Sound and Recording

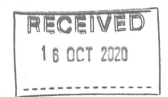

Sound and Recording
Applications and Theory

Seventh edition

Francis Rumsey and Tim McCormick

Focal Press
Taylor & Francis Group

NEW YORK AND LONDON

First published 1992 by Focal Press
This edition published 2014
by Focal Press
70 Blanchard Road, Suite 402, Burlington, MA 01803

and by Focal Press
2 Park Square, Milton Park, Abingdon, Oxon OX14 4RN

Focal Press is an imprint of the Taylor & Francis Group, an informa business

Notices
Knowledge and best practice in this field are constantly changing. As new
research and experience broaden our understanding, changes in research methods,
professional practices, or medical treatment may become necessary.

Practitioners and researchers must always rely on their own experience and
knowledge in evaluating and using any information, methods, compounds, or
experiments described herein. In using such information or methods they should
be mindful of their own safety and the safety of others, including parties for whom
they have a professional responsibility.

Product or corporate names may be trademarks or registered trademarks, and are
used only for identification and explanation without intent to infringe.

Library of Congress Cataloging in Publication Data
Rumsey, Francis.
Sound and recording : applications and theory / Francis Rumsey and
Tim McCormick. — 7th edition.
pages cm
Originally published: 1992.
Includes index.
1. Sound—Recording and reproducing. I. Title.
TK7881.4.R858 2014
621.389'3—dc23   2013030814

ISBN: 978-0-415-84337-9 (pbk)
ISBN: 978-0-415-84340-9 (hbk)
ISBN: 978-0-203-75623-2 (ebk)

Typeset in Kuenstler 480 BT
By MPS Limited, Chennai, India
Printed and bound by CPI Group (UK) Ltd, Croydon, CR0 4YY

Contents

Contents **ix**

Fact File Directory

Preface to the Second Edition

One of the greatest dangers in writing a book at an introductory level is to sacrifice technical accuracy for the sake of simplicity. In writing *Sound and Recording: An Introduction* we have gone to great lengths not to fall into this trap, and have produced a comprehensive introduction to the field of audio, intended principally for the newcomer to the subject, which is both easy to understand and technically precise. We have written the book that we would have valued when we first entered the industry, and as such it represents a readable reference, packed with information. Many books stop after a vague overview, just when the reader wants some clear facts about a subject, or perhaps assume too much knowledge on the reader's behalf. Books by contributed authors often suffer from a lack of consistency in style, coverage and technical level. Furthermore, there is a tendency for books on audio to be either too technical for the beginner or, alternatively, subjectively biased towards specific products or operations. There are also quite a number of American books on sound recording which, although good, tend to ignore European trends and practices. We hope that we have steered a balanced course between these extremes, and have deliberately avoided any attempt to dictate operational practice.

Sound and Recording: An Introduction is definitely biased towards an understanding of 'how it works', as opposed to 'how to work it', although technology is never discussed in an abstract manner but related to operational reality. Although we have included a basic introduction to acoustics and the nature of sound perception, this is not a book on acoustics or musical acoustics (there are plenty of those around). It is concerned with the principles of audio recording and reproduction, and has a distinct bias towards the professional rather than the consumer end of the market. The coverage of subject matter is broad, including chapters on digital audio, timecode synchronization and MIDI, amongst other more conventional subjects, and there is comprehensive coverage of commonly misunderstood subjects such as the decibel, balanced lines, reference levels and metering systems.

This second edition of the book has been published only two years after the first, and the subject matter has not changed significantly enough in the interim to warrant major modifications to the existing chapters.

The key difference between the second and first editions is the addition of a long chapter on stereo recording and reproduction. This important topic is covered in considerable detail, including historical developments, principles of stereo reproduction, surround sound and stereo microphone techniques. Virtually every recording or broadcast happening today is made in stereo, and although surround sound has had a number of notable 'flops' in the past it is likely to become considerably more important in the next ten years. Stereo and surround sound are used extensively in film, video and television production, and any new audio engineer should be familiar with the principles.

Since this is an introductory book, it will be of greatest value to the student of sound recording or music technology, and to the person starting out on a career in sound engineering or broadcasting. The technical level has deliberately been kept reasonably low for this reason, and those who find this frustrating probably do not need the book! Nonetheless, it is often valuable for the seasoned audio engineer to go back to basics. Further reading suggestions have been made in order that the reader may go on to a more in-depth coverage of the fields introduced here, and some of the references are considerably more technical than this book. Students will find these suggestions valuable when planning a course of study.

Francis Rumsey and
Tim McCormick

Preface to the Third Edition

Since the first edition of *Sound and Recording* some of the topics have advanced quite considerably, particularly the areas dependent on digital and computer technology. Consequently I have rewritten the chapters on digital recording and MIDI (Chapters 10 and 15), and have added a larger section on mixer automation in Chapter 7. Whereas the first edition of the book was quite 'analogue', I think that there is now a more appropriate balance between analogue and digital topics. Although analogue audio is by no means dead (sound will remain analogue for ever!), most technological developments are now digital.

I make no apologies for leaving in the chapter on record players, although some readers have commented that they think it is a waste of space. People still use record players, and there is a vast store of valuable material on LP record. I see no problem with keeping a bit of history in the book – you never know, it might come in useful one day when everyone has forgotten (and some may never have known) what to do with vinyl discs. It might even appease the faction of our industry that continues to insist that vinyl records are the highest fidelity storage medium ever invented.

Francis Rumsey,
Guildford

Preface to the Fourth Edition

The fourth edition is published ten years after *Sound and Recording* was first published, which is hard to believe. The book has been adopted widely by students and tutors on audio courses around the world. In that time audio technology and techniques have changed in some domains but not in others. All the original principles still apply but the emphasis has gradually changed from predominantly analogue to quite strongly digital, although many studios still use analogue mixers and multitrack tape recorders for a range of purposes and we do not feel that the death-knell of analogue recording has yet been sounded. Readers of earlier editions will notice that the chapter on record players has finally been reduced in size and relegated to an appendix. While we continue to believe that information about the LP should remain in the literature as the format lingers on, it is perhaps time to remove it from the main part of the book.

In this edition a new chapter on surround sound has been added, complemented by a reworked chapter preceding it that is now called 'Two-channel stereo'. Surround sound was touched upon in the previous edition but a complete chapter reflects the increased activity in this field with the coming of new multichannel consumer replay formats.

The chapter on auditory perception has been reworked to include greater detail on spatial perception and the digital audio chapter has been updated to include DVD-A and SACD, with information about Direct Stream Digital (DSD), the MiniDisc, computer-based editing systems and their operation. Chapter 5 on loudspeakers now includes information about distributed-mode loudspeakers (DML) and a substantial section on directivity and the various techniques used to control it. Finally a glossary of terms has now been provided, with some additional material that supports the main text.

Francis Rumsey and
Tim McCormick

Preface to the Fifth Edition

The fifth edition of *Sound and Recording* includes far greater detail on digital audio than the previous editions, reflecting the growing 'all-digital' trend in audio equipment and techniques. In place of the previous single chapter on the topic there are now three chapters (Chapters 8, 9 and 10) covering principles, recording and editing systems, and applications. This provides a depth of coverage of digital audio in the fifth edition that should enable the reader to get a really detailed understanding of the principles of current audio systems. We believe, however, that the detailed coverage of analogue recording should remain in its current form, at least for this iteration of the book. We have continued the trend, begun in previous new editions, of going into topics in reasonable technical depth but without using unnecessary mathematics. It is intended that this will place *Sound and Recording* slightly above the introductory level of the many broad-ranging textbooks on recording techniques and audio, so that those who want to understand how it works a bit better will find something to satisfy them here.

The chapter previously called 'A guide to the audio signal chain' has been removed from this new edition, and parts of that material have now found their way into other chapters, where appropriate. For example, the part dealing with the history of analogue recording has been added to the start of Chapter 6. Next, the material dealing with mixers has been combined into a single chapter (it is hard to remember why we ever divided it into two) and now addresses both analogue and digital systems more equally than before. Some small additions have been made to Chapters 12 and 13 and Chapter 14 has been completely revised and extended, now being entitled 'MIDI and synthetic audio control'.

**Francis Rumsey and
Tim McCormick**

Preface to the Sixth Edition

When we first wrote this book it was our genuine intention to make it an introduction to the topic of sound and recording that would be useful to students starting out in the field. However, we readily admit that over the years the technical level of the book has gradually risen in a number of chapters, and that there are now many audio and music technology courses that do not start out by covering the engineering aspects of the subject at this level. For this reason, and recognizing that many courses use the book as a somewhat more advanced text, we have finally allowed the book's subtitle, 'An Introduction', to fall by the wayside.

In this edition we have overhauled many of the chapters, continuing the expansion and reorganization of the digital audio chapters to include more recent details of pitch correction, file formats, interfaces and Blu-Ray disk. The coverage of digital tape formats has been retained in reduced form, partly for historical reasons. Chapters 6 and 7, covering analog recording and noise reduction, have been shortened but it is felt that they still justify inclusion given that such equipment is still in use in the field. As fewer and fewer people in the industry continue to be familiar with such things as bias, replay equalization, azimuth and noise reduction line-up, we feel it is important that such information should continue to be available in the literature while such technology persists. Likewise, the appendix on record players survives, it being surprising how much this equipment is still used.

The chapter on mixers has been thoroughly reworked and updated, as it had become somewhat disorganized during its evolution through various editions, and the chapter on MIDI has been expanded to include more information on sequencing principles. Chapter 15 on synchronization has been revised to include substantially greater coverage of digital audio synchronization topics, and the information about MIDI sync has also been moved here. The section on digital plug-ins has been moved into the chapter on outboard equipment. A number of other additions have also been made to the book, including an introduction to loudspeaker design parameters, further information on Class D amplifiers, and updated information on wireless microphone frequencies.

Finally, a new chapter on sound quality has been added at the end of the book, which incorporates some of the original appendix dealing with equipment specifications. This chapter introduces some of the main concepts relating to the perception and evaluation of sound quality, giving examples of relationships to simple aspects of audio equipment performance.

Francis Rumsey and
Tim McCormick
January 2009

Preface to the Seventh Edition

Since the sixth edition appeared in 2009 there have been significant developments in several fields, and revisions, updates and deletions have taken place in a number of areas. The digital audio chapters have been substantially revised and include such topics as parametric and high-resolution audio coding, recent interfaces, file formats and networks, the latest workstation audio processing technology, and issues concerning mixing 'in the box' (that is, entirely within the computer) and 'out of the box'. Digital mastering issues such as loudness normalization and initiatives such as Apple's Mastered for iTunes are also included. Audio network requirements and protocols for IP-based communication, for instance, RAVENNA, X-192, AVB and Q-LAN are covered.

After CD sales peaked in about 2000 there has been a year-by-year decline, counterbalanced somewhat by a growth in downloading activity. DVD sales burgeoned, and surround sound developments have been driven not by audio-only formats such as the quadraphonics and ambisonics of yesterday or the SACD of today, but by the audio-visual industry: film, DVD and the higher definition TV formats. Chapter 17 introduces an overview of some of the advanced immersive audio systems recently introduced and to an extent still under development. Those principally involved only with sound will inevitably find themselves working in the audio-visual field from time to time, and discussions of Wave Field Synthesis and Dolby Atmos in particular underline the issues concerning surround formats suitable for the professional or public arena and their relevance to the domestic environment.

Information specifically about analog recording, principally covered in Chapters 6 and 7, now occupies about 6% of the whole book. For those who have left analog behind for ever, or indeed have never even encountered it, its continuing presence for those who still find the information useful should not trouble them. When the first edition of this book appeared in 1992, no one would have predicted that vinyl would still be in use this deep into the present century nearly 40 years after LP sales peaked in 1978, and so an appendix has been allowed to remain which helps people to get the best from the format, and indeed to help avoid damage of literally irreplaceable hardware merely by playing it on poorly aligned equipment.

Chapter 3 includes a section on digital radio microphones, and Chapter 4 now includes a section on developments in loudspeaker sensitivity and the issues which confine this parameter within certain bounds. Also included is information about the highly directional 'audio spotlight' techniques.

Chapter 14 has been re-named 'MIDI and Remote Control' and reflects developments in the usage of computer networks to enable conventional-looking mixer control surfaces to communicate with computer-based recording and editing processes. It includes information about current thinking and co-operation by those developing and specifying systems, and looks specifically at the Open Control Architecture and AES64-2012 proposals. The Avid EuCon format is also covered which has been in the field for some years. These topics reflect the ever greater part the computer is playing in the sound industry.

For information on all Focal Press publications visit our website at: www.focalpress.com

Francis Rumsey and
Tim McCormick

What is Sound?

A VIBRATING SOURCE

Sound is produced when an object (the source) vibrates and causes the air around it to move. Consider the sphere shown in Figure 1.1. It is a pulsating sphere which could be imagined as something like a squash ball, and it is pulsating regularly so that its size oscillates between being slightly larger than normal and then slightly smaller than normal. As it pulsates it will alternately compress and then rarefy the surrounding air, resulting in a series of compressions and rarefactions traveling away from the sphere, rather like a three-dimensional version of the ripples which travel away from a stone dropped into a pond. These are known as longitudinal waves since the air particles move in the same dimension as the direction of wave travel. The alternative to longitudinal wave motion is transverse wave motion (see Figure 1.2), such as is found in vibrating strings, where the motion of the string is at right angles to the direction of apparent wave travel.

1

FIGURE 1.1

*(a) A simple sound source
can be imagined to be like
a pulsating sphere radiating
spherical waves. (b) The
longitudinal wave thus created
is a succession of compressions
and rarefactions of the air.*

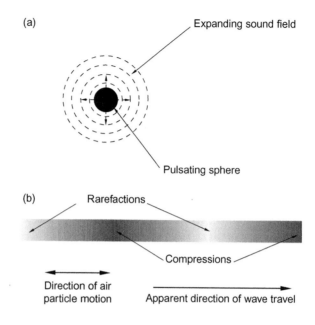

(a)

Expanding sound field

Pulsating sphere

(b)   Rarefactions

Compressions

Direction of air
particle motion

Apparent direction of wave travel

FIGURE 1.2

*In a transverse wave the
motion of any point on the
wave is at right angles to
the apparent direction of
motion of the wave.*

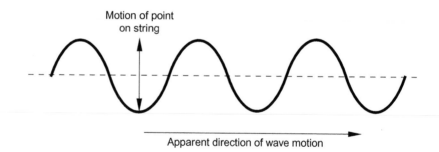

Motion of point
on string

Apparent direction of wave motion

CHARACTERISTICS OF A SOUND WAVE

The rate at which the source oscillates is the frequency of the sound wave
it produces, and is quoted in hertz (Hz) or cycles per second (cps). 1000
hertz is termed 1 kilohertz (1 kHz). The amount of compression and rarefaction of the air which results from the sphere's motion is the amplitude
of the sound wave, and is related to the loudness of the sound when it is
finally perceived by the ear (see Chapter 2). The distance between two adjacent peaks of compression or rarefaction as the wave travels through the
air is the wavelength of the sound wave, and is often represented by the
Greek letter lambda (λ). The wavelength depends on how fast the sound

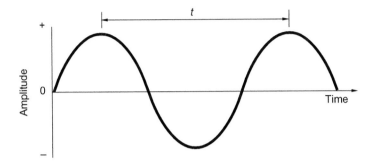

FIGURE 1.3

A graphical representation of a sinusoidal sound waveform. The period of the wave is represented by t, and its frequency by 1/t.

wave travels, since a fast-traveling wave would result in a greater distance between peaks than a slow-traveling wave, given a fixed time between compression peaks (i.e. a fixed frequency of oscillation of the source).

As shown in Figure 1.3, the sound wave's characteristics can be represented on a graph, with amplitude plotted on the vertical axis and time plotted on the horizontal axis. It will be seen that both positive and negative ranges are shown on the vertical axis: these represent compressions (+) and rarefactions (−) of the air. This graph represents the waveform of the sound. For a moment, a source vibrating in a very simple and regular manner is assumed, in so-called simple harmonic motion, the result of which is a simple sound wave known as a sine wave. The simplest vibrating systems oscillate in this way, such as a mass suspended from a spring, or a swinging pendulum (see also 'Phase', p. 8). It will be seen that the frequency (f) is the inverse of the time between peaks or troughs of the wave $(f=1/t)$. So the shorter the time between oscillations of the source, the higher the frequency. The human ear is capable of perceiving sounds with frequencies between approximately 20 Hz and 20 kHz (see 'Frequency perception', Chapter 2); this is known as the audio frequency range or audio spectrum.

HOW SOUND TRAVELS IN AIR

Air is made up of gas molecules and has an elastic property (imagine putting a thumb over the end of a bicycle pump and compressing the air inside – the air is springy). Longitudinal sound waves travel in air in somewhat the same fashion as a wave travels down a row of up-ended dominoes after the first one is pushed over. The half-cycle of compression created by the vibrating source causes successive air particles to be moved in a knock-on effect, and this is normally followed by a balancing rarefaction which causes a similar motion of particles in the opposite direction.

It may be appreciated that the net effect of this is that individual air particles do not actually travel – they oscillate about a fixed point – but the result is that a wave is formed which appears to move away from the source. The speed at which it moves away from the source depends on the density and elasticity of the substance through which it passes, and in air the speed is relatively slow compared with the speed at which sound travels through most solids. In air the speed of sound is approximately 340 meters per second (m s^{-1}), although this depends on the temperature of the air. At freezing point the speed is reduced to nearer 330 m s^{-1}. In steel, to give an example of a solid, the speed of sound is approximately 5100 m s^{-1}.

The frequency and wavelength of a sound wave are related very simply if the speed of the wave (usually denoted by the letter c) is known:

$$c = f\lambda \quad \text{or} \quad \lambda = c/f$$

To show some examples, the wavelength of sound in air at 20 Hz (the low-frequency or LF end of the audio spectrum), assuming normal room temperature, would be:

$$\lambda = 340/20 = 17 \text{ meters}$$

whereas the wavelength of 20 kHz (at the high-frequency or HF end of the audio spectrum) would be 1.7 cm. Thus it is apparent that the wavelength of sound ranges from being very long in relation to most natural objects at low frequencies, to quite short at high frequencies. This is important when considering how sound behaves when it encounters objects – whether the object acts as a barrier or whether the sound bends around it (see Fact File 1.5).

SIMPLE AND COMPLEX SOUNDS

In the foregoing example, the sound had a simple waveform – it was a sine wave or sinusoidal waveform – the type which might result from a very simple vibrating system such as a weight suspended on a spring. Sine waves have a very pure sound because they consist of energy at only one frequency, and are often called pure tones. They are not heard very commonly in real life (although they can be generated electrically) since most sound sources do not vibrate in such a simple manner. A person whistling or a recorder (a simple wind instrument) produces a sound which approaches a sinusoidal waveform. Most real sounds are made up of a combination of vibration patterns which result in a more complex waveform. The more complex the waveform, the more like noise the sound becomes, and when the waveform has a highly random pattern the sound is said to be noise (see 'Frequency spectra of non-repetitive sounds', p. 7).

The important characteristic of sounds which have a definite pitch is that they are repetitive: that is, the waveform, no matter how complex, repeats its pattern in the same way at regular intervals. All such waveforms can be broken down into a series of components known as harmonics, using a mathematical process called Fourier analysis (after the mathematician Joseph Fourier). Some examples of equivalent line spectra for different waveforms are given in Figure 1.4. This figure shows another way of depicting the characteristics of the sound graphically – that is, by drawing a so-called line spectrum which shows frequency along the horizontal axis and amplitude up the vertical axis. The line spectrum shows the relative strengths of different frequency components which make up a sound. Where there is a

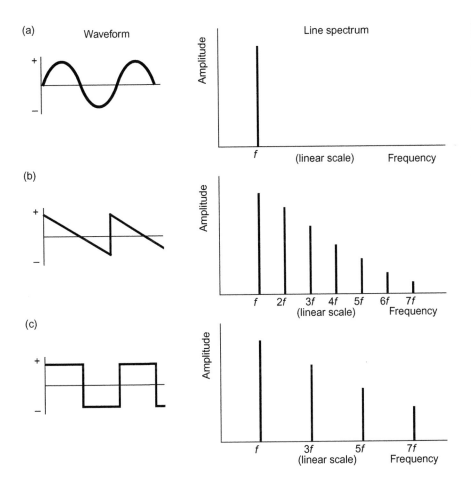

FIGURE 1.4

Equivalent line spectra for a selection of simple waveforms. (a) The sine wave consists of only one component at the fundamental frequency f. (b) The sawtooth wave consists of components at the fundamental and its integer multiples, with amplitudes steadily decreasing. (c) The square wave consists of components at odd multiples of the fundamental frequency.

line there is a frequency component. It will be noticed that the more complex the waveform the more complex the corresponding line spectrum.

For every waveform, such as that shown in Figure 1.3, there is a corresponding line spectrum: waveforms and line spectra are simply two different ways of showing the characteristics of the sound. Figure 1.3 is called a time-domain plot, whilst the line spectrum is called a frequency-domain plot. Unless otherwise stated, such frequency-domain graphs in this book will cover the audio-frequency range from 20 Hz at the lower end to 20 kHz at the upper end.

In a reversal of the above breaking-down of waveforms into their component frequencies it is also possible to construct or synthesize waveforms by adding together the relevant components.

FREQUENCY SPECTRA OF REPETITIVE SOUNDS

As will be seen in Figure 1.4, the simple sine wave has a line spectrum consisting of only one component at the frequency of the sine wave. This is known as the fundamental frequency of oscillation. The other repetitive waveforms, such as the square wave, have a fundamental frequency as well as a number of additional components above the fundamental. These are known as harmonics, but may also be referred to as overtones or partials.

Harmonics are frequency components of a sound which occur at integer multiples of the fundamental frequency, that is at twice, three times, four times and so on. Thus a sound with a fundamental of 100 Hz might also contain harmonics at 200 Hz, 400 Hz and 600 Hz. The reason for the existence of these harmonics is that most simple vibrating sources are capable of vibrating in a number of harmonic modes at the same time. Consider a stretched string, as shown in Figure 1.5. It may be made to vibrate in any of a number of modes, corresponding to integer multiples of the fundamental frequency of vibration of the string (the concept of 'standing waves' is introduced below). The fundamental corresponds to the mode in which the string moves up and down as a whole, whereas the harmonics correspond to modes in which the vibration pattern is divided into points of maximum and minimum motion along the string (these are called antinodes and nodes). It will be seen that the second mode involves two peaks of vibration, the third mode three peaks, and so on.

In accepted terminology, the fundamental is also the first harmonic, and thus the next component is the second harmonic, and so on. Confusingly, the second harmonic is also known as the first overtone. For the waveforms shown in Figure 1.4, the fundamental has the highest

(a)

(b)

Antinode

Node

(c)

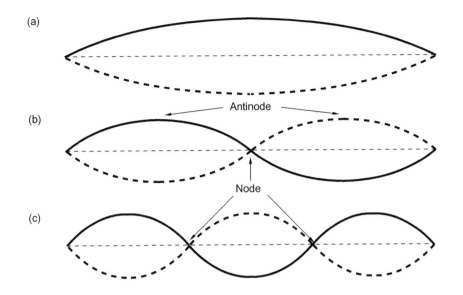

FIGURE 1.5
Modes of vibration of a stretched string. (a) Fundamental. (b) Second harmonic. (c) Third harmonic.

amplitude, and the amplitudes of the harmonics decrease with increasing frequency, but this will not always be the case with real sounds since many waveforms have line spectra which show the harmonics to be higher in amplitude than the fundamental. It is also quite feasible for there to be harmonics missing in the line spectrum, and this depends entirely on the waveform in question.

It is also possible for there to be overtones in the frequency spectrum of a sound which are not related in a simple integer-multiple fashion to the fundamental. These cannot correctly be termed harmonics, and they are more correctly referred to as overtones or inharmonic partials. They tend to arise in vibrating sources which have a complicated shape, and which do not vibrate in simple harmonic motion but have a number of repetitive modes of vibration. Their patterns of oscillation are often unusual, such as might be observed in a bell or a percussion instrument. It is still possible for such sounds to have a recognizable pitch, but this depends on the strength of the fundamental. In bells and other such sources, one often hears the presence of several strong inharmonic overtones.

FREQUENCY SPECTRA OF NON-REPETITIVE SOUNDS

Non-repetitive waveforms do not have a recognizable pitch and sound noise-like. Their frequency spectra are likely to consist of a collection of components at unrelated frequencies, although some frequencies may be

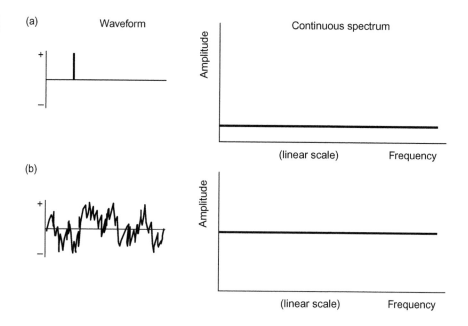

FIGURE 1.6

Frequency spectra of non-repetitive waveforms. (a) Pulse. (b) Noise.

more dominant than others. The analysis of such waves to show their frequency spectra is more complicated than with repetitive waves, but is still possible using a mathematical technique called Fourier transformation, the result of which is a frequency-domain plot of a time-domain waveform.

Single, short pulses can be shown to have continuous frequency spectra which extend over quite a wide frequency range, and the shorter the pulse the wider its frequency spectrum but usually the lower its total energy (see Figure 1.6). Random waveforms will tend to sound like hiss, and a completely random waveform in which the frequency, amplitude and phase of components are equally probable and constantly varying is called white noise. A white noise signal's spectrum is flat, when averaged over a period of time, right across the audio-frequency range (and theoretically above it). White noise has equal energy for a given bandwidth, whereas another type of noise, known as pink noise, has equal energy per octave. For this reason white noise sounds subjectively to have more high-frequency energy than pink noise.

PHASE

Two waves of the same frequency are said to be 'in phase' when their compression (positive) and rarefaction (negative) half-cycles coincide exactly in time and space (see Figure 1.7). If two in-phase signals of equal amplitude are added together, or superimposed, they will sum to produce another

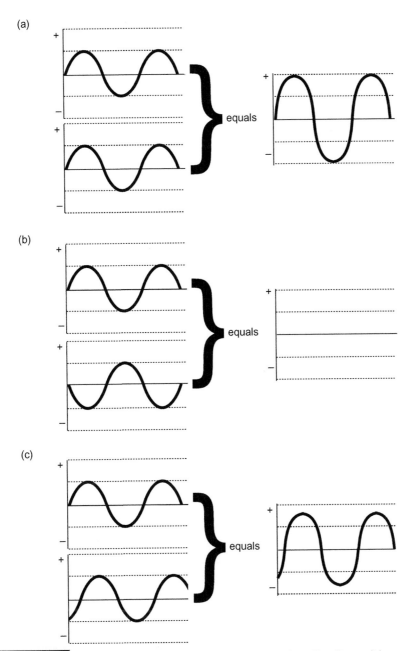

FIGURE 1.7 *(a) When two identical in-phase waves are added together, the result is a wave of the same frequency and phase but twice the amplitude. (b) Two identical out-of-phase waves add to give nothing. (c) Two identical waves partially out of phase add to give a resultant wave with a phase and amplitude which is the point-by-point sum of the two.*

FIGURE 1.8

If the two loudspeakers in the drawing emit the same wave at the same time, the phase difference between the waves at the listener's ear will be directly related to the delay t_1–t_2.

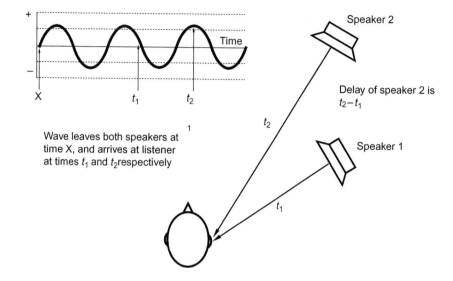

Wave leaves both speakers at time X, and arrives at listener at times t_1 and t_2 respectively

signal of the same frequency but twice the amplitude. Signals are said to be out of phase when the positive half-cycle of one coincides with the negative half-cycle of the other. If these two signals are added together they will cancel each other out, and the result will be no signal.

Clearly, these are two extreme cases, and it is entirely possible to superimpose two sounds of the same frequency which are only partially in phase with each other. The resultant wave in this case will be a partial addition or partial cancelation, and the phase of the resulting wave will lie somewhere between that of the two components (see Figure 1.7c).

Phase differences between signals can be the result of time delays between them. If two identical signals start out at sources equidistant from a listener at the same time as each other then they will be in phase by the time they arrive at the listener. If one source is more distant than the other then it will be delayed, and the phase relationship between the two will depend upon the amount of delay (see Figure 1.8). A useful rule-of-thumb is that sound travels about 30 cm (1 foot) per millisecond, so if the second source in the above example were 1 meter (just over 3 ft) more distant than the first it would be delayed by just over 3 ms. The resulting phase relationship between the two signals, it may be appreciated, would depend on the frequency of the sound, since at a frequency of around 330 Hz the 3 ms delay would correspond to one wavelength and thus the delayed signal would be in phase with the undelayed signal. If the delay had been half this (1.5 ms) then the two signals would have been out of phase at 330 Hz.

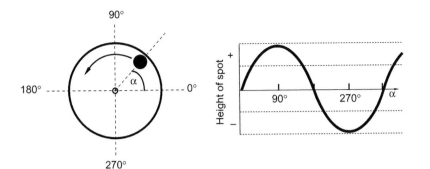

FIGURE 1.9

The height of the spot varies sinusoidally with the angle of rotation of the wheel. The phase angle of a sine wave can be understood in terms of the number of degrees of rotation of the wheel.

Phase is often quoted as a number of degrees relative to some reference, and this must be related back to the nature of a sine wave. A diagram is the best way to illustrate this point, and looking at Figure 1.9 it will be seen that a sine wave may be considered as a graph of the vertical position of a rotating spot on the outer rim of a disc (the amplitude of the wave), plotted against time. The height of the spot rises and falls regularly as the circle rotates at a constant speed. The sine wave is so called because the spot's height is directly proportional to the mathematical sine of the angle of rotation of the disc, with zero degrees occurring at the origin of the graph and at the point shown on the disc's rotation in the diagram. The vertical amplitude scale on the graph goes from minus one (maximum negative amplitude) to plus one (maximum positive amplitude), passing through zero at the halfway point. At 90° of rotation the amplitude of the sine wave is maximum positive (the sine of 90° is 1), and at 180° it is zero ($\sin 180° = 0$). At 270° it is maximum negative ($\sin 270° = -1$), and at 360° it is zero again. Thus in one cycle of the sine wave the circle has passed through 360° of rotation.

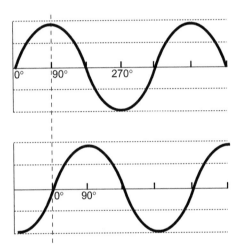

FIGURE 1.10 *The lower wave is 90° out of phase with the upper wave.*

It is now possible to go back to the phase relationship between two waves of the same frequency. If each cycle is considered as corresponding to 360°, then one can say just how many degrees one wave is ahead of or behind another by comparing the 0° point on one wave with the 0° point on the other (see Figure 1.10). In the example wave 1 is 90° out of phase with wave 2. It is important to realize that phase is only a relevant concept in the case of continuous repetitive waveforms, and has little meaning in the case of impulsive or transient sounds where time difference is the more relevant quantity. It can be deduced from the foregoing discussion that

(a) the higher the frequency, the greater the phase difference which would result from a given time delay between two signals, and (b) it is possible for there to be more than 360° of phase difference between two signals if the delay is great enough to delay the second signal by more than one cycle. In the latter case it becomes difficult to tell how many cycles of delay have elapsed unless a discontinuity arises in the signal, since a phase difference of 360° is indistinguishable from a phase difference of 0°.

SOUND IN ELECTRICAL FORM

Although the sound that one hears is due to compression and rarefaction of the air, it is often necessary to convert sound into an electrical form in order to perform operations on it such as amplification, recording and mixing. As detailed in Fact File 3.1 and Chapter 3, it is the job of the microphone to convert sound from an acoustical form into an electrical form. The process of conversion will not be described here, but the result is important because if it can be assumed for a moment that the microphone is perfect then the resulting electrical waveform will be exactly the same shape as the acoustical waveform which caused it.

The equivalent of the amplitude of the acoustical signal in electrical terms is the voltage of the electrical signal. If the voltage at the output of a microphone were to be measured whilst the microphone was picking up an acoustical sine wave, one would measure a voltage which changed sinusoidally as well. Figure 1.11 shows this situation, and it may be seen that an acoustical compression of the air corresponds to a positive-going voltage, whilst an acoustical rarefaction of the air corresponds to a negative-going voltage. (This is the norm, although some sound reproduction systems introduce an absolute phase reversal in the relationship between acoustical phase and electrical phase, such that an acoustical compression becomes equivalent to a negative voltage. Some people claim to be able to hear the difference.)

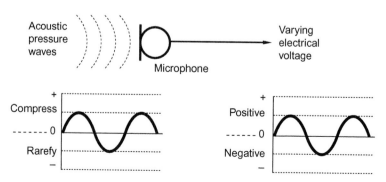

FIGURE 1.11

A microphone converts variations in acoustical sound pressure into variations in electrical voltage. Normally, a compression of the air results in a positive voltage and a rarefaction results in a negative voltage.

The other important quantity in electrical terms is the current flowing down the wire from the microphone. Current is the electrical equivalent of the air particle motion discussed in 'How sound travels in air', p. 3. Just as the acoustical sound wave was carried in the motion of the air particles, so the electrical sound wave is carried in the motion of tiny charge carriers which reside in the metal of a wire (these are called electrons). When the voltage is positive the current moves in one direction, and when it is negative the current moves in the other direction. Since the voltage generated by a microphone is repeatedly alternating between positive and negative, in sympathy with the sound wave's compression and rarefaction cycles, the current similarly changes direction each half cycle. Just as the air particles in 'Characteristics of a sound wave', p. 2, did not actually go anywhere in the long term, so the electrons carrying the current do not go anywhere either – they simply oscillate about a fixed point. This is known as alternating current or AC.

A useful analogy to the above (both electrical and acoustical) exists in plumbing. If one considers water in a pipe fed from a header tank, as shown in Figure 1.12, the voltage is equivalent to the pressure of water which results from the header tank, and the current is equivalent to the rate of flow of water through the pipe. The only difference is that the diagram is concerned with a direct current situation in which the direction of flow is not repeatedly changing. The quantity of resistance should be introduced here, and is analogous to the diameter of the pipe. Resistance impedes the flow of water through the pipe, as it does the flow of electrons through a wire and the flow of acoustical sound energy through a substance. For a fixed voltage (or water pressure in this analogy), a high resistance (narrow pipe) will result in a small current (a trickle of water), whilst a low resistance (wide pipe) will result in a large current. The relationship between voltage, current and resistance was established by Ohm, in the form of Ohm's law, as described in Fact File 1.1. There is also a relationship between power and voltage, current and resistance.

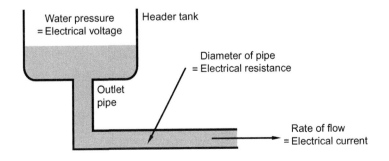

FIGURE 1.12

There are parallels between the flow of water in a pipe and the flow of electricity in a wire, as shown in this drawing.

FACT FILE 1.1 OHM'S LAW

Ohm's law states that there is a fixed and simple relationship between the current flowing through a device (I), the voltage across it (V), and its resistance (R), as shown in the diagram:

$$V = IR$$

or:

$$I = V/R$$

or:

$$R = V/I$$

Thus if the resistance of a device is known, and the voltage dropped across it can be measured, then the current flow may be calculated, for example.

There is also a relationship between the parameters above and the power in watts (W) dissipated in a device:

$$W = I^2 R = V^2/R$$

In AC systems, resistance is replaced by impedance, a complex term which contains both resistance and reactance components. The reactance part varies with the frequency of the signal; thus the impedance of an electrical device also varies with the frequency of a signal. Capacitors (basically two conductive plates separated by an insulator) are electrical devices which present a high impedance to low-frequency signals and a low impedance to high-frequency signals. They will not pass direct current. Inductors (basically coils of wire) are electrical devices which present a high impedance to high-frequency signals and a low impedance to low-frequency signals. Capacitance is measured in farads, inductance in henrys.

DISPLAYING THE CHARACTERISTICS OF A SOUND WAVE

Two devices can be introduced at this point which illustrate graphically the various characteristics of sound signals so far described. It would be useful to (a) display the waveform of the sound, and (b) display the frequency spectrum of the sound. In other words (a) the time-domain signal and (b) the frequency-domain signal.

An oscilloscope is used for displaying the waveform of a sound, and a spectrum analyzer is used for showing which frequencies are contained in the signal and their amplitudes. Examples of such devices are pictured in Figure 1.13. Both devices accept sound signals in electrical form and display their analyses of the sound on a screen. The oscilloscope displays a moving spot which scans horizontally at one of a number of fixed speeds

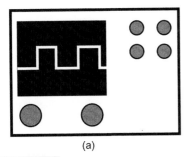 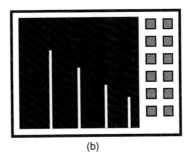

(a)   (b)

FIGURE 1.13 *(a) An oscilloscope displays the waveform of an electric signal by means of a moving spot which is deflected up by a positive signal and down by a negative signal. (b) A spectrum analyzer displays the frequency spectrum of an electrical waveform in the form of lines representing the amplitudes of different spectral components of the signal.*

from left to right and whose vertical deflection is controlled by the voltage of the sound signal (up for positive, down for negative). In this way it plots the waveform of the sound as it varies with time. Many oscilloscopes have two inputs and can plot two waveforms at the same time, and this can be useful for comparing the relative phases of two signals (see 'Phase', p. 8).

The spectrum analyzer works in different ways depending on the method of spectrum analysis. A real-time analyzer displays a constantly updating line spectrum, similar to those depicted earlier in this chapter, and shows the frequency components of the input signal on the horizontal scale together with their amplitudes on the vertical scale.

THE DECIBEL

The unit of the decibel is used widely in sound engineering, often in preference to other units such as volts, watts, or other such absolute units, since it is a convenient way of representing the ratio of one signal's amplitude to another's. It also results in numbers of a convenient size which approximate more closely to one's subjective impression of changes in the amplitude of a signal, and it helps to compress the range of values between the maximum and minimum sound levels encountered in real signals. For example, the range of sound intensities (see next section) which can be handled by the human ear covers about 14 powers of ten, from 0.000 000 000 001 W m^{-2} to around 100 W m^{-2}, but the equivalent range in decibels is only from 0 to 140 dB.

Some examples of the use of the decibel are given in Fact File 1.2. The relationship between the decibel and human sound perception is discussed

FACT FILE 1.2 THE DECIBEL

Basic Decibels

The decibel is based on the logarithm of the ratio between two numbers. It describes how much larger or smaller one value is than the other. It can also be used as an absolute unit of measurement if the reference value is fixed and known. Some standardized references have been established for decibel scales in different fields of sound engineering (see below).

The decibel is strictly ten times the logarithm to the base ten of the ratio between the powers of two signals:

$$dB = 10\log_{10}(P_1/P_2)$$

For example, the difference in decibels between a signal with a power of 1 watt and one of 2 watts is $10\log(2/1) = 3$.

If the decibel is used to compare values other than signal powers, the relationship to signal power must be taken into account. Voltage has a square relationship to power (from Ohm's law: $W = V^2/R$); thus to compare two voltages:

$$dB = 10\log(V_1^2/V_2^2), \quad \text{or} \quad 10\log(V_1/V_2)^2, \quad \text{or} \quad 20\log(V_1/V_2)$$

For example, the difference in decibels between a signal with a voltage of 1 volt and one of 2 volts is $20\log(2/1) = 6\,dB$. So a doubling in voltage gives rise to an increase of 6 dB, and a doubling in power gives rise to an increase of 3 dB. A similar relationship applies to acoustical sound pressure (analogous to electrical voltage) and sound power (analogous to electrical power).

Decibels with a Reference

If a signal level is quoted in decibels, then a reference must normally be given, otherwise the figure means nothing; e.g. 'Signal level = 47 dB' cannot have a meaning unless one knows that the signal is 47 dB above a known point. '+ 8 dB ref. 1 volt' has a meaning since one now knows that the level is 8 dB higher than 1 volt, and thus one could calculate the voltage of the signal.

There are exceptions in practice, since in some fields a reference level is accepted as implicit. Sound pressure levels (SPLs) are an example, since the reference level is defined worldwide as $2 \times 10^{-5}\,\mathrm{N\,m^{-2}}$ (20 μPa). Thus to state 'SPL = 77 dB' is probably acceptable, although confusion can still arise due to misunderstandings over such things as weighting curves (see Fact File 1.4). In sound recording, 0 dB or 'zero level' is a nominal reference level used for aligning equipment and setting recording levels, often corresponding to 0.775 volts (0 dBu) although this is subject to variations in studio centers in different locations. (Some studios use 4 dBu as their reference level, for example.) '0 dB' does not mean 'no signal', it means that the signal concerned is at the same level as the reference.

Often a letter is placed after 'dB' to denote the reference standard in use (e.g. 'dBm'), and a number of standard abbreviations are in use, some examples of which are given below. Sometimes the suffix denotes a particular frequency weighting characteristic used in the measurement of noise (e.g. 'dBA').

Abbrev	Ref. level
dBV	1 volt
dBu	0.775 volt (Europe)
dBv	0.775 volt (USA)
dBm	1 milliwatt (see Chapter 12)
dBA	dB SPL, A-weighted response

Useful Decibel Ratios to Remember (Voltages or SPLs)

It is more common to deal in terms of voltage or SPL ratios than power ratios in audio systems. Here are some useful dB equivalents of different voltage or SPL relationships and multiplication factors:

dB	Multiplication factor
0	1
+3	$\sqrt{2}$
+6	2
+20	10
+60	1000

in more detail in Chapter 2. Operating levels in recording equipment are discussed further in 'Metering systems', Chapter 5 and 'Magnetic recording levels', Chapter 6.

Decibels are not only used to describe the ratio between two signals, or the level of a signal above a reference, they are also used to describe the voltage gain of a device. For example, a microphone amplifier may have a gain of 60 dB, which is the equivalent of multiplying the input voltage by a factor of 1000, as shown in the example below:

$$20 \log 1000/1 = 60 \, dB$$

SOUND POWER AND SOUND PRESSURE

A simple sound source, such as the pulsating sphere used at the start of this chapter, radiates sound power omnidirectionally – that is, equally in all directions, rather like a three-dimensional version of the ripples moving away from a stone dropped in a pond. The sound source generates a certain amount of power, measured in watts, which is gradually distributed over an increasingly large area as the wavefront travels further from the source; thus the amount of power per square meter passing through the surface of the imaginary sphere surrounding the source gets smaller with increasing distance (see Fact File 1.3). For practical purposes the intensity of the direct sound from a source drops by 6 dB for every doubling in distance from the source (see Figure 1.14).

The amount of acoustical power generated by real sound sources is surprisingly small, compared with the number of watts of electrical power involved in lighting a light bulb, for example. An acoustical source radiating 20 watts would produce a sound pressure level close to the threshold of pain if a listener was close to the source. Most everyday sources generate fractions of a watt of sound power, and this energy is eventually dissipated into heat by absorption (see below). The amount of heat produced by the dissipation of acoustic energy is relatively insignificant – the chances of increasing the temperature of a room by shouting are slight, at least in the physical sense.

Acoustical power is sometimes confused with the power output of an amplifier used to drive a loudspeaker, and audio engineers will be familiar with power outputs from amplifiers of many hundreds of watts. It is important to realize that loudspeakers are very inefficient devices – that is, they only convert a small proportion of their electrical input power into acoustical power. Thus, even if the input to a loudspeaker was to be, say,

FACT FILE 1.3 THE INVERSE-SQUARE LAW

The law of decreasing power per unit area (intensity) of a wavefront with increasing distance from the source is known as the inverse-square law, because intensity drops in proportion to the inverse square of the distance from the source. Why is this? It is because the sound power from a point source is spread over the surface area of a sphere (S), which from elementary math is given by:

$$S = 4\pi r^2$$

where r is the distance from the source or the radius of the sphere, as shown in the diagram.

If the original power of the source is W watts, then the intensity, or power per unit area (I) at distance r, is:

$$I = W/4\pi r^2$$

For example, if the power of a source was 0.1 watt, the intensity at 4 m distance would be:

$$I = 0.1/(4 \times 3.14 \times 16) = 0.0005\,Wm^{-2}$$

The sound intensity level (SIL) of this signal in decibels can be calculated by comparing it with the accepted reference level of $10^{-12}\,Wm^{-2}$:

$$SIL(dB) = 10\log((5 \times 10^{-4})/(10^{-12})) = 87\,dB$$

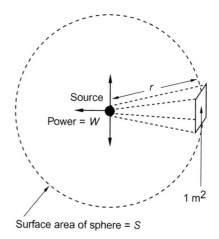

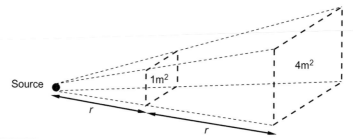

FIGURE 1.14   *The sound power which had passed through $1\,m^2$ of space at distance r from the source will pass through $4\,m^2$ at distance 2r, and thus will have one quarter of the intensity.*

100 watts electrically, the acoustical output power might only be perhaps 1 watt, suggesting a loudspeaker that is only 1% efficient. The remaining power would be dissipated as heat in the voice coil.

Sound pressure is the effect of sound power on its surroundings. To use a central heating analogy, sound power is analogous to the heat energy generated by a radiator into a room, whilst sound pressure is analogous to the temperature of the air in the room. The temperature is what a person entering the room would feel, but the heat-generating radiator is the source of power. Sound pressure level (SPL) is measured in newtons per square meter (Nm^{-2}). A convenient reference level is set for sound pressure and intensity measurements, this being referred to as 0 dB. This level of 0 dB is approximately equivalent to the threshold of hearing (the quietest sound perceivable by an average person) at a frequency of 1 kHz, and corresponds to an SPL of $2 \times 10^{-5} Nm^{-2}$, which in turn is equivalent to an intensity of approximately $10^{-12} Wm^{-2}$ in the free field (see below).

Sound pressure levels are often quoted in dB (e.g. SPL = 63 dB means that the SPL is 63 dB above $2 \times 10^{-5} Nm^{-2}$). The SPL in dB may not accurately represent the loudness of a sound, and thus a subjective unit of loudness has been derived from research data, called the phon. This is discussed further in Chapter 2. Some methods of measuring sound pressure levels are discussed in Fact File 1.4.

FREE AND REVERBERANT FIELDS

The free field in acoustic terms is an acoustical area in which there are no reflections. Truly free fields are rarely encountered in reality, because there are nearly always reflections of some kind, even if at a very low level. If the reader can imagine the sensation of being suspended out of doors, way above the ground, away from any buildings or other surfaces, then he or she will have an idea of the experience of a free-field condition. The result is an acoustically 'dead' environment. Acoustic experiments are sometimes performed in anechoic chambers, which are rooms specially treated so as to produce almost no reflections at any frequency – the surfaces are totally absorptive – and these attempt to create near free-field conditions.

In the free field all the sound energy from a source is radiated away from the source and none is reflected; thus the inverse-square law (Fact File 1.3) entirely dictates the level of sound at any distance from the source. Of course the source may be directional, in which case its directivity factor must be taken into account. A source with a directivity factor of 2 on its axis of maximum radiation radiates twice as much power in this direction as it would have if it had been radiating omnidirectionally. The directivity index of a source is measured in dB, giving the above example a directivity index of 3 dB. If calculating the intensity at a given distance from a directional source (as shown in Fact File 1.3), one must take into account its directivity factor on the axis concerned by multiplying the power of the source by the directivity factor before dividing by $4\pi r^2$.

FACT FILE 1.4 MEASURING SPLS

Typically, a sound pressure level (SPL) meter is used to measure the level of sound at a particular point. It is a device that houses a high-quality omnidirectional (pressure) microphone (see 'Omnidirectional pattern', Chapter 3) connected to amplifiers, filters and a meter (see diagram).

Weighting Filters

The microphone's output voltage is proportional to the SPL incident upon it, and the weighting filters may be used to attenuate low and high frequencies according to a standard curve such as the 'A'-weighting curve, which corresponds closely to the sensitivity of human hearing at low levels (see Chapter 2). SPLs quoted simply in dB are usually unweighted – in other words all frequencies are treated equally – but SPLs quoted in dBA will have been A-weighted and will correspond more closely to the perceived loudness of the signal. A-weighting was originally designed to be valid up to a loudness of 55 phons, since the ear's frequency response becomes flatter at higher levels; between 55 and 85 phons the 'B' curve was intended to be used; above 85 phons the 'C' curve was used. The 'D' curve was devised particularly for measuring aircraft engine noise at very high levels.

Now most standards suggest that the 'A' curve may be used for measuring noise at any SPL, principally for ease of comparability of measurements, but there is still disagreement in the industry about the relative merits of different curves. The 'A' curve attenuates low and high frequencies and will therefore under-read quite substantially for signals at these frequencies. This is an advantage in some circumstances and a disadvantage in others. The 'C' curve is recommended in the USA and Japan for aligning sound levels using noise signals in movie theaters, for example. This only rolls off the very extremes of the audio spectrum and is therefore quite close to an unweighted reading. Some researchers have found that the 'B' curve produces results that more closely relate measured sound signal levels to subjective loudness of those signals.

Noise Criterion or Rating (NC or NR)

Noise levels are often measured in rooms by comparing the level of the noise across the audible range with a standard set of curves called the noise criteria (NC) or noise rating (NR) curves. These curves set out how much noise is acceptable in each of a number of narrow frequency bands for the noise to meet a certain criterion. The noise criterion is then that of the nearest curve above which none of the measured results rises. NC curves are used principally in the USA, whereas NR curves are used principally in Europe. They allow considerably higher levels in low-frequency bands than in middle- and high-frequency bands, since the ear is less sensitive at low frequencies.

In order to measure the NC or NR of a location it is necessary to connect the measuring microphone to a set of filters or a spectrum analyzer which is capable of displaying the SPL in one octave or one-third octave bands.

Further Reading

British Standard 5969. Specification for sound level meters.
British Standard 6402. Sound exposure meters.

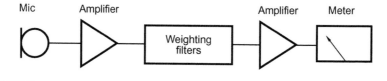

In a room there is both direct and reflected sound. At a certain distance from a source contained within a room the acoustic field is said to be diffuse or reverberant, since reflected sound energy predominates over direct sound. A short time after the source has begun to generate sound a diffuse pattern of reflections will have built up throughout the room, and the reflected sound energy will become roughly constant at any point in the room. Close to the source the direct sound energy is still at quite a high level, and thus the reflected sound makes a smaller contribution to the total. This region is called the near field. (It is popular in sound recording to make use of so-called 'near-field monitors', which are loudspeakers mounted quite close to the listener, such that the direct sound predominates over the effects of the room.)

The exact distance from a source at which a sound field becomes dominated by reverberant energy depends on the reverberation time of the room, and this in turn depends on the amount of absorption in the room, and the room's volume (see Fact File 1.5). Figure 1.15 shows how the SPL changes as distance increases from a source in three different rooms. Clearly, in the acoustically 'dead' room, the conditions approach that of the free field (with sound intensity dropping at close to the expected 6 dB per doubling in distance), since the amount of reverberant energy is very small. The critical distance at which the contribution from direct sound equals that from reflected sound is further from the source than when the room is very reverberant. In the reverberant room the sound pressure level does not change much with distance from the source because reflected sound energy predominates after only a short distance. This is important in room design, since although a short reverberation time may be desirable in a recording control room, for example, it has the disadvantage that the change in SPL with distance from the speakers will be quite severe, requiring very highly powered amplifiers and heavy-duty speakers to provide the necessary level. A slightly longer reverberation time makes the room less disconcerting to work in, and relieves the requirement on loudspeaker power.

STANDING WAVES

The wavelength of sound varies considerably over the audible frequency range, as indicated in Fact File 1.5. At high frequencies, where the wavelength is small, it is appropriate to consider a sound wavefront rather like light – as a ray. Similar rules apply, such as the angle of incidence of a sound wave to a wall is the same as the angle of reflection. At low frequencies where the wavelength is comparable with the dimensions of the room it is necessary to consider other factors, since the room behaves more as a complex resonator, having certain frequencies at which strong pressure peaks and dips are set up in various locations.

Standing waves or eigentones (sometimes also called room modes) may be set up when half the wavelength of the sound or a multiple is equal to one of the dimensions of the room (length, width or height). In such

FACT FILE 1.5 ABSORPTION, REFLECTION AND RT

Absorption

When a sound wave encounters a surface some of its energy is absorbed and some reflected. The absorption coefficient of a substance describes, on a scale from 0 to 1, how much energy is absorbed. An absorption coefficient of 1 indicates total absorption, whereas 0 represents total reflection. The absorption coefficient of substances varies with frequency.

The total amount of absorption present in a room can be calculated by multiplying the absorption coefficient of each surface by its area and then adding the products together. All of the room's surfaces must be taken into account, as must people, chairs and other furnishings. Tables of the performance of different substances are available in acoustics references (see 'Recommended further reading'). Porous materials tend to absorb high frequencies more effectively than low frequencies, whereas resonant membrane- or panel-type absorbers tend to be better at low frequencies. Highly tuned artificial absorbers (Helmholtz absorbers) can be used to remove energy in a room at specific frequencies. The trends in absorption coefficient are shown in the diagram below.

Reflection

The size of an object in relation to the wavelength of a sound is important in determining whether the sound wave will bend round it or be reflected by it. When an object is large in relation to the wavelength the object will act as a partial barrier to the sound, whereas when it is small the sound will bend or diffract around it. Since sound wavelengths in air range from approximately 18 meters at low frequencies to just over 1 cm at high frequencies, most commonly encountered objects will tend to act as barriers to sound at high frequencies but will have little effect at low frequencies.

Reverberation Time

W.C. Sabine developed a simple and fairly reliable formula for calculating the reverberation time (RT_{60}) of a room, assuming that absorptive material is distributed evenly around the surfaces. It relates the volume of the room (V) and its total absorption (A) to the time taken for the sound pressure level to decay by 60 dB after a sound source is turned off.

$$RT_{60} = (0.16V)/A \text{ seconds}$$

In a large room where a considerable volume of air is present, and where the distance between surfaces is large, the absorption of the air becomes more important, in which case an additional component must be added to the above formula:

$$RT_{60} = (0.16V)/(A + xV) \text{ seconds}$$

where x is the absorption factor of air, given at various temperatures and humidities in acoustics references.

The Sabine formula has been subject to modifications by such people as Eyring, in an attempt to make it more reliable in extreme cases of high absorption, and it should be realized that it can only be a guide.

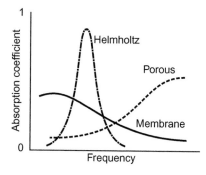

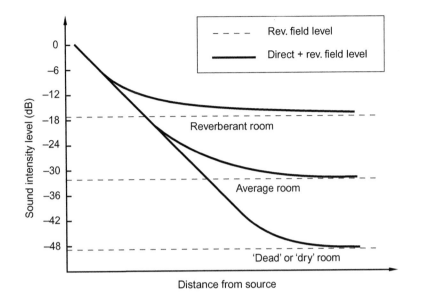

FIGURE 1.15

As the distance from a source increases direct sound level drops but reverberant sound level remains roughly constant. The resultant sound level experienced at different distances from the source depends on the reverberation time of the room, since in a reverberant room the level of reflected sound is higher than in a 'dead' room.

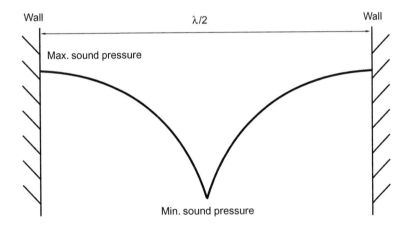

FIGURE 1.16

When a standing wave is set up between two walls of a room there arise points of maximum and minimum pressure. The first simple mode or eigentone occurs when half the wavelength of the sound equals the distance between the boundaries, as illustrated, with pressure maxima at the boundaries and a minimum in the center.

a case (see Figure 1.16) the reflected wave from the two surfaces involved is in phase with the incident wave and a pattern of summations and cancelations is set up, giving rise to points in the room at which the sound pressure is very high, and other points where it is very low. For the first mode (pictured), there is a peak at the two walls and a trough in the center of the room. It is easy to experience such modes by generating a low-frequency sine tone into a room from an oscillator connected to an amplifier and loudspeaker placed in a corner. At selected low frequencies the room will resonate strongly and the pressure peaks may be experienced by walking around the room. There are always peaks towards the boundaries of the room, with troughs distributed at regular intervals between them. The positions of these depend on whether the mode has been created between the walls or between the floor and ceiling. The frequencies (f) at which the strongest modes will occur is given by:

$$f = (c/2) \times (n/d)$$

where c is the speed of sound, d is the dimension involved (distance between walls or floor and ceiling), and n is the number of the mode.

A more complex formula can be used to predict the frequencies of all the modes in a room, including those secondary modes formed by reflections between four and six surfaces (oblique and tangential modes). The secondary modes typically have lower amplitudes than the primary modes (the axial modes) since they experience greater absorption. The formula is:

$$f = (c/2)\sqrt{((p/L)^2 + (q/W)^2 + (r/H)^2)}$$

where p, q and r are the mode numbers for each dimension (1, 2, 3 {...}) and L, W and H are the length, width and height of the room. For example, to calculate the first axial mode involving only the length, make $p = 1$, $q = 0$ and $r = 0$. To calculate the first oblique mode involving all four walls, make $p = 1$, $q = 1$, $r = 0$, and so on.

Some quick sums will show, for a given room, that the modes are widely spaced at low frequencies and become more closely spaced at high frequencies. Above a certain frequency, there arise so many modes per octave that it is hard to identify them separately. As a rule-of-thumb, modes tend only to be particularly problematical up to about 200 Hz. The larger the room the more closely spaced the modes. Rooms with more than one dimension equal will experience so-called degenerate modes in which modes between two dimensions occur at the same frequency, resulting in an even stronger resonance at a particular frequency than otherwise. This is to be avoided.

Since low-frequency room modes cannot be avoided, except by introducing total absorption, the aim in room design is to reduce their effect by adjusting the ratios between dimensions to achieve an even spacing. A number of 'ideal' mode-spacing criteria have been developed by acousticians, but there is not the space to go into these in detail here. Larger rooms are generally more pleasing than small rooms, since the mode spacing is closer at low frequencies, and individual modes tend not to stick out so prominently, but room size has to be traded off against the target reverberation time. Making walls non-parallel does not prevent modes from forming (since oblique and tangential modes are still possible); it simply makes their frequencies more difficult to predict.

The practical difficulty with room modes results from the unevenness in sound pressure throughout the room at mode frequencies. Thus a person sitting in one position might experience a very high level at a particular frequency whilst other listeners might hear very little. A room with prominent LF modes will 'boom' at certain frequencies, and this is unpleasant and undesirable for critical listening. The response of the room modifies the perceived frequency response of a loudspeaker, for example, such that even if the loudspeaker's own frequency response may be acceptable it may become unacceptable when modified by the resonant characteristics of the room.

Room modes are not the only results of reflections in enclosed spaces, and some other examples are given in Fact File 1.6.

FACT FILE 1.6 ECHOES AND REFLECTIONS

Early Reflections

Early reflections are those echoes from nearby surfaces in a room which arise within the first few milliseconds (up to about 50 ms) of the direct sound arriving at a listener from a source (see the diagram). It is these reflections which give the listener the greatest clue as to the size of a room, since the delay between the direct sound and the first few reflections is related to the distance of the major surfaces in the room from the listener. Artificial reverberation devices allow for the simulation of a number of early reflections before the main body of reverberant sound decays, and this gives different reverberation programs the characteristic of different room sizes.

Echoes

Echoes may be considered as discrete reflections of sound arriving at the listener after about 50 ms from the direct sound. These are perceived as separate arrivals, whereas those up to around 50 ms are normally integrated by the brain with the first arrival, not being perceived consciously as echoes. Such echoes are normally caused by more distant surfaces which are strongly reflective, such as a high ceiling or distant rear wall. Strong echoes are usually annoying in critical listening situations and should be suppressed by dispersion and absorption.

Flutter Echoes

A flutter echo is sometimes set up when two parallel reflective surfaces face each other in a room, whilst the other surfaces are absorbent. It is possible for a wavefront to become 'trapped' into bouncing back and forth between these two surfaces until it decays, and this can result in a 'buzzing' or 'ringing' effect on transients (at the starts and ends of impulsive sounds such as hand claps).

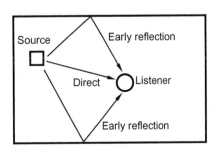

RECOMMENDED FURTHER READING

GENERAL ACOUSTICS

Alton Everest, F., 2009. The Master Handbook of Acoustics, fifth edition. Tab Electronics.

Campbell, M., Greated, C., 2001. The Musician's Guide to Acoustics. Oxford University Press.

Howard, D., Angus, J., 2009. Acoustics and Psychoacoustics, fourth edition. Focal Press.

Kleiner, M., 2011. Acoustics and Audio Technology. J Ross Publishing.

Long, M., 2006. Architectural Acoustics. Academic Press.

Rossing, T.D., 2001. The Science of Sound, third edition. Addison-Wesley.

Auditory Perception

In this chapter the mechanisms by which sound is perceived will be introduced. The human ear often modifies the sounds presented to it before they are presented to the brain, and the brain's interpretation of what it receives from the ears will vary depending on the information contained in the nervous signals. An understanding of loudness perception is important when considering such factors as the perceived frequency balance of a reproduced signal, and an understanding of directional perception is relevant to the study of stereo recording techniques. Below, a number of aspects of the hearing process will be related to the practical world of sound recording and reproduction.

THE HEARING MECHANISM

Although this is not intended to be a lesson in physiology, it is necessary to investigate the basic components of the ear, and to look at how information about sound signals is communicated to the brain. Figure 2.1

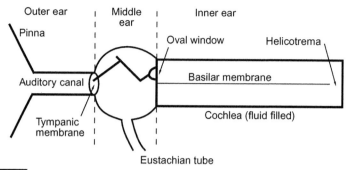

FIGURE 2.1   *A simplified mechanical diagram of the ear.*

shows a diagram of the ear mechanism, not anatomically accurate but showing the key mechanical components. The outer ear consists of the pinna (the visible skin and bone structure) and the auditory canal, and is terminated by the tympanic membrane or 'ear drum'. The middle ear consists of a three-bone lever structure which connects the tympanic membrane to the inner ear via the oval window (another membrane). The inner ear is a fluid-filled bony spiral device known as the cochlea, down the center of which runs a flexible membrane known as the basilar membrane. The cochlea is shown here as if 'unwound' into a straight chamber for the purposes of description. At the end of the basilar membrane, furthest from the middle ear, there is a small gap called the helicotrema which allows fluid to pass from the upper to the lower chamber. There are other components in the inner ear, but those noted above are the most significant.

The ear drum is caused to vibrate in sympathy with the air in the auditory canal when excited by a sound wave, and these vibrations are transferred via the bones of the middle ear to the inner ear, being subject to a multiplication of force of the order of 15:1 by the lever arrangement of the bones. The lever arrangement, coupled with the difference in area between the tympanic membrane and the oval window, helps to match the impedances of the outer and inner ears so as to ensure optimum transfer of energy. Vibrations are thus transferred to the fluid in the inner ear in which pressure waves are set up. The basilar membrane is not uniformly stiff along its length (it is narrow and stiff at the oval window end and wider and more flexible at the far end), and the fluid is relatively incompressible; thus a high-speed pressure wave travels through the fluid and a pressure difference is created across the basilar membrane.

FREQUENCY PERCEPTION

The motion of the basilar membrane depends considerably on the frequency of the sound wave, there being a peak of motion which moves closer towards the oval window the higher the frequency (see Figure 2.2).

At low frequencies the membrane has been observed to move as a whole, with the maximum amplitude of motion at the far end, whilst at higher frequencies there arises a more well-defined peak. It is interesting to note that for every octave (i.e. for every doubling in the frequency) the position of this peak of maximum vibration moves a similar length up the membrane, and this may explain the human preference for displaying frequency-related information on a logarithmic frequency scale, which represents an increase in frequency by showing octaves as equal increments along a frequency axis.

Frequency information is transmitted to the brain in two principal ways. At low frequencies hair cells in the inner ear are stimulated by the vibrations of the basilar membrane, causing them to discharge small electrical impulses along the auditory nerve fibers to the brain. These impulses are found to be synchronous with the sound waveform, and thus the period of the signal can be measured by the brain. Not all nerve fibers are capable of discharging once per cycle of the sound waveform (in fact most have spontaneous firing rates of a maximum of 150 Hz with many being much lower than this). Thus at all but the lowest frequencies the period information is carried in a combination of nerve fiber outputs, with at least a few firing on every cycle (see Figure 2.3). There is evidence to suggest that nerve fibers may re-trigger faster if they are 'kicked' harder – that is, the louder the sound the more regularly they may be made to fire. Also, whilst some fibers will trigger with only a low level of stimulation, others will only fire at high sound levels.

The upper frequency limit at which nerve fibers appear to cease firing synchronously with the signal is around 4 kHz, and above this frequency

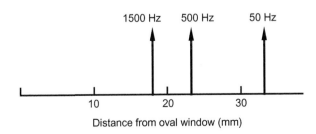

FIGURE 2.2

The position of maximum vibration on the basilar membrane moves towards the oval window as frequency increases.

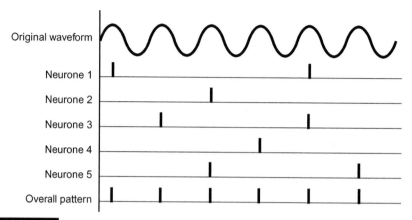

Original waveform

Neurone 1

Neurone 2

Neurone 3

Neurone 4

Neurone 5

Overall pattern

FIGURE 2.3 *Although each neurone does not normally fire on every cycle of the causatory sound wave, the outputs of a combination of neurones firing on different cycles represent the period of the wave.*

the brain relies increasingly on an assessment of the position of maximum excitation of the membrane to decide on the pitch of the signal. There is clearly an overlap region in the middle-frequency range, from about 200 Hz upwards, over which the brain has both synchronous discharge information and 'position' information on which to base its measurement of frequency. It is interesting to note that one is much less able to determine the precise musical pitch of a note when its frequency is above the synchronous discharge limit of 4 kHz.

The frequency selectivity of the ear has been likened to a set of filters, and this concept is described in more detail in Fact File 2.1. It should be noted that there is an unusual effect whereby the perceived pitch of a note is related to the loudness of the sound, such that the pitch shifts slightly with increasing sound level. This is sometimes noticed as loud sounds decay, or when removing headphones, for example. The effect of 'beats' may also be noticed when two pure tones of very similar frequency are sounded together, resulting in a pattern of addition and cancelation as they come in and out of phase with each other. The so-called 'beat frequency' is the difference frequency between the two signals, such that signals at 200 Hz and 201 Hz would result in a cyclic modulation of the overall level, or beat, at 1 Hz. Combined signals slightly further apart in frequency result in a 'roughness' which disappears once the frequencies of the two signals are further than a critical band apart.

FACT FILE 2.1 CRITICAL BANDWIDTH

The basilar membrane appears to act as a rough mechanical spectrum analyzer, providing a spectral analysis of the incoming sound to an accuracy of between one-fifth and one-third of an octave in the middle frequency range (depending on which research data is accepted). It acts rather like a bank of overlapping filters of a fixed bandwidth. This analysis accuracy is known as the critical bandwidth, which is the range of frequencies passed by each notional filter.

The critical band concept is important in understanding hearing because it helps to explain why some signals are 'masked' in the presence of others (see Fact File 2.3). Fletcher, working in the 1940s, suggested that only signals lying within the same critical band as the wanted signal would be capable of masking it, although other work on masking patterns seems to suggest that a signal may have a masking effect on frequencies well above its own.

With complex signals, such as noise or speech, for example, the total loudness of the signal depends to some extent on the number of critical bands covered by a signal. It can be demonstrated by a simple experiment that the loudness of a constant power signal does not begin to increase until its bandwidth extends over more than the relevant critical bandwidth, which appears to support the previous claim. (A useful demonstration of this phenomenon is to be found on the Compact Disc entitled 'Auditory Demonstrations' described at the end of this chapter.)

Although the critical band concept helps to explain the first level of frequency analysis in the hearing mechanism, it does not account for the fine frequency selectivity of the ear which is much more precise than one-third of an octave. One can detect changes in pitch of only a few hertz, and in order to understand this it is necessary to look at the ways in which the brain 'sharpens' the aural tuning curves. For this the reader is referred to Moore (2003), as detailed at the end of this chapter.

LOUDNESS PERCEPTION

The subjective quantity of 'loudness' is not directly related to the SPL of a sound signal (see 'Sound power and sound pressure', Chapter 1). The ear is not uniformly sensitive at all frequencies, and a set of curves has been devised which represents the so-called equal-loudness contours of hearing (see Fact File 2.2). This is partially due to the resonances of the outer ear which have a peak in the middle-frequency region, thus increasing the effective SPL at the ear drum over this range.

The unit of loudness is the phon. If a sound is at the threshold of hearing (just perceivable) it is said to have a loudness of 0 phons, whereas if a sound is at the threshold of pain it will probably have a loudness of around 140 phons. Thus the ear has a dynamic range of approximately 140 phons, representing a range of sound pressures with a ratio of around 10 million to one between the loudest and quietest sounds perceivable. As indicated in Fact File 1.4, the 'A'-weighting curve is often used when measuring sound levels because it shapes the signal spectrum to represent more closely the

FACT FILE 2.2 EQUAL-LOUDNESS CONTOURS

Fletcher and Munson devised a set of curves to show the sensitivity of the ear at different frequencies across the audible range. They derived their results from tests on a large number of subjects who were asked to adjust the level of test tones until they appeared equally as loud as a reference tone with a frequency of 1 kHz. The test tones were spread across the audible spectrum. From these results could be drawn curves of average 'equal loudness', indicating the SPL required at each frequency for a sound to be perceived at a particular loudness level (see diagram).

Loudness is measured in phons, the zero phon curve being that curve which passes through 0 dB SPL at 1 kHz – in other words, the threshold of hearing curve. All points along the 0 phon curve will sound equally loud, although clearly a higher SPL is required at extremes of the spectrum than in the middle. The so-called Fletcher–Munson curves are not the only equal-loudness curves in existence – Robinson and Dadson, amongst others, have published revised curves based upon different test data. The shape of the curves depends considerably on the type of sound used in the test, since filtered noise produces slightly different results to sine tones.

It will be seen that the higher-level curves are flatter than the low-level curves, indicating that the ear's frequency response changes with signal level. This is important when considering monitoring levels in sound recording (see text).

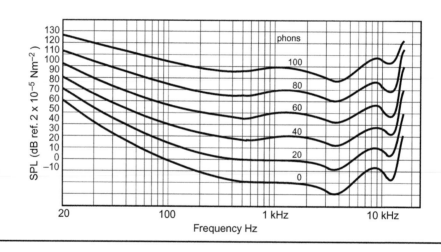

subjective loudness of low-level signals. A noise level quoted in dBA is very similar to a loudness level in phons.

To give an idea of the loudnesses of some common sounds, the background noise of a recording studio might be expected to measure at around 20 phons, a low-level conversation perhaps at around 50 phons, a busy office at around 70 phons, shouted speech at around 90 phons, and a full symphony orchestra playing loudly at around 120 phons. These figures of course depend on the distance from the sound source, but are given as a guide.

The loudness of a sound depends to a great extent on its nature. Broadband sounds tend to appear louder than narrow-band sounds, because they cover more critical bands (see Fact File 2.1), and distorted sounds appear psychologically to be louder than undistorted sounds, perhaps because one associates distortion with system overload. If two music signals are played at identical levels to a listener, one with severe distortion and the other without, the listener will judge the distorted signal to be louder.

A further factor of importance is that the threshold of hearing is raised at a particular frequency in the presence of another sound at a similar frequency. In other words, one sound may 'mask' another – a principle described in more detail in Fact File 2.3.

FACT FILE 2.3 MASKING

Most people have experienced the phenomenon of masking, although it is often considered to be so obvious that it does not need to be stated. As an example: it is necessary to raise your voice in order for someone to hear you if you are in noisy surroundings. The background noise has effectively raised the perception threshold so that a sound must be louder before it can be heard. If one looks at the masking effect of a pure tone, it will be seen that it raises the hearing threshold considerably for frequencies which are the same as or higher than its own (see diagram). Frequencies below the masking tone are less affected. The range of frequencies masked by a tone depends mostly on the area of the basilar membrane set into motion by the tone, and the pattern of motion of this membrane is more extended towards the HF end than towards the LF end. If the required signal produces more motion on the membrane than the masking tone produces at that point then it will be perceived.

The phenomenon of masking has many practical uses in audio engineering. It is used widely in noise reduction systems, since it allows the designer to assume that low-level noise which exists in the same frequency band as a high-level music signal will be effectively masked by the music signal. It is also used in digital audio data compression systems, since it allows the designer to use lower resolution in some frequency bands where the increased noise will be effectively masked by the wanted signal.

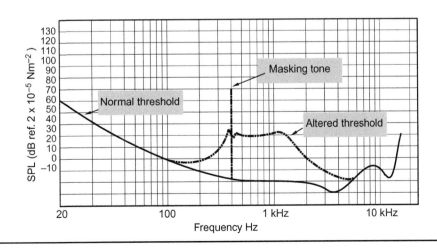

In order to give the impression of a doubling in perceived loudness, an increase of some 9–10 dB is required. Although 6 dB represents a doubling of the actual sound pressure, the hearing mechanism appears to require a greater increase than this for the signal to appear to be twice as loud. Another subjective unit, rarely used in practice, is that of the sone: 1 sone is arbitrarily aligned with 40 phons, and 2 sones is twice as loud as 1 sone, representing approximately 49 phons; 3 sones is three times as loud, and so on. Thus the sone is a true indication of the relative loudness of signals on a linear scale, and sone values may be added together to arrive at the total loudness of a signal in sones.

The ear is by no means a perfect transducer; in fact it introduces considerable distortions into sound signals due to its non-linearity. At high signal levels, especially for low-frequency sounds, the amount of harmonic and intermodulation distortion (see Chapter 18) produced by the ear can be high.

PRACTICAL IMPLICATIONS OF EQUAL-LOUDNESS CONTOURS

The non-linear frequency response of the ear presents the sound engineer with a number of problems. First, the perceived frequency balance of a recording will depend on how loudly it is replayed, and thus a balance made in the studio at one level may sound different when replayed in the home at another. In practice, if a recording is replayed at a much lower level than that at which it was balanced it will sound lacking in bass and extreme treble – it will sound thin and lack warmth. Conversely, if a signal is replayed at a higher level than that at which it was balanced it will have an increased bass and treble response, sounding boomy and overbright.

A 'loudness' control is often provided on hi-fi amplifiers to boost low and high frequencies for low-level listening, but this should be switched out at higher levels. Rock-and-roll and heavy-metal music often sounds lacking in bass when replayed at moderate sound levels because it is usually balanced at extremely high levels in the studio.

Some types of noise will sound louder than others, and hiss is usually found to be most prominent due to its considerable energy content at middle–high frequencies. Rumble and hum may be less noticeable because the ear is less sensitive at low frequencies, and a low-frequency noise which causes large deviations of the meters in a recording may not sound particularly loud in reality. This does not mean, of course, that rumble and hum are acceptable.

Recordings equalized to give a strong mid-frequency content often sound rather 'harsh', and listeners may complain of listening fatigue, since the ear is particularly sensitive in the range between about 1 and 5 kHz.

SPATIAL PERCEPTION

Spatial perception principles are important when considering stereo sound reproduction (see Chapters 16 and 17) and when designing PA rigs for large auditoria, since an objective in both these cases is to give the illusion of directionality and spaciousness.

Sound source localization

Most research into the mechanisms underlying directional sound perception conclude that there are two primary mechanisms at work, the importance of each depending on the nature of the sound signal and the conflicting environmental cues that may accompany discrete sources. These broad mechanisms involve the detection of timing or phase differences between the ears, and of amplitude or spectral differences between the ears. The majority of spatial perception is dependent on the listener having two ears, although certain monoaural cues have been shown to exist – in other words it is mainly the differences in signals received by the two ears that matter.

Time-based cues

A sound source located off the 0° (center front) axis will give rise to a time difference between the signals arriving at the ears of the listener which is related to its angle of incidence, as shown in Figure 2.4. This rises to a maximum for sources at the side of the head, and enables the brain to localize sources in the direction of the earlier ear. The maximum time delay between the ears is of the order of 650 μs or 0.65 ms and is called the binaural delay. It is apparent that humans are capable of resolving direction down to a resolution of a few degrees by this method. There is no obvious way of distinguishing between front and rear sources or of detecting elevation by this method, but one way of resolving this confusion is by taking into account the effect of head movements. Front and rear sources at the same angle of offset from center to one side, for example, will result in opposite changes in time of arrival for a given direction of head turning.

Time difference cues are particularly registered at the starts and ends of sounds (onsets and offsets) and seem to be primarily based on the low-frequency content of the sound signal. They are useful for monitoring the

FIGURE 2.4

The interaural time difference (ITD) for a listener depends on the angle of incidence of the source, as this affects the additional distance that the sound wave has to travel to the more distant ear. In this model the ITD is given by r(θ + sin θ)/c (where c = 340m/s, the speed of sound, and θ is in radians.

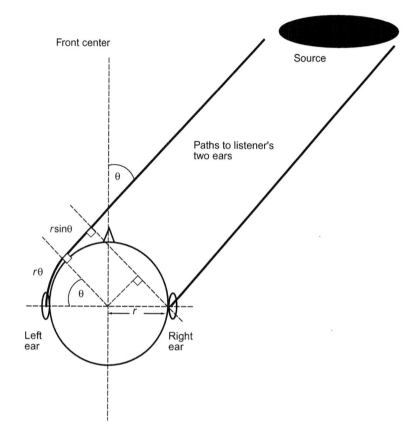

differences in onset and offset of the overall envelope of sound signals at higher frequencies.

Timing differences can be expressed as phase differences when considering sinusoidal signals. The ear is sensitive to interaural phase differences only at low frequencies and the sensitivity to phase begins to deteriorate above about 1 kHz. At low frequencies the hair cells in the inner ear fire regularly at specific points in the phase of the sound cycle, but at high frequencies this pattern becomes more random and not locked to any repeatable point in the cycle. Sound sources in the lateral plane give rise to phase differences between the ears that depend on their angle of offset from the 0° axis (center front). Because the distance between the ears is constant, the phase difference will depend on the frequency and location of the source. (Some sources also show a small difference in the time delay between the ears at LF and HF.) Such a phase difference model of directional perception is only really relevant for continuous sine waves auditioned in anechoic environments, which are rarely heard except in laboratories. It also gives ambiguous information above about 700 Hz where the distance between the ears is equal to half a wavelength of the sound, because it is impossible to tell which ear is lagging and which is leading. Also there arise frequencies where the phase difference is zero. Phase differences can also be confusing in reflective environments where room modes and other effects of reflections may modify the phase cues present at the ears.

When two or more physically separated sources emit similar sounds the precedence effect is important in determining the apparent source direction, as explained in Fact File 2.4.

Amplitude and spectral cues

The head's size makes it an appreciable barrier to sound at high frequencies but not at low frequencies. Furthermore, the unusual shape of the pinna (the visible part of the outer ear) gives rise to reflections and

---

FACT FILE 2.4 THE PRECEDENCE EFFECT

The precedence effect is important for understanding sound localization when two or more sources are emitting essentially the same sound (e.g. a person speaking and a loudspeaker in a different place emitting an amplified version of their voice). It is primarily a feature of transient sounds rather than continuous sounds. In such an example both ears hear both the person and the loudspeaker. The brain tends to localize based on the interaural delay arising from the earliest arriving wavefront, the source appearing to come from a direction towards that of the earliest arriving signal (within limits).

This effect operates over delays between the sources that are somewhat greater than the interaural delay, of the order of a few milliseconds. Similar sounds arriving within up to 50 ms of each other tend to be perceptually fused together, such that one is not perceived as an echo of the other. The time delay over which this fusing effect obtains depends on the source, with clicks tending to separate before complex sounds like music or speech. The timbre and spatial qualities of this 'fused sound', though, may be affected. One form of precedence effect is sometimes referred to as the Haas effect after the Dutch scientist who conducted some of the original experiments. It was originally identified in experiments designed to determine what would happen to the perception of speech in the presence of a single echo. Haas determined that the delayed 'echo' could be made substantially louder than the earlier sound before it was perceived to be equally loud, as shown in the approximation below. The effect depends considerably on the spatial separation of the two or more sources involved. This has important implications for recording techniques where time and intensity differences between channels are used either separately or combined to create spatial cues.

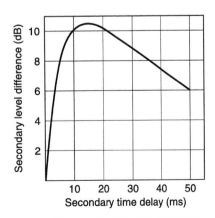

resonances that change the spectrum of the sound at the eardrum depending on the angle of incidence of a sound wave. Reflections off the shoulders and body also modify the spectrum to some extent. A final amplitude cue that may be relevant for spherical wave sources close to the head is the level difference due to the extra distance traveled between the ears by off-center sources. For sources at most normal distances from the head this level difference is minimal, because the extra distance traveled is negligible compared with that already traveled.

The sum of all of these effects is a unique head-related transfer function or HRTF for every source position and angle of incidence, including different elevations and front–back positions. Some examples of HRTFs at different angles are shown in Figure 2.5. It will be seen that there are numerous spectral peaks and dips, particularly at high frequencies, and common features have been found that characterize certain source positions. This, therefore, is a unique form of directional encoding that the brain can learn. Typically, sources to the rear give rise to a reduced high-frequency response in both ears compared to those at the front, owing to the slightly forward-facing shape of the pinna. Sources to one side result in an increased high-frequency difference between the ears, owing to the shadowing effect of the head.

These HRTFs are superimposed on the natural spectra of the source themselves. It is therefore hard to understand how the brain might use the monoaural spectral characteristics of sounds to determine their positions as it would be difficult to separate the timbral characteristics of sources from those added by the HRTF. Monaural cues are likely to be more detectable with moving sources, because moving sources allow the brain to track changes in the spectral characteristics that should be independent of a source's own spectrum. For lateralization it is most likely to be differences in HRTFs between the ears that help the brain to localize sources, in conjunction with the associated interaural time delay. Monaural cues may be more relevant for localization in the median plane where there are minimal differences between the ears.

There are remarkable differences in HRTFs between individuals, although common features can be found. Figure 2.6 shows just two HRTF curves measured by Begault for different subjects, illustrating the problem of generalization in this respect.

The so-called concha resonance (that created by the main cavity in the center of the pinna) is believed to be responsible for creating a sense of externalization – in other words a sense that the sound emanates from outside the head rather than within. Sound-reproducing systems that disturb or distort this resonance, such as certain headphone types, tend to create in-the-head localization as a result.

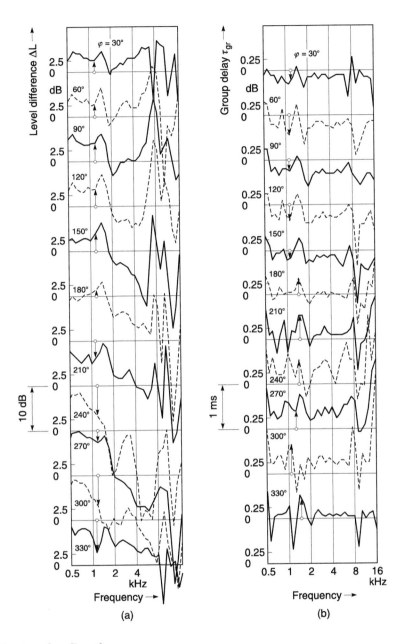

FIGURE 2.5

Monaural transfer functions of the left ear for several directions in the horizontal plane, relative to sound incident from the front; anechoic chamber, 2m loudspeaker distance, impulse technique, 25 subjects, complex averaging (Blauert, 1997). (a) Level difference; (b) time difference. (Courtesy of MIT Press.)

Effects of reflections

Reflections arising from sources in listening spaces affect spatial perception significantly, as discussed in Fact File 2.5. Reflections in the early time period after direct sound (up to 50–80 ms) typically have the effect of broadening or deepening the spatial attributes of a source. They are unlikely to

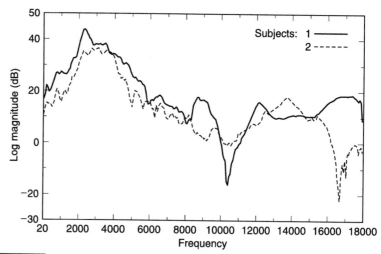

FIGURE 2.6   *HRTFs of two subjects for a source at 0° azimuth and elevation. Note considerable HF differences. (Begault, 1991.)*

be individually localizable. In the period up to about 20 ms they can cause severe timbral coloration if they are at high levels. After 80 ms they tend to contribute more to the sense of envelopment or spaciousness of the environment.

Interaction between hearing and other senses

Some spatial cues are context dependent and may be strongly influenced by the information presented by other senses, particularly vision. Learned experience leads the brain to expect certain cues to imply certain spatial conditions, and if this is contradicted then confusion may arise. For example, it is unusual to experience the sound of a plane flying along beneath one, but the situation can occasionally arise when climbing mountains. Generally one expects planes to fly above, and most people will look up or duck when played loud binaural recordings of planes flying over, even if the spectral cues do not imply this direction.

It is normal to rely quite heavily on the visual sense for information about events within the visible field, and it is interesting to note that most people, when played binaural recordings (see Chapter 16) of sound scenes without accompanying visual information or any form of head tracking, localize the scene primarily behind them rather than in front. In fact obtaining front images from any binaural system using headphones is surprisingly difficult. This may be because one is used to using the hearing sense to localize things where they cannot be seen, and that if something

FACT FILE 2.5 REFLECTIONS AFFECT SPACIOUSNESS

The subjective phenomenon of apparent or auditory source width (ASW) has been studied for a number of years, particularly by psychoacousticians interested in the acoustics of concert halls. ASW relates to the issue of how large a space a source appears to occupy from a sonic point of view (ignoring vision for the moment), as shown below. Individual source width should be distinguished from overall 'sound stage width' (in other words, the distance perceived between the left and right limits of a stereophonic scene).

Early reflected energy in a space (up to about 80 ms) appears to modify the ASW of a source by broadening it somewhat, depending on the magnitude and time delay of early reflections. Concert hall experiments seem to show that subjects prefer larger amounts of ASW, but it is not clear what is the optimum degree of ASW (presumably sources that appeared excessively large would be difficult to localize and unnatural).

Envelopment, spaciousness and sometimes 'room impression' are typically spatial features of a reverberant environment rather than individual sources, and are largely the result of late reflected sound (particularly lateral reflections after about 80 ms). Spaciousness is used most often to describe the sense of open space or 'room' in which the subject is located, usually as a result of some sound sources such as musical instruments playing in that space. It is also related to the sense of 'externalization' perceived – in other words whether the sound appears to be outside the head rather than constrained to a region close to or inside it. Envelopment is a similar term and is used to describe the sense of immersivity and involvement in a (reverberant) soundfield, with that sound appearing to come from all around. It is regarded as a positive quality that is experienced in good concert halls.

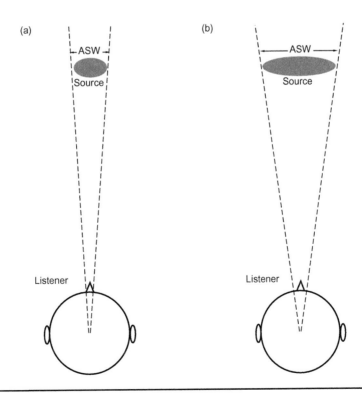

cannot be seen it is likely to be behind. In the absence of the ability to move the head to resolve front–back conflicts the brain tends to assume a rear sound image. So-called 'reversals' in binaural audio systems are consequently very common.

Resolving conflicting cues

In environments where different cues conflict in respect of the implied location of sound sources, the hearing process appears to operate on a sort of majority decision logic basis. In other words it evaluates the available information and votes on the most likely situation, based on what it can determine. Auditory perception has been likened to a hypothesis generation and testing process, whereby likely scenarios are constructed from the available information and tested against subsequent experience (often over a very short time interval). Context-dependent cues and those from other senses are quite important here. Since there is a strong precedence effect favoring the first-arriving wavefront, the direct sound in a reflective environment (which arrives at the listener first) will tend to affect localization most, while subsequent reflections may be considered less important. Head movements will also help to resolve some conflicts, as will visual cues. Reflections from the nearest surfaces, though, particularly the floor, can aid the localizing process in a subtle way. Moving sources also tend to provide more information than stationary ones, allowing the brain to measure changes in the received information that may resolve some uncertainties.

Distance and depth perception

Apart from lateralization of sound sources, the ability to perceive distance and depth of sound images is crucial to our subjective appreciation of sound quality. Distance is a term specifically related to how far away an individual source appears to be, whereas depth can describe the overall front–back distance of a scene and the sense of perspective created. Individual sources may also appear to have depth.

A number of factors appear to contribute to distance perception, depending on whether one is working in reflective or 'dead' environments. Considering for a moment the simple differences between a sound source close to a listener and the same source further away, the one further away will have the following differences:

- Quieter (extra distance traveled)
- Less high-frequency content (air absorbtion)
- More reverberant (in reflective environment)

- Less difference between time of direct sound and first-floor reflection
- Attenuated ground reflection

Numerous studies have shown that absolute distance perception, using the auditory sense alone, is very unreliable in non-reflective environments, although it is possible for listeners to be reasonably accurate in judging relative distances (since there is then a reference point with known distance against which other sources can be compared). In reflective environments, on the other hand, there is substantial additional information available to the brain. The ratio of direct to reverberant sound is directly related to source distance. The reverberation time and the early reflection timing tells the brain a lot about the size of the space and the distance to the surfaces, thereby giving it boundaries beyond which sources could not reasonably be expected to lie.

Naturalness in spatial hearing

The majority of spatial cues received in reproduced sound environments are similar to those received in natural environments, although their magnitudes and natures may be modified somewhat. There are, nonetheless, occasional phenomena that might be considered as specifically associated with reproduced sound, being rarely or never encountered in natural environments. The one that springs most readily to mind is the 'out-of-phase' phenomenon, in which two sound sources such as loudspeakers or headphones are oscillating exactly 180° out of phase with each other – usually the result of a polarity inversion somewhere in the signal chain. This creates an uncomfortable sensation with a strong but rather unnatural sense of spaciousness, and makes phantom sources hard to localize. The out-of-phase sensation never arises in natural listening and many people find it quite disorientating and uncomfortable. Its unfamiliarity makes it hard to identify for naïve listeners, whereas for expert audio engineers its sound is unmistakable. Naïve listeners may even quite like the effect, and extreme phase effects have sometimes been used in low-end audio products to create a sense of extra stereo width.

Audio engineers also often refer to problems with spatial reproduction as being 'phasy' in quality. Usually this is a negative term that can imply abnormal phase differences between the channels, or an unnatural degree of phase difference that may be changing with time. Anomalies in signal processing or microphone technique can create such effects and they are unique to reproduced sound, so there is in effect no natural anchor or reference point against which to compare these experiences.

RECOMMENDED FURTHER READING

Blauert, J., 1997. Spatial Hearing, second edition. Translated by J.S. Allen. MIT Press.

Bregman, A., 1994. Auditory Scene Analysis: The Perceptual Organisation of Sound. MIT Press.

Howard, D., Angus, J., 2009. Acoustics and Psychoacoustics, fourth edition. Focal Press.

Moore, B.C.J., 2003. An Introduction to the Psychology of Hearing, fifth edition. Academic Press.

RECOMMENDED LISTENING

Auditory Demonstrations (Compact Disc). Philips Cat. No. 1126-061. Available from the Acoustical Society of America: http://asa.aip.org/discs.html.

Microphones

CHAPTER CONTENTS

A microphone is a transducer that converts acoustical sound energy into electrical energy, based on the principle described in Fact File 3.1. It performs the opposite function to a loudspeaker, which converts electrical energy into acoustical energy. The three most common principles of operation are the moving coil or 'dynamic', the ribbon, and the capacitor or condenser. The principles of these are described in Fact Files 3.2–3.4

FACT FILE 3.1 ELECTROMAGNETIC TRANSDUCERS

Electromagnetic transducers facilitate the conversion of acoustic signals into electrical signals. They also act to convert electrical signals back into acoustic sound waves. The principle is very simple: if a wire can be made to move in a magnetic field, perpendicular to the lines of flux linking the poles of the magnet, then an electric current is induced in the wire (see diagram). The direction of motion governs the direction of current flow in the wire. If the wire can be made to move back and forth then an alternating current can be induced in the wire, related in frequency and amplitude to the motion of the wire. Conversely, if a current is made to flow through a wire that cuts the lines of a magnetic field then the wire will move.

It is a short step from here to see how acoustic sound signals may be converted into electrical signals and vice versa. A simple moving-coil microphone, as illustrated in Fact File 3.2, involves a wire moving in a magnetic field, by means of a coil attached to a flexible diaphragm that vibrates in sympathy with the sound wave. The output of the microphone is an alternating electrical current, whose frequency is the same as that of the sound wave that caused the diaphragm to vibrate. The amplitude of the electrical signal generated depends on the mechanical characteristics of the transducer, but is proportional to the velocity of the coil.

Vibrating systems, such as transducer diaphragms, with springiness (compliance) and mass, have a resonant frequency (a natural frequency of free vibration). If the driving force's frequency is below this resonant frequency then the motion of the system depends principally on its stiffness; at resonance the motion is dependent principally on its damping (resistance); and above resonance it is mass controlled. Damping is used in transducer diaphragms to control the amplitude of the resonant response peak, and to ensure a more even response around resonance. Stiffness and mass control are used to ensure as flat a frequency response as possible in the relevant frequency ranges. A similar, but reversed process, occurs in a loudspeaker, where an alternating current is fed into a coil attached to a diaphragm, there being a similar magnet around the coil. This time the diaphragm moves in sympathy with the frequency and magnitude of the incoming electrical audio signal, causing compression and rarefaction of the air.

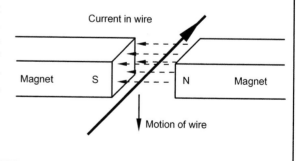

FACT FILE 3.2 DYNAMIC MICROPHONE – PRINCIPLES

The moving-coil microphone functions like a moving-coil speaker in reverse. As shown in the diagram, it consists of a rigid diaphragm, typically 20–30 mm in diameter, which is suspended in front of a magnet. A cylindrical former is attached to the diaphragm on to which is wound a coil of very fine-gauge wire. This sits in the gap of a strong permanent magnet. When the diaphragm is made to vibrate by sound waves the coil in turn moves to and fro in the magnet's gap, and an alternating current flows in the coil, producing the electrical output (see Fact File 3.1). Some models have sufficient windings on the coil to produce a high enough output to be fed directly to the output terminals, whereas other models use fewer windings, the lower output then being fed to a step-up transformer in the microphone casing and then to the output. The resonant frequency of dynamic microphone diaphragms tends to be in the middle frequency region.

The standard output impedance of professional microphones is 200 ohms. This value was chosen because it is high enough to allow useful step-up ratios to be employed in the output transformers, but low enough to allow a microphone to drive long lines of 100 meters or so. It is possible, though, to encounter dynamic microphones with output impedances between 50 and 600 ohms. Some moving-coil models have a transformer that can be wired to give a high-level, high-impedance output suitable for feeding into the lower-sensitivity inputs found on guitar amplifiers and some PA amplifiers. High-impedance outputs can, however, only be used to drive cables of a few meters in length, otherwise severe high-frequency loss results. (This is dealt with fully in Chapter 12.)

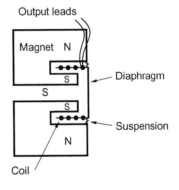

FACT FILE 3.3 RIBBON MICROPHONE – PRINCIPLES

The ribbon microphone consists of a long thin strip of conductive metal foil, pleated to give it rigidity and 'spring', lightly tensioned between two end clamps, as shown in the diagram. The opposing magnetic poles create a magnetic field across the ribbon such that when it is excited by sound waves a current is induced into it (see Fact File 3.1). The electrical output of the ribbon is very small, and a transformer is built into the microphone which steps up the output. The step-up ratio of a particular ribbon design is chosen so that the resulting output impedance is the standard 200 ohms, this also giving an electrical output level comparable with that of moving-coil microphones. The resonant frequency of ribbon microphones is normally at the bottom of the audio spectrum.

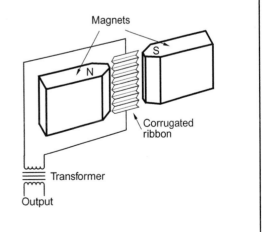

FACT FILE 3.4 CAPACITOR MICROPHONE – PRINCIPLES

The capacitor (or condenser) microphone operates on the principle that if one plate of a capacitor is free to move with respect to the other, then the capacitance (the ability to hold electrical charge) will vary. As shown in the diagram, the capacitor consists of a flexible diaphragm and a rigid back plate, separated by an insulator, the diaphragm being free to move in sympathy with sound waves incident upon it. The 48 volts DC phantom power (see 'Microphone powering options', p. 64) charges the capacitor via a very high resistance. A DC blocking capacitor simply prevents the phantom power from entering the head amplifier, allowing only audio signals to pass.

When sound waves move the diaphragm the capacitance varies, and thus the voltage across the capacitor varies proportionally, since the high resistance only allows very slow leakage of charge from the diaphragm (much slower than the rate of change caused by audio frequencies). This voltage modulation is fed to the head amplifier (via the blocking capacitor) which converts the very high impedance output of the capacitor capsule to a much lower impedance. The output transformer balances this signal (see 'Balanced lines', Chapter 12) and conveys it to the microphone's output terminals. The resonant frequency of a capacitor mic diaphragm is normally at the upper end of the audio spectrum.

The head amplifier consists of a field-effect transistor (FET) which has an almost infinitely high input impedance. Other electronic components are also usually present which perform tasks such as voltage regulation and output stage duties. Earlier capacitor microphones had valves built into the housing, and were somewhat more bulky affairs than their modern counterparts. Additionally, extra wiring had to be incorporated in the mic leads to supply the valves with HT (high-tension) and valve-heater voltages. They were thus not particularly convenient to use, but such is the quality of sound available from capacitor mics that they quickly established themselves. Today, the capacitor microphone is the standard top-quality type; other types being used for relatively specialized applications. The electrical current requirement of capacitor microphones varies from model to model, but generally lies between 0.5 mA and 8 mA, drawn from the phantom power supply.

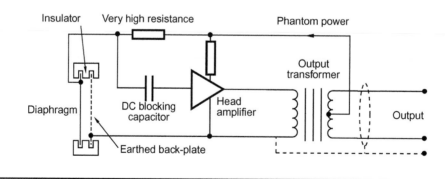

THE MOVING-COIL OR DYNAMIC MICROPHONE

The moving-coil microphone is widely used in the sound reinforcement industry, its robustness making it particularly suitable for hand-held vocal use. Wire-mesh bulbous wind shields are usually fitted to such models,

FACT FILE 3.5 BASS TIP-UP

Pressure-gradient microphones are susceptible to a phenomenon known as bass tip-up, meaning that if a sound source is close to the mic (less than about a meter) the low frequencies become unnaturally exaggerated. In normal operation, the driving force on a pressure-gradient microphone is related almost totally to the phase difference of the sound wave between front and rear of the diaphragm (caused by the extra distance traveled by the wave). For a fixed path-length difference between front and rear, therefore, the phase difference increases with frequency. At LF the phase difference is small and at MF to HF it is larger.

Close to a small source, where the microphone is in a field of roughly spherical waves, sound pressure drops as distance from the source increases (see Fact File 1.3).

Thus, in addition to the phase difference between front and rear of the mic's diaphragm, there is a pressure difference due to the natural level drop with distance from the source. Since the driving force on the diaphragm due to phase difference is small at LF, this pressure drop makes a significant additional contribution, increasing the overall output level at LF. At HF the phase difference is larger, and thus the contribution made by pressure difference is smaller as a proportion of the total driving force.

At greater distances from the source, the sound field approximates more closely to one of plane waves, and the pressure drop over the front–back distance may be considered insignificant as a driving force on the diaphragm, making the mic's output related only to front–back phase difference.

and contain foam material which attenuates wind noise and 'p-blasting' from the vocalist's mouth. Built-in bass attenuation is also often provided to compensate for the effect known as bass tip-up or proximity effect, a phenomenon whereby sound sources at a distance of less than 50 cm or so are reproduced with accentuated bass if the microphone has a directional response (see Fact File 3.5). The frequency response of the moving-coil mic tends to show a resonant peak of several decibels in the upper-mid frequency or 'presence' range, at around 5 kHz or so, accompanied by a fairly rapid fall-off in response above 8 or 10 kHz. This is due to the fact that the moving mass of the coil–diaphragm structure is sufficient to impede the diaphragm's rapid movement necessary at high frequencies. The shortcomings have actually made the moving coil a good choice for vocalists since the presence peak helps to lift the voice and improve intelligibility. Its robustness has also meant that it is almost exclusively used as a bass drum mic in the rock industry. Its sound quality is restricted by its slightly uneven and limited frequency response, but it is extremely useful in applications such as vocals, drums and the micing-up of guitar amplifiers.

One or two high-quality moving-coil mics have appeared with an extended and somewhat smoother frequency response, and one way of achieving this has been to use what are effectively two mic capsules in one housing, one covering mid and high frequencies, one covering the bass.

THE RIBBON MICROPHONE

The ribbon microphone at its best is capable of very high-quality results. The comparatively 'floppy' suspension of the ribbon gives it a low-frequency resonance at around 40 Hz, below which its frequency response fairly quickly falls away. At the high-frequency end the frequency response remains smooth. However, the moving mass of the ribbon itself means that it has difficulty in responding to very high frequencies, and there is generally a roll-off above 14 kHz or so. Reducing the size (therefore the mass) of the ribbon reduces the area for the sound waves to work upon and its electrical output becomes unacceptably low. One manufacturer has adopted a 'double-ribbon' principle which goes some way towards removing this dilemma. Two ribbons, each half the length of a conventional ribbon are mounted one above the other and are connected in series. They are thus analogous to a conventional ribbon that has been 'clamped' in the center. Each ribbon now has half the moving mass and thus a better top-end response. Both of them working together still maintain the necessary output.

The ribbon mic is rather more delicate than the moving coil, and it is better suited to applications where its smooth frequency response comes into its own, such as the micing of acoustic instruments and classical ensembles. There are, however, some robust models which look like moving-coil vocal mics and can be interchanged with them. Micing a rock bass drum with one is still probably not a good idea, due to the very high transient sound pressure levels involved.

THE CAPACITOR OR CONDENSER MICROPHONE

Basic capacitor microphone

The great advantage of the capacitor mic's diaphragm over moving-coil and ribbon types is that it is not attached to a coil and former, and it does not need to be of a shape and size which makes it suitable for positioning along the length of a magnetic field. It therefore consists of an extremely light disc, typically 12–25 mm in diameter, frequently made from polyester coated with an extremely thin vapor-deposited metal layer so as to render it conductive. Sometimes the diaphragm itself is made of a metal such as titanium. The resonant frequency of the diaphragm is typically in the 12–20 kHz range, but the increased output here is rather less prominent than with moving coils due to the diaphragm's very light weight.

Occasionally capacitor microphones are capable of being switched to give a line level output, this being simple to arrange since an amplifier is built into the mic anyway. The high-level output gives the signal rather more immunity to interference when very long cables are employed, and it also removes the need for microphone amplifiers at the mixer or tape recorder. Phantom power does, however, still need to be provided (see 'Phantom power', p. 64).

Electret designs

A much later development was the so-called 'electret' or 'electret condenser' principle. The need to polarize the diaphragm with 48 volts is dispensed with by introducing a permanent electrostatic charge into it during manufacture. In order to achieve this the diaphragm has to be of a more substantial mass, and its audio performance is therefore closer to a moving-coil than to a true capacitor type. The power for the head amplifier is supplied either by a small dry-cell battery in the stem of the mic or by phantom power. The electret principle is particularly suited to applications where compact size and light weight are important, such as in small portable cassette machines (all built-in mics are now electrets) and tie-clip microphones which are ubiquitous in television work. They are also made in vast quantities very cheaply.

Later on, the so-called 'back electret' technique was developed. Here, the diaphragm is the same as that of a true capacitor type, the electrostatic charge being induced into the rigid back plate instead. Top-quality examples of back electrets are therefore just as good as conventional capacitor mics with their 48 volts of polarizing voltage.

RF capacitor microphone

Still another variation on the theme is the RF (radio frequency) capacitor mic, in which the capacitor formed by the diaphragm and back plate forms part of a tuned circuit to generate a steady carrier frequency which is much higher than the highest audio frequency. The sound waves move the diaphragm as before, and this now causes modulation of the tuned frequency. This is then demodulated by a process similar to the process of FM radio reception, and the resulting output is the required audio signal. (It must be understood that the complete process is carried out within the housing of the microphone and it does not in itself have anything to do with radio microphone systems, as discussed in 'Radio microphones', p. 68.)

DIRECTIONAL RESPONSES AND POLAR DIAGRAMS

Microphones are designed to have a specific directional response pattern, described by a so-called 'polar diagram'. The polar diagram is a form of two-dimensional contour map, showing the magnitude of the

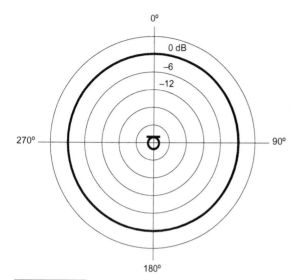

FIGURE 3.1 *Idealized polar diagram of an omnidirectional microphone.*

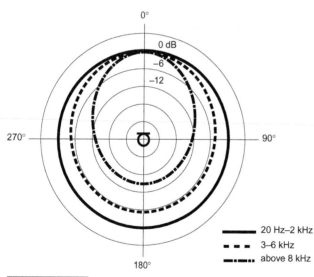

FIGURE 3.2 *Typical polar diagram of an omnidirectional microphone at a number of frequencies.*

microphone's output at different angles of incidence of a sound wave. The distance of the polar plot from the center of the graph (considered as the position of the microphone diaphragm) is usually calibrated in decibels, with a nominal 0 dB being marked for the response at zero degrees at 1 kHz. The further the plot is from the center, the greater the output of the microphone at that angle.

Omnidirectional pattern

Ideally, an omnidirectional or 'omni' microphone picks up sound equally from all directions. The omni polar response is shown in Figure 3.1, and is achieved by leaving the microphone diaphragm open at the front, but completely enclosing it at the rear, so that it becomes a simple pressure transducer, responding only to the change of air pressure caused by the sound waves. This works extremely well at low and mid frequencies, but at high frequencies the dimensions of the microphone capsule itself begin to be comparable with the wavelength of the sound waves, and a shadowing effect causes high frequencies to be picked up rather less well to the rear and sides of the mic. A pressure increase also results for high-frequency sounds from the front. Coupled with this is the possibility for cancelations to arise when a high-frequency wave, whose wavelength is comparable with the diaphragm diameter, is incident from the side of the diaphragm. In such a case positive and negative peaks of the wave may result in opposing forces on the diaphragm.

Figure 3.2 shows the polar response plot which can be expected from a real omnidirectional microphone with a capsule half an inch (13 mm) in diameter. It is perfectly omnidirectional up to around 2 kHz, but then it begins to lose sensitivity at the rear; at 3 kHz its sensitivity at 180° will typically be 6 dB down compared with lower frequencies. Above 8 kHz, the 180° response could be as much as 15 dB down, and the response at 90° and 270° could show perhaps a 10 dB loss. As a consequence, sounds which are being picked up significantly off axis from the microphone will be reproduced with considerable treble loss, and will sound dull. It is at its best on axis and up to 45° either side of the front of the microphone.

High-quality omnidirectional microphones are characterized by their wide, smooth frequency response extending both to the lowest bass frequencies and the high treble with minimum resonances or coloration. This is due to the fact that they are basically very simple in design, being just a capsule which is open at the front and completely enclosed at the rear. (In fact a very small opening is provided to the rear of the diaphragm in order to compensate for overall changes in atmospheric pressure which would otherwise distort the diaphragm.) The small tie-clip microphones which one sees in television work are usually omnidirectional electret types which are capable of very good performance. The smaller the dimensions of the mic, the better the polar response at high frequencies, and mics such as these have quarter-inch diaphragms which maintain a very good omnidirectional response right up to 10 kHz.

Omni microphones are usually the most immune to handling and wind noise of all the polar patterns, since they are only sensitive to absolute sound pressure. Patterns such as figure-eight (especially ribbons) and cardioid, described below, are much more susceptible to handling and wind noise than omnis because they are sensitive to the large pressure difference created across the capsule by low-frequency movements such as those caused by wind or unwanted diaphragm motion. A pressure-gradient microphone's mechanical impedance (the diaphragm's resistance to motion) is always lower at LF than that of a pressure (omni) microphone, and thus it is more susceptible to unwanted LF disturbances.

Figure-eight or bidirectional pattern

The figure-eight or bidirectional polar response is shown in Figure 3.3. Such a microphone has an output proportional to the mathematical cosine of the angle of incidence. One can quickly draw a figure-eight plot on a piece of graph paper, using a protractor and a set of cosine tables or pocket calculator. Cos 0° = 1, showing a maximum response on the forward axis (this will be termed the 0 dB reference point). Cos 90° = 0, so at 90° off axis no sound is picked up. Cos 180° is −1, so the output produced by a sound which is picked up by the rear lobe of the microphone will be 180° out of phase compared with an identical sound picked up by the front lobe. The

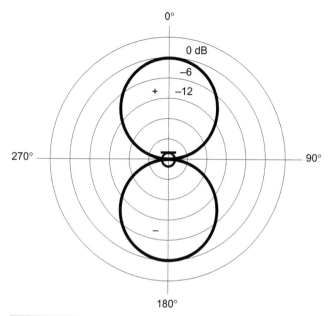

FIGURE 3.3 *Idealized polar diagram of a figure-eight microphone.*

phase is indicated by the + and − signs on the polar diagram. At 45° off axis, the output of the microphone is 3 dB down (cos 45° represents 0.707 or $1/\sqrt{2}$ times the maximum output) compared with the on-axis output.

Traditionally the ribbon microphone has sported a figure-eight polar response, and the ribbon has been left completely open both to the front and to the rear. Such a diaphragm operates on the pressure-gradient principle, responding to the difference in pressure between the front and the rear of the microphone. Consider a sound reaching the mic from a direction 90° off axis to it. The sound pressure will be of equal magnitude on both sides of the diaphragm and so no movement will take place, giving no output. When a sound arrives from the 0° direction a phase difference arises between the front and rear of the ribbon, due to the small additional distance traveled by the wave. The resulting difference in pressure produces movement of the diaphragm and an output results.

At very low frequencies, wavelengths are very long and therefore the phase difference between front and rear of the mic is very small, causing a gradual reduction in output as the frequency gets lower. In ribbon microphones this is compensated for by putting the low-frequency resonance of the ribbon to good use, using it to prop up the bass response. Single-diaphragm capacitor mic designs which have a figure-eight polar response do not have this option, since the diaphragm resonance is at a very high frequency, and a gradual roll-off in the bass can be expected unless other means such as electronic frequency correction in the microphone design have been employed. Double-diaphragm switchable types which have a figure-eight capability achieve this by combining a pair of back-to-back cardioids (see next section) that are mutually out of phase.

Like the omni, the figure-eight can give very clear uncolored reproduction. The polar response tends to be very uniform at all frequencies, except for a slight narrowing above 10 kHz or so, but it is worth noting that a ribbon mic has a rather better polar response at high frequencies in the horizontal plane than in the vertical plane, due to the fact that the ribbon is long and thin. A high-frequency sound coming from a direction somewhat above the plane of the microphone will suffer partial cancelation, since at frequencies where the wavelength begins to be comparable with the length of the ribbon the wave arrives partially out of phase at the lower portion compared with the upper portion, therefore reducing the effective acoustical drive of the ribbon compared with mid frequencies. Ribbon figure-eight microphones should therefore be orientated either upright or upside-down with their stems vertical so as to obtain the best polar response in the horizontal plane, vertical polar response usually being less important.

Although the figure-eight picks up sound equally to the front and to the rear, it must be remembered that the rear pickup is out of phase with the front, and so correct orientation of the mic is required.

Cardioid or unidirectional pattern

The cardioid pattern is described mathematically as $1 + \cos\theta$, where θ is the angle of incidence of the sound. Since the omni has a response of 1 (equal all round) and the figure-eight has a response represented by $\cos\theta$, the cardioid may be considered theoretically as a product of these two responses. Figure 3.4a illustrates its shape. Figure 3.4b shows an omni and a figure-eight superimposed, and one can see that adding the two produces the cardioid shape: at 0°, both polar responses are of equal amplitude and phase, and so they reinforce each other, giving a total output which

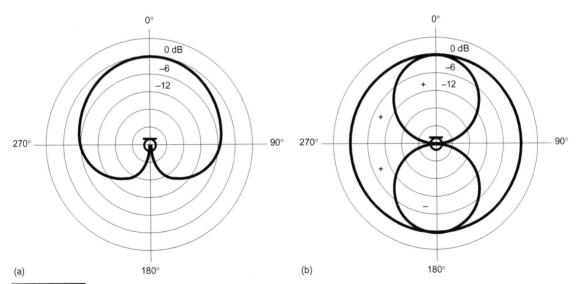

(a)

(b)

FIGURE 3.4 *(a) Idealized polar diagram of a cardioid microphone. (b) A cardioid microphone can be seen to be the mathematical equivalent of an omni and a figure-eight response added together.*

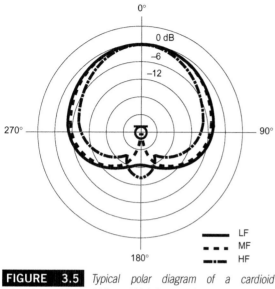

LF
MF
HF

FIGURE 3.5 *Typical polar diagram of a cardioid microphone at low, middle and high frequencies.*

is actually twice that of either separately. At 180°, however, the two are of equal amplitude but opposite phase, and so complete cancelation occurs and there is no output. At 90° there is no output from the figure-eight, but just the contribution from the omni, so the cardioid response is 6 dB down at 90°. It is 3 dB down at 65° off axis.

One or two early microphone designs actually housed a figure-eight and an omni together in the same casing, electrically combining their outputs to give a resulting cardioid response. This gave a rather bulky mic, and also the two diaphragms could not be placed close enough together to produce a good cardioid response at higher frequencies due to the fact that at these frequencies the wavelength of sound became comparable with the distance between the diaphragms. The designs did, however, obtain a cardioid from first principles. The BBC type 4033 was one such example.

The cardioid response is now obtained by leaving the diaphragm open at the front, but introducing various acoustic labyrinths at the rear which cause sound to reach the back of the diaphragm in various combinations of phase and amplitude to produce a resultant cardioid response. This is difficult to achieve at all frequencies simultaneously, and Figure 3.5 illustrates the polar pattern of a typical cardioid mic with a three-quarter-inch diaphragm. As can be seen, at mid frequencies the polar response is very good. At low frequencies it tends to degenerate towards omni, and at very high frequencies it becomes rather more directional than is desirable. Sound arriving from, say, 45° off axis will be reproduced with treble loss, and sounds arriving from the rear will not be completely attenuated, the low frequencies being picked up quite uniformly.

The above example is very typical of moving-coil cardioids, and they are in fact very useful for vocalists due to the narrow pickup at high frequencies helping to exclude off-axis sounds, and also the relative lack of pressure-gradient component at the bass end helping to combat bass tip-up. High-quality capacitor cardioids with half-inch diaphragms achieve a rather more ideal cardioid response. Owing to the presence of acoustic labyrinths, coloration of the sound is rather more likely, and it is not unusual to find that a relatively cheap electret omni will sound better than a fairly expensive cardioid.

Hypercardioid pattern

The hypercardioid, sometimes called 'cottage loaf' because of its shape, is shown in Figure 3.6. It is described mathematically by the formula 0.5 + cos θ, i.e. it is a combination of an omni attenuated by 6 dB, and a figure-eight. Its response is in between the cardioid and figure-eight patterns, having a relatively small rear lobe which is out of phase with the front lobe. Its sensitivity is 3 dB down at 55° off axis. Like the cardioid, the polar response is obtained by introducing acoustic labyrinths to the rear of the diaphragm. Because of the large pressure-gradient component it too is fairly susceptible to bass tip-up. Practical examples of hypercardioid microphones tend to have polar responses which are tolerably close to the ideal. The hypercardioid has the highest direct-to-reverberant ratio of the patterns described, which means that the ratio between the level of on-axis sound and the level of reflected sounds picked up from other angles is very high, and so it is good for excluding unwanted sounds such as excessive room ambience or unwanted noise.

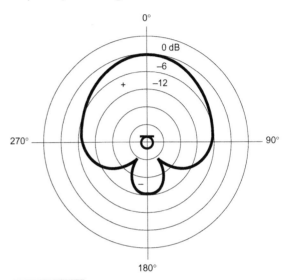

FIGURE 3.6 *Idealized polar diagram of a hypercardioid microphone.*

SPECIALIZED MICROPHONE TYPES

Rifle microphone

The rifle microphone is so called because it consists of a long tube of around three-quarters of an inch (1.9 cm) in diameter and perhaps 2 feet (61 cm) in length, and looks rather like a rifle barrel. The design is effectively an ordinary cardioid microphone to which has been attached a long barrel along which slots are cut in such a way that a sound arriving off axis enters the slots along the length of the tube and thus various versions of the sound arrive at the diaphragm at the bottom of the tube in relative phases which tend to result in cancelation. In this way, sounds arriving off axis are greatly attenuated compared with sounds arriving on axis. Figure 3.7

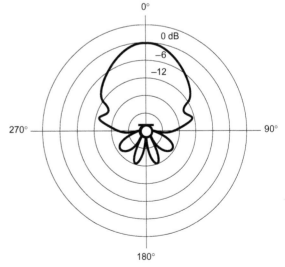

FIGURE 3.7 *Typical polar diagram of a highly directional microphone.*

illustrates the characteristic club-shaped polar response. It is an extremely directional device, and is much used by news sound crews where it can be pointed directly at a speaking subject, excluding crowd noise. It is also used for wildlife recording, sports broadcasts, along the front of theater stages in multiples, and in audience participation discussions where a particular speaker can be picked out. For outside use it is normally completely enclosed in a long, fat wind shield, looking like a very big cigar. Half-length versions are also available which have a polar response midway between a club shape and a hypercardioid. All versions, however, tend to have a rather wider pickup at low frequencies.

Parabolic microphone

An alternative method of achieving high directionality is to use a parabolic dish, as shown in Figure 3.8. The dish has a diameter usually of between 0.5 and 1 meter, and a directional microphone is positioned at its focal point. A large 'catchment area' is therefore created in which the sound is concentrated at the head of the mic. An overall gain of around 15 dB is typical, but at the lower frequencies where the wavelength of sound becomes comparable with the diameter of the dish the response falls away. Because this device actually concentrates the sound rather than merely rejecting off-axis sounds, comparatively high outputs are achieved from distant sound sources. They are very useful for capturing bird song, and they are also sometimes employed around the boundaries of cricket pitches. They are, however, rather cumbersome in a crowd, and can also produce a rather colored sound.

Boundary or 'pressure-zone' microphone

The so-called boundary or pressure-zone microphone (PZM) consists basically of an omnidirectional microphone capsule mounted on a plate usually of around 6 inches (15 cm) square or 6 inches in diameter such that the capsule points directly at the plate and is around 2 or 3 millimeters away from it. The plate is intended to be placed on a large flat surface such as a wall or floor, and it can also be placed on the underside of

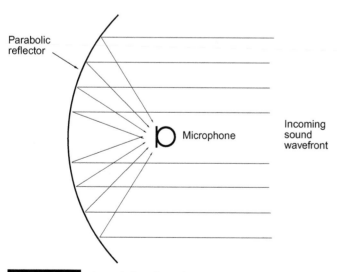

Parabolic reflector

Microphone

Incoming sound wavefront

FIGURE 3.8 *A parabolic reflector is sometimes used to 'focus' the incoming sound wavefront at the microphone position, thus making it highly directional.*

a piano lid, for instance. Its polar response is hemispherical. Because the mic capsule is a simple omni, quite good-sounding versions are available with electret capsules fairly cheaply, and so if one wishes to experiment with this unusual type of microphone one can do so without parting with a great deal of money. It is important to remember though that despite its looks it is not a contact mic – the plate itself does not transduce surface vibrations – and it should be used with the awareness that it is equivalent to an ordinary omnidirectional microphone pointing at a flat surface, very close to it. The frequency response of such a microphone is rarely as flat as that of an ordinary omni, but it can be unobtrusive in use.

SWITCHABLE POLAR PATTERNS

The double-diaphragm capacitor microphone, such as the commercial example shown in Figure 3.9, is a microphone in which two identical diaphragms are employed, placed each side of a central rigid plate in the manner of a sandwich. Perforations in the central plate give both diaphragms an essentially cardioid response. When the polarizing voltage on both diaphragms is the same, the electrically combined output gives an omnidirectional response due to the combination of the back-to-back cardioids in phase. When the polarizing voltage of one diaphragm is opposite to that of the other, and the potential of the rigid central plate is midway between the two, the combined output gives a figure-eight response (back-to-back cardioids mutually out of phase). Intermediate combinations give cardioid and hypercardioid polar responses. In this way the microphone is given a switchable polar response which can be adjusted either by switches on the microphone itself or via a remote control box. Some microphones with switchable polar patterns achieve this by employing a conventional single diaphragm around which are placed appropriate mechanical labyrinths which can be switched to give the various patterns.

Another method manufacturers have used is to make the capsule housing on the end of the microphone detachable, so that a cardioid capsule, say, can be unscrewed and removed to be replaced with, say, an omni. This also facilitates the use of extension tubes whereby a long thin pipe of around a meter or so in length with suitably threaded terminations is inserted between the main microphone body and the capsule. The body of the microphone is mounted on a short floor stand and the thin tube now brings the capsule up to the required height, giving a visually unobtrusive form of microphone stand.

FIGURE 3.9 *A typical double-diaphragm condenser microphone with switchable polar pattern: the AKG C4141B-ULS. (Courtesy of AKG Acoustics GmbH.)*

FIGURE 3.10 *A typical stereo microphone: the Neumann SM69. (Courtesy of FWO Bauch Ltd.)*

FIGURE 3.11 *A typical 'sum-and-difference' stereo microphone: the Shure VP88. (Courtesy of HW International.)*

FIGURE 3.12 *The SoundField ST250 microphone is based on sound-field principles, and can be operated either end- or side-fire, or upside-down, using electrical matrixing of the capsules within the control unit. (Courtesy of SoundField Ltd.)*

STEREO MICROPHONES

Stereo microphones, such as the example shown in Figure 3.10, are available in which two microphones are built into a single casing, one capsule being rotatable with respect to the other so that the angle between the two can be adjusted. Also, each capsule can be switched to give any desired polar response. One can therefore adjust the mic to give a pair of figure-eight microphones angled at, say, 90°, or a pair of cardioids at 120°, and so on. Some stereo mics, such as that pictured in Figure 3.11, are configured in a sum-and-difference arrangement, instead of as a left–right pair, with a 'sum' capsule pointing forwards and a figure-eight 'difference' capsule facing sideways. The sum-and-difference or 'middle and side' (M and S) signals are combined in a matrix box to produce a left–right stereo signal by adding M and S to give the left channel and subtracting M and S to give the right channel. This is discussed in more detail in Fact File 3.6.

A sophisticated stereo microphone is the Soundfield Research microphone. In this design, four 'subcardioid' capsules (i.e. between omni and cardioid) are arranged in a tetrahedral array such that their outputs can be combined in various ways to give four outputs, termed 'B format'. The raw output from the four capsules is termed 'A format'. The four B-format signals consist of a forward-facing figure-eight ('X'), a sideways-facing figure-eight ('Y'), an up-and-down-facing figure-eight ('Z'), and an omnidirectional output ('W'). These are then appropriately combined to produce any configuration of stereo microphone output, each channel being fully adjustable from omni through cardioid to figure-eight, the angles between the capsules also being fully adjustable. The tilt angle of the microphone, and also the 'dominance' (the front-to-back pickup ratio) can also be controlled. All of this is achieved electronically by a remotely sited control unit. Additionally, the raw B-format signals can be recorded on a four-channel recorder, later to be replayed through the control unit where all of the above parameters can be chosen after the recording session.

The ST250 is a second generation stereo microphone based on soundfield principles, designed to be smaller and to be usable either 'end-fire' or 'side-fire' (see Figure 3.12). It can be electronically inverted and polar patterns and capsule angles are variable remotely.

FACT FILE 3.6 SUM AND DIFFERENCE PROCESSING

MS signals may be converted to conventional stereo very easily, either using three channels on a mixer, or using an electrical matrix. M is the mono sum of two conventional stereo channels, and S is the difference between them.

Thus:

$$M = (L + R)/2$$

$$S = (L - R)/2$$

and

$$L = (M + S)/2$$

$$R = (M - S)/2$$

A pair of transformers may be used wired as shown in the diagram to obtain either MS from LR, or vice versa. Alternatively, a pair of summing amplifiers may be used, with the M and S (or L and R) inputs to one being wired in phase (so that they add) and to the other out of phase (so that they subtract).

The mixer configuration shown in the diagram may also be used. Here the M signal is panned centrally (feeding L and R outputs), whilst the S signal is panned left (M + S = L). A post-fader insertion feed is taken from the S channel to a third channel which is phase reversed to give − S. The gain of this channel is set at 0dB and is

panned right (M − S = R). If the S fader is varied in level, the width of the stereo image and the amount of rear pickup can be varied.

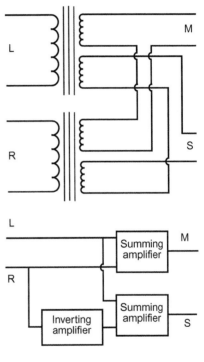

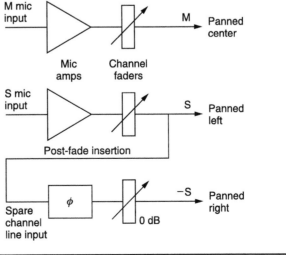

MICROPHONE PERFORMANCE

Professional microphones have a balanced low-impedance output usually via a three-pin XLR-type plug in their base. The impedance, which is usually around 200 ohms but sometimes rather lower, enables long microphone leads to be used. Also, the balanced configuration, discussed in 'Balanced lines', Chapter 12, gives considerable immunity from interference. Other parameters which must be considered are sensitivity (see Fact File 3.7) and noise (see Fact File 3.8).

FACT FILE 3.7 MICROPHONE SENSITIVITY

The sensitivity of a microphone is an indication of the electrical output which will be obtained for a given acoustical sound pressure level (SPL). The standard SPL is either 74 dB (= 1 µB) or 94 dB (= 1 pascal or 10 µB) (µB = microbar). One level is simply ten times greater than the other, so it is easy to make comparisons between differently specified models. 74 dB is roughly the level of moderately loud speech at a distance of 1 meter. 94 dB is 20 dB or ten times higher than this, so a microphone yielding 1 mV µB^{-1}, will yield 10 mV in a soundfield of 94 dB. Other ways of specifying sensitivity include expressing the output as being so many decibels below a certain voltage for a specified SPL. For example, a capacitor mic may have a sensitivity figure of − 60 dBV Pa^{-1} meaning that its output level is 60 dB below 1 volt for a 94 dB SPL, which is 1 mV (60 dB = times 1000).

Capacitor microphones are the most sensitive types, giving values in the region of 5–15 mV Pa^{-1}, i.e. a sound pressure level of 94 dB will give between 5 and 15 millivolts of electrical output. The least sensitive microphones are ribbons, having typical sensitivities of 1–2 mV Pa^{-1}, i.e. around 15 or 20 dB lower than capacitor types. Moving coils are generally a little more sensitive than ribbons, values being typically 1.5–3 mV Pa^{-1}.

FACT FILE 3.8 MICROPHONE NOISE SPECIFICATIONS

All microphones inherently generate some noise. The common way of expressing capacitor microphone noise is the 'A'-weighted equivalent self-noise. A typical value of 'A'-weighted self-noise of a high-quality capacitor microphone is around 18 dBA. This means that its output noise voltage is equivalent to the microphone being placed in a soundfield with a loudness of 18 dBA. A self-noise in the region of 25 dBA from a microphone is rather poor, and if it were to be used to record speech from a distance of a couple of meters or so the hiss would be noticeable on the recording. The very best capacitor microphones achieve self-noise values of around 12 dBA.

When comparing specifications one must make sure that the noise specification is being given in the same units. Some manufacturers give a variety of figures, all taken using different weighting systems and test meter characteristics, but the 'A'-weighted self-noise discussed will normally be present amongst them. Also, a signal-to-noise ratio is frequently quoted for a 94 dB reference SPL, being 94 minus the self-noise, so a mic with a self-noise of 18 dBA will have a signal-to-noise ratio of 76 dBA for a 94 dB SPL, which is also a very common way of specifying noise.

Microphone sensitivity in practice

The consequence of mics having different sensitivity values is that rather more amplification is needed to bring ribbons and moving coils up to line level than is the case with capacitors. For example, speech may yield, say, 0.15 mV from a ribbon. To amplify this up to line level (775 mV) requires a gain of around x5160 or 74 dB. This is a lot, and it taxes the noise performance of the equipment and will also cause considerable amplification of any interference that manages to get into the microphone cables.

Consider now the same speech recording, made using a capacitor microphone of $1\,mV\,\mu B^{-1}$ sensitivity. Now only x775 or 57 dB of gain is needed to bring this up to line level, which means that any interference will have a rather better chance of being unnoticed, and also the noise performance of the mixer will not be so severely taxed. This does not mean that high-output capacitor microphones should always be used, but it illustrates that high-quality mixers and microphone cabling are required to get the best out of low-output mics.

Microphone noise in practice

The noise coming from a capacitor microphone is mainly caused by the head amplifier. Since ribbons and moving coils are purely passive devices one might think that they would therefore be noiseless. This is not the case, since a 200 ohm passive resistance at room temperature generates a noise output between 20 Hz and 20 kHz of 0.26 μV (μV = microvolts). Noise in passive microphones is thus due to thermal excitation of the charge carriers in the microphone ribbon or voice coil, and the output transformer windings. To see what this means in equivalent self-noise terms so that ribbons and moving coils can be compared with capacitors, one must relate this to sensitivity.

Take a moving coil with a sensitivity of $0.2\,mV\,\mu B^{-1}$, which is 2 mV for 94 dB SPL. The noise is 0.26 μV or 0.00026 mV. The signal-to-noise ratio is given by dividing the sensitivity by the noise:

$$2 \div 0.00026 = 7600$$

and then expressing this in decibels:

$$dB = 20\log 7600 = 77\,dB$$

This is an unweighted figure, and 'A' weighting will usually improve it by a couple of decibels. However, the microphone amplifier into which the mic needs to be plugged will add a bit of noise, so it is a good idea to leave this figure as it is to give a fairly good comparison with the capacitor example. (Because the output level of capacitor mics is so much higher than that of moving coils, the noise of a mixer's microphone amplifier does not figure in the noise discussion as far as these are concerned. The noise generated by a capacitor mic is far higher than noise generated by good microphone amplifiers and other types of microphone.)

A 200 ohm moving-coil mic with a sensitivity of $0.2\,\text{mV}\,\mu\text{B}^{-1}$ thus has a signal-to-noise ratio of about 77 dB, and therefore an equivalent self-noise of $94 - 77 = 17\,\text{dB}$ which is comparable with high-quality capacitor types, providing that high-quality microphone amplifiers are also used. A low-output 200 ohm ribbon microphone could have a sensitivity of $0.1\,\text{mV}\,\mu\text{B}^{-1}$, i.e. 6 dB less than the above moving-coil example. Because its 200 ohm thermal noise is roughly the same, its equivalent self-noise is therefore 6 dB worse, i.e. 23 dB. This would probably be just acceptable for recording speech and classical music if an ultra-low-noise microphone amplifier were to be used which did not add significantly to this figure.

The discussion of a few decibels here and there may seem a bit pedantic, but in fact self-noises in the low twenties are just on the borderline of being acceptable if one wishes to record speech or the quieter types of classical music. Loud music, and mic positions close to the sound sources such as is the practice with rock music, generate rather higher outputs from the microphones and here noise is rarely a problem. But the high output levels generated by close micing of drums, guitar amps and the like can lead to overload in the microphone amplifiers. For example, if a high-output capacitor microphone is used to pick up a guitarist's amplifier, outputs as high as 150 mV or more can be generated. This would overload some fixed-gain microphone input stages, and an in-line attenuator which reduces the level by an appropriate amount such as 10–20 dB would have to be inserted at the mixer or recorder end of the microphone line. Attenuators are available built into a short cylindrical tube which carries an XLR-type plug at one end and a socket at the other end. It is simply inserted between the mixer or recorder input and the mic lead connector. It should not be connected at the microphone end because it is best to leave the level of signal along the length of the mic lead high to give it greater immunity from interference.

MICROPHONE POWERING OPTIONS

Phantom power

Consideration of capacitor microphones reveals the need for supplying power to the electronics which are built into the casing, and also the need for a polarizing voltage across the diaphragm of many capacitor types. It would obviously be inconvenient and potentially troublesome to incorporate extra wires in the microphone cable to supply this power, and so an ingenious method was devised whereby the existing wires in the cable which carry the audio signal could also be used to carry the DC voltage necessary for the operation of capacitor mics – hence the term 'phantom power', since it is invisibly carried over the audio wires. Furthermore, this system does not preclude the connection of a microphone not requiring power to a powered circuit. The principle is outlined in Fact File 3.9.

It will be appreciated that if, for instance, a ribbon microphone is connected to the line in place of a capacitor mic, no current will flow into the microphone because there will be no center tap provided on the microphone's output transformer. Therefore, it is perfectly safe to connect other types of balanced microphone to this line. The two 6k8 resistors are necessary for the system because if they were replaced simply by two wires directly connected to the audio lines, these wires would short-circuit the lines together and so no audio signal would be able to pass. The phantom

FACT FILE 3.9 PHANTOM POWERING

The diagram below illustrates the principle of phantom powering. Arrows indicate the path of the phantom power current. (Refer to Chapter 12 for details of the balanced line system.) Here 48 volts DC are supplied to the capacitor microphone as follows: the voltage is applied to each of the audio lines in the microphone cable via two equal value resistors, 6800 (6k8) ohms being the standard value. The current then travels along both audio lines and into the microphone. The microphone's output transformer secondary has either a 'center tap' – that is, a wire connected half-way along the transformer winding, as shown in the diagram – or two resistors as in the arrangement shown at the other end of the line. The current thus travels towards the center of the winding from each end, and then via the center tap to the electronic circuit and diaphragm of the microphone. To complete the circuit, the return path for the current is provided by the screening braid of the microphone cable.

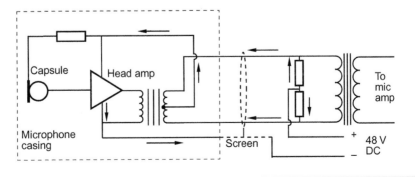

power could be applied to a center tap of the input transformer, but if a short circuit were to develop along the cabling between one of the audio wires and the screen, potentially large currents could be drawn through the transformer windings and the phantom power supply, blowing fuses or burning out components. Two 6k8 resistors limit the current to around 14 mA, which should not cause serious problems. The 6k8 value was chosen so as to be high enough not to load the microphone unduly, but low enough for there to be only a small DC voltage drop across them so that the microphone still receives nearly the full 48 volts. This is known as the P48 standard. Two real-life examples will be chosen to investigate exactly how much voltage drop occurs due to the resistors.

First, the current flows through both resistors equally and so the resistors are effectively 'in parallel'. Two equal-value resistors in parallel behave like a single resistor of half the value, so the two 6k8 resistors can be regarded as a single 3k4 resistor as far as the 48 V phantom power is concerned. Ohm's law (see Fact File 1.1) states that the voltage drop across a resistor is equal to its resistance multiplied by the current passing through it. Now a Calrec 1050C microphone draws 0.5 milliamps (=0.0005 amps) through the resistors, so the voltage drop is $3400 \times 0.0005 = 1.7$ volts. Therefore the microphone receives $48 - 1.7$ volts, i.e. 46.3 volts. The Schoeps CMC-5 microphone draws 4 mA so the voltage drop is $3400 \times 0.004 = 13.6$ volts. Therefore the microphone receives 48–13.6 volts, i.e. 34.4 volts. The manufacturer normally takes this voltage drop into account in the design of the microphone, although examples exist of mics which draw so much current that they load down the phantom voltage of a mixer to a point where it is no longer adequate to power the mics. In such a case some mics become very noisy, some will not work at all, and yet others may produce unusual noises or oscillation. A stand-alone dedicated power supply or internal battery supply may be the solution in difficult cases.

The universal standard is 48 volts, but some capacitor microphones are designed to operate on a range of voltages down to 9 volts, and this can be advantageous, for instance, when using battery-powered equipment on location, or out of doors away from a convenient source of mains power.

Figure 3.13 illustrates the situation with phantom powering when electronically balanced circuits are used, as opposed to transformers. Capacitors are used to block the DC voltage from the power supply, but they present a very low impedance to the audio signal.

A–B powering

Another form of powering for capacitor microphones which is sometimes encountered is A–B powering. Figure 3.14 illustrates this system

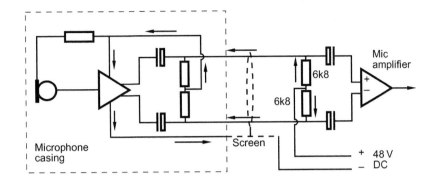

FIGURE 3.13

A typical 48 volt phantom powering arrangement in an electronically balanced circuit.

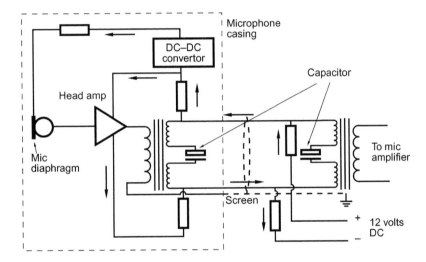

FIGURE 3.14

A typical 12 volt A–B powering arrangement

schematically. Here, the power is applied to one of the audio lines via a resistor and is taken to the microphone electronics via another resistor at the microphone end. The return path is provided by the other audio line as the arrows show. The screen is not used for carrying any current. There is a capacitor at the center of the winding of each transformer. A capacitor does not allow DC to pass, and so these capacitors prevent the current from short-circuiting via the transformer windings. The capacitors have a very low impedance at audio frequencies, so as far as the audio signal is concerned they are not there. The usual voltage used in this system is 12 volts.

Although, like phantom power, the existing microphone lines are used to carry the current, it is dangerous to connect another type of microphone in place of the one illustrated. If, say, a ribbon microphone were to be connected, its output transformer would short-circuit the applied current.

Therefore 12 volt A–B powering should be switched off before connecting any other type of microphone, and this is clearly a disadvantage compared with the phantom powering approach. It is encountered most commonly in location film sound recording equipment.

CONNECTORS

The 3-pin XLR connector (see Fact File 12.3) is the industry standard for balanced microphones. Increasing activity in the surround sound field has given rise to an AES standard for a 5.1 microphone multiway output socket with pins as outlined in AES65-2012. This consists of a single connector to replace the six XLR outputs of the control box, simplifying connection and greatly reducing panel space needed. Table 3.1 outlines the pin designations. A standard circular 19-pin socket, M16 screw-locking, is employed, one pin being left unused.

RADIO MICROPHONES

Radio microphones are widely used in film, broadcasting, theater and other industries, and it is not difficult to think of circumstances in which freedom from trailing microphone cables can be a considerable advantage in all of the above.

Principles

The radio microphone system consists of a microphone front end (which is no different from an ordinary microphone); an FM (frequency modulation) transmitter, either built into the housing of the mic or housed in a separate case into which the mic plugs; a short aerial via which the signal

Table 3.1   AES Standard for Multipin Surround Sound Microphone Connector

Audio channel	Connector contact			Surround channel	Colour code
	+	–	Shield		
1	A	B	N	LEFT	yellow
2	C	D	O	RIGHT	red
3	E	F	P	CENTRE	orange
4	G	H	R	LFE	black/grey
5	I	K	T	LEFT SURROUND	blue
6	L	U	M	RIGHT SURROUND	green

is transmitted; and a receiver which is designed to receive the signal from a particular transmitter. Only one specified transmission frequency is picked up by a given receiver. The audio output of the receiver then feeds a mixer or tape machine in the same manner as any orthodox microphone or line level source would. The principle is illustrated in Figure 3.15.

The transmitter can be built into the stem of the microphone, or it can be housed in a separate case, typically the size of a packet of cigarettes, into which the microphone or other signal source is plugged. A small battery which fits inside the casing of the transmitter provides the power, and this can also supply power to those capacitor mics which are designed to operate at the typical 9 volts of the transmitter battery. The transmitter is of the FM type (see Fact File 3.10), as this offers high-quality audio performance.

Frequently, two or more radio microphones need to be used. Each transmitter must transmit at a different frequency, and the spacing between each adjacent frequency must not be too close otherwise they will interfere with each other. In practice, channels with a minimum spacing of 0.2 MHz are used. Although only one transmitter can be used at a given frequency, any number of receivers can of course be used, as is the case with ordinary radio reception.

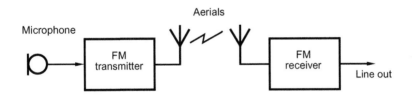

FIGURE 3.15

A radio microphone incorporates an FM transmitter, resulting in no fixed link between microphone and mixer

FACT FILE 3.10 FREQUENCY MODULATION

In FM systems the transmitter radiates a high-frequency radio wave (the carrier) whose frequency is modulated by the amplitude of the audio signal. The positive-going part of the audio waveform causes the carrier frequency to deviate upwards, and the negative-going part causes it to deviate downwards. At the receiver, the modulated carrier is demodulated, converting variations in carrier frequency back into variations in the amplitude of an audio signal.

Audio signals typically have a wide dynamic range, and this affects the degree to which the carrier frequency is modulated. The carrier deviation must be kept within certain limits, and manufacturers specify the maximum deviation permitted. The standard figure for a transmitter with a carrier frequency of around 175 MHz is $\pm$ 75 kHz, meaning that the highest-level audio signal modulates the carrier frequency between 175.075 MHz and 174.925 MHz. The transmitter incorporates a limiter to ensure that these limits are not exceeded.

Facilities

Transmitters are often fitted with facilities which enable the operator to set the equipment up for optimum performance. A 1 kHz line-up tone is sometimes encountered which sends a continuous tone to the receiver to check continuity. Input gain controls are useful, with an indication of peak input level, so that the transmitter can be used with mics and line level sources of widely different output levels. It is important that the optimum setting is found, as too great an input level may cause a limiter (see 'The compressor/limiter', Chapter 13) to come into action much of the time, which can cause compression and 'pumping' noises as the limiter operates. Too weak a signal gives insufficient drive, and poor signal-to-noise ratios can result.

The receiver will have a signal strength indicator. This can be very useful for locating 'dead spots'; transmitter positions which cause unacceptably low meter readings should be avoided, or the receiving aerial should be moved to a position which gives better results. Another useful facility is an indicator which tells the condition of the battery in the transmitter. When the battery voltage falls below a certain level, the transmitter sends out an inaudible warning signal to the receiver which will then indicate this condition. The operator then has a warning that the battery will soon fail, which is often within 15 minutes of the indication.

The 'wireless guide' download available at www.lectrosonics.com provides a huge amount of information about radio microphones.

Licenses

Transmitting equipment usually requires a license for its operation, and governments normally rigidly control the frequency bands over which a given user can operate. This ensures that local and network radio transmitters do not interfere with police, ambulance and fire brigade equipment, etc. In the UK the frequency band for which radio mics do not have to be licensed is between 173.8 MHz and 175 MHz. Each radio mic transmitter needs to be spaced at least 0.2 MHz apart, and commonly used frequencies are 173.8, 174.1, 174.5, 174.8 and 175.0 MHz. An additional requirement is that the frequencies must be crystal controlled, which ensures that they cannot drift outside tightly specified limits. Maximum transmitter power is limited to 10 milliwatts, which gives an effective radiated power (ERP) at the aerial of 2 milliwatts which is very low, but adequate for the short ranges over which radio mics are operated.

In recent years the VHF frequencies have been largely replaced by those in the UHF band, and these have proved to be notably free from interference problems. Aerials are correspondingly shorter, and thus more convenient to wear. Frequencies are mainly in the 800 MHz band, and frequencies for which a UK license is not required are between 863.1 and 864.9 MHz. The Joint Frequency Management Group Ltd is the body which administers licenses in the UK, and their website www.jfmg.co.uk provides much useful information on the subject. The situation in the USA is considerably more complicated, with both licensed and widespread unlicensed use of wireless microphones, often in the 'white space' between other services such as local television channels. There is increased competition for the bandwidth previously used by wireless microphones, from services such as commercial data networking and homeland security. The digital television transition in 2009 also forced licensed wireless microphones onto frequencies below 698 MHz. However, this is accompanied by an FCC ruling that allows continued use of wireless microphones on unused television band channels provided that they meet very strict technical criteria that will limit their potential for interference.

Digital radio microphones

Virtually everything in the audio chain is now digital, but radio mics have been slow to adopt this technology. It has been partly due to difficulties in design, ensuring that bandwidth for a given frequency is sufficiently small to enable a large number of channels to be accommodated across a given frequency range, with any data reduction techniques being used maintaining high sound quality and low latency. Also, because high quality UHF analog systems have performed excellently, the need to go digital has been rather less compelling than within other applications. Nonetheless, digital radio microphone systems are now increasingly prevalent.

Analog FM transmission involves the frequency modulation of a carrier wave, the modulation index (how far above and below the carrier wave frequency the signal can deviate) defining the minimum possible frequency spacing of adjacent channels, and hence the maximum number of channels that can be fitted into an allocated frequency band. Other issues such as inter-channel interference further limit the practical number of usable channels. Digital transmission can be accomplished in a variety of ways. The carrier wave can be modulated by the digital code using simple amplitude shift keying (ASK) where a 1 modulates the carrier wave up in level, and a 0 modulates it down in level. This is equivalent to analog AM radio. With frequency shift keying (FSK) the 1 s and 0 s modulate the frequency of the carrier wave up or down. This is equivalent to the analog FM system. Phase shift keying (PSK) modulates the phase of the carrier wave. Quadrature phase shift keying (QPSK) is a more complex technique which allows a high bit rate to be accommodated in a manageably small bandwidth, and this is the technique normally used for digital radio microphones. Required bandwidth for a particular channel is further kept to a minimum by using data reduction techniques similar to MP3 and other broadcasting systems so that a usefully large number of channels can be fitted into the allowable frequency band. As with analog systems, certain combinations of frequency are best avoided. Bandwidth and data reduction techniques cause latency (delay) of up to 4 or 5 milliseconds in commercial systems. This in itself is insufficient to cause noticeable loss of synchronization between sound and vision, but it can become significant if the signal passes through other equipment such as digital mixers and processors which themselves have latency, and the cumulative effect can become noticeable in some situations.

Other approaches have been marketed. One manufacturer uses uncompressed coding in its SpectraPulse system (Audio-Technica). It operates in a band above 6 GHz using UWB (Ultra Wide Band) transmission. 24 kHz sampling with 16-bit resolution gives 14 channels of uncompressed audio with a data rate of 8 Mbit/s^{-1}. The penalty is an audio bandwidth of 12 kHz. Lectrosonics' 'digital hybrid' system encodes the audio digitally but transmits it as an analog data signal via conventional FM transmission. A compressed error signal is sent to the receiver using FM, representing amplitude-compressed deviations between the input signal and a computed extrapolation of that signal. In the receiver a similar predictor reconstructs the signal by using the transmitted error signal. Conventional digital transmission mutes abruptly when signal strength falls below a certain threshold, but such a hybrid system tends to fade into noise in a similar manner to an analog FM system.

The Joint Frequency Management Group (JFMG) (www.jfmg.co.uk) regulates permitted frequencies and issues appropriate licenses; currently, the frequency allocation for digital radio microphones across Europe is 470–865 MHz, enabling up to at least 32 digital radio mic channels to be accommodated, the exact number depending on how the manufacturer has specified the system with regard to required bandwidth for each channel and the combinations of frequencies that will work well together. The user is however advised to consult the JFMG's website for the latest situation as frequency allocations can be altered as situations develop. In the USA, permitted frequencies occupy a similar range, but again the user is advised to enquire which band of frequencies is appropriate for a particular region or city; the Shure website (www.shure.com) for instance, gives helpful information.

For licensing, it does not matter in principle whether the radio mics are analog or digital, it is the frequencies over which they operate that is the issue. The following information on aerials, diversity and the like applies equally to both types.

Aerials

The dimensions of the transmitting aerial are related to the wavelength of the transmitted frequency. The wavelength (λ) in an electrical conductor at a frequency of 174.5 MHz is approximately 64 inches (160 cm). To translate this into a suitable aerial length, it is necessary to discuss the way in which a signal resonates in a conductor. It is convenient to consider a simple dipole aerial, as shown in Figure 3.16. This consists of two conducting rods, each a quarter of a wavelength long, fed by the transmitting signal as shown. The center of the pair is the nodal point and exhibits a characteristic impedance of about 70 ohms. For a radio mic, we need a total length of $\lambda/2$, i.e. 64/2 = 32 inches (80 cm).

A 32-inch dipole will therefore allow the standard range of radio mic frequencies to resonate along its length to give efficient radiation, the precise length not being too critical. Consideration also has to be given to the radiated polar response (this is not the same as the microphone's polar response). Figure 3.17 shows the polar response for a dipole. As can be seen, it is a figure-eight with no radiation in the directions in which the two halves are pointing. Another factor is polarization of the signal.

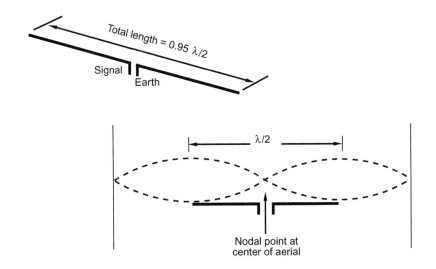

FIGURE 3.16

A simple dipole aerial configuration.

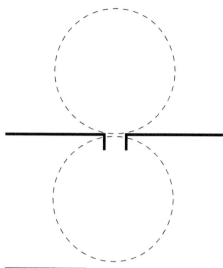

FIGURE 3.17 *The dipole has a figure-eight radiation pattern.*

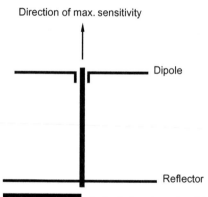

FIGURE 3.18 *A simple two-element aerial incorporates a dipole and a reflector for greater directionality than a dipole.*

Electromagnetic waves consist of an electric wave plus a magnetic wave radiating at right angles to each other, and so if a transmitting aerial is orientated vertically, the receiving aerial should also be orientated vertically. This is termed vertical polarization.

The radio mic transmitter therefore has a transmitting aerial of about 16 inches long: half of a dipole. The other half is provided by the earth screen of the audio input lead, and will be in practice rather longer than 16 inches. The first-mentioned half is therefore looked upon as being the aerial proper, and it typically hangs vertically downwards. The screened signal input cable will generally be led upwards, but other practical requirements tend to override its function as part of the aerial system.

Another type which is often used for hand-held radio mics is the helical aerial. This is typically rather less than half the length of the 16 inch aerial, and has a diameter of a centimeter or so. It protrudes from the base of the microphone. It consists of a tight coil of springy wire housed in a plastic insulator, and has the advantage of being both smaller and reasonably tolerant of physical abuse. Its radiating efficiency is, however, less good than the 16-inch length of wire. At the receiver, a similar aerial is required. The helical aerial is very common here, and its short stubby form is very convenient for outside broadcast and film crews. A 16-inch length of metal tubing, rather like a short car aerial, can be a bit unwieldy although it is a more efficient receiver.

Other aerial configurations exist, offering higher gain and directionality. In the two-element aerial shown in Figure 3.18 the reflector is slightly larger than the dipole, and is spaced behind it at a distance which causes reflection of signal back on to it. It increases the gain, or strength of signal output, by 3 dB. It also attenuates signals approaching from the rear and sides. The three-element 'Yagi', named after its Japanese inventor and shown in Figure 3.19, uses the presence of a director and reflector to increase the gain of a conventional dipole, and a greatly elongated rectangle called a folded dipole is used, which itself has a characteristic impedance of about 300 ohms. The other elements are positioned such that the final impedance is reduced to the standard 50 ohms. The three-element Yagi is even more directional than the dipole, and has increased gain. It can be useful in very difficult reception conditions, or where longer distances are involved such as receiving the signal from a transmitter carried by a rock climber for running commentary! The multi-element, high-gain, highly directional UHF television aerial is of course a familiar sight on our roof-tops.

These aerials can also be used for transmitting, the principles being exactly the same. Their increased directionality also helps to combat multipath problems. The elements should be vertically orientated, because the transmitting aerial will normally be vertical, and the 'direction of maximum sensitivity' arrows on the figures show the direction in which the aerials should be pointed.

Another aerial type which has proved very useful where a large number of radio mics are being used, for example in big musical productions, is the log-periodic type. It covers a considerably wider bandwidth than the aerial types previously discussed. This type is rather like a dipole but with an array of graduated dipole elements along a boom, the longer ones to the rear, progressing down to shorter ones at the front. The appropriate pair then resonate according to the frequency of the transmission, and by these means a single aerial (or, more usually, a pair in a diversity system) can cover the wide band of frequencies over which the numerous transmitters operate. The presence of other elements behind and/or in front of a particular resonating pair gives the aerial a cardioid-like polar response, which is useful for pointing the aerial to the area of desired coverage. Superficially it resembles a Yagi but is fundamentally different in two respects. First, the pairs of elements are somewhat different in length from their neighbors; and second, each pair is an active dipole in its own right, the two halves being insulated from each other. Usually such aerials are fabricated within a solid flat plate rather like the paddle on the end of a rowing oar.

Another technique for improving the signal-to-noise ratio under difficult reception conditions is noise reduction, which operates as follows. Inside the transmitter there is an additional circuit which compresses the incoming audio signal, thus reducing its overall dynamic range. At the receiver, a reciprocal circuit expands the audio signal, after reception and demodulation, and as it pushes the lower-level audio signals back down to their correct level it also therefore pushes the residual noise level down. Previously unacceptable reception conditions will often yield usable results when such transmitters and receivers are employed. It should be noted though that the system does not increase signal strength, and all the problems of transmission and reception still apply. (Noise reduction systems are covered further in Chapter 7.)

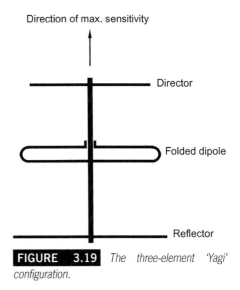

Direction of max. sensitivity

Director

Folded dipole

Reflector

FIGURE   3.19 *The three-element 'Yagi' configuration.*

Aerial siting and connection

It is frequently desirable to place the receiving aerial itself closer to the transmitter than the receiver, in order to pick up a strong signal. To do this an aerial is rigged at a convenient position close to the transmitter, for example in the wings of a theater stage, or on the front of a balcony, and then an aerial lead is run back to the receiver. A helical dipole aerial is frequently employed. In such a situation, characteristic impedance must be considered. As discussed in 'Principles', Chapter 12, when the wavelength of the electrical signal in a conductor is similar to the length of the conductor, reflections can be set up at the receiving end unless the cable is properly terminated. Therefore, impedance matching must be employed between the aerial and the transmitter or receiver, and additionally the connecting lead needs to have the correct characteristic impedance.

The standard value for radio microphone equipment is 50 ohms, and so the aerial, the transmitter, the receiver, the aerial lead and the connectors must all be rated at this value. This cannot be measured using a simple test meter, but an aerial and cable can be tuned using an SWR (standing wave ratio) meter to detect the level of the reflected signal. The aerial lead should be a good-quality, low-loss type, otherwise the advantage of siting the aerial closer to the transmitter will be wasted by signal loss along the cable. Poor signal reception causes noisy performance, because the receiver has a built-in automatic gain control (AGC), which sets the amplification of the carrier frequency to an appropriate value. Weak signals simply require higher amplification and therefore higher noise levels result.

The use of several radio microphones calls for a complementary number of receivers which all need an aerial feed. It is common practice to use just one aerial which is plugged into the input of an aerial distribution amplifier. This distribution unit has several outputs which can be fed into each receiver. It is not possible simply to connect an aerial to all the inputs in parallel due to the impedance mismatch that this would cause.

Apart from obvious difficulties such as metallic structures between transmitter and receiver, there are two phenomena which cause the reception of the radio signal to be less than perfect. The first phenomenon is known as multipath (see Figure 3.20). When the aerial transmits, the signal reaches the receiving aerial by a number of routes. First, there is the direct path from aerial to aerial. Additionally, signals bounce off the walls of the building and reach the receiving aerial via a longer route. So the receiving aerial is faced with a number of signals of more or less random phase and strength, and these will sometimes combine to cause severe signal cancelation and consequently very poor reception. The movement

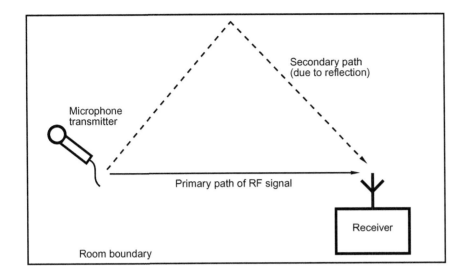

FIGURE 3.20

Multipath distortion can arise between source and receiver due to reflections.

of the transmitter along with the person wearing it will alter the relationship between these multipath signals, and so 'dead spots' are sometimes encountered where particular combinations of multipath signals cause signal 'drop-out'. The solution is to find out where these dead spots are by trial and error, and re-siting the receiving aerial until they are minimized or eliminated. It is generally good practice to site the aerial close to the transmitter so that the direct signal will be correspondingly stronger than many of the signals arriving from the walls. Metal structures should be kept clear of wherever possible due to their ability to reflect and screen RF signals. Aerials can be rigged on metal bars, but at right angles to them, not parallel.

The other phenomenon is signal cancelation from other transmitters when a number of channels are in use simultaneously. Because the transmitting frequencies of the radio mics will be quite close together, partial cancelation of all the signals takes place. The received signals are therefore weaker than for a single transmitter on its own. Again, siting the receiving aerial close to the transmitters is a good idea. The 'sharpness' or 'Q' of the frequency tuning of the receivers plays a considerable part in obtaining good reception in the presence of a number of signals. A receiver may give a good performance when only one transmitter is in use, but a poor Q will vastly reduce the reception quality when several are used. This should be checked for when systems are being evaluated, and the testing of one channel on its own will not of course show up these kinds of problems.

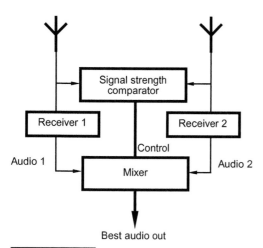

FIGURE   3.21 *A diversity receiver incorporates two aerials spaced apart and two receivers. The signal strength from each aerial is used to determine which output will have the higher quality.*

Diversity reception

A technique known as 'spaced diversity' goes a good way towards combatting the above problems. In this system, two aerials feed two identical receivers for each radio channel. A circuit continuously monitors the signal strength being received by each receiver and automatically selects the one which is receiving the best signal (see Figure 3.21). When they are both receiving a good signal, the outputs of the two are mixed together. A crossfade is performed between the two as one RF signal fades and the other becomes strong.

The two aerials are placed some distance apart, in practice several meters gives good results, so that the multipath relationships between a given transmitter position and each aerial will be somewhat different. A dead spot for one aerial is therefore unlikely to coincide with a dead spot for the other one. A good diversity system overcomes many reception problems, and the considerable increase in reliability of performance is well worth the extra cost. The point at which diversity becomes desirable is when more than two radio microphones are to be used, although good performance from four channels in a non-diversity installation is by no means out of the question. Good radio microphones are very expensive, a single channel of a quality example costing over a thousand pounds today. Cheaper ones exist, but experience suggests that no radio microphone at all is vastly preferable to a cheap one.

RECOMMENDED FURTHER READING

AES, 1979. Microphones: An Anthology. Audio Engineering Society.

Rayburn, R., 2011. Eargle's Microphone Book: From Mono to Stereo to Surround – A Guide to Microphone Design and Application, third edition. Focal Press.

See also 'General further reading' at the end of this book.

USEFUL WEBSITES

www.lectrosonics.com: The 'wireless guide' download provides a huge amount of information about radio microphones.

www.jfmg.co.uk: The UK's radio microphone licensing authority; contains useful information about frequencies and other issues.

Loudspeakers

A loudspeaker is a transducer which converts electrical energy into acoustical energy. A loudspeaker must therefore have a diaphragm of some sort which is capable of being energized in such a way that it vibrates to produce sound waves which are recognizably similar to the original sound from which the energizing signal was derived. To ask a vibrating plastic loudspeaker cone to reproduce the sound of, say, a violin is to ask a great deal, and it is easy to take for granted how successful the best examples have become. Continuing development and refinement of the loudspeaker has brought about a more or less steady improvement in its general performance, but it is a sobering thought that one very rarely mistakes a sound coming from a speaker for the real sound itself, and that one nevertheless has to use these relatively imperfect devices to assess the results of one's work. Additionally, it is easy to hear significant differences between one model and another. Which is right? It is important not to tailor a sound to suit a particular favorite model. There are several principles by which loudspeakers can function, and the commonly employed ones will be briefly discussed.

A word or two must be said about the loudspeaker enclosure. The box can have as big an influence on the final sound of a speaker system as can the drivers themselves. At first sight surprising, this fact can be more readily appreciated when one remembers that a speaker cone radiates virtually the same amount of sound into the cabinet as out into the room. The same amount of acoustical energy that is radiated is therefore also being concentrated in the cabinet, and the sound escaping through the walls and also back out through the speaker cone has a considerable influence upon the final sound of the system.

THE MOVING-COIL LOUDSPEAKER

The moving-coil principle is by far the most widely used, as it can be implemented in very cheap transistor radio speakers, PA (public address) systems, and also top-quality studio monitors, plus all performance levels and applications in between. Figure 4.1 illustrates a cutaway view of a typical moving-coil loudspeaker. Such a device is also known as a drive unit or driver, as it is the component of a complete speaker system which actually produces the sound or 'drives' the air. Basically, the speaker consists of a powerful permanent magnet which has an annular gap to accommodate a coil of wire wound around a cylindrical former. This former is attached to the cone or diaphragm which is held in its rest position by a suspension system which usually consists of a compliant, corrugated, doped (impregnated) cloth material and a compliant surround around the edge of the cone which can be made of a type of rubber, doped fabric, or it can even be an extension of the cone itself, suitably treated to allow the required amount of movement of the cone.

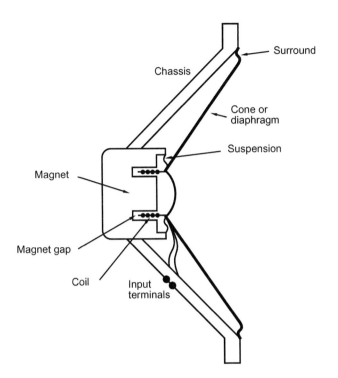

FIGURE 4.1

Cross-section through a typical moving-coil loudspeaker.

Chassis

Surround

Cone or diaphragm

Suspension

Magnet

Magnet gap

Coil

Input terminals

The chassis usually consists either of pressed steel or a casting, the latter being particularly desirable where large heavy magnets are employed, since the very small clearance between the coil and the magnet gap demands a rigid structure to maintain the alignment, and a pressed steel chassis can sometimes be distorted if the loudspeaker is subject to rough handling as is inevitably the case with portable PA systems and the like. (A properly designed pressed steel chassis should not be overlooked though.) The cone itself can in principle be made of almost any material, common choices being paper pulp (as used in many PA speaker cones for its light weight, giving good efficiency), plastics of various types (as used in many hi-fi speaker cones due to the greater consistency achievable than with paper pulp, and the potentially lower coloration of the sound, usually at the expense of increased weight and therefore lower efficiency which is not crucially important in a domestic loudspeaker), and sometimes metal foil.

The principle of operation is based on the principle of electromagnetic transducers described in Fact File 3.1, and is the exact reverse of the process involved in the moving-coil microphone (see Fact File 3.2). The cone vibration sets up sound waves in the air which are an acoustic analog of the electrical input signal. Thus in principle the moving-coil speaker is a very crude and simple device, but the results obtained today are incomparably superior to the original 1920s Kellog and Rice design. It is, however, a great tribute to those pioneers that the principle of operation of what is still today's most widely used type of speaker is still theirs.

OTHER LOUDSPEAKER TYPES

The electrostatic loudspeaker first became commercially viable in the 1950s, and is described in Fact File 4.1. The electrostatic principle is far less commonly employed than is the moving coil, since it is difficult and expensive to manufacture and will not produce the sound levels available from moving-coil speakers. The sound quality of the best examples, such as the Quad ESL 63 pictured in Figure 4.2, is, however, rarely equaled by other types of speakers.

Another technique in producing a panel-type speaker membrane has been to employ a light film on which is attached a series of conductive strips which serve as the equivalent of the coil of a moving-coil cone speaker. The panel is housed within a system of strong permanent

---

FACT FILE 4.1 ELECTROSTATIC LOUDSPEAKER–PRINCIPLES

The electrostatic loudspeaker's drive unit consists of a large, flat diaphragm of extremely light weight, placed between two rigid plates. The diagram shows a side view. There are parallels between this loudspeaker and the capacitor microphone described in Chapter 3.

The diaphragm has a very high resistance, and a DC polarizing voltage in the kilovolt (kV) range is applied to the center tap of the secondary of the input transformer, and charges the capacitor formed by the narrow gap between the diaphragm and the plates. The input signal appears (via the transformer) across the two rigid plates and thus modulates the electrostatic field. The diaphragm, being the other plate of the capacitor, thus experiences a force which alters according to the input signal. Being free to move within certain limits with respect to the two rigid plates, it thus vibrates to produce the sound.

There is no cabinet as such to house the speaker, and sound radiates through the holes of both plates. Sound therefore emerges equally from the rear and the front of the speaker, but not from the sides. Its polar response is therefore a figure-eight, similar to a figure-eight microphone with the rear lobe being out of phase with the front lobe.

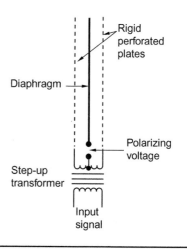

magnets, and the drive signal is applied to the conductive strips. Gaps in the magnets allow the sound to radiate. Such systems tend to be large and expensive like the electrostatic models, but again very high-quality results are possible. In order to get adequate bass response and output level from such panel speakers the diaphragm needs to be of considerable area.

The ribbon loudspeaker principle has sometimes been employed in high-frequency applications ('tweeters') and has recently also been employed in large full-range models. Figure 4.3 illustrates the principle. A light corrugated aluminum ribbon, clamped at each end, is placed between two magnetic poles, one north, one south. The input signal is applied, via a step-down transformer, to each end of the ribbon. The alternating nature of the signal causes an alternating magnetic field around the ribbon, which behaves like a single turn of a coil in a moving-coil speaker. The magnets each side thus cause the ribbon to vibrate, producing sound waves. The impedance of the ribbon is often extremely low, and an amplifier cannot drive it directly. A transformer is therefore used which steps up the impedance of the ribbon. The ribbon itself produces a very low acoustic output and often has a horn in front of it to improve its acoustical matching with the air, giving a higher output for a given electrical input. Some ribbons are, however, very long–half a meter or more–and drive the air directly.

A recent panel type of speaker is the so-called 'distributed mode loudspeaker' (DML), developed by the NXT company following the UK's Defence Evaluation and Research Agency's discovery that certain lightweight composite panels used in military aircraft could act as efficient sound radiators (Figure 4.4). Its operating principle is the antithesis of conventional wisdom: whereas it is normal practice to strive for 'pistonic' motion of a cone driver or panel, the complete area of the radiating surface moving backwards and forwards as a whole with progressively smaller areas of the surface moving as frequency increases, the DML panel is deliberately made very flexible so that a multiplicity of bending modes or resonances, equally distributed in frequency, are set up across its surface.

FIGURE 4.2 *The Quad ESL63 electrostatic loudspeaker. (Courtesy of Quad Electroacoustics Ltd.)*

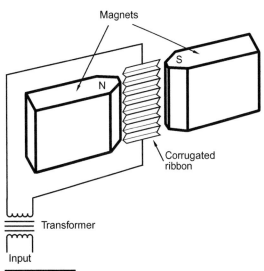

FIGURE 4.3 *A ribbon loudspeaker mechanism.*

FIGURE 4.4   *DML loudspeaker. (Courtesy of New Transducers Ltd.)*

This creates a large number of small radiating areas which are virtually independent of each other, giving an uncorrelated set of signals but summing to give a resultant output. The panel is driven not across its whole area but usually at a strategically placed point by a moving-coil transducer. Because of the essentially random-phase nature of the radiating areas, the panel is claimed not to suffer from the higher-frequency beaming effects of conventional panels, and also there is not the global 180° out-of-phase radiation from the rear.

Further research into DML materials has brought the promise of integrated audio-visual panels, a single screen radiating both sound and vision simultaneously.

There are a few other types of speaker in use, but these are sufficiently uncommon for descriptions not to be merited in this brief outline of basic principles.

MOUNTING AND LOADING DRIVE UNITS

'Infinite baffle' systems

The moving-coil speaker radiates sound equally in front of and to the rear of the diaphragm or cone. As the cone moves forward it produces a compression of the air in front of it but a rarefaction behind it, and vice versa. The acoustical waveforms are therefore 180° out of phase with each other and when they meet in the surrounding air they tend to cancel out, particularly at lower frequencies where diffraction around the cone occurs. A cabinet is therefore employed in which the drive unit sits, which has the job of preventing the sound radiated from the rear of the cone from reaching the open air. The simplest form of cabinet is the sealed box (commonly, but wrongly, known as the 'infinite baffle') which will usually have some sound-absorbing material inside it such as plastic foam or fiber wadding. A true 'infinite baffle' would be a very large flat piece of sheet material with a circular hole cut in the middle into which the drive unit would be mounted. Diffraction around the baffle would then only occur at frequencies below that where the wavelength approached the size of the baffle, and thus cancelation of the two mutually out-of-phase signals would not occur over most of the range, but for this to be effective at the lowest frequencies the baffle would have to measure at least 3 or 4 meters square. The only practical means of employing this type of loading is to mount the speaker in the dividing wall between two rooms, but this is rarely encountered for obvious reasons.

Bass reflex systems

Another form of loading is the bass reflex system, as shown in Figure 4.5. A tunnel, or port, is mounted in one of the walls of the cabinet, and the various parameters of cabinet internal volume, speaker cone weight, speaker cone suspension compliance, port dimensions, and thus mass of air inside the port are chosen so that at a specified low frequency the air inside the port will resonate, which reduces the movement of the speaker cone at that frequency. The port thus produces low-frequency output of its own, acting in combination with the driver. In this manner increased low-frequency output, increased efficiency, or a combination of the two can be achieved. However, it is worth remembering that at frequencies lower than the resonant frequency the driver is acoustically unloaded because the port now behaves simply as an open window. If extremely low frequencies from, say, mishandled microphones or record player arms reach the speaker they will cause considerable excursion of the speaker cone which can cause damage. The air inside a closed box system, however, provides a mechanical supporting 'spring' right down to the lowest frequencies.

A device known as an auxiliary bass radiator (ABR) is occasionally used as an alternative to a reflex port, and takes the form of a further bass unit without its own magnet and coil. It is thus undriven electrically. Its cone mass acts in the same manner as the air plug in a reflex port, but has the advantage that mid-range frequencies are not emitted, resulting in lower coloration.

A further form of bass loading is described in Fact File 4.2.

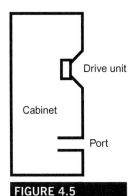

FIGURE 4.5

A ported bass reflex cabinet construction.

Coupled cavity systems

A form of bass loading that has found much favor in small domestic surround sound subwoofers is the coupled cavity, although the technique has been in use even in large sound reinforcement subwoofers for many years. The simplest arrangement of the loading is shown in Figure 4.6a. The drive unit looks into a second or 'coupled' enclosure which is fitted with a port. Whereas reflex ports are tuned to a specific low frequency, here the port is tuned to a frequency above the pass band of the system, e.g. above 120 Hz or so, and the port therefore radiates all sound below that. Above the tuned frequency, it exhibits a 12 dB/octave roll-off. However, secondary resonances in the system require that a low pass filter must still be employed. The increased loading on the driver–it now drives an air cavity coupled to the air plug in the port–produces greater efficiency. Figures 4.6b and 4.6c show other arrangements. In (b) two drivers look into a common central cavity. In (c), the two drivers are also reflex loaded.

FACT FILE 4.2 TRANSMISSION LINE SYSTEM

A form of bass loading is the acoustic labyrinth or 'transmission line', as shown in the diagram. A large cabinet houses a folded tunnel the length of which is chosen so that resonance occurs at a specified low frequency. Above that frequency, the tunnel, which is filled or partially filled with acoustically absorbent material, gradually absorbs the rear-radiated sound energy along its length. At resonance, the opening, together with the air inside the tunnel, behaves like the port of a bass reflex design. An advantage of this type of loading is the very good bass extension achievable, but a large cabinet is required for its proper functioning.

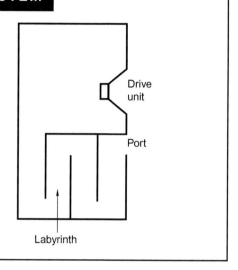

Drive unit

Port

Labyrinth

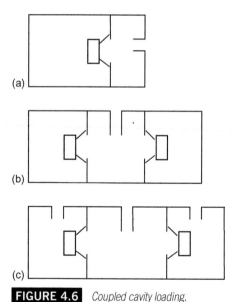

(a)

(b)

(c)

FIGURE 4.6 *Coupled cavity loading.*

Horn loading

Horn loading is a technique commonly employed in large PA loudspeaker systems, as described in Fact File 4.3. Here, a horn is placed in front of the speaker diaphragm. The so-called 'long-throw' horn tends to beam the sound over an included angle of perhaps 90° horizontally and 40° vertically. The acoustical energy is therefore concentrated principally in the forward direction, and this is one reason for the horn's high efficiency. The sound is beamed forwards towards the rear of the hall with relatively little sound reaching the side walls. The 'constant directivity' horn aims to achieve a consistent spread of sound throughout the whole of its working frequency range, and this is usually achieved at the expense of an uneven frequency response. Special equalization is therefore often applied to compensate for this.

The long-throw horn does not do much for those members of an audience who are close to the stage between the speaker stacks, and an acoustic lens is often employed, which, as its name suggests, diffracts the sound, such that the higher frequencies are spread out over a wider angle to give good coverage at the front. Figure 4.7 shows a typical acoustic lens. It consists of a number of metal plates which are

FACT FILE 4.3 HORN LOUDSPEAKER–PRINCIPLES

A horn is an acoustic transformer, that is, it helps to match the air impedance at the throat of the horn (the throat is where the speaker drive unit is) with the air impedance at the mouth. Improved acoustic efficiency is therefore achieved, and for a given electrical input a horn can increase the acoustical output of a driver by 10 dB or more compared with the driver mounted in a conventional cabinet. A horn functions over a relatively limited frequency range, and therefore relatively small horns are used for the high frequencies, larger ones for upper mid frequencies, and so on. This is very worthwhile where high sound levels need to be generated in large halls, rock concerts and open-air events.

Each design of horn has a natural lower cut-off frequency which is the frequency below which it ceases to load the driver acoustically. Very large horns indeed are needed to reproduce low frequencies, and one technique has been to fold the horn up by building it into a more conventional-looking cabinet. The horn principle is rarely employed at bass frequencies due to the necessarily large size. It is, however, frequently employed at mid and high frequencies, but the higher coloration of the sound it produces tends to rule it out for hi-fi and studio monitoring use other than at high frequencies if high sound levels are required. Horns tend to be more directional than conventional speakers, and this has further advantages in PA applications.

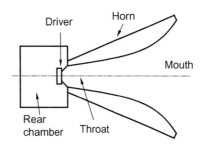

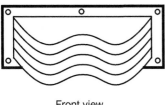

Front view

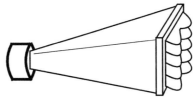

Side view

FIGURE 4.7
An example of an acoustic lens.

shaped and positioned with respect to each other in such a manner as to cause outward diffraction of the high frequencies. The downward slope of the plates is incidental to the design requirements and it is not incorporated to project the sound downwards. Because the available acoustic output is spread out over a wider area than is the case with the long-throw horn, the on-axis sensitivity tends to be lower.

The high efficiency of the horn has also been much exploited in those PA applications which do not require high sound quality, and their use for outdoor events such as fêtes, football matches and the like, as well as on railway station platforms, will have been noticed. Often, a contrivance

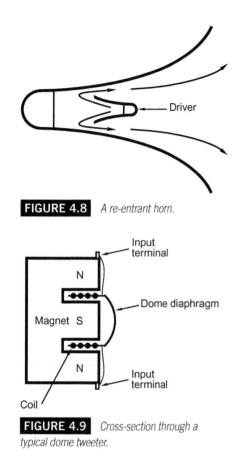

FIGURE 4.8 *A re-entrant horn.*

FIGURE 4.9 *Cross-section through a typical dome tweeter.*

known as a re-entrant horn is used, as shown in Figure 4.8. It can be seen that the horn has been effectively cut in half, and the half which carries the driver is turned around and placed inside the bell of the other. Quite a long horn is therefore accommodated in a compact structure, and this method of construction is particularly applicable to hand-held loudhailers.

The high-frequency horn is driven not by a cone speaker but by a 'compression driver' which consists of a dome-shaped diaphragm usually with a diameter of 1 or 2 inches (2.5 or 5 cm). It resembles a hi-fi dome tweeter but with a flange or thread in front of the dome for fixing on to the horn. The compression driver can easily be damaged if it is driven by frequencies below the cut-off frequency of the horn it is looking into.

COMPLETE LOUDSPEAKER SYSTEMS

Two-way systems

It is a fact of life that no single drive unit can adequately reproduce the complete frequency spectrum from, say, 30 Hz to 20 kHz. Bass frequencies require large drivers with relatively high cone excursions so that adequate areas of air can be set in motion. Conversely, the same cone could not be expected to vibrate at 15 kHz–15,000 times a second to reproduce very high frequencies. A double bass is much larger than a flute, and the strings of a piano which produce the low notes are much fatter and longer than those for the high notes.

The most widely used technique for reproducing virtually the whole frequency spectrum is the so-called two-way speaker system, which is employed at many quality levels from fairly cheap audio packages to very high-quality studio monitors. It consists of a bass/mid driver which handles frequencies up to around 3 kHz, and a high-frequency unit or 'tweeter' which reproduces frequencies from 3 kHz to 20 kHz or more. Figure 4.9 shows a cutaway view of a tweeter. Typically of around 1 inch (2.5 cm) in diameter, the dome is attached to a coil in the same way that a cone is in a bass/mid driver. The dome can be made of various materials, 'soft' or 'hard', and metal domes are also frequently employed. A bass/mid driver cannot adequately reproduce high frequencies as has been said. Similarly, such a small dome tweeter would actually be damaged if bass frequencies were fed to it; thus a crossover network is required to feed each drive unit with frequencies in the correct range, as described in Fact File 4.4.

In a basic system the woofer would typically be of around 8 inches (20 cm) in diameter for a medium-sized domestic speaker, mounted in a cabinet having several cubic feet internal volume. Tweeters are usually sealed at the rear, and therefore they are simply mounted in an appropriate hole cut in the front baffle of the enclosure. This type of speaker is commonly encountered at the cheaper end of the price range, but its simplicity makes it

FACT FILE 4.4 A BASIC CROSSOVER NETWORK

A frequency-dividing network or 'crossover' is fitted into the speaker enclosure which divides the incoming signal into high frequencies (above about 3 kHz) and lower frequencies, sending the latter to the bass/mid unit or 'woofer' and the former to the tweeter. A simple example of the principle involved is illustrated in the diagram. In practical designs additional account should be taken of the fact that speaker drive units are not pure resistances.

The tweeter is fed by a capacitor. A capacitor has an impedance which is inversely proportional to frequency, that is, at high frequencies its impedance is very low and at low frequencies its impedance is relatively high. The typical impedance of a tweeter is 8 ohms, and so for signals below the example of 3 kHz (the 'crossover frequency') a value of capacitor is chosen which exhibits an impedance of 8 ohms also at 3 kHz, and due to the nature of the voltage/current phase relationship of the signal across a capacitor the power delivered to the tweeter is attenuated by 3 dB at that frequency. It then falls at a rate of 6 dB per octave thereafter (i.e. the tweeter's output is 9 dB down at 1.5 kHz, 15 dB down at 750 Hz and so on) thus protecting the tweeter from lower frequencies. The formula which contains the value of the capacitor for the chosen 3 kHz frequency is:

$$f = 1/(2\pi RC)$$

where R is the resistance of the tweeter, and C is the value of the capacitor in farads.

The capacitor value will more conveniently be expressed in microfarads (millionths of a farad) and so the final formula becomes:

$$C = 159155 \div (8\,\text{ohms} \times 3000\,\text{Hz}) = 6.7\,\mu\text{F}$$

Turning now to the woofer, it will be seen that an inductor is placed in series with it. An inductor has an impedance which rises with frequency; therefore, a value is chosen that gives an impedance value similar to that of the woofer at the chosen crossover frequency. Again, the typical impedance of a woofer is 8 ohms. The formula which contains the value of the inductor is:

$$f = R/(2\pi L)$$

where L = inductance in henrys, R = speaker resistance, f = crossover frequency. The millihenry (one-thousandth of a henry, mH) is more appropriate, so this gives:

$$L = 8000/(2\pi \times 3000) = 0.42\,\text{mH}$$

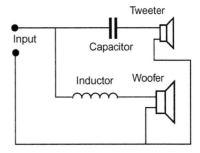

well worth study since it nevertheless incorporates the basic features of many much more costly designs. The latter differ in that they make use of more advanced and sophisticated drive units, higher-quality cabinet materials and constructional techniques, and a rather more sophisticated crossover which usually incorporates both inductors and capacitors in the treble and bass sections as well as resistors which together give much steeper filter slopes than our 6 dB/octave example. Also, the overall frequency response can be adjusted by the crossover to take account of, say, a woofer which gives more acoustic output in the mid-range than in the bass: some attenuation of the mid-range can give a flatter and better-balanced frequency response.

Three-way systems

Numerous three-way loudspeaker systems have also appeared where a separate mid-range driver is incorporated along with additional crossover components to restrict the frequencies feeding it to the mid-range, for example between 400 Hz and 4 kHz. It is an attractive technique due to the fact that the important mid frequencies where much of the detail of music and speech resides are reproduced by a dedicated driver designed specially for that job. But the increased cost and complexity does not always bring about a proportional advance in sound quality.

ACTIVE LOUDSPEAKERS

So far, only 'passive' loudspeakers have been discussed, so named because simple passive components–resistors, capacitors and inductors–are used to divide the frequency range between the various drivers. 'Active' loudspeakers are also encountered, in which the frequency range is divided by active electronic circuitry at line level, after which each frequency band is sent to a separate power amplifier and thence to the appropriate speaker drive unit. The expense and complexity of active systems has tended to restrict the active technique to high-powered professional PA applications where four-, five- and even six-way systems are employed, and to professional studio monitoring speakers, such as the Rogers LS5/8 system pictured in Figure 4.10. Active speakers are still comparatively rare in domestic audio.

Each driver has its own power amplifier, which of course immediately increases the cost and

FIGURE 4.10 *Rogers LS5/8 high-quality active studio loudspeaker. (Courtesy of Swisstone Electronics Ltd.)*

complexity of the speaker system, but the advantages include: lower distortion (due to the fact that the signal is now being split at line level, where only a volt or so at negligible current is involved, as compared with the tens of volts and several amps that passive crossovers have to deal with); greater system-design flexibility due to the fact that almost any combination of speaker components can be used because their differing sensitivities, impedances and power requirements can be compensated for by adjusting the gains of the separate power amplifiers or electronic crossover outputs; better control of final frequency response, since it is far easier to incorporate precise compensating circuitry into an electronic crossover design than is the case with a passive crossover; better clarity of sound and firmer bass simply due to the lack of passive components between power amplifiers and drivers; and an improvement in power amplifier performance due to the fact that each amplifier now handles a relatively restricted band of frequencies.

In active systems amplifiers can be better matched to loudspeakers, and the system can be designed as a whole, without the problems which arise when an unpredictable load is attached to a power amplifier. In passive systems, the designer has little or no control over which type of loudspeaker is connected to which type of amplifier, and thus the design of each is usually a compromise between adaptability and performance. Some active speakers have the electronics built into the speaker cabinet which simplifies installation.

Electronic equalization has also been used to extract a level of bass performance from relatively small-sized enclosures which would not normally be expected to extend to very low frequencies. For example, looking at Figure 4.16a, the response of this speaker can be seen to be about 6 dB down at 55 Hz compared with the mid frequencies, with a roll-off of about 12 dB per octave. Applying 6 dB of boost at 55 Hz with an appropriate filter shape would extend the bass response of the speaker markedly. However, a 6 dB increase corresponds to a four times increase in input power at low frequencies, causing a large increase in speaker cone excursion. For these reasons, such a technique can only be implemented if special high-powered long-throw bass drivers are employed, designed specifically for this kind of application.

SUBWOOFERS

Good bass response from a loudspeaker requires a large internal cabinet volume so that the resonant frequency of the system can be correspondingly low, the response of a given speaker normally falling away below this resonant point. This implies the use of two large enclosures which are likely to be visually obtrusive in a living room, for instance. A way around this problem is to incorporate a so-called 'subwoofer' system. A separate speaker cabinet is employed which handles only the deep bass frequencies, and it is usually driven by its own power amplifier. The signal to drive the power amp comes from an electronic crossover which subtracts the low bass frequencies from the feed to the main stereo amplifier and speakers, and sends the mono sum of the deep bass to the subwoofer system.

Freed from the need to reproduce deep bass, the main stereo speakers can now be small high-quality systems; the subwoofer can be positioned anywhere in the room according to the manufacturers of such systems since it only radiates frequencies below around 100 Hz or so, where sources tend to radiate only omnidirectionally anyway. Degradation of the stereo image has sometimes been noted when the subwoofer is a long way from the stereo pair, and a position close to one of these is probably a good idea.

Subwoofers are also employed in concert and theater sound systems. It is difficult to achieve both high efficiency and a good bass response at the same time from a speaker intended for public address use, and quite large and loud examples often have little output below 70 Hz or so. Subwoofer systems, if properly integrated into the system as a whole, can make a large difference to the weight and scale of live sound.

LOUDSPEAKER PERFORMANCE

Impedance

The great majority of loudspeaker drive units and systems are labeled 'Impedance = 8 ohms'. This is, however, a nominal figure, the impedance in practice varying widely with frequency (see 'Sound in electrical form', Chapter 1). A speaker system may indeed have an 8 ohm impedance at, say, 150 Hz, but at 50 Hz it may well be 30 ohms, and at 10 kHz it could be 4 ohms. Figure 4.11 shows the impedance plot of a typical two-way, sealed box, domestic hi-fi speaker.

The steep rise in impedance at a certain low frequency is indicative of the low-frequency resonance of the system. Other undulations are indicative of the reactive nature of the speaker due to capacitive and inductive elements in the crossover components and the drive units themselves. Also, the driver/box interface has an effect, the most obvious place being at the already-mentioned LF resonant frequency.

Figure 4.12 shows an impedance plot of a bass reflex design. Here we see the characteristic 'double hump' at the bass end. The high peak at

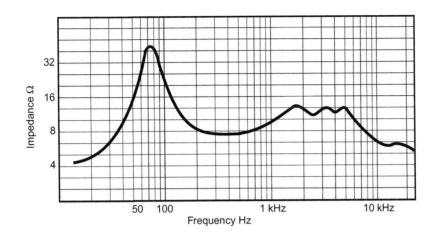

FIGURE 4.11

Impedance plot of a typical two-way sealed-box domestic loudspeaker.

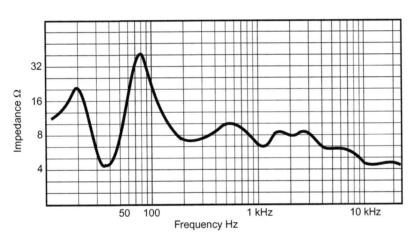

FIGURE 4.12

Impedance plot of a typical bass reflex design.

about 70 Hz is the bass driver/cabinet resonance point. The trough at about 40 Hz is the resonant frequency of the bass reflex port where maximum LF sound energy is radiated from the port itself and minimum energy is radiated from the bass driver. The low peak at about 20 Hz is virtually equal to the free-air resonance of the bass driver itself because at very low frequencies the driver is acoustically unloaded by the cabinet due to the presence of the port opening. A transmission-line design exhibits a similar impedance characteristic.

The DC resistance of an 8 ohm driver or speaker system tends to lie around 7 ohms, and this simple measurement is a good guide if the impedance of an unlabeled speaker is to be estimated. Other impedances encountered include 15 ohm and 4 ohm models. The 4 ohm speakers are harder to drive because for a given amplifier output voltage they draw twice as much current. The 15 ohm speaker is an easy load, but its higher imped-ance means that less current is drawn from the amplifier and so the power (volts x amps) driving the speaker will be correspondingly less. So a power amplifier may not be able to deliver its full rated power into this higher impedance. Thus 8 ohms has become virtually standard, and compe-tently designed amplifiers can normally be expected to drive competently designed speakers. Higher-powered professional power amplifiers can also be expected to drive two 8 ohm speakers in parallel, giving a resultant nom-inal impedance of 4 ohms.

Sensitivity

A loudspeaker's sensitivity is a measure of how efficiently it converts electrical sound energy into acoustical sound energy. The principles are described in Fact File 4.5. Loudspeakers are very inefficient devices indeed. A typical high-quality domestic speaker system has an efficiency of less than 1%, and therefore if 20 watts is fed into it the resulting acoustic out-put will be less than 0.2 acoustical watts. Almost all of the rest of the power is dissipated as heat in the voice coils of the drivers. Horn-loaded systems can achieve a much better efficiency, figures of around 10% being typical. An efficiency figure is not in itself a very helpful thing to know, parameters such as sensitivity and power handling being much more use-ful. But it is as well to be aware that most of the power fed into a speaker has to be dissipated as heat, and prolonged high-level drive causes high voice-coil temperatures.

FACT FILE 4.5 LOUDSPEAKER SENSITIVITY

Sensitivity is defined as the acoustic sound output for a given voltage input. The standard conditions are an input of 2.83 volts (corresponding to 1 watt into 8 ohms) and an acoustic SPL measurement at a distance of 1 meter in front of the speaker. The input signal is pink noise which contains equal sound energy per octave (see 'Frequency spectra of non-repetitive sounds', Chapter 1). A single frequency may correspond with a peak or dip in the speaker's response, leading to an inaccurate overall assessment. For example, a domestic speaker may have a quoted sensitivity of 86 dB W^{-1}, that is, 1 watt of input will produce 86 dB output at 1 meter.

Sensitivities of various speakers differ quite widely and this is not an indication of the sound quality. A high-level professional monitor speaker may have a sensitivity of 98 dB W^{-1} suggesting that it will be very much louder than its domestic cousin, and this will indeed be the case. High-frequency PA horns sometimes achieve a value of 118 dB for just 1 watt input. Sensitivity is thus a useful guide when considering which types of speaker to choose for a given application. A small speaker having a quoted sensitivity of 84 dB W^{-1} and 40 watts power handling will not fill a large hall with sound. The high sound level capability of large professional models will be wasted in a living room.

It has been suggested that sensitivity is not an indication of quality. In fact, it is often found that lower-sensitivity models tend to produce a better sound. This is because refinements in sound quality usually come at the expense of reduced acoustical output for a given input, and PA speaker designers generally have to sacrifice absolute sound quality in order to achieve the high sensitivity and sound output levels necessary for the intended purpose.

Sensitivity: practical design limitations

Designers have always had to work with the conflicting requirements of good sound quality and sensitivity. In the early twentieth century when valve (tube) amplifiers offered only a few watts of power, drivers had to be horn loaded to obtain adequate levels in sound reinforcement and cinema applications, and domestic speakers were often horn loaded too. The early drive units of the 1920s in fact incorporated electromagnets ('energizing coils') because the permanent magnets available at the time were of inadequate strength to give useful sensitivity. By the 1930s, permanent magnets had been developed with the necessary strength, and in particular an alloy of aluminum, cobalt, iron and nickel known as Alnico was offering high magnetic flux in a magnet structure that was quite compact. The addition of titanium and copper gave an alloy known as Ticonal which also gave high strength for a given weight and size.

The speaker cones of the time were invariably made of very thin, flimsy paper pulp, its light weight combined with the inherent stiffness of a conical form enabling high sensitivity to be achieved in conjunction with the new high powered permanent magnets and lightweight voice coils and formers. Such cones suffer from high coloration due to their lack of rigidity; the coil moving too and fro at the apex of the cone controlled its movement in the near vicinity, but further away severe 'breakup modes' appear due to cone flexure where areas of the surface vibrate at various different amplitudes, frequencies and phases causing frequency response irregularities and distortions of several types. Such lively cone behavior proved a good match for the electric guitar during its development in the 1930s and 1940s, and even in the twenty-first century one still finds thin, flimsy paper pulp cones in the best sounding guitar speakers. Alnico magnets shot up in price during the 1970s because of deteriorating political situations in the African countries from which cobalt was sourced, and many 'vintage' guitar speaker models are now fitted with the much cheaper, and much larger for a given magnetic strength, ceramic or ferrite magnets. The latter type of magnet has for many years been used widely in hi-fi and sound reinforcement drive units.

Later developments in magnetic materials have included samarium cobalt, and then an alloy of neodymium, iron and boron was found to give a magnetic strength of about a factor of ten greater than a ceramic magnet of comparable dimensions. Such magnets are notable for their physically small size, and they allow a very efficient magnetic circuit to be designed around the coil as a consequence. The material is however very expensive, and its principal advantage for sound reinforcement speakers apart from efficiency is its somewhat lighter weight for a given strength compared with other magnetic materials. It also comes into its own where high strength is needed in a very small device such as the in-ear headphone and similar applications.

During the course of the twentieth century when higher and higher powered amplifiers became the norm, it was possible to sacrifice sensitivity in the interests of improved sound quality and accuracy, and domestic hi-fi and studio monitor speakers began to be fitted with cones made of various plastic materials (e.g. Bextrene, Polypropylene and other co-polymers and materials, such as that shown in Figure 4.13) which, together with appropriately designed cone flare profiles, gave a much more predictable and consistent series of break-up modes, usually beginning at a somewhat higher frequency than was the case with paper pulp. Even metal has been used in some high quality examples, as shown in Figure 4.14. In partnership with advanced measuring techniques–anechoic chambers, high quality measuring microphones, spectrum and distortion analyzers, laser interferometry to give a visual picture of cone behavior and the like–considerable advances were made in the pursuit of accuracy; but usually at the expense of sensitivity.

FIGURE 4.13   *A carbon fibre cone can possess a high stiffness/mass ratio.*

FIGURE 4.14   *An aluminium cone (Jordan JX53).*

In the 1980s the typical sensitivity of a good domestic hi-fi speaker was about 86 dB/watt, this being perfectly adequate given that amplifiers of 30–100 watts and more were becoming commonplace and relatively cheap. The 1970s had seen an almost wholesale move from valves to transistors, except among guitarists and certain hi-fi enthusiasts. But since then there has been a trend towards improving domestic speaker sensitivity, and today the average is more like 89 dB/watt. The extra 3 dB is worthwhile for several reasons. It indicates developments in cone materials and profiles which can be lower in mass than older designs whilst improving

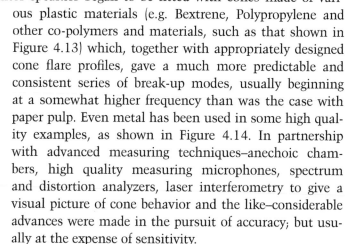

on their performance. Other things being equal, a lighter cone will store less energy and give potentially lower coloration than its heavier counterpart, and improved voice coil, suspension and magnet designs have also made their contributions. One technique has been to increase the diameter of the voice coil considerably so that it now drives the cone a substantial way towards the centre of its area, which helps it to control cone behavior and break-up modes more effectively. Also, speaker distortion tends to be a function of power input rather than acoustical output, and less power needed for a given sound level brings the promise of a lower distortion design. A 3 dB increase in sensitivity after all represents a halving of power needed for a given SPL.

The sensitivity trend for guitar and sound reinforcement drive units was rather different. In fact, little less than a sensitivity war was being waged between competing manufacturers during the course of the 1970s, and this had reached a plateau by 1980, arriving at values of about 102 dB/watt for the most sensitive 12-inch models and in the region of 105 dB for 15-inch drivers. That was about as far as the conventional cone speaker design could be taken, and this can be appreciated by considering the coil/magnet gap relationship. Figure 4.1 shows how the coil sits in the magnet's annular slot between its poles. When signal is applied, the coil moves to and fro to drive the cone. The most sensitive speakers employ a gap length which is equal to the coil length, so that the whole of the coil is immersed in the magnetic field, giving maximum efficiency for the system. Many guitar speakers are of this type (as are most high frequency drivers); voice coil and cone excursion for this application is minimal because the electric guitar frequency spectrum is weak in fundamentals, and the coil stays almost entirely within the magnetic field even for quite high power inputs. Increased input causes the ends of the coil to move to and fro beyond the magnet gap, introducing compression and distortion artefacts which have in fact played their part in electric guitar sound. Drive units intended for lower distortion sound reinforcement and hi-fi therefore employ a coil that is slightly longer than the magnet gap such that a small percentage of its length extends beyond the gap to both front and rear when no drive is being applied, as shown in Figure 4.15a. Such a design is slightly less sensitive because not all of the coil is immersed in the magnetic field, but the coil can now move a significant distance to and fro whilst keeping the same percentage of its length within the gap. Higher output for a given value of distortion is thereby achieved. The peak-to-peak linear excursion, equal to the total excess length of the coil compared to the magnet gap length, is the X_{max} value given in a drive unit's specification.

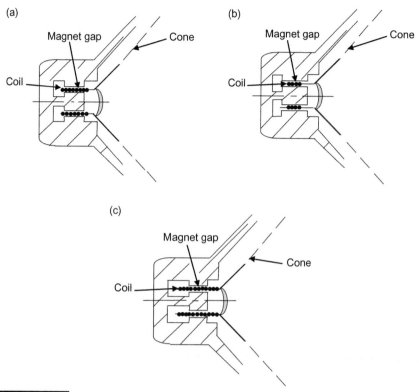

FIGURE 4.15   *a) Many bass-mid drivers employ a coil which is slightly longer than the magnet gap. b) A short coil in a long magnet gap can be employed for high linear excursion whilst still operating into the mid range. c) A long coil, much longer than the magnet gap, gives high linear excursion in a dedicated bass driver.*

It becomes clear then that a point will be reached where increasing coil length even further in pursuit of higher power handling and greater excursion brings with it reduced sensitivity, as a greater percentage of the coil now lies outside of the magnetic field. This portion of the coil still dissipates power, but it can contribute no driving force. Increases beyond a certain point therefore become self-defeating. The optimum was reached by about 1980, and drive unit sensitivities have not increased since then. Another, less-encountered alternative is to employ a short coil in a long magnet gap, considerable excursion being possible before the coil reaches either end of the gap, as shown in Figure 4.15b. The short, relatively low mass coil is capable of extending the frequency response well into the upper mid range as well as providing good bass extension by virtue of the long excursion. The X_{max} in such a design now specifies the total permitted travel of the coil whilst still remaining fully confined within that gap. The over-long coil is used successfully in high powered low frequency drivers where large linear excursions are required whilst retaining consistent immersion of the coil in the magnetic field. Somewhat reduced sensitivity has to be the trade-off, and the relatively large, high mass coil combined with a stiff straight-sided cone does not have an extended frequency response, this being a dedicated bass driver technique. Some domestic sub-woofer drivers have X_{max} values as high as 25 mm, with corresponding sensitivities in the 80 dB/watt range, as shown in Figure 4.15c.

Since the 1980s, development of high temperature glues and coil formers such as polyimide (trade name Kapton) and efficient venting systems for them have allowed coil operating temperatures of up to 300°C or so to be withstood, considerably increasing power handling capacities and therefore output levels of drive units.

Distortion

Distortion in loudspeaker systems is generally an order of magnitude or more higher than in other audio equipment. Much of it tends to be second-harmonic distortion whereby the loudspeaker will add frequencies an octave above the legitimate input signal. This is especially manifest at low frequencies where speaker diaphragms have to move comparatively large distances to reproduce them. When output levels of greater than 90 dB for domestic systems and 105 dB or so for high-sensitivity systems are being produced, low-frequency distortion of around 10% is quite common, this consisting mainly of second-harmonic and partly of third-harmonic distortion.

At mid and high frequencies distortion is generally below 1%, this being confined to relatively narrow bands of frequencies which correspond to areas such as crossover frequencies or driver resonances. Fortunately, distortion of this magnitude in a speaker does not indicate impending damage, and it is just that these transducers are inherently non-linear to this extent. Much of the distortion is at low frequencies where the ear is comparatively insensitive to it, and also the predominantly second-harmonic character is subjectively innocuous to the ear. Distortion levels of 10–15% are fairly common in the throats of high-frequency horns.

Frequency response

The frequency response of a speaker also indicates how linear it is. Ideally, a speaker would respond equally well to all frequencies, producing a smooth 'flat' output response to an input signal sweeping from the lowest to the highest frequencies at a constant amplitude. In practice, only the largest speakers produce a significant output down to 20 Hz or so, but even the smallest speaker systems can respond to 20 kHz. The 'flatness' of the response, i.e. how evenly a speaker responds to all frequencies, is a rather different matter. High-quality systems achieve a response that is within 6 dB of the 1 kHz level from 80 Hz to 20 kHz, and such a frequency response might look like Figure 4.16(a). Figure 4.16(b) is an example of a rather lower-quality speaker which has a considerably more ragged response and an earlier bass roll-off.

The frequency response can be measured using a variety of different methods, some manufacturers taking readings under the most favorable conditions to hide inadequacies. Others simply quote something like '± 3 dB from 100 Hz to 15 kHz'. This does at least give a fairly good idea of the smoothness of the response. These specifications do not, however,

FIGURE 4.16

Typical loudspeaker frequency response plots. (a) A high-quality unit. (b) A lower-quality unit.

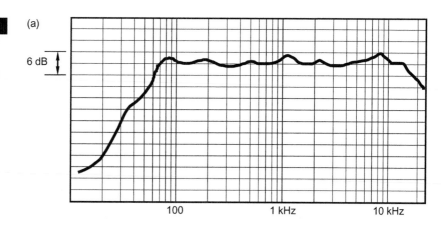

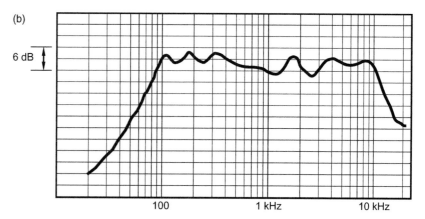

tell you how a system will sound, and they must be used only as a guide. They tell nothing of coloration levels, or the ability to reproduce good stereo depth, or the smoothness of the treble, or the 'tightness' of the bass.

Power handling

Power handling is the number of watts a speaker can handle before unacceptable amounts of distortion ensue. It goes hand-in-hand with sensitivity in determining the maximum sound level a speaker can deliver. For example, a domestic speaker may be rated at 30 watts and have a sensitivity of $86\,dB\,W^{-1}$. The decibel increase of 30 watts over 1 watt is given by:

$$dB\ increase\ =\ 10\log 30\ =\ 15\,dB$$

Therefore, the maximum output level of this speaker is $86 + 15 = 101\,dB$ at 1 meter for 30 watts input. This is loud, and quite adequate for domestic use. Consider now a PA speaker with a quoted sensitivity of $99\,dB\,W^{-1}$. 30 watts input now produces $99 + 15 = 114\,dB$, some $13\,dB$ more than with the previous example for the same power input. To get $114\,dB$ out of the $86\,dB\,W^{-1}$ speaker one would need to drive it with no less than 500 watts, which would of course be way beyond its capabilities. This dramatically demonstrates the need to be aware of the implications of sensitivity and power handling.

A 30 watt speaker can, however, safely be driven even by a 500 watt amplifier providing that sensible precautions are taken with respect to how hard the amplifier is driven. Occasional peaks of more than 30 watts will be quite happily tolerated; it is sustained high-level drive which will damage a speaker. It is perfectly all right to drive a high-power speaker with a low-power amplifier, but care must be taken that the latter is not overdriven otherwise the harsh distortion products can easily damage high-frequency horns and tweeters even though the speaker system may have quoted power handling well in excess of the amplifier. The golden rule is to listen carefully. If the sound is clean and unstressed, all will be well.

Directivity

Directivity, or dispersion, describes the angle of coverage of a loudspeaker's output. Very low frequencies radiated from a speaker are effectively omnidirectional, because the wavelength of the sound is large compared with the dimensions of the speaker and its enclosure, and efficient diffraction of sound around the latter is the result. As the frequency increases, wavelengths become comparable to the dimensions of the speaker's front surface, diffraction is curtailed, and the speaker's output is predominantly in the forwards direction. At still higher frequencies, an even narrower dispersion angle results as a further effect comes into play: off-axis phase cancelation. If one listens, say, 30° off-axis from the front of a speaker, a given upper frequency (with a short wavelength) arrives which has been radiated both from the closest side of the speaker cone to the listener and from the furthest side of the cone, and these two sound sources will not therefore be in phase with each other because of the different distances they are away from one another. Phase cancelation therefore occurs, perceived output level falls, and the effect becomes more severe as frequencies increase. The phenomenon is mitigated by designing for progressively smaller radiating areas of the speaker cone to be utilized as the frequency increases, finally crossing over to a tweeter of very small dimensions. By these means, fairly even dispersion of sound, at least in the mid and lower treble regions, can be maintained.

Various other methods have been used to control directivity (the acoustic lens has been covered) and one or two will be described. Low frequencies, which are normally omnidirectional, have been given a cardioid-like dispersion pattern by mounting large speaker drivers on essentially open baffles which by themselves give a figure-of-eight polar response, the output falling with falling frequency. To these was added a considerable amount of absorbent material to the rear, and together with appropriate bass boost to flatten the frequency response of the speakers, predominantly forward-radiation of low frequencies was achieved. A more elegant technique has been to mount essentially open-baffle speakers (the rear radiation therefore being 180° out of phase with the front producing a figure-of-eight polar pattern, and with bass boost applied to flatten the frequency response) adjacent to closed-box omnidirectional speakers. Their combined acoustical outputs thereby produce a cardioid dispersion pattern, useful for throwing low frequencies forwards into an auditorium rather than across a stage where low-frequency feedback with microphones can be a problem.

A more sophisticated technique has been to use a conventional forward-facing subwoofer, adding to it another which is behind, and facing rearward. Using DSP processing to give appropriate delay and phase change with frequency, the rear-facing driver's output can be made to cancel the sound which reaches the rear of the enclosure from the front-facing driver. A very consistent cardioid response can thereby be achieved over the limited range of frequencies across which the subwoofer operates.

Another fascinating technique, introduced by Philips in 1983, is the Bessel Array. It was developed to counteract the beaming effects of multiple-speaker systems. Essentially it makes use of Bessel coefficients to specify phase relationships and output level requirements from each of a horizontal row of speakers necessary to obtain an overall dispersion pattern from the row which is the same as one speaker on its own. Normally, path-length differences between off-axis listeners and the various speaker drivers result in phase cancelations and consequent loss of level, particularly in the upper frequency range. For a horizontal five-speaker row, labeled A, B, C, D and E, the Bessel function gives:

$$A:B:C:D:E = 1:2:2:-2:1$$

In other words, speakers A and E are required to draw half the current of speakers B, C and D; and speaker D must be connected out of phase. A practical implementation would be to connect speakers A and E in series, with speakers B, C and D each connected straight across the system's input terminals but with D wired out of phase. The speaker drivers are mounted side by side very close together to give good results across the frequency range.

For a seven-speaker row, the Bessel function gives:

$$A:B:C:D:E:F:G = 1:2:2:0:-2:2:-1$$

Speaker D can therefore be omitted, but a space in the row must be left in its position so as to preserve the correct distance relationships between the others.

Both horizontal and vertical rows of speakers can be combined into a square arrangement so that an array of, for example, 25 speakers, together having potentially very high power handling and output level capability, can, however, give the same dispersion characteristics of one speaker on its own. The amplitude and phase relationships necessary in such an array are given by the numbers in the circles representing the speakers in Figure 4.17.

It is worth mentioning that the same technique can also be applied to microphones, offering potential for a high output, very low noise array whilst still maintaining a good polar response.

A highly directional speaker incorporating a parabolic reflector of about 1.3 meters in diameter has been developed by the Meyer loudspeaker company as their type SB-1. Designed to work between 500 Hz and 15 kHz, the system comprises an outrigger supporting a small horn and compression driver at the focus of the dish which fires into it, and a small hole at the

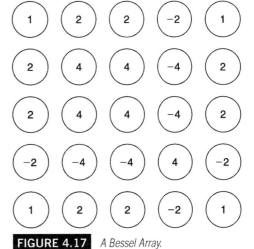

FIGURE 4.17 *A Bessel Array.*

dish's center admits sound from a 12-inch cone driver. Claimed dispersion (−6 dB points) is 10° vertical and 10° horizontal, and maximum peak
output at 100 meters distance is 110 dB. The Watkins Electronic Music
company (WEM) used this parabolic reflector technique at the Isle of
Wight Music Festival in 1970.

'Modulated Ultrasound'

The idea of directing sound via a tightly controlled beam in the manner
of a spotlight was first investigated by both the USA and Soviet militaries
in the 1960s in connection with sonar. Some decades later the idea was
developed for propagating sound through the air, but considerable technical difficulties have meant that commercial designs have appeared only
in recent years. These have included the Audio Spotlight (Holosonics),
Hypersonic Sound and Sennheiser's Audio Beam among others. The
design concept is very simple. Both dispersion of sound from a conventional loudspeaker and reflections of sound from walls and other objects
mean that sound can be heard throughout the listening space to a greater
or lesser extent, and the prospect of directing a tight ultrasonic beam of
sound, amplitude modulated by the audio signal, to deliver audio to a
clearly defined location suggested itself as a viable technique analagous
to the AM radio system. Those familiar with the principle of AM radio
transmission will recall that a high frequency carrier wave is amplitude
modulated by the audio signal, and at the receiving end the carrier wave
is separated from the much lower audio frequencies, the latter comprising
the program content. With a sound beam, no demodulation of the arriving sound is required because the ultrasonic carrier wave, typically about
50 kHz in commercial systems, cannot be heard by human ears. Only the
audio frequencies are perceived.

There were two principal problems to overcome. First, a speaker system
had to be developed capable of directing a tight ultrasonic beam at a high
sound pressure level to a destination some meters away. This was achieved
using a large number of ultrasonic piezoelectric transducers–more than
one hundred in practical examples–mounted on a surface of about 30 cm
square, or alternatively on a disc of comparable area. The wavelength
of sound at 50 kHz is just under a centimeter, and this ensures that any
sound which is directed significantly off axis is subjected to efficient phase
cancellations as the outputs from the variously spaced transducers will be
substantially out of phase, their outputs only reinforcing each other in a
tight forward beam. The many transducers provide the necessary high output, in the order of 130 dB, of the ultrasonic carrier wave. High output is
necessary because the air absorbs high frequencies to a somewhat greater

extent than the lower frequencies, and also because the level of the modu-
lating audio frequencies is somewhat lower than that of the carrier wave.
Health and safety issues have to be considered with such sound levels; for
instance the USA's Occupational Safety and Health Administration sets a
top limit of 145 dB for ultrasonic sound.

The second problem was that of distortion. A high level ultrasonic beam alters the speed of sound along its path and causes the air
to behave non-linearly, and it also creates an environment of very short
wave compressions and rarefactions of the air from which the contained
audio frequencies have to emerge. The ensuing distortions are rather
more complex than the familiar harmonic distortions of audio systems
in general, and reciprocal distortion conditioning has had to be developed
and added to the signal beforehand in order to bring distortion down to
acceptable levels.

Applications have included trade fairs and theme parks where sound
spillage can be a problem between events and exhibitions, the tightly
controlled beam delivering sound only to the area where the listener is
positioned. Museums and art galleries are other examples where their particular properties would be appropriate.

Panel speaker dispersion

The previous section describes how a tight beam of sound can be delivered
to a specific location. Conventional panel speakers such as electrostatics
have suffered in the past from inadequate dispersion of sound in the listening environment, particularly at the higher frequencies, for the same basic
reason: that a relatively large vibrating diaphragm generates the sound from
a large area rather than from a near-point source. Off-access phase cancellation is the result. The original Quad ESL63 electrostatic design and its
successors such as the current model 2912 deal with the problem by the
following means. Imagine the sound source coming not from the panel
itself but from a virtual point source positioned about 30 cm behind the
panel. The sound coming straight from the point source to the middle of
the panel can be considered to continue through to the listener. But the
sound travelling from the point source at an angle to a position elsewhere
on the panel takes a little longer to reach it because the distance is now
slightly greater. The sound coming from the panel, to recreate this mechanism, will need to be delayed incrementally as one moves towards its
edges. The Quad designs achieve this in the horizontal plane by dividing
the panel up into a series of concentric arcs of circles laterally, an array of
inductors and capacitors providing the necessary incremental delays to feed
the sections. By these means, the panel as a whole simulates a point source

30 cm *behind* it over the critical upper mid and high frequencies, somewhat improving lateral dispersion.

The DML panel loudspeakers do not however suffer from poor dispersion because the panel does not vibrate as a conventional single diaphragm would. Instead, a multitude of breakup modes exist across its surface which create a large number of small radiating areas, their outputs de-correlated with each other with respect to phase such that off-axis phase cancellations do not occur in a systematic and predictable way as is the case with conventional panels. In contrast with other loudspeaker types and electrostatic panels that do not have the Quad-style dispersion feature, the DML's dispersion pattern does not therefore change significantly with panel size.

SETTING UP LOUDSPEAKERS

Phase

Phase is a very important consideration when wiring up speakers. A positive-going voltage will cause a speaker cone to move in a certain direction, which is usually forwards, although at least two American and two British manufacturers have unfortunately adopted the opposite convention. It is essential that both speakers of a stereo pair, or all of the speakers of a particular type in a complete sound rig, are 'in phase', that is, all the cones are moving in the same direction at any one time when an identical signal is applied. If two stereo speakers are wired up out of phase, this produces vague 'swimming' sound images in stereo, and cancelation of bass frequencies. This can easily be demonstrated by temporarily connecting one speaker in opposite phase and then listening to a mono signal source–speech from the radio is a good test. The voice will seem to come from nowhere in particular, and small movements of the head produce sudden large shifts in apparent sound source location. Now reconnect the speakers in phase and the voice will come from a definite position in between the speakers. It will also be quite stable when you move a few feet to the left or to the right.

Occasionally it is not possible to check the phase of an unknown speaker by listening. An alternative method is to connect a 1.5 V battery across the input terminals and watch which way the cone of the bass driver moves. If it moves forwards, then the positive terminal of the battery corresponds to the positive input terminal of the speaker. If it moves backwards as the battery is connected, then the positive terminal of the battery is touching the negative input terminal of the speaker. The terminals can then be labeled + and – .

Positioning

Loudspeaker positioning has a significant effect upon the performance. In smaller spaces such as control rooms and living rooms the speakers are likely to be positioned close to the walls, and 'room gain' comes into effect whereby the low frequencies are reinforced. This happens because at these frequencies the speaker is virtually omnidirectional, i.e. it radiates sound equally in all directions. The rear- and side-radiated sound is therefore reflected off the walls and back into the room to add more bass power. As we move higher in frequency, a point is reached whereby the wavelength of lower mid frequencies starts to become comparable with the distance between the speaker and a nearby wall. At half wavelengths the reflected sound is out of phase with the original sound from the speaker and some cancelation of sound is caused. Additionally, high-frequency 'splash' is often caused by nearby hard surfaces, this often being the case in control rooms where large consoles, tape machines, outboard processing gear, etc. can be in close proximity to the speakers. Phantom stereo images can thus be generated which distort the perspective of the legitimate sound. A loudspeaker which has an encouragingly flat frequency response can therefore often sound far from neutral in a real listening environment. It is therefore essential to give consideration to loudspeaker placement, and a position such that the speakers are at head height when viewed from the listening position (high-frequency dispersion is much narrower than at lower frequencies, and therefore a speaker should be listened to on axis) and also away from room boundaries will give the most tonally accurate sound.

Some speakers, however, are designed to give their best when mounted directly against a wall, the gain in bass response from such a position being allowed for in the design. A number of professional studio monitors are designed to be let into a wall such that their drivers are then level with the wall's surface. The manufacturers' instructions should be heeded, in conjunction with experimentation and listening tests. Speech is a good test signal. Male speech is good for revealing boominess in a speaker, and female speech reveals treble splash from hard-surfaced objects nearby. Electronic music is probably the least helpful since it has no real-life reference by which to assess the reproduced sound. It is worth emphasizing that the speaker is the means by which the results of previous endeavor are judged, and that time spent in both choosing and siting is time well spent.

Speakers are of course used in audio-visual work, and one frequently finds that it is desirable to place a speaker next to a video monitor screen. But the magnetic field from the magnets can affect the picture quality by pulling the internal electron beams off course. Some speakers are specially magnetically screened so as to avoid this.

Loudspeaker positioning issues affecting two-channel stereo and surround sound reproduction are covered in greater detail in Chapters 16 and 17.

THIELE–SMALL PARAMETERS AND ENCLOSURE VOLUME CALCULATIONS

Low-frequency performance of a driver/box combination is one of the few areas of loudspeaker design where the performance of the practical system closely resembles the theoretical design aims. This is because at low frequencies the speaker cone acts as a pure piston, and wavelengths are long, minimizing the effects of enclosure dimensions and objects close to the speaker. Nearby boundaries, e.g. walls and the floor, have a significant effect at very low frequencies, but these are predictable and easily allowed for.

It was A.N. Thiele and Richard Small, working largely independently of each other in Australia mainly during the 1960s, who modeled driver and enclosure behavior in terms of simple electrical circuits. They substituted, for instance, the DC resistance of the coil with a resistor, its inductance with an inductor of the same value, and the places where the speaker's impedance rose with decreasing frequency could be represented by a capacitor of an appropriate value in the circuit model. The latter could also represent the 'stiffness' of the air enclosed in the box. Electrical formulae could then be applied to these models to predict the behavior of particular drive units in various sizes and types of enclosure. A series of 'Thiele–Small' parameters are therefore associated with a particular drive unit and they enable systems to be designed, the low frequency performance of which can be predicted with considerable accuracy, something which had previously been a largely empirical affair before their work. A host of parameters are specified by the manufacturers, and include such things as magnet flux density, sensitivity, diaphragm moving mass, mechanical resistance of suspension, force factor, equivalent volume of suspension compliance, mechanical Q, electrical Q, and free air resonance. The list is a comprehensive description of the drive unit in question, but fortunately only three need to be considered when designing an enclosure for the target low-frequency performance. These are the free-air resonance, represented by the symbol f_0; the equivalent air volume of the suspension compliance, V_{AS}; and the total Q of the driver, Q_T. They will be considered in turn.

f_0 is the free-air resonance of the driver, determined by taking an impedance plot and noting the frequency at which a large, narrow peak in the impedance takes place. Such a rise can be seen in Figure 4.12 at about 20 Hz.

V_{AS} can be explained as follows. Imagine a bass driver with an infinitely compliant suspension, its cone being capable of being moved to and fro with the fingers with no effort. Now mount the driver in a closed box of say 70 liters internal volume. Push the cone with the fingers again, and the spring of the enclosed air now supports the cone, and one feels an impedance as one pushes against that spring. The suspension of the drive unit is thus specified as an equivalent air volume. The enclosure volume of both closed-box and reflex systems is always smaller than the drive unit's V_{AS} in order that the air stiffness loads the cone of the speaker adequately, as its own suspension system is insufficient to control low frequency excursion alone.

Q_T, the total Q of the driver, is the average between the electrical Q, Q_E, and the mechanical Q, Q_M. Briefly, Q_E is determined using a formula containing total moving mass of the diaphragm, the resistance of the coil and the Bl factor (flux density multiplied by a length of coil wire immersed in the magnetic field, indicating the force with which the coil pushes the cone for a given input level). Q_M is determined by calculating the Q of the low-frequency peak in the impedance plot, dividing the center peak frequency by the bandwidth at the $-3\,\mathrm{dB}$ points each side of this. Q_T is always quoted, so one does not need to calculate it from the other parameters.

Before moving on to discuss how the above is used in calculations, system Q must be looked at. This is explained in Fact File 4.6.

For the following discussions, a Q of 0.7 will be assumed. Different Q values can be substituted by the reader to explore the effect this has on enclosure volume.

For a closed box ('infinite baffle') system, a three-stage process is involved. The following formula is used first:

$$Q_{TC}/Q_T = x = f_3/f_0$$

where Q_{TC} is the chosen system Q (assumed to be 0.7), and f_3 is the resonant frequency of the system. x is the ratio between the two quantities. For example, if the driver's Q_T is 0.35, then x is 2. If the driver's f_0 is 25 Hz, then f_3 is 50 Hz. Therefore, such a driver in a box giving a Q of 0.7 will have a resonant frequency of 50 Hz.

The next stage is to calculate the box volume required to achieve this performance. For this, x needs first to be converted into α, the *compliance ratio*. This is the ratio between the drive unit's V_{AS} and the box volume.

FACT FILE 4.6 LOW-FREQUENCY Q

The graph shows a family of curves for various possible low-frequency alignments. A system Q of 0.7 for a closed box is usually the target figure for medium-sized enclosures in both the domestic and the sound reinforcement contexts. The roll-off is a smooth 12 dB per octave, and there is no emphasis at any frequency. A Q of 0.7 means that the response is 3 dB down compared with the flat part of the frequency response above it. (Refer to Q in the 'Glossary of terms' for a discussion of how Q value relates to dB of attenuation at the point of resonance.) A Q of 0.6 has an earlier but gentler bass roll-off, and the frequency response is about 4.5 dB down at resonance. Q values above 0.7 progressively overemphasize the response at resonance, producing an undesirable 'hump'. Large enclosures for domestic use have to be designed with room boundary gain taken into consideration. A model with an impressively extended response down to very low frequencies in an anechoic chamber will often sound 'boomy' in a room because the low frequencies, which are omnidirectional, reflect off the rear wall and floor to reinforce the primary sound from the speaker, adding to its output. A Q value of 0.6 or even 0.5 is therefore best chosen for large domestic enclosures so that the total combined response in a real listening environment is more even and natural.

For very small speakers, a Q value of greater than 0.7, say 0.9 or slightly more, can be chosen which gives slight bass emphasis and helps give the impression of a 'fuller' sound from a small enclosure. Such an alignment is used judiciously so as to avoid overemphasizing the upper bass frequencies.

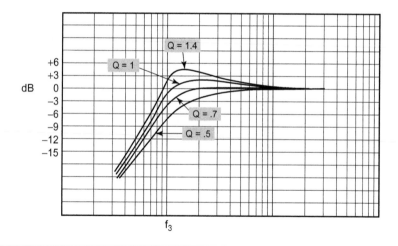

The latter is always smaller than the former. This is done using the simple formula:

$$x^2 - 1 = \alpha$$

In the present example, this gives an α of 3. To calculate box size, the following formula is used:

$$\alpha = V_{AS}/V_B$$

where V_B is the box volume. If the driver has a V_{AS} of 80 liters, this then gives a box size of about 27 liters. These results are quite typical of a medium-sized domestic speaker system.

The bass reflex design is rather more complex. This consists of a box with a small port or 'tunnel' mounted in one of its walls, the air plug in the port together with the internal air volume resonating at a particular low frequency. At this frequency, maximum sound output is obtained from the port, and the drive unit's cone movement is reduced compared with a closed box system. First, we will look at the formula for the enclosure:

$$f_c = \frac{344.8\,R}{2\sqrt{(\pi V_b[L + 1.7\,R])}}$$

where R is the port radius (assuming a circular port), L is the port length, and V_b is the enclosure volume. All dimensions are in meters; V_b is in cubic meters. f_c is the resonant frequency of the system. 344.8 is the speed of sound in meters per second at normal temperatures and pressures. The port can in principle have any cross-sectional shape, and more than one port can be used. It is the total cross-sectional area combined with the total length of the port or ports which are required for the calculation. Note that nowhere does the drive unit in question appear in the calculation. The box with port is a resonant system alone, and the design must be combined with drive unit calculations assuming a closed box system with a target Q rather lower than is customary for a closed box, usually much nearer to 0.5 so that the driver has a slow low-frequency roll-off as the output from the port rises, producing a smooth transition. The reflex enclosure volume is therefore typically in the order of about 80% of the drive unit's V_{AS}. Looking again at Figure 4.12, we see two low-frequency peaks in the impedance with a trough in between. The lowest one is the free-air resonance of the driver (altered slightly by air loading), the trough is the reflex port resonant frequency, and the upper peak is the box/driver resonant frequency. The designer must ensure that the design arrived at with the particular chosen driver ensures these conditions. One does not, for instance, design for a port resonant frequency of 30 Hz when the drive unit's free-air resonance is 40 Hz.

The design procedure is normally to calculate for a closed-box system with a chosen drive unit, the target Q being closer to 0.5 than to 0.7. (Reflex systems are larger than closed-box systems for a given driver.) One notes the driver/box resonant frequency, and then chooses port dimensions which give a port resonance which is midway between the driver/box resonance and the driver's free-air resonance. Final dimensions will be chosen during prototyping for optimum subjective results in a real listening environment.

Above port resonance, the output from the port falls at the rate of 12 dB/octave, and the design aim is to give a smooth transition between the port's output and the driver's output. The port gives a useful degree of bass extension. Below resonance, the speaker cone simply pumps air in and out of the port, and the latter's output is therefore 180° out of phase with the former's, producing a rapid 24 dB/octave roll-off in the response. Furthermore, the drive unit is not acoustically loaded below resonance, and particularly in sound reinforcement use where very high powers are involved care must be taken to curtail very low-frequency drive to reflex enclosures, otherwise excessive cone excursions combined with large currents drawn from the amplifiers can cause poor performance and premature failure.

The abrupt 24 dB/octave roll-off of the reflex design means that it interfaces with room boundaries less successfully than does a closed-box system with its more gradual roll-off. However, some reflex designs are deliberately 'de-tuned' to give a less rapid fall in response, helping to avoid a 'boomy' bass quality when the speaker is placed close to walls.

Drive units with Q_T values of 0.2 and below are well suited to bass reflex use. Drivers with Q_T values of 0.3 and above are better suited to closed-box designs. If one runs calculations for bass reflex designs using drive units with high Q_T values, one finds that the drivers' free air resonances and the driver/box resonances are uncomfortably close together, leaving little room to place the port resonances in between. If one runs calculations for closed box designs with drivers having low Q_T values, one finds that the system resonances are disappointingly high even though the drivers' free air resonances may be encouragingly low.

In the above discussions, no mention has been made of sound absorbent material in the box, either in the form of foam lining or wadding filling the volume of the enclosure. This should not be necessary for low-frequency considerations, and it is usually included to absorb mid frequency energy from the rear of the speaker cone to prevent it from re-emerging back through the cone and the enclosure walls, coloring the sound. However, the presence particularly of volume-filling material has the effect of reducing the speed of sound in the enclosure, making it apparently larger, sometimes by as much as 15% for some types of filling. This must be taken into consideration in the final design. An over-dense filling tends to behave in a manner more like a solid mass, and the box volume is apparently reduced. There is no reason in principle to include sound absorbent material in a box intended purely for low-frequency use, unless one wishes to increase its apparent acoustic volume for economic or space-saving reasons.

The website www.thielesmall.com provides a large data base for drive unit parameters, both obsolete and current, and it helps one to find replacement drivers with closely matching specifications which will give comparable performances in given enclosures.

DIGITAL SIGNAL PROCESSING IN LOUDSPEAKERS

Digital signal processing (DSP) is used increasingly in loudspeakers to compensate for a range of linear and non-linear distortion processes that typically arise. DSP can also be used in crossover design and for controlling the spatial radiation characteristics of loudspeakers or loudspeaker arrays. With the help of such technology it may be possible to get better performance out of smaller loudspeaker units by using electronics to counteract physical inadequacies. Some such processes can make use of psychoacoustical phenomena, such as a means of extending the perceived bass response without actually reproducing the relevant low frequencies, and it may also be possible to modify the way in which the loudspeaker interacts with the listening room. Finally, there are various ways by which it may be possible to engineer an 'all-digital' signal chain, even using digital forms of representation right up to the point where the binary data is converted into an acoustical waveform.

RECOMMENDED FURTHER READING

Beranek, L., Mellow, T., 2012. Acoustics: Sound Fields and Transducers. Academic Press.

Borwick, J. (Ed.), 2001. Loudspeaker and Headphone Handbook. Focal Press.

Colloms, M., 2005. High Performance Loudspeakers, sixth edition. Wiley.

Eargle, J., 2003. Loudspeaker Handbook. Kluwer Academic Publishers.

Toole, F., 2008. Sound Reproduction: The Acoustics and Psychoacoustics of Loudspeakers and Rooms. Focal Press.

USEFUL WEBSITES

www.thielesmall.com: Among other things, has a valuable drive unit database which helps one to compare parameters of obsolete drivers with current ones.

www.xlrtechs.com/dbkeele.com/papers.htm: Contains papers covering a wide variety of design and performance issues, with further information about Bessel Arrays.

Mixers

This chapter describes the principles and basic operation of audio mixers. It begins with a description of a simple system and moves on to consider the facilities of large-scale multitrack systems. Because many design and layout concepts of analog mixers find a place in more recent digital mixers, these aspects are covered here in a fairly generic way. Those features found more commonly only in digital systems are described towards the end of the chapter.

In its simplest form an audio mixer combines several incoming signals into a single output signal. This cannot be achieved simply by connecting all the incoming signals in parallel and then feeding them into a single input because they may influence each other. The signals need to be isolated from each other. Individual control of at least the level of each signal is also required.

In practice, mixers do more than simply mix. They can provide phantom power for capacitor microphones (see 'The capacitor or condenser microphone', Chapter 3); pan control (whereby each signal can be placed in any desired position in a stereo image); filtering and equalization; routing facilities; and monitoring facilities, whereby one of a number of sources can be routed to loudspeakers for listening, often without affecting the mixer's main output.

A SIMPLE SIX-CHANNEL ANALOG MIXER

Overview

By way of example, a simple six-channel analog mixer will be considered, having six inputs and two outputs (for stereo). Figure 5.1 illustrates such a notional six-into-two mixer with basic facilities. It also illustrates the back panel. The inputs illustrated are via XLR-type three-pin latching connectors, and are of a balanced configuration. Separate inputs are provided for microphone and line level signals, although it is possible to encounter systems which simply use one socket switchable to be either mic or line. Many cheap mixers have unbalanced inputs via quarter-inch jack sockets, or even 'phono' sockets such as are found on hi-fi amplifiers. Some mixers employ balanced XLR inputs for microphones, but unbalanced jack or phono inputs for line level signals, since the higher-level line signal is less susceptible to noise and interference, and will probably have traveled a shorter distance.

On some larger mixers a relatively small number of multipin connectors are provided, and multicore cables link these to a large jackfield which consists of rows of jack sockets mounted in a rack, each being individually labeled. All inputs and outputs will appear on this jackfield, and patch cords of a meter or so in length with GPO-type jack plugs at each end enable the inputs and outputs to be interfaced with other equipment and tie-lines in any appropriate combination. (The jackfield is more fully described in 'Patchbay or jackfield', p. 142, and 'Jackfields (patchbays)', Chapter 12.)

The outputs are also on three-pin XLR-type connectors. The convention for these audio connections is that inputs have sockets or holes, outputs have pins. This means that the pins of the connectors 'point' in the direction of the signal, and therefore one should never be confused as

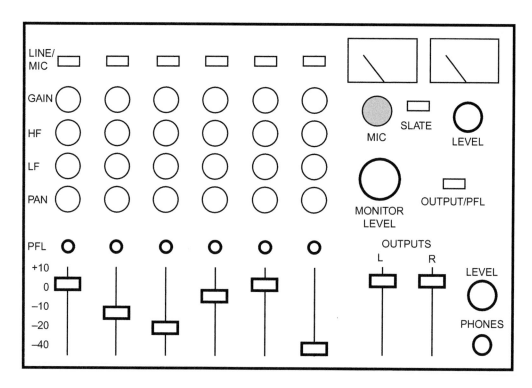

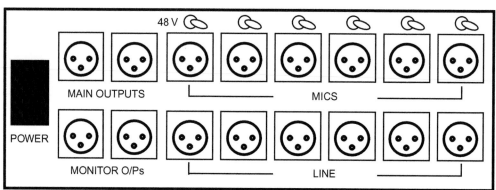

FIGURE 5.1 *Front panel and rear connectors of a typical simple six-channel mixer.*

to which connectors are inputs and which are outputs. The microphone inputs also have a switch each for supplying 48 V phantom power to the microphones if required. Sometimes this is found on the input module itself, or sometimes on the power supply, switching 48 V for all the inputs at once.

The term 'bus' is frequently used to describe a signal path within the mixer to which a number of signals can be attached and thus combined. For instance, routing some input channels to the 'stereo bus' conveys those channels to the stereo output in the manner of a bus journey in the conventional everyday sense. A bus is therefore a mixing path to which signals can be attached.

Input channels

All the input channels in this example are identical, and so only one will be described. The first control in the signal chain is input gain or sensitivity. This control adjusts the degree of amplification provided by the input amplifier, and is often labeled in decibels, either in detented steps or continuously variable. Inputs are normally switchable between mic and line. In 'mic' position, depending on the output level of the microphone connected to the channel (see 'Microphone performance', Chapter 3), the input gain is adjusted to raise the signal to a suitable line level, and up to 80 dB or so of gain is usually available here (see 'Miscellaneous features', p. 121). In 'line' position little amplification is used and the gain control normally provides adjustment either side of unity gain (0 dB), perhaps ± 20 dB either way, allowing the connection of high-level signals from such devices as CD players, tape machines and musical keyboards.

The equalization or EQ section which follows (see 'Equalizer section', p. 134) has only two bands in this example – treble and bass – and these provide boost and cut of around ±12 dB over broad low- and high-frequency bands (e.g. centered on 100 Hz and 10 kHz). This section can be used like the tone controls on a hi-fi amplifier to adjust the spectral balance of the signal. The fader controls the overall level of the channel, usually offering a small amount of gain (up to 12 dB) and infinite attenuation. The law of the fader is specially designed for audio purposes (see Fact File 5.1). The pan control divides the mono input signal between left and right mixer outputs, in order to position the signal in a virtual stereo sound stage (see Fact File 5.2).

Output section

The two main output faders (left and right) control the overall level of all the channel signals which have been summed on the left and right mix buses, as shown in the block diagram (Figure 5.2). The outputs of these faders (often called the group outputs) feed the main output connectors on the rear panel, and an internal feed is taken from the main outputs to the monitor selector. The monitor selector on this simple example

FACT FILE 5.1 FADER FACTS

Fader Law

Channel and output faders, and also rotary level controls, can have one of two laws: linear or logarithmic (the latter sometimes also termed 'audio taper'). A linear law means that a control will alter the level of a signal (or the degree of cut and boost in a tone control circuit) in a linear fashion: that is, a control setting midway between maximum and minimum will attenuate a signal by half its voltage, i.e. −6 dB. But this is not a very good law for an audio level control because a 6 dB drop in level does not produce a subjective halving of loudness. Additionally, the rest of the scaling (−10 dB, −20 dB, −30 dB and so on) has to be accommodated within the lower half of the control's travel, so the top half gives control over a mere 6 dB, the bottom half all the rest.

For level control, therefore, the logarithmic or 'log' law is used whereby a non-linear voltage relationship is employed in order to produce an approximately even spacing when the control is calibrated in decibels, since the decibel scale is logarithmic. A log fader will therefore attenuate a signal by 10 dB at a point approximately a quarter of the way down from the top of its travel. Equal dB increments will then be fairly evenly spaced below this point. A rotary log pot ('pot' is short for potentiometer) will have its maximum level usually set at the 5 o'clock position and, the −10 dB point will be around the 2 o'clock position. An even subjective attenuation of volume level is therefore produced by the log law as the control is

gradually turned down. A linear law causes very little to happen subjectively until one reaches the lowest quarter of the range, at which point most of the effect takes place.

The linear law is, however, used where a symmetrical effect is required about the central position; for example, the cut and boost control of a tone control section will have a central zero position about which the signal is cut and boosted to an equal extent either side of this.

Electrical Quality

There are two types of electrical track in use in analog faders, along which a conductive 'wiper' runs as the fader is moved to vary its resistance. One type of track consists of a carbon element, and is cheap to manufacture. The quality of such carbon tracks is, however, not very consistent and the 'feel' of the fader is often scrapy or grainy, and as it is moved the sound tends to jump from one level to another in a series of tiny stages rather than in a continuous manner. The carbon track wears out rather quickly, and can become unreliable.

The second type employs a conductive plastic track. Here, an electrically conductive material is diffused into a strip of plastic in a controlled manner to give the desired resistance value and law (linear or log). Much more expensive than the carbon track, the conductive plastic track gives smooth, continuous operation and maintains this standard over a long period of time. It is the only serious choice for professional-quality equipment.

can be switched to route either the main outputs or the PFL bus (see Fact File 5.3) to the loudspeakers. The monitor gain control adjusts the loudspeaker output level without affecting the main line output level, but of course any changes made to the main fader gain will affect the monitor output.

The slate facility on this example allows for a small microphone mounted in the mixer to be routed to the main outputs, so that comments from the engineer (such as take numbers) can be recorded on a tape machine connected to the main outputs. A rotary control adjusts the slate level.

FACT FILE 5.2 PAN CONTROL

The pan control on a mixer is used for positioning a signal somewhere between left and right in the stereo mix image. It does this by splitting a single signal from the output of a fader into two signals (left and right), setting the position in the image by varying the level difference between left and right channels. It is thus not the same as the balance control on a stereo amplifier, which takes in a stereo signal and simply varies the relative levels between the two channels. A typical pan-pot law would look similar to that shown in the diagram, and ensures a roughly constant perceived level of sound as the source is panned from left to right in stereo. The output of the pan-pot usually feeds the left and right channels of the stereo mix bus (the two main summation lines which combine the outputs of all channels on the mixer), although on mixers with more than two mix buses the pan-pot's output may be switched to pan between any pair of buses, or perhaps simply between odd and even groups (see Fact File 5.4).

On some older consoles, four-way routing is provided to a quadraphonic mix bus, with a left–right pot and a front–back pot. These are rare now. Many stereo pan-pots use a dual-gang variable resistor which follows a law giving a 4.5 dB level drop to each channel when panned centrally, compared with the level sent to either channel at the extremes. The 4.5 dB figure is a compromise between the −3 dB and −6 dB laws. Pan-pots which only drop the level by 3 dB in the center cause a rise in level of any centrally panned signal if a mono sum is derived from the left and right outputs of that channel, since two identical signals summed together will give a rise in level of 6 dB. A pot which gives a 6 dB drop in the center results in no level rise for centrally panned signals in the mono sum. Unfortunately, the 3 dB drop works best for stereo reproduction, resulting in no perceived level rise for centrally panned signals.

Only about 18 dB of level difference is actually required between left and right channels to give the impression that a source is either fully left or fully right in a loudspeaker stereo image, but most pan-pots are designed to provide full attenuation of one channel when rotated fully towards the other. This allows for the two buses between which signals are panned to be treated independently, such as when a pan control is used to route a signal either to odd or even channels of a multi-track bus (see 'Routing section', p. 132).

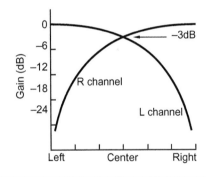

Miscellaneous features

Professional-quality microphones have an output impedance of around 200 ohms, and the balanced microphone inputs will have an input impedance of between 1000 and 2000 ohms ('2 kΩ', k = thousand). The outputs should have an impedance of around 200 ohms or lower. The headphone output impedance will typically be 100 ohms or so. Small mixers usually have a separate power supply which plugs into the mains. This typically contains a mains transformer, rectifiers and regulating circuitry, and

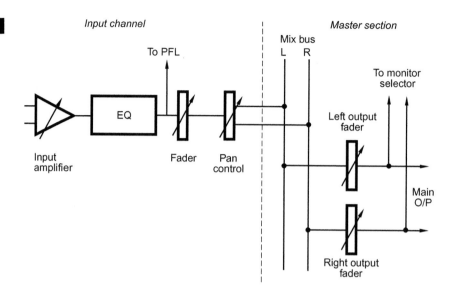

FIGURE 5.2

Block diagram of a typical signal path from channel input to main output on a simple mixer.

FACT FILE 5.3 PRE-FADE LISTEN (PFL)

Pre-fade listen, or PFL, is a facility which enables a signal to be monitored without routing it to the main outputs of the mixer. It also provides a means for listening to a signal in isolation in order to adjust its level or EQ.

Normally, a separate mono mixing bus runs the length of the console picking up PFL outputs from each channel. A PFL switch on each channel routes the signal from before the fader of that channel to the PFL bus (see diagram), sometimes at the same time as activating internal logic which switches the mixer's monitor outputs to monitor the PFL bus. If no such logic exists, the mixer's monitor selector will allow for the selection of PFL, in which position the monitors will reproduce any channel currently with its PFL button pressed. On some broadcast and live consoles a separate small PFL loudspeaker is provided on the mixer itself, or perhaps on a separate output, in order that selected sources can be checked without affecting the main monitors.

Sometimes PFL is selected by 'overpressing' the channel fader concerned at the bottom of its travel (i.e. pushing it further down). This activates a microswitch which performs the same functions as above. PFL has great advantages in live work and broadcasting, since it allows the engineer to listen to sources before they are faded up (and thus routed to the main outputs which would be carrying the live program). It can also be used in studio recording to isolate sources from all the others without cutting all the other channels, in order to adjust equalization and other processing with greater ease.

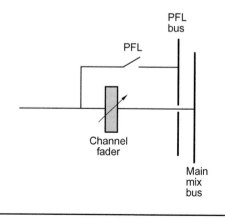

it supplies the mixer with relatively low DC voltages. The main advantage of a separate power supply is that the mains transformer can be sited well away from the mixer, since the alternating 50 Hz mains field around the former can be induced into the audio circuits. This manifests itself as 'mains hum' which is only really effectively dealt with by increasing the distance between the mixer and the transformer. Large mixers usually have separate rack-mounting power supplies.

The above-described mixer is very simple, offering few facilities, but it provides a good basis for the understanding of more complex models. A typical commercial example of a compact mixer is shown in Figure 5.3.

FIGURE 5.3 *A compact stereo mixer: the Seemix 'Seeport'. (Courtesy of Seemix Sound AS.)*

A MULTITRACK MIXER

Overview

The stereo mixer outlined in the previous section forms only half the story in a multitrack recording environment. Conventionally, popular music recording involves at least two distinct stages: the 'track-laying' phase and the 'mixdown' phase. In the former, musical tracks are layed down on a multitrack recorder in stages, with backing tracks and rhythm tracks being recorded first, followed by lead tracks and vocals. In the mixdown phase, all the previously recorded tracks are played back through the mixer and combined into a stereo or surround mix to form the finished product which goes to be made into a commercial release. Since the widespread adoption of electronic instruments and MIDI-controlled equipment (see Chapter 14), MIDI-sequenced sound sources are often played directly into the mix in the second stage.

For these reasons, as well as requiring mixdown signal paths from many inputs to a stereo bus the mixer also requires signal paths for routing many input signals to a multitrack recorder. Often it will be necessary to perform both of these functions simultaneously – that is, recording microphone signals to multitrack whilst also mixing the return from multitrack into stereo, so that the engineer and producer can hear what the finished result will sound like, and so that any musicians who may be overdubbing additional tracks can be given a mixed feed of any previously recorded tracks in headphones. The latter is known as the monitor mix

FIGURE 5.4

In multitrack recording two signal paths are needed – one from mic or line input to the multitrack recorder, and one returning from the recorder to contribute to a 'monitor' mix.

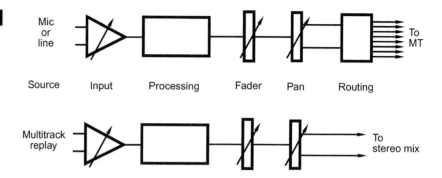

and this often forms the basis for the stereo mixdown when the tracklaying job is finished.

So there are two signal paths in this case: one from the microphone or line source to the multitrack recorder, and one from the multitrack recorder back to the stereo mix, as shown in Figure 5.4. The path from the microphone input which usually feeds the multitrack machine will be termed the channel path, whilst the path from the line input or tape return which usually feeds the stereo mix will be termed the monitor path.

It is likely that some basic signal processing such as equalization will be required in the feed to the multitrack recorder (see below), but the more comprehensive signal processing features are usually applied in the mixdown path. The situation used to be somewhat different in the American market where there was a greater tendency to record on multitrack 'wet', that is with all effects and EQ, rather than applying the effects on mixdown.

In-line and split configurations

As can be seen from Figure 5.4, there are two complete signal paths, two faders, two sets of EQ, and so on. This takes up space, and there are two ways of arranging this physically, one known as the split-monitoring, or European-style console, the other as the in-line console. The split console is the more obvious of the two, and its physical layout is shown in Figure 5.5. It contains the input channels on one side (usually the left), a master control section in the middle, and the monitor mixer on the other side. So it really is two consoles in one frame. It is necessary to have as many monitor channels as there are tracks on the tape, and these channels are likely to need some signal processing. The monitor mixer is used during track laying for mixing a stereo version of the material that is being recorded, so that everyone can hear a rough mix of what the end result will

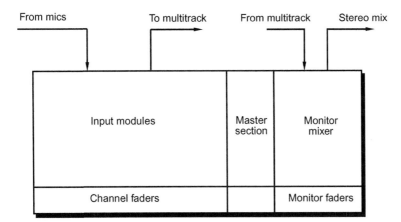

From mics   To multitrack   From multitrack   Stereo mix

Input modules	Master section	Monitor mixer
Channel faders		Monitor faders

FIGURE 5.5

A typical 'split' or 'European-style' multitrack mixer has input modules on one side and monitor modules on the other: two separate mixers in effect.

sound like. On mixdown every input to the console can be routed to the stereo mix bus so as to increase the number of inputs for outboard effects, etc. and so that the comprehensive facilities provided perhaps only on the left side of the console are available for the multitrack returns.

This layout has advantages in that it is easily assimilated in operation, and it makes the channel module less cluttered than the in-line design (described below), but it can make the console very large when a lot of tracks are involved. It can also increase the build cost of the console because of the near doubling in facilities and metalwork required, and it lacks flexibility, especially when switching over from track laying to remixing.

The in-line layout involves the incorporation of the monitor paths from the right-hand side of the split console (the monitor section) into the left side, rather as if the console were sawn in half and the right side merged with the left, as shown in Figure 5.6. In this process a complete monitor signal path is fitted into the module of the same-numbered channel path, making it no more than a matter of a few switches to enable facilities to be shared between the two paths. In such a design each module will contain two faders (one for each signal path), but usually only one EQ section, one set of auxiliary sends (see below), one dynamics control section, and so on, with switches to swap facilities between paths. (A simple example showing only the switching needed to swap one block of processing is shown in Figure 5.7.) Usually this means that it is not possible to have EQ in both the multitrack recording path and the stereo mix path, but some more recent designs have made it possible to split the equalizer so that some frequency-band controls are in the channel path whilst others are in the monitor path. The band ranges are then made to overlap considerably which makes the arrangement quite flexible.

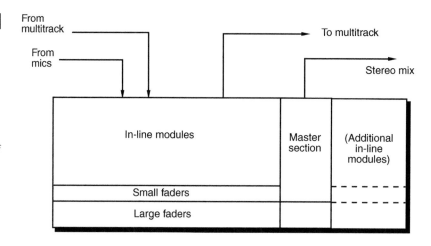

A typical 'in-line' mixer incorporates two signal paths in one module, providing two faders per module (one per path). This has the effect of reducing the size of the mixer for a given number of channels, when compared with a split design.

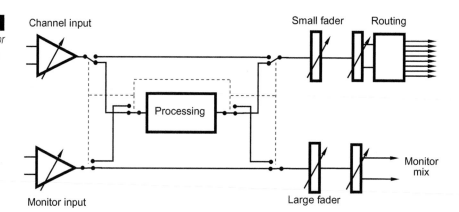

The in-line design allows for sound processing facilities such as EQ and dynamics to be shared or switched between the signal paths.

Further aspects of the in-line design

It has already been stated that there will be two main faders associated with each channel module in an in-line console: one to control the gain of each signal path. Sometimes the small fader is not a linear slider but a rotary knob. It is not uniformly agreed as to whether the large fader at the bottom of the channel module should normally control the monitor level of the like-numbered tape track or whether it should control the channel output level to multitrack tape. Convention originally had it that American consoles made the large fader the monitor fader in normal operation, while British consoles tended to make it the channel fader. Normally their functions can be swapped over, depending on whether one is mixing down or track laying, either globally (for the whole console), in which case the fader swap will probably happen automatically when switching the console from 'recording' to 'remix' mode, or on individual channels, in which case the operation is usually performed using a control labeled something like 'fader flip', 'fader reverse' or 'changeover'. The process of fader swapping is mostly used for convenience, since more precise control can be exercised over a large fader near the operator than over a small fader which is further away, so the large fader is assigned to the function that is being used most in the current operation. This is coupled with the fact that in an automated console it is almost invariably the large fader that is automated, and the automation is required most in the mixdown process.

Confusion can arise when operating in-line mixers, such as when a microphone signal is fed into, say, mic input 1 and is routed to track 13 on the tape. In such a case the operator will control the monitor level of that track (and therefore the level of that microphone's signal in the stereo mix) on monitor fader 13, whilst the channel fader on module 1 will control the multitrack record level for that mic signal.

If a 24-track recorder is in use with the mixer, then monitor faders higher than number 24 will not normally carry a tape return, but will be free for other sources. More than one microphone signal can be routed to each track on the tape (as each multitrack output on the mixer has its own mix bus), so there will be a number of level controls that affect each source's level in the monitor mix, each of which has a different purpose:

- MIC LEVEL TRIM – adjusts the gain of the microphone pre-amplifier at the channel input. Usually located at the top of the module.
- CHANNEL FADER – comes next in the chain and controls the individual level of the mic (or line) signal connected to that module's input before it goes to tape. Located on the same-numbered module as the input. (May be switched to be either the large or small fader, depending on configuration.)
- BUS TRIM or TRACK SUBGROUP – will affect the overall level of all signals routed to a particular track. Usually located with the track routing buttons at the top of the module. Sometimes a channel fader can be made to act as a subgroup master.
- MONITOR FADER – located in the return path from the multitrack recorder to the stereo mix. Does not affect the recorded level on the multitrack tape, but affects the level of this track in the mix. (May be switched to be either the large or small fader, depending on configuration.)

A typical in-line multitrack mixer is shown in the photograph in Figure 5.8.

CHANNEL GROUPING

Grouping is a term that refers to the simultaneous control of more than one signal at a time. It usually means that one fader controls the levels of a number of slave channels. Two types of channel grouping are currently common: audio grouping and 'control' grouping. The latter is often called VCA grouping, but there are other means of control grouping that are not quite the same as the direct VCA control method. The two approaches have very different results, although initially they may appear to be similar because one fader appears to control a number of signal levels. The primary reason for using group faders of any kind is in order to reduce the number of faders that the engineer has to handle at a time. This can be done in a situation where a number of channels are carrying audio signals that can be faded up and down together. These signals do not all have to be at the same initial level, and indeed one is still free to adjust levels individually within a group. A collection of channels carrying drum sounds, or carrying an orchestral string section, would be examples of suitable groups. The two approaches are described in Fact Files 5.4 and 5.5.

FIGURE 5.8 *A typical in-line mixer: the Soundcraft 'Sapphyre'. (Courtesy of Soundcraft Electronics Ltd.)*

AN OVERVIEW OF TYPICAL MIXER FACILITIES

Most mixing consoles provide a degree of sound signal processing on board, as well as routing to external processing devices. The very least of these facilities is some form of equalization (a means of controlling the gain at various frequencies), and there are few consoles which do not include this. As well as signal processing, there will be a number of switches that make changes to the signal path or operational mode of the console. These may operate on individual channels, or they may function globally (affecting the whole console at once). The following section is a guide to the facilities commonly found on multitrack consoles. Figure 5.9 shows the typical location of these sections on an in-line console module.

FACT FILE 5.4 AUDIO GROUPS

Audio groups are so called because they create a single audio output which is the sum of a number of channels. A single fader controls the level of the summed signal, and there will be a group output from the console which is effectively a mix of the audio signals in that group, as shown in the diagram. The audio signals from each input to the group are fed via equal-value resistors to the input of a summing or virtual-earth amplifier.

The stereo mix outputs from an in-line console are effectively audio groups, one for the left, one for the right, as they constitute a sum of all the signals routed to the stereo output and include overall level control. In the same way, the multitrack routing buses on an in-line console are also audio groups, as they are sums of all the channels routed to their respective tracks. More obviously, some smaller or older consoles will have routing buttons on each channel module for, say, four audio group destinations, these being really the only way of routing channels to the main outputs.

The master faders for audio groups will often be in the form of four or eight faders in the central section of the console. They may be arranged such that one may pan a channel between odd and even groups, and it would be common for two of these groups (an odd and an even one) to be used as the stereo output in mixdown. It is also common for perhaps eight audio group faders to be used as 'subgroups', themselves having routing to the stereo mix, so that channel signals can be made more easily manageable by routing them to a subgroup (or panning between two subgroups) and thence to the main mix via a single level control (the subgroup fader), as shown in the diagram. (Only four subgroups are shown in the diagram, without pan controls. Subgroups one and three feed the left mix bus, and two and four feed the right mix bus. Sometimes subgroup outputs can be panned between left and right main outputs.)

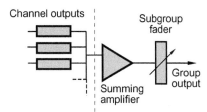

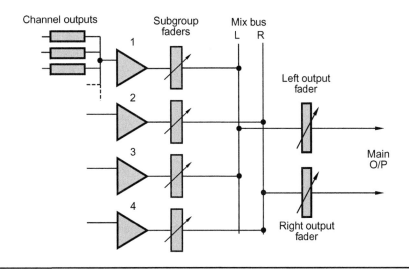

FACT FILE 5.5 CONTROL GROUPS

Control grouping differs from audio grouping primarily because it does not give rise to a single summed audio output for the group: the levels of the faders in the group are controlled from one fader, but their outputs remain separate. Such grouping can be imagined as similar in its effect to a large hand moving many faders at the same time, each fader maintaining its level in relation to the others.

The most common way of achieving control grouping is to use VCAs (voltage-controlled amplifiers), whose gain can be controlled by a DC voltage applied to a control pin. In the VCA fader, audio is not passed through the fader itself but is routed through a VCA, whose gain is controlled by a DC voltage derived from the fader position, as shown in the diagram. So the fader now carries DC instead of audio, and the audio level is controlled indirectly. A more recent alternative to the VCA is the DCA, or digitally controlled attenuator, whose gain is controlled by a binary value instead of a DC voltage. This can be easier to implement in digitally controlled mixers.

Indirect gain control opens up all sorts of new possibilities. The gain of the channel could be controlled externally from a variety of sources, either by combining the voltage from an external controller in an appropriate way with the fader's voltage so that it would still be possible to set the relative level of the channel, or by breaking the direct connection between the DC fader and the VCA so that an automation system could intervene, as discussed in 'Automation', p. 154. It becomes possible to see that group faders could be DC controls which could be connected to a number of channel VCAs such that their gains would go up and down together. Further to this,

a channel VCA could be assigned to any of the available groups simply by selecting the appropriate DC path: this is often achieved by means of thumbwheel switches on each fader, as shown in the diagram.

Normally, there are dedicated VCA group master faders in a non-automated system. They usually reside in the central section of a mixer and will control the overall levels of any channel faders assigned to them by the thumbwheels by the faders. In such a system, the channel audio outputs would normally be routed to the main mix directly, the grouping affecting the levels of the individual channels in this mix.

In an automated system grouping may be achieved via the automation processor which will allow any fader to be designated as the group master for a particular group. This is possible because the automation processor reads the levels of all the faders, and can use the position of the designated master to modify the data sent back to the other faders in the group (see 'Automation', p. 154).

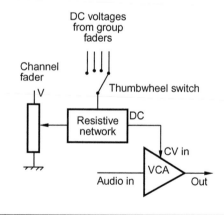

Input section

- *Input gain control*
  Sets the microphone or line input amplifier gain to match the level of the incoming signal. This control is often a coarse control in 10 dB steps, sometimes accompanied by a fine trim. Opinion varies as to whether this control should be in detented steps or continuous.

Detented steps of 5 or 10 dB make for easy reset of the control to an exact gain setting, and precise gain matching of channels.

- *Phantom power*
  Many professional mics require 48 volts phantom powering (see 'Microphone powering options', Chapter 3). There is sometimes a switch on the module to turn it on or off, although most balanced mics which do not use phantom power will not be damaged if it is accidentally left on. Occasionally this switch is on the rear of the console, by the mic input socket, or it may be in a central assignable switch panel. Other methods exist: for example, one console requires that the mic gain control is pulled out to turn on the phantom power.

- *MIC/LINE switch*
  Switches between the channel's mic input and line input. The line input could be the playback output from a tape machine, or another line level signal such as a synth or effects device.

- *PAD*
  Usually used for attenuating the mic input signal by something like 20 dB, for situations when the mic is in a field of high sound pressure. If the mic is in front of a kick drum, for example, its output may be so high as to cause the mic input to clip. Also, capacitor mics tend to produce a higher output level than dynamic mics, requiring that the pad be used on some occasions.

- *Phase reverse or 'Φ'*
  Sometimes located after the mic input for reversing the phase of the signal, to compensate for a reversed directional mic, a mis-wired lead, or to create an effect. This is often left until later in the signal path.

- *HPF/LPF*
  Filters can sometimes be switched in at the input stage, which will usually just be basic high- and

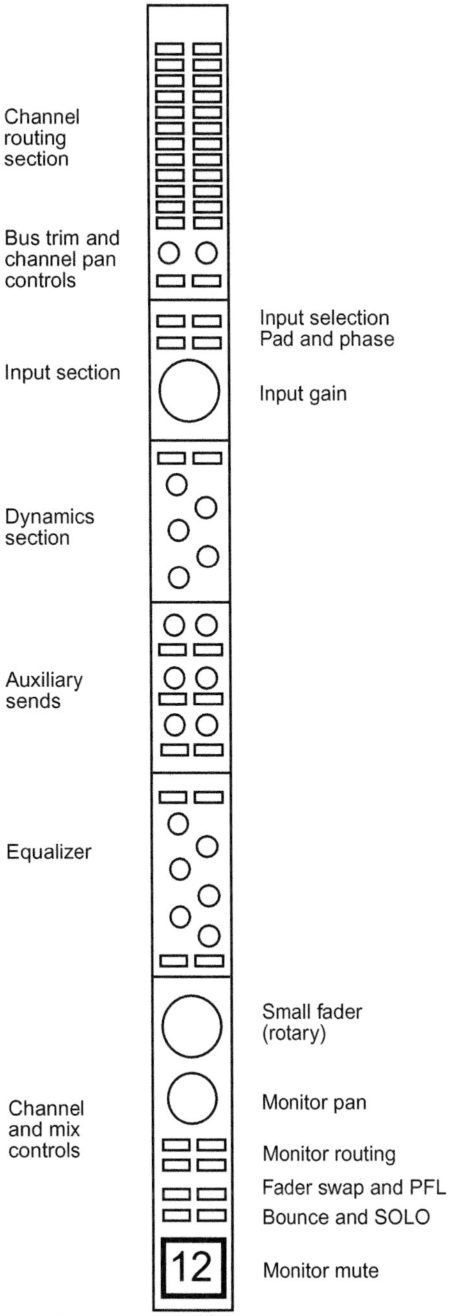

FIGURE 5.9 *Typical layout of controls on an in-line mixer module (for description see text).*

low-pass filters which are either in or out, with no frequency adjustment. These can be used to filter out unwanted rumble or perhaps hiss from noisy signals. Filtering rumble at this stage can be an advantage because it saves clipping later in the chain.

Routing section

■ *Track routing switches*
The number of routing switches depends on the console: some will have 24, some 32 and some 48. The switches route the channel path signal to the multitrack machine, and it is possible to route a signal to more than one track. The track assignment is often arranged as pairs of tracks, so that odd and even tracks can be assigned together, with a pan-pot used to pan between them as a stereo pair, e.g. tracks three and four could be a stereo pair for background vocals, and each background vocal mic could be routed to three and four, panned to the relevant place in the image. In an assignable console these controls may be removed to a central assignable routing section.

It is common for there to be fewer routing switches than there are tracks, so as to save space, resulting in a number of means of assigning tracks. Examples are rotary knobs to select the track, one button per pair of tracks with 'odd/even/both' switch, and a 'shift' function to select tracks higher than a certain number. The multitrack routing may be used to route signals to effects devices during mixdown, when the track outputs are not being used for recording. In this case one would patch into the track output on the patchfield (see below) and take the relevant signal to an effects input somewhere else on the patchfield. In order to route monitor path signals to the track routing buses it may be necessary to use a switch which links the output of the monitor fader to the track assignment matrix.

In theater sound mixers it is common for output routing to be changed very frequently, and thus routing switches may be located close to the channel fader, rather than at the top of the module as in a music mixer. On some recent mixers, track routing is carried out on a matrix which resides in the central section above the main faders. This removes unnecessary clutter from the channel modules and reduces the total number of switches required. It may also allow the storing of routing configurations in memory for later recall.

- *Mix routing switches*
  Sometimes there is a facility for routing the channel path output sig-nal to the main monitor mix, or to one of perhaps four output groups, and these switches will often be located along with the track routing.

- *Channel pan*
  Used for panning channel signals between odd and even tracks of the multitrack, in conjunction with the routing switches.

- *Bus trim*
  Used for trimming the overall level of the send to multitrack for a particular bus. It will normally trim the level sent to the track which corresponds to the number of the module.

- *Odd/Even/Both*
  Occasionally found when fewer routing buttons are used than there are tracks. When one routing button is for a pair of tracks, this switch will determine whether the signal is sent to the odd channel only, the even channel only, or to both (in which case the pan control is operative).

- *DIRECT*
  Used for routing the channel output directly to the corresponding track on the multitrack machine without going via the summing buses. This can reduce the noise level from the console since the summing procedure used for combining a number of channel outputs to a track bus can add noise. If a channel is routed directly to a track, no other signals can be routed to that track.

Dynamics section

Some advanced consoles incorporate dynamics control on every module, so that each signal can be treated without resorting to external devices. The functions available on the best designs rival the best external devices, incorporating compressor and expander sections which can act as limiters and gates respectively if required. One system allows the EQ to be placed in the side-chain of the dynamics unit, providing frequency-sensitive limit-ing, among other things, and it is usually possible to link the action of one channel's dynamics to the next in order to 'gang' stereo channels so that the image does not shift when one channel has a sudden change in level while the other does not.

When dynamics are used on stereo signals it is important that left and right channels have the same settings, otherwise the image may be affected. If dynamics control is not available on every module, it is sometimes offered on the central section with inputs and outputs on the patchbay. Dynamics control will not be covered further here, but is discussed in more detail in 'The compressor/limiter', p. 394.

Equalizer section

The EQ section is usually split into three or four sections, each operating on a different frequency band. As each band tends to have similar functions these will be described in general. The principles of EQ are described in greater detail in 'EQ explained', p. 142.

- *HF, MID 1, MID 2, LF*
  A high-frequency band, two mid-frequency bands, and a low-frequency band are often provided. If the EQ is parametric these bands will allow continuous variation of frequency (over a certain range), 'Q', and boost/cut. If it is not parametric, then there may be a few switched frequencies for the mid band, and perhaps a fixed frequency for the LF and HF bands.

- *Peaking/shelving or BELL*
  Often provided on the upper and lower bands for determining whether the filter will provide boost/cut over a fixed band (whose width will be determined by the Q), or whether it will act as a shelf, with the response rising or rolling off above or below a certain frequency (see 'EQ explained', p. 142).

- *Q*
  The Q of a filter is defined as its center frequency divided by its bandwidth (the distance between frequencies where the output of the filter is 3 dB lower than the peak output). In practice this affects the 'sharpness' of the filter peak or notch, high Q giving the sharpest response, and low Q giving a very broad response. Low Q would be used when boost or cut over a relatively wide range of frequencies is required, while high Q is used to boost or cut one specific region (see Fact File 5.6).

- *Frequency control*
  Sets the center frequency of a peaking filter, or the turnover frequency of a shelf.

- *Boost/cut*
  Determines the amount of boost or cut applied to the selected band, usually up to a maximum of around ±15 dB.

FACT FILE 5.6 VARIABLE *Q*

Some EQ sections provide an additional control whereby the *Q* of the filter can be adjusted. This type of EQ section is termed a parametric EQ since all parameters, cut/boost, frequency, and *Q* can be adjusted. The diagram below illustrates the effect of varying the *Q* of an EQ section. High *Q* settings affect very narrow bands of frequencies, low *Q* settings affect wider bands. The low *Q* settings sound 'warmer' because they have gentle slopes and therefore have a more gradual and natural effect on the sound. High *Q* slopes are good for a rather more overt emphasis of a particular narrow band, which of course can be just as useful in the appropriate situation. Some EQ sections are labeled parametric even though the *Q* is not variable. This is a misuse of the term, and it is wise to check whether or not an EQ section is truly parametric even though it may be labeled as such.

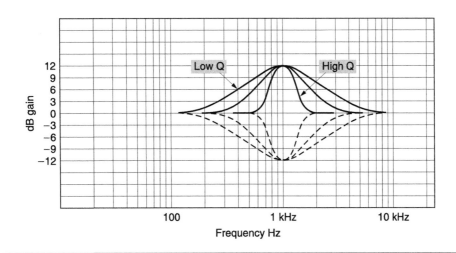

- *HPF/LPF*
  Sometimes the high- and low-pass filters are located here instead of at the input, or perhaps in addition. They normally have a fixed frequency turnover point and a fixed roll-off of either 12 or 18 dB per octave. Often these will operate even if the EQ is switched out.

- *CHANNEL*
  The American convention is for the main equalizer to reside normally in the monitor path, but it can be switched so that it is in the channel path. Normally the whole EQ block is switched at once, but on some recent models a section of the EQ can be switched separately. This would be used to equalize the signal which is being recorded on multitrack tape. If the EQ is in the monitor path then it will only affect the replayed signal. The traditional European

convention is for EQ to reside normally in the channel path, so as to allow recording with EQ.

- *IN/OUT*
  Switches the EQ in or out of circuit. Equalization circuits can introduce noise and phase distortion, so they are best switched out when not required.

Channel and mix controls

- *Pan*
  This control is a continuous rotary knob, and is used to place the signal of that channel in any desired position in the stereo picture. See Fact File 5.2.

- *Fader reverse*
  Swaps the faders between mix and channel paths, so that the large fader can be made to control either the mix level or the channel level. Some systems defeat any fader automation when the large fader is put in the channel path. Fader reverse can often be switched globally, and may occur when the console mode is changed from recording to mixdown.

- *Line/Tape or Bus/Tape*
  Switches the source of the input to the monitor path between the line output of the same-numbered channel and the return from multitrack tape. Again it is possible that this may be switched globally. In 'line' or 'bus' mode the monitor paths are effectively 'listening to' the line output of the console's track assignment buses, while in 'tape' mode the monitor paths are listening to the off-tape signal (unless the tape machine's monitoring is switched to monitor the line input of the tape machine, in which case 'line' and 'tape' will effectively be the same thing!). If a problem is suspected with the tape machine, switching to monitor 'line' will bypass the tape machine entirely and allow the operator to check if the console is actually sending anything.

- *Broadcast, or 'mic to mix, or 'simulcast'*
  Used for routing the mic signal to both the channel and monitor paths simultaneously, so that a multitrack recording can be made while a stereo mix is being recorded or broadcasted. The configuration means that any alterations made to the channel path will not affect the stereo mix, which is important when the mix output is live (see Figure 5.10).

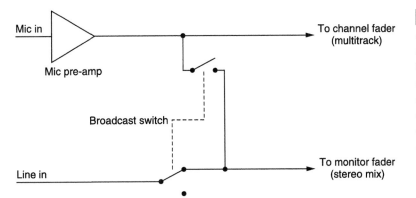

FIGURE 5.10

A 'broadcast mode' switch in an in-line console allows the microphone input to be routed to both signal paths, such that a live stereo mix may be made independent of any changes to multitrack recording levels.

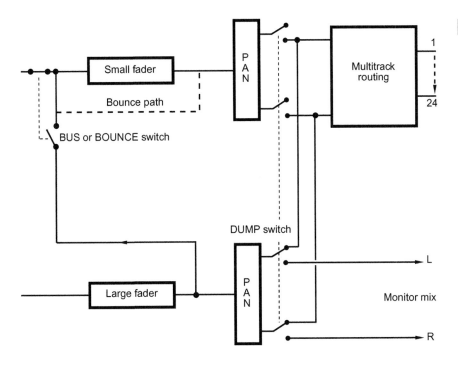

FIGURE 5.11

Signal routings for 'bounce', 'bus' and 'dump' modes (see text).

- *BUS or 'monitor-to-bus'*
  Routes the output of the monitor fader to the input of the channel path (or the channel fader) so that the channel path can be used as a post-fader effects send to any one of the multitrack buses (used in this case as aux sends), as shown in Figure 5.11. If a BUS TRIM control is provided on each multitrack output this can be used as the master effects-send level control.

- *DUMP*
  Incorporated (rarely) on some consoles to route the stereo panned mix output of a track (i.e. after the monitor path pan-pot) to the multitrack assignment switches. In this way, the mixed version of a group of tracks can be 'bounced down' to two tracks on the multitrack, panned and level-set as in the monitor mix (see Figure 5.11).

- *BOUNCE*
  A facility for routing the output of the monitor fader to the multitrack assignment matrix, before the pan control, in order that tracks can be 'bounced down' so as to free tracks for more recording by mixing a group of tracks on to a lower number of tracks. BOUNCE is like a mono version of DUMP (see Figure 5.11).

- *MUTE or CUT*
  Cuts the selected track from the mix. There may be two of these switches, one for cutting the channel signal from the multitrack send, the other for cutting the mix signal from the mix.

- *PFL*
  See Fact File 5.3.

- *AFL*
  After fade listen (AFL) is similar to PFL, except that it is taken from after the fader. This is sometimes referred to as SOLO, which routes a panned version of the track to the main monitors, cutting everything else. These functions are useful for isolating signals when setting up and spotting faults. On many consoles the AFL bus will be stereo. Solo functions are useful when applying effects and EQ, in order that one may hear the isolated sound and treat it individually without hearing the rest of the mix. Often a light is provided to show that a solo mode is selected, because there are times when nothing can be heard from the loudspeakers due to a solo button being down with no signal on that track. A solo safe control may be provided centrally, which prevents this feature from being activated.

- *In-place solo*
  On some consoles, solo functions as an 'in-place' solo, which means that it actually changes the mix output, muting all tracks which are not solo'ed and picking out all the solo'ed tracks. This may be preferable to AFL as it reproduces the exact contribution of each

channel to the mix, at the presently set master mix level. Automation systems often allow the solo functions to be automated in groups, so that a whole section can be isolated in the mix. In certain designs, the function of the automated mute button on the monitor fader may be reversed so that it becomes solo.

Auxiliary sends

The number of aux(iliary) sends depends on the console, but there can be up to ten on an ordinary console, and sometimes more on assignable models. Aux sends are 'take-off points' for signals from either the channel or mix paths, and they appear as outputs from the console which can be used for foldback to musicians, effects sends, cues, and so on. Each module will be able to send to auxiliaries, and each numbered auxiliary output is made up of all the signals routed to that aux send. So they are really additional mix buses. Each aux will have a master gain control, usually in the center of the console for adjusting the overall gain of the signal sent from the console, and may have basic EQ. Aux sends are often a combination of mono and stereo buses. Mono sends are usually used as routes to effects, while stereo sends may have one level control and a pan control per channel for mixing a foldback source.

- *Aux sends 1–n*
  Controls for the level of each individual channel in the numbered aux mix.

- *Pre/post*
  Determines whether the send is taken off before or after the fader. If it is before then the send will still be live even when the fader is down. Generally, 'cue' feeds will be pre-fade, so that a mix can be sent to foldback which is independent of the monitor mix. Effects sends will normally be taken post-fade, in order that the effect follows a track's mix level.

- *Mix/channel*
  Determines whether the send is taken from the mix or channel paths. It will often be sensible to take the send from the channel path when effects are to be recorded on to multitrack rather than on to the mix. This function has been labeled 'WET' on some designs.

- *MUTE*
  Cuts the numbered send from the aux mix.

Master control section

The master control section usually resides in the middle of the console, or near the right-hand end. It will contain some or all of the following facilities:

■ *Monitor selection*
A set of switches for selecting the source to be monitored. These will include recording machines, aux sends, the main stereo mix, and perhaps some miscellaneous external sources like CD players, cassette machines, etc. They only select the signal going to the loudspeakers, not the mix outputs. This may be duplicated to some extent for a set of additional studio loudspeakers, which will have a separate gain control.

■ *DIM*
Reduces the level sent to the monitor loudspeakers by a considerable amount (usually around 40 dB), for quick silencing of the room.

■ *MONO*
Sums the left and right outputs to the monitors into mono so that mono compatibility can be checked.

■ *Monitor phase reverse*
Phase reverses one channel of the monitoring so that a quick check on suspected phase reversals can be made.

■ *TAPE/LINE*
Usually a global facility for switching the inputs to the mix path between the tape returns and the console track outputs. Can be reversed individually on modules.

■ *FADER REVERSE*
Global swapping of small and large faders between mix and channel paths.

■ *Record/Overdub/Mixdown*
Usually globally configures mic/line input switching, large and small faders and auxiliary sends depending on mode of operation. (Can be overridden on individual channels.)

■ *Auxiliary level controls*
Master controls for setting the overall level of each aux send output.

■ *Foldback and Talkback*
There is often a facility for selecting which signals are routed to the stereo foldback which the musicians hear on their headphones.

Sometimes this is as comprehensive as a cue mixer which allows mixing of aux sends in various amounts to various stereo cues, while often it is more a matter of selecting whether foldback consists of the stereo mix, or one of the aux sends. Foldback level is controllable, and it is sometimes possible to route left and right foldback signals from different sources. Talkback is usually achieved using a small microphone built into the console, which can be routed to a number of destinations. These destinations will often be aux sends, multitrack buses, mix bus, studio loudspeakers and foldback.

- *Oscillator*
  Built-in sine-wave oscillators vary in quality and sophistication, some providing only one or two fixed frequencies, while others allow the generation of a whole range. If the built-in oscillator is good it can be used for lining up the tape machine, as it normally can be routed to the mix bus or the multitrack outputs. The absolute minimum requirement is for accurate 1 kHz and 10 kHz tones, the 10 kHz being particularly important for setting the bias of an analog tape machine. The oscillator will have an output level control.

- *Slate*
  Provides a feed from the console talkback mic to the stereo output, often superimposing a low-frequency tone (around 50 Hz) so that the slate points can be heard when winding a tape at high speed. Slate would be used for recording take information on to tape.

- *Master faders*
  There may be either one stereo fader or left and right faders to control the overall mix output level. Often the group master faders will reside in this section.

Effects returns

Effects returns are used as extra inputs to the mixer, supplied specifically for inputs from external devices such as reverberation units. These are often located in the central section of the console and may be laid out like reduced-facility input channels. Returns sometimes have EQ, perhaps more basic than on channels, and they may have aux sends. Normally they will feed the mix, although sometimes facilities are provided to feed one or more returns to the multitrack via assignment switches. A small fader or rotary level control is provided, as well as a pan-pot for a mono return. Occasionally, automated faders may be assigned to the return channels so as to allow automated control of their levels in the mix.

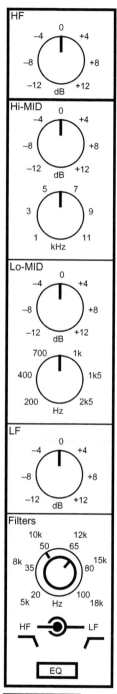

FIGURE 5.12 *Typical layout of an EQ section.*

Patchbay or jackfield

Most large consoles employ a built-in jackfield or patchbay for routing signals in ways which the console switching does not allow, and for sending signals to and from external devices. Just about every input and output on every module in the console comes up on the patchbay, allowing signals to be cross-connected in virtually any configuration. The jackfield is usually arranged in horizontal rows, each row having an equal number of jacks. Vertically, it tries to follow the signal path of the console as closely as possible, so the mic inputs are at the top and the multitrack outputs are nearer the bottom. In between these there are often insert points which allow the engineer to 'break into' the signal path, often before or after the EQ, to insert an effects device, compressor, or other external signal processor. Insert points usually consist of two rows, one which physically breaks the signal chain when a jack is inserted, and one which does not. Normally it is the lower row which breaks the chain, and should be used as inputs. The upper row is used as an output or send. Normaling is usually applied at insert points, which means that unless a jack is inserted the signal will flow directly from the upper row to the lower.

At the bottom of the patchfield will be all the master inputs and outputs, playback returns, perhaps some parallel jacks, and sometimes some spare rows for connection of one's own devices. Some consoles bring the microphone signals up to the patchbay, but there are some manufacturers who would rather not do this unless absolutely necessary as it is more likely to introduce noise, and phantom power may be present on the jackfield. Jackfields are covered in further detail in 'Jackfields (patchbays)', Chapter 12.

EQ EXPLAINED

The tone control or EQ (5equalization) section provides mid-frequency controls in addition to bass and treble. A typical comprehensive EQ section may have first an HF (high-frequency) control similar to a treble control but operating only at the highest frequencies. Next would come a hi-mid control, affecting frequencies from around 1 kHz to 10 kHz, the center frequency being adjusted by a separate control. Lo-mid controls would come next, similar to the hi-mid but operating over a range of say 200 Hz to 2 kHz. Then would come an LF (low-frequency) control. Additionally, high- and low-frequency filters can be provided. The complete EQ section looks something like that shown in Figure 5.12. An EQ section takes up quite a bit of space, and so it is quite common for dual concentric or even assignable controls (see below) to be used. For instance, the cut/boost controls of the hi- and lo-mid sections can be surrounded by annular skirts which select the frequency.

Principal EQ bands

The HF section affects the highest frequencies and provides up to 12 dB of boost or cut. This type of curve is called a shelving curve because it gently boosts or cuts the frequency range towards a shelf where the level remains relatively constant (see Figure 5.13a). Next comes the hi-mid section. Two controls are provided here, one to give cut or boost, the other to select the desired center frequency. The latter is commonly referred to as a 'swept mid' because one can sweep the setting across the frequency range.

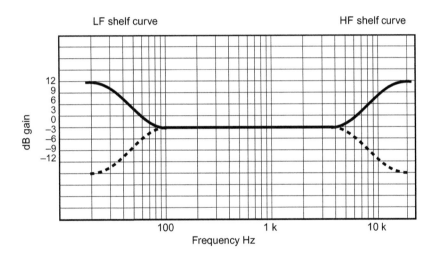

FIGURE 5.13A

Typical HF and LF shelf EQ characteristics shown at maximum settings.

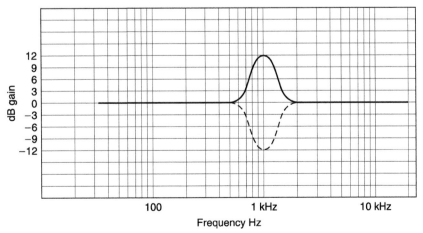

FIGURE 5.13B

Typical MF peaking filter characteristic.

MF peaking filter characteristics at 1, 5 and 10kHz.

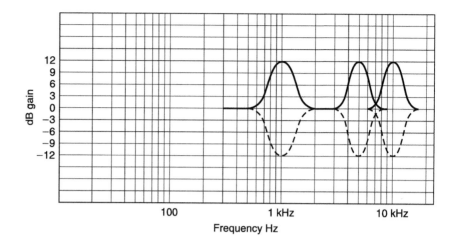

High-pass filters with various turnover frequencies.

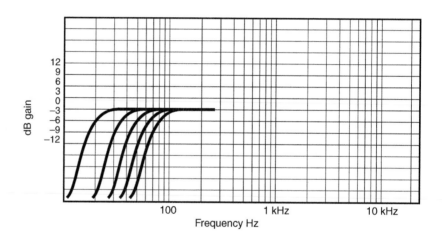

Figure 5.13b shows the result produced when the frequency setting is at the 1 kHz position, termed the center frequency. Maximum boost and cut affects this frequency the most, and the slopes of the curve are considerably steeper than those of the previous shelving curves. This is often referred to as a 'bell' curve due to the upper portion's resemblance to the shape of a bell. It has a fairly high 'Q', that is, its sides are steep. Q is defined as:

$$Q = \text{center frequency} \div \text{bandwidth}$$

where the bandwidth is the spacing in hertz between the two points at which the response of the filter is 3 dB lower than that at the center frequency. In the example shown the center frequency is 1 kHz and the bandwidth is 400 Hz, giving $Q = 2.5$.

MF EQ controls are often used to hunt for trouble-spots; if a particular instrument (or microphone) has an emphasis in its spectrum somewhere which does not sound very nice, some mid cut can be introduced, and the frequency control can be used to search for the precise area in the frequency spectrum where the trouble lies. Similarly, a dull sound can be given a lift in an appropriate part of the spectrum which will bring it to life in the overall mix. Figure 5.13c shows the maximum cut and boost curves obtained with the frequency selector at either of the three settings of 1, 5 and 10 kHz. The high Q of the filters enables relatively narrow bands to be affected. Q may be varied in some cases, as described in Fact File 5.6.

The lo-mid section is the same as the hi-mid section except that it covers a lower band of frequencies. Note though that the highest frequency setting overlaps the lowest setting of the hi-mid section. This is quite common, and ensures that no 'gaps' in the frequency spectrum are left uncovered.

Filters

High- and low-cut filters provide fixed attenuation slopes at various frequencies. Figure 5.13d shows the responses at LF settings of 80, 65, 50, 35 and 20 Hz. The slopes are somewhat steeper than is the case with the HF and LF shelving curves, and slope rates of 18 or 24 dB per octave are typical. This enables just the lowest, or highest, frequencies to be rapidly attenuated with minimal effect on the mid band. Very low traffic rumble could be removed by selecting the 20 or 35 Hz setting. More serious low-frequency noise may require the use of one of the higher turnover frequencies. High-frequency hiss from, say, a noisy guitar amplifier or air escaping from a pipe organ bellows can be dealt with by selecting the turnover frequency of the HF section which attenuates just sufficient HF noise without unduly curtailing the HF content of the wanted sound.

STEREO LINE INPUT MODULES

In broadcast situations it is common to require a number of inputs to be dedicated to stereo line level sources, such as CD players, electronic musical instruments, etc. Such modules are sometimes offered as an option for multitrack consoles, acting as replacements for conventional I/O modules and allowing two signals to be faded up and down together with one fader. Often the EQ on such modules is more limited, but the module may provide for the selection of more than one stereo source, and routing to the main mix as well as the multitrack. It is common to require that stereo modules always reside in special slots on the console, as they may require special wiring. Such modules are also used to provide facilities for handling LP turntable outputs, offering RIAA equalization (see 'RIAA equalization', Appendix 1).

Stereo microphone inputs can also be provided, with the option for MS (middle and side) format signals as well as AB (conventional left and right) format (see 'Stereo microphones', Chapter 3). A means of control over stereo width can be offered on such modules.

DEDICATED MONITOR MIXER

A dedicated monitor mixer is often used in live sound reinforcement work to provide a separate monitor mix for each musician, in order that each artist may specify his or her precise monitoring requirements. A comprehensive design will have, say, 24 inputs containing similar facilities to a conventional mixer, except that below the EQ section there will be a row of rotary or short-throw faders which individually send the signal from that channel to the group outputs, in any combination of relative levels. Each group output will then provide a separate monitor mix to be fed to headphones or amplifier racks.

TECHNICAL SPECIFICATIONS

This section contains some guidance concerning the meanings and commonly encountered values of technical specifications for mixers.

Input noise

The output from a microphone is in the millivolt range, and so needs considerable amplification to bring it up to line level. Amplification of the signal also brings with it amplification of the microphone's own noise output (discussed in 'Microphone noise in practice', Chapter 3), which one can do nothing about, as well as amplification of the mixer's own input noise. The latter must therefore be as low as possible so as not to compromise the noise performance unduly. A 200 ohm source resistance on its own generates $0.26\,\mu V$ of noise (20 kHz bandwidth). Referred to the standard line level of 775 mV (0 dBu) this is -129.6 dBu. A microphone amplifier will add its own noise to this, and so manufacturers quote an 'equivalent input noise' (EIN) value which should be measured with a 200 ohm source resistance across the input.

An amplifier with a noise contribution equal to that of the 200 ohm resistor will degrade the theoretically 'perfect' noise level by 3 dB, and so the quoted equivalent input noise will be $-129.6 + 3 = -126.6$ dBm. (Because noise contributions from various sources sum according to their power content, not their voltage levels, dBm is traditionally used to express input noise level.) This value is quite respectable, and good-quality mixers should not be noisier than this. Values of around -128 dBm are sometimes encountered, which are excellent, indicating that the input resistance is generating more noise than the amplifier. Make sure that the EIN is quoted with a 200 ohm source, and a bandwidth up to 20 kHz, unweighted. A 150 ohm source, sometimes specified, will give an apparently better EIN simply because this resistor is itself quieter than a 200 ohm one, resistor noise being proportional to ohmic value. Also, weighting gives a flattering result, so one always has to check the measuring conditions. Make sure that EIN is quoted in dBm or dBu. Some manufacturers quote EIN in dBV (i.e. ref. 1 volt) which gives a result that looks 2.2 dB better. An input should have high common mode rejection as well as low noise, as discussed in Fact File 5.7.

Output noise

The output residual noise of a mixer, with all faders at minimum, should be at most -90 dBu. There is no point in having a very quiet microphone amplifier if a noisy output stage ruins it. With all channels routed to the output, and all faders at the 'zero' position, output noise (or 'mixing' noise) should be at least -80 dBu with the channel inputs switched to 'line' and set for unity gain. Switching these to 'mic' inevitably increases noise levels

---

FACT FILE 5.7 COMMON MODE REJECTION

As discussed in 'Balanced lines', Chapter 12, common mode rejection is the ability of a balanced input to reject interference which can be induced into the signal lines. A microphone input should have a CMRR (common mode rejection ratio) of 70 dB or more; i.e. it should attenuate the interference by 70 dB. But look at how this measurement is made. It is relatively easy to achieve 70 dB at, say, 500 Hz, but rejection is needed most at high frequencies – between 5 and 20 kHz – and so a quoted CMRR of '70 dB at 15 kHz' or '70 dB between 100 Hz and 10 kHz' should be sought. Line level CMRR can be allowed to be rather lower since the signal voltage level is a lot higher than in microphone cabling. CMRRs of as low as 30 dB at 10 kHz are deemed to be adequate.

Common mode rejection is a property of a balanced input, and so it is not applicable to a balanced output. However, output balance is sometimes quoted which gives an indication of how closely the two legs of a balanced output are matched. If the two legs were to be combined in antiphase total cancelation would ideally be achieved. In practice, around 70 dB of attenuation should be looked for.

because this increases the gain of the input amplifier. It underlines the reason why all unused channels should be unrouted from the mix buses, or their faders brought down to a minimum. Digital mixers with 'scene' memories tend to be programmed by copying a particular scene to another vacant scene, then modifying it for the new requirements. When doing this, one needs to ensure that all unwanted inputs and routing from the copied scene are removed so as to maintain the cleanest possible signal. Make sure that the aux outputs have a similarly good output noise level.

Impedance

A microphone input should have a minimum impedance of $1\,k\Omega$. A lower value than this degrades the performance of many microphones. A line level input should have a minimum impedance of $10\,k\Omega$. Whether it is balanced or unbalanced should be clearly stated, and consideration of the type of line level equipment that the mixer will be partnered with will determine the importance of balanced line inputs. All outputs should have a low impedance, below 200 ohms, balanced. Check that the aux outputs are also of very low impedance, as sometimes they are not. If insert points are provided on the input channels and/or outputs, these also should have very low output and high input impedances.

Frequency response

A frequency response that is within $0.2\,dB$ between $20\,Hz$ and $20\,kHz$ for all combinations of input and output is desirable. The performance of audio transformers varies slightly with different source and load impedances, and a specification should state the range of loads between which a 'flat' frequency response will be obtained. Above $20\,kHz$, and probably below $15\,Hz$ or so, the frequency response should fall away so that unwanted out-of-band frequencies are not amplified, for example radio-frequency breakthrough or subsonic interference.

Distortion

With an analog mixer, distortion should be quoted at maximum gain through the mixer and a healthy output level of, say, $10\,dBu$ or more. This will produce a typical worst case, and should normally be less than 0.1% THD (total harmonic distortion). The distortion of the low-gain line level inputs to outputs can be expected to be lower: around 0.01%. The outputs should be loaded with a fairly low impedance which will require more current from the output stages than a high impedance will, this helping to reveal any shortcomings. A typical value is 600 ohms.

Clipping and overload margins are discussed in Fact File 5.8.

FACT FILE 5.8 CLIPPING

A good mixer will be designed to provide a maximum electrical output level of at least +20dBu. Many will provide +24dBu. Above this electrical level clipping will occur, where the top and bottom of the audio waveform are chopped off, producing sudden and excessive distortion (see diagram). Since the nominal reference level of 0dBu usually corresponds to a meter indication of PPM 4 or −4VU, it is very difficult to clip the output stages of a mixer. The maximum meter indication on a PPM would correspond in this case to an electrical output of around +12dBu, and thus one would have to be severely bending the meter needles to cause clipping.

Clipping, though, may occur at other points in the signal chain, especially when large amounts of EQ boost have been added. If, say, 12dB of boost has been applied on a channel, and the fader is set well above the 0dB mark, clipping on the mix bus may occur, depending on overload margins here. Large amounts of EQ boost should not normally be used without a corresponding overall gain reduction of the channel for this reason.

An input pad or attenuator is often provided to prevent the clipping of mic inputs in the presence of high-level signals (see 'Input section', p. 130).

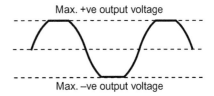

Max. +ve output voltage

Max. −ve output voltage

Crosstalk

In analog mixers, a signal flowing along one path may induce a small signal in another, and this is termed 'crosstalk'. Crosstalk from adjacent channels should be well below the level of the legitimate output signal, and a figure of −80 dB or more should be looked for at 1 kHz. Crosstalk performance tends to deteriorate at high frequencies due to capacitive coupling in wiring harnesses, for instance, but a crosstalk of at least −60 dB at 15 kHz should still be sought. Similarly, very low-frequency crosstalk often deteriorates due to the power supply source impedance rising here, and a figure of −50 dB at 20 Hz is reasonable.

Ensure that crosstalk between all combinations of input and output is of a similarly good level. Sometimes crosstalk between channel auxiliaries is rather poorer than that between the main outputs.

METERING SYSTEMS

Metering systems are provided on audio mixers to indicate the levels of audio signals entering and leaving the mixer. Careful use of metering is vital for optimizing noise and distortion, and to the recording of the correct audio level on tape. In this section the merits of different metering systems are examined.

Mechanical metering

Two primary types of mechanical meters have been used: the VU (volume unit) meter (Figure 5.14) and the PPM (peak program meter), as shown in Figure 5.15. These are very different to each other, the only real similarity being that they both have swinging needles. The British, or BBC-type, PPM is distinctive in styling in that it is black with numbers ranging from 1 to 7 equally spaced across its scale, there being a 4 dB level difference between each gradation, except between 1 and 2 where there is usually a 6 dB change in level. The EBU PPM (also shown in Figure 5.15) has a scale calibrated in decibels. The VU, on the other hand, is usually white or cream, with a scale running from −20 dB up to 3 dB, ranged around a zero point which is usually the studio's electrical reference level. Originally the VU meter was associated with a variable attenuator which could vary the electrical alignment level for 0 VU up to +24 dBu, although it is common for this to be fixed these days at 0 VU= +4 dBu.

It is important to know how meter readings relate to the line-up standard in use in a particular environment, and to understand that these standards may vary between establishments and areas of work. Fact File 5.9 discusses the relationship between meter indication and signal levels, considering practical issues such as the onset of distortion.

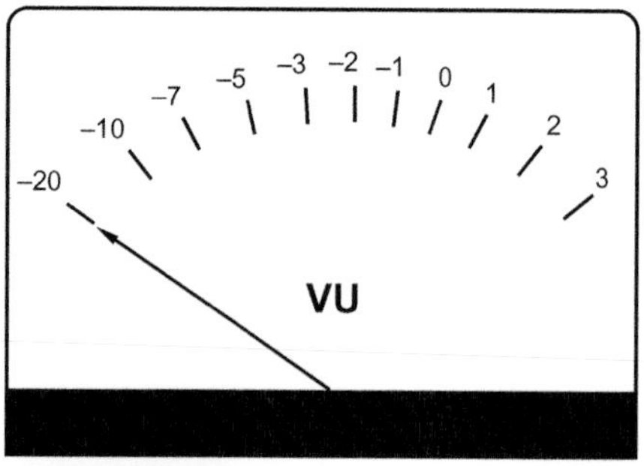

Problems with mechanical meters

PPMs respond well to signal peaks, that is, they have a fast rise-time, whereas VUs are quite the opposite: they have a very slow rise-time. This means that VUs do not give a true representation of the peak level going on to tape, especially in cases when a signal with a high

FIGURE 5.14   *Typical VU meter scale.*

FIGURE 5.15

(Left) BBC-type peak programme meter (PPM). (Right) European-type PPM.

FACT FILE 5.9   METERING, SIGNAL LEVELS AND DISTORTION

Within a studio there is usually a 'reference level' and a 'peak recording level'. In the broadcast domain these are usually referred to as 'alignment level' and 'permitted maximum level (PML)' as shown in Figure 5.17. The reference or alignment level usually relates to the level at which a 1 kHz line-up tone should play back on the console's meters. In analog mixers this may correspond to PPM 4 on a BBC-type PPM or 'Test' on a European PPM. Electrically PPM 4 usually corresponds to a level of 0 dBu, although the German and Nordic metering standards traditionally had it as −3 dBu. In the digital domain, line-up level usually corresponds to either −20 dBFS (SMPTE) or −18 dBFS (EBU), depending on the area of the world and standard concerned. A relationship is therefore established between meter reading and signal level in each domain.

In digital audio systems, where compatibility with other systems is not an issue, it is possible to peak close to 0 dBFS without incurring increases in distortion, and many recording engineers use all this 'headroom' in order to maximize dynamic range. Digital systems clip hard at 0 dBFS whereas analog tape tends to give rise to gradually increasing distortion and level compression as levels rise. Because of the typical relationship between analog and digital levels, peak digital level can correspond to electrical levels as high as +18 to +22 dBu, which can be inconvenient in mixed format systems sharing the same meters. In broadcasting it is normal to peak no more than 8–9 dB above line-up level (PPM 6 in the UK) as higher levels than this can have serious effects on analog transmitter distortion. Limiters are normally used in broadcasting systems, which start to take effect rapidly above this level. Because of the standardized relationship between analog and digital levels in broadcasting, the PML is lower than digital peak level, leading to a degree of undermodulation of the digital system that is considered acceptable in the interests of compatibility.

transient content, such as a harpsichord, is being recorded, often showing as much as 10–15 dB lower than a peak-reading meter. This can result in overmodulation of the recording, especially with digital recorders where the system is very sensitive to peak overload. Nonetheless, many people are used to working with VUs, and have learned to interpret them. They are good for measuring continuous signals such as tones, but their value for monitoring program material is dubious in the age of digital recording.

VUs have no control over the fall-time of the needle, which is much the same as the rise-time, whereas PPMs are engineered to have a fast rise-time and a longer fall-time, which tends to be more subjectively useful. The PPM was designed to indicate peaks that would cause audible distortion, but does not measure the absolute peak level of a signal. Mechanical meters take up a lot of space on a console, and it can be impossible to find space for one meter per channel in the case of a multitrack console. In this case there are often only meters on the main outputs, and perhaps some auxiliary outputs, these being complemented on more expensive consoles by electronic bargraph meters, usually consisting of LED or liquid crystal displays, or some form of 'plasma' display.

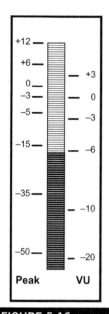

FIGURE 5.16

Typical peak-reading bargraph meter with optional VU scale.

Electronic bargraph metering

Unlike mechanical meters, electronic bargraphs have no mechanical inertia to overcome, so they can effectively have an infinitely fast rise-time although this may not be the ideal in practice. Cheaper bargraphs are made out of a row of LEDs (light emitting diodes), and the resolution accuracy depends on the number of LEDs used. This type of display is sometimes adequate, but unless there are a lot of gradations it is difficult to use them for line-up purposes. Plasma and liquid crystal displays look almost continuous from top to bottom, and do not tend to have the glare of LEDs, being more comfortable to work with for any length of time. Such displays often cover a dynamic range far greater than any mechanical meter, perhaps from −50 dB up to +12 dB, and so can be very useful in showing the presence of signals which would not show up on a mechanical PPM. Such a meter is illustrated in Figure 5.16.

There may be a facility provided to switch the peak response of these meters from PEAK to VU mode, where they will imitate the scale and ballistic response of a VU meter. On more up-market designs it may be possible to use the multitrack bargraphs as a spectrum analyzer display, indicating perhaps a one-third octave frequency-band analysis of the signal fed to it. Occasionally, bargraph displays incorporate a peak-hold facility. A major advantage of these vertical bargraphs is that they take up very little horizontal space on a meter bridge and can thus be used for providing one meter for every channel of the console: useful for monitoring the record levels on a multitrack tape machine. In this case, the feed to the meter is usually taken off at the input to the monitor path of an in-line module.

Miscellaneous meters may also be provided on the aux send outputs for giving some indication of the level being sent to auxiliary devices such as effects. These are commonly smaller than the main meters, or may consist of LED bargraphs with lower resolution. A phase meter or correlation meter is another option, this usually being connected between the left and right main monitor outputs to indicate the degree of phase correlation between these signals. This can be either mechanical or electronic. In broadcast environments, sum and difference (or M and S) meters may be provided to show the level of the mono-compatible and stereo difference signals in stereo broadcasting. These often reside alongside a stereo meter for left and right output levels.

Relationship between different metering standards

Figure 5.17 shows a number of common meter scales and the relationship between them. The relationship between meter indication and electrical level varies depending on the type of meter and the part of the world

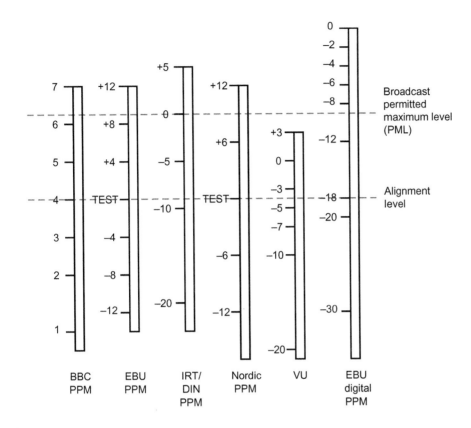

FIGURE 5.17

Graphical comparison of commonly encountered meter scalings.

concerned. As introduced in Fact File 5.9, there is a further relationship to be concerned with, this being that between the electrical output level of the mixer and the recording or transmitted level, and between the analog and digital domains.

Meter take-off point

Output level meter-driving circuits should normally be connected directly across the outputs so that they register the real output levels of the mixer. This may seem self-evident but there are certain models in which this is not the case, the meter circuit taking its drive from a place in the circuit just before the output amplifiers. In such configurations, if a faulty lead or piece of equipment connected to the mixer places, say, a short-circuit across the output the meter will nevertheless read normal levels, and lack of signal reaching a destination will be attributed to other causes. The schematic circuit diagrams of the mixer can be consulted to ascertain whether such an arrangement has been employed. If it is not clear, a steady test tone can be sent to the mixer's output, giving a high meter reading. Then a short-circuit can be deliberately applied across the output (the output amplifier will not normally be harmed by several seconds of short-circuit) and the meter watched. If the indicated level drastically reduces then the meter is correctly registering the real output. If it stays high then the meter is taking its feed from elsewhere.

AUTOMATION

Background

The original and still most common form of mixer automation is a means of storing fader positions dynamically against time for reiteration at a later point in time, synchronous with recorded material. The aim of automation has been to assist an engineer in mixdown when the number of faders that need to be handled at once becomes too great for one person. Fader automation has resulted in engineers being able to concentrate on sub-areas of a mix at each pass, gradually building up the finished product and refining it.

MCI first introduced VCA (voltage controlled amplifier) automation for their JH500 series of mixing consoles in the mid-1970s, and this was soon followed by imitations with various changes from other manufacturers. Moving fader automation systems, such as Neve's NECAM, were introduced slightly later and tended to be more expensive than VCA systems. During the mid-1980s, largely because of the falling cost of microprocessor hardware, console automation enjoyed further advances resulting in developments such as snapshot storage, total dynamic automation, retrofit automation packages, and MIDI-based automation. The rise of digital mixers and digitally controlled analog mixers with integral automation has continued the trend towards total automation of most mixer controls as a standard feature of many new products.

In the following sections a number of different approaches to console automation will be presented and discussed.

Fader automation

There are two common means of memorizing and controlling the gain of a channel: one which stores the positions of the fader and uses this data to control the gain of a VCA or digitally controlled attenuator (DCA), the other which also stores fader movements but uses this information actually to drive the fader's position using a motor. The former is cheaper to implement than the latter, but is not so ergonomically satisfactory because the fader's physical position may not always correspond to the gain of the channel.

It is possible to combine elements of the two approaches in order that gain control can be performed by a VCA but with the fader being moved mechanically to display the gain. This allows for rapid changes in level which might be impossible using physical fader movements, and also allows for dynamic gain offsets of a stored mix whilst retaining the previous gain profile (see below). In the following discussion the term 'VCA faders' may be taken to refer to any approach where indirect gain control of the channel is employed, and many of the concepts apply also to DCA implementations.

With VCA faders it is possible to break the connection between a fader and the corresponding means of level control, as was described in Fact File 5.5. It is across this breakpoint that an automation system will normally be connected. The automation processor then reads a digital value corresponding to the position of the fader and can return a value to control the gain of the channel (see Figure 5.18).

The information sent back to the VCA would depend on the operational mode of the system at the time, and might or might not correspond directly to the fader position. Common operational modes are:

■ WRITE: channel level corresponds directly to the fader position

■ READ: channel level controlled by data derived from a previously stored mix

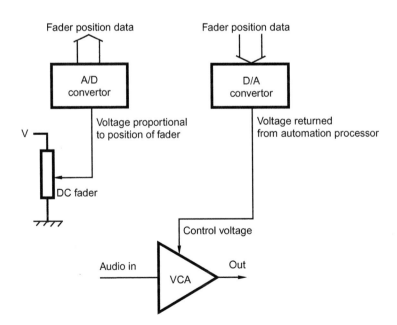

FIGURE 5.18

Fader position is encoded so that it can be read by an automation computer. Data returned from the computer is used to control a VCA through which the audio signal flows.

- UPDATE: channel level controlled by a combination of previously stored mix data and current fader position

- GROUP: channel level controlled by a combination of the channel fader's position and that of a group master

In a VCA implementation the fader position is measured by an analog-to-digital convertor (see Chapter 8), which turns the DC value from the fader into a binary number (usually eight or ten bits) which the microprocessor can read. An eight bit value suggests that the fader's position can be represented by one of 256 discrete values, which is usually enough to give the impression of continuous movements, although professional systems tend to use ten bit representation for more precise control (1024 steps). The automation computer 'scans' the faders many times a second and reads their values. Each fader has a unique address and the information obtained from each address is stored in a different temporary memory location by the computer. A generalized block diagram of a typical system is shown in Figure 5.19.

The disadvantage of such a system is that it is not easy to see what the level of the channel is. During a read or update pass the automation

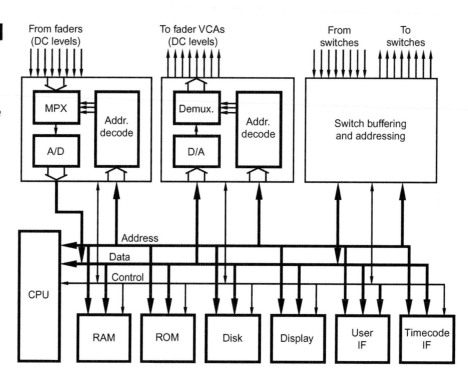

FIGURE 5.19

Generalized block diagram of a mixer automation system handling switches and fader positions. The fader interfaces incorporate a multiplexer (MPX) and demultiplexer (Demux) to allow one convertor to be shared between a number of faders. RAM is used for temporary mix data storage; ROM may hold the operating software program. The CPU is the controlling microprocessor.

computer is in control of the channel gain, rather than the fader. The fader can be half way to the bottom of its travel whilst the gain of the VCA is near the top. Sometimes a mixer's bargraph meters can be used to display the value of the DC control voltage which is being fed from the automation to the VCA, and a switch is sometimes provided to change their function to this mode. Alternatively a separate display is provided for the automation computer, indicating fader position with one marker and channel gain with another.

Such faders are commonly provided with 'null' LEDs: little lights on the fader package which point in the direction that the fader must be moved to make its position correspond to the stored level. When the lights go out (or when they are both on), the fader position is correct. This can sometimes be necessary when modifying a section of the mix by writing over the original data. If the data fed from the automation is different to the position of the fader, then when the mode is switched from READ to WRITE there will be a jump in level as the fader position takes over from the stored data. The null lights allow the user to move the fader towards the position dictated by the stored data, and most systems only switch from read to write when the null point is crossed, to ensure a smooth transition. The same procedure is followed when coming out of rewrite mode, although it can be bypassed in favor of a sudden jump in level.

Update mode involves using the relative position of the fader to modify the stored data. In this mode, the fader's absolute position is not important because the system assumes that its starting position is a point of unity gain, thereafter adding the changes in the fader's position to the stored data. So if a channel was placed in update mode and the fader moved up by 3 dB, the overall level of the updated passage would be increased by 3 dB (see Figure 5.20). For fine changes in gain the fader can be preset near the top of its range before entering update mode, whereas larger changes can be introduced nearer the bottom (because of the gain law of typical faders).

Some systems make these modes relatively invisible, anticipating which mode is most appropriate in certain situations. For example, WRITE mode is required for the first pass of a new mix, where the absolute fader positions are stored, whereas subsequent passes might require all the faders to be in UPDATE.

A moving fader system works in a similar fashion, except that the data which is returned to the fader is used to set the position of a drive mechanism which physically moves the fader to the position in which it was when the mix was written. This has the advantage that the fader is its own means of visual feedback from the automation system and will always represent the gain of the channel.

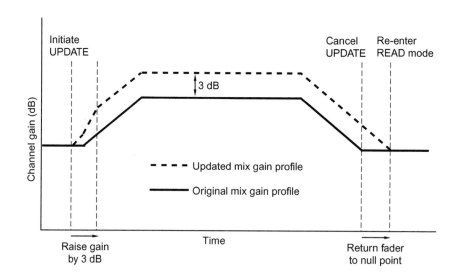

If the fader were to be permanently driven, there would be a problem when both the engineer and the automation system wanted to control the gain. Clutches or other forms of control are employed to remove the danger of a fight between fader and engineer in such a situation, and the fader is usually made touch-sensitive to detect the presence of a hand on it.

Such faders are, in effect, permanently in UPDATE mode, as they can at any time be touched and the channel gain modified, but there is usually some form of relative mode which can be used for offsetting a complete section by a certain amount. The problem with relative offsets and moving faders is that if there is a sudden change in the stored mix data while the engineer is holding the fader, it will not be executed. The engineer must let go for the system to take control again. This is where a combination of moving fader and VCA-type control comes into its own.

Grouping automated faders

Conventional control grouping (Fact File 5.5) is normally achieved by using dedicated master faders. In an automated console it may be possible to do things differently. The automation computer has access to data representing the positions of all the main faders on the console, so it may allow any fader to be designated a group master for a group of faders assigned to it. It can do this by allowing the user to set up a fader as a group master (either by pressing a button on the fader panel, or from a central control panel). It will then use the level from this fader to modify the data sent back to all the other faders in that group, taking into account their individual positions as well. This idea means that a master fader can reside physically within the group of faders to which it applies, although this may not always be the most desirable way of working.

Sometimes the computer will store automation data relating to groups in terms of the motions of the individual channels in the group, without storing the fact that a certain fader was the master, whereas other systems will store the data from the master fader, remembering the fact that it was a master originally.

Mute automation

Mutes are easier to automate than faders because they only have two states. Mute switches associated with each fader are also scanned by the automation computer, although only a single bit of data is required to represent the state of each switch. A simple electronic switch can be used to effect the mute, and in analog designs this often takes the form of a FET (field effect transistor) in the signal path, which has very high attenuation in its 'closed' position (see Figure 5.21). Alternatively, some more basic systems effect mutes by a sudden change in channel gain, pulling the fader down to maximum attenuation.

Storing the automation data

Early systems converted the data representing the fader positions and mute switches into a modulated serial data stream which could be recorded alongside the audio to which it related on a multitrack tape. In order to allow updates of the data, at least two tracks were required: one to play back the old data, and one to record the updated data, these usually being the two outside tracks of the tape (1 and 24 in the case of a 24-track machine). This was limiting, in that only the two most recent mixes were ever available for comparison (unless more tracks were set aside for automation), whole tracks had to be mixed at a time (because otherwise the updated track would be incomplete), and at least two audio tracks were lost on the tape. Yet it meant that the mix data was always available alongside the music, eliminating the possibility of losing a disk with the mix data stored separately.

More recent systems use computer hardware to store mix data, in RAM and on disks. Data is synchronized to the audio by recording time-code alongside audio which uniquely identifies any point in time, this being read by the automation system and used to relate recording position

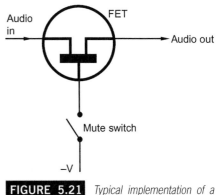

FIGURE 5.21 *Typical implementation of a FET mute switch.*

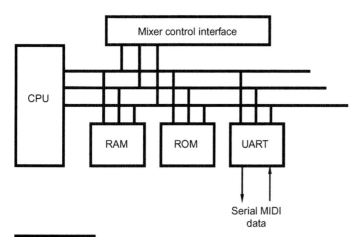

FIGURE 5.22 *A UART is used to route MIDI data to and from the automation computer.*

to stored data. This method gives almost limitless flexibility in the modification of a mix, allowing one to store many versions, of which sections can be joined together 'off-line' (that is, without the recorder running) or on-line, to form the finished product. The finished mix can be dumped to a disk for more permanent storage, and this disk could contain a number of versions of the mix.

It is becoming quite common for some automation systems to use MIDI or MIDI over Ethernet (ipMIDI) for the transmission of automation data. A basic automation computer associated with the mixer converts control positions into MIDI information and transmits/receives it using a device known as a UART which generates and decodes serial data at the appropriate rate for the MIDI standard, as shown in Figure 5.22. MIDI data can then be stored on a conventional sequencer or using dedicated software, as described in Chapter 14.

Integrating machine control

Control of recording machines is a common feature of modern mixers. It may only involve transport remotes being mounted in the center panel somewhere, or it may involve a totally integrated autolocator/synchronizer associated with the rest of the automation system. On top-flight desks, controls are provided on the channel modules for putting the relevant tape track into record-ready mode, coupled with the record function of the transport remotes. This requires careful interfacing between the console and the recording machine, but means that it is not necessary to work with a separate recording machine remote unit by the console.

It is very useful to be able to address the automation in terms of the mix in progress: in other words, 'go back to the second chorus', should mean something to the system, even if abbreviated. The alternative is to have to address the system in terms of timecode locations. Often, keys are provided which allow the engineer to return to various points in the mix, both from a mix data point-of-view and from the recording machines' point-of-view, so that the automation system locates the recorder to the position described in the command, ready to play.

Total automation systems

SSL originally coined the term 'Total Recall' for its system, which was a means of telling the operator where the controls should be and leaving him or her to reset them him/herself. This saved an enormous amount of time in the resetting of the console in between sessions, because it saved having to write down the positions of every knob and button.

True Total Reset is quite a different proposition and requires an interface between the automation system and every control on the console, with some means of measuring the position of the control, some means of resetting it, and some means of displaying what is going on. A number of options exist, for example one could:

- motorize all the rotary pots

- make all the pots continuously rotating and provide a display

- make the pots into up/down-type incrementers with display

- provide assignable controls with larger displays

Of these, the first is impractical in most cases due to the space that motorized pots would take up, the reliability problem and the cost, although it does solve the problem of display. The second would work, but again there is the problem that a continuously rotating pot would not have a pointer because it would merely be a means of incrementing the level from wherever it was at the time, so extra display would be required and this takes up space. Nonetheless some ingenious solutions have been developed, including incorporating the display in the head of rotary controls (see Figure 5.23). The third is not ergonomically very desirable, as the human prefers analog interfaces rather than digital ones, and there is no room on a conventional console for all the controls to be of this type with their associated displays. Most of the designs which have implemented total automation have adopted a version of the fourth option: that is, to use fewer controls than there are functions, and to provide larger displays.

The concept of total automation is inherent in the principles of an assignable mixing console, as described below in the section on digital mixers. In such a console, few of the controls carry audio directly as they are only interfaces to the control system, so one knob may control the HF EQ for any channel to which it is assigned, for example. Because of this indirect control, usually via a microprocessor, it is relatively easy to implement a means of storing the switch closures and settings in memory for reiteration at a later date.

(a) (b)

FIGURE 5.23 *Two possible options for positional display with continuously rotating knobs in automated systems. (a) Lights around the rim of the knob itself; (b) lights around the knob's base.*

Dynamic and static systems

Many analog assignable consoles use the modern equivalent of a VCA: the digitally controlled attenuator, also to control the levels of various functions such as EQ, aux sends, and so on. Full dynamic automation requires regular scanning of all controls so as to ensure smooth operation, and a considerable amount of data is generated this way. Static systems exist which do not aim to store the continuous changes of all the functions, but they will store 'snapshots' of the positions of controls which can be recalled either manually or with respect to timecode. This can often be performed quite regularly (many times a second) and in these cases we approach the dynamic situation, but in others the reset may take a second or two which precludes the use of it during mixing. Changes must be silent to be useful during mixing.

Other snapshot systems merely store the settings of switch positions, without storing the variable controls, and this uses much less processing time and memory. Automated routing is of particular use in theater work where sound effects may need to be routed to a complex combination of destinations. A static memory of the required information is employed so that a single command from the operator will reset all the routing ready for the next set of sound cues.

DIGITAL MIXERS

Much of what has been said in this chapter applies equally to both analog and digital mixers, at least in conceptual and operational terms. To complete the picture some features specific to digital mixers will now be described, although the principles of digital audio processing are explained in more detail in Chapter 8. Digital mixing has now reached the point where it can be implemented cost effectively, and there are a number of reasonably priced digital mixers with full automation. At the other end of the scale companies are manufacturing large-scale studio mixers with an emphasis on ultra-high sound quality and an ergonomically appropriate control interface. At the low cost end of the scale, digital mixers are implemented within computer-based workstations and represented graphically on the computer display. Faders and other controls are moved using a mouse.

Audio handling

In a digital mixer incoming analog signals are converted to the digital domain as early as possible so that all the functions are performed entirely in the digital domain, with as much as 64-bit internal processing resolution to cope with extremes of signal level, summing, EQ settings and other effects. The advantage of this is that once the signal is in the digital domain it is inherently more robust than its analog counterpart: it is virtually immune from crosstalk, and is unaffected by lead capacitance, electromagnetic fields from mains wiring, additional circuit distortion and noise, and other forms of interference. Digital inputs and outputs can be provided to connect recording devices and other digital equipment without conversion to analog. Inputs can be a mixture of analog and digital (the latter configurable via plug-in modules for Tascam, ADAT, Yamaha and AES/EBU formats, for example) with digital and analog main and monitoring outputs. Functions such as gain, EQ, delay, phase, routing, and effects such as echo, reverb, compression and limiting, can all be carried out in the digital domain precisely and repeatably using digital signal processing.

Inputs and outputs, digital and analog, are often provided by a series of outboard rack-mounting units which incorporate the D/A and A/D convertors, microphone amplifiers and phantom power, and these can be positioned where needed, the gain still being adjusted from the mixer. In a recording studio one rack could be by the control surface itself, one or more in the recording studio area, and one by the recording machines. In a theater, units would be placed next to power amplifiers, and in off-stage areas where musicians play. These units are connected in a daisy-chain loop to the main control surface via coaxial BNC cabling, MADI interface, or proprietary fiber optic links, the latter being preferable for longer distances.

Assignable control surfaces

Operationally a digital mixer can remain similar to its analog counterpart, although commercial examples have tended at least partially to use assignable control surface designs as they can be easier to implement in digital mixers. With assignable designs many of the controls of the traditional console such as pan, EQ, aux send and group assign are present only as single assignable sections or multi-function controls, many facilities can be packed into a unit of modest dimensions and quite modest cost.

A fully assignable digital mixing console is ergonomically quite different from its analog counterpart. Typically, the control surface consists of many input channel faders and/or rotary knobs, each channel having 'active' and 'select' buttons. Much smaller areas of the control surface are given over to single sections of EQ, routing (aux and group) and processing: these sections are automatically assigned to one particular channel when its 'select' button is active before adjustments can take place. Thus many processes which affect the signals are not continuously on view or at the fingertips of the operator as is the case with the traditional analog desk. The assignable design is therefore more suitable to recording work (particularly post-session mixdowns) where desk states can be built up gradually and saved to scene memories, rather than to live performance and primary recording work where continuous visual indication of and access to controls remains desirable.

Facilities such as channel delay, effects processing, moving fader automation and fader ganging, scene memories offering total recall of all settings, MIDI (including memory load and dump via separate MIDI data filers), and timecode interface are typically offered, and a display screen shows the status of all controls: either in simple global formats for the whole console for parameters such as routing, channel delay, scene memory details and the like, or in much greater detail for each individual channel. Metering can also be shown. Cursors facilitate both navigation around the screen displays and adjustments of the various parameters.

FIGURE 5.24   *Digico D5T. (Courtesy of the RSC.)*

FIGURE 5.25   *Detail of input channels display. (Digico D5T, courtesy of the RSC.)*

Digital mixers – a case study

The difficulty and expense of implementing true 'total recall' of an analog mixer, that is automated resetting of all surface controls, has already been discussed. Digital mixers can incorporate such a feature routinely, and the Digico D5T illustrated in Figure 5.24 is a typical example of a console in which all set-up parameters including such things as input gain and phantom power switching are recallable in seconds when a particular project or show is loaded into the mixer or recalled from its memory store. Such mixers are essentially versions of computer mixing systems but with a hardware control surface to provide a more traditional mode of hands-on operation, still an essential feature for live mixing work and many types of recording and broadcast session. Ergonomically, the mixer combines traditional 'analog' facilities of channel faders, aux send and EQ knobs, 'VCA' and group faders, with a considerable degree of assignability using large touch-sensitive screens and several selectable 'layers' across which banks of inputs can be displayed and accessed (Figure 5.25).

A master output screen can display a variety of things such as group outputs, automation parameters, scene memory information, matrix settings and the like. A console such as this can offer 96 input channels, 20 aux sends, 24 group sends, and in a theater version a 32-output matrix section. It is not difficult to appreciate the huge size and cost of an equivalent analog console. A QWERTY key pad facilitates the labeling of all sections.

Typically, adjacent to each bank of channel faders will be a row of buttons enabling access to different control layers. Layer 1 could be input channels 1 to 8, the accompanying screen display showing such things as input gain, phantom power, routing, aux send levels and EQ. Touching the appropriate area of the display expands that area for ease of viewing and adjustment, e.g. touching the EQ section of a channel displays the settings in much more detail and assigns the EQ controls adjacent to the screen to that channel. Layer 2 could display channels 25 to 30 (channels 9 to 24 being provided by adjacent banks of faders). Layer 3 of all the fader banks could give fader control of all the matrix outputs, or all the group outputs, or all the aux master outputs, or a combination. All of these things are chosen by the operator and set up to his or her requirements. The top layer would normally be assigned to inputs which need to be continuously on view, e.g. musicians' microphones and vocal mics, radio mics and DI inputs. The lower layers would be assigned to things such as CD players, sampler and other replay machine outputs, and probably to some of the effects returns. These inputs do not normally need to be accessed quickly. Other features such as digital delay and EQ on inputs and outputs (the latter particularly useful in live sound work), compressors and limiters, and internal effects processors, are routinely available. This reduces the number of outboard effects processors needed. The settings for these are all programmable and recordable along with the rest of the console settings.

Two main observations can be made regarding the operation of such consoles compared with their analog counterparts. First, a good deal of initial setting up, assigning and labeling needs to be carried out before a session can begin. Input and output channels need to be assigned to appropriate sockets on the outboard units around the building; the various layers have to be assigned to inputs/outputs/auxs/VCAs as appropriate, and labeled; and a series of scene memories has to be created in anticipation of what will be required for the show or recording session. Second, the operation of the console often requires a two-stage thinking process. Although channel faders and some other facilities for a particular layer will be instantly available for adjustment, many other

FIGURE 5.26   *The Midas Heritage 3000. (Courtesy of Klark Teknik.)*

facilities will need to be accessed either on a different layer or by touching an area of a screen before adjustments can be made. Additionally, adjustments need to be stored in a scene memory. Normally, storing changes such as input gain, EQ and aux send levels in a particular scene will automatically store those changes to the other scene memories. Channel fader adjustments will be stored only to that particular scene. Just what adjustments are stored to the present scene, and which ones are automatically stored to a bank of scenes, can be chosen by the operator. The complete project then needs to be stored to the mixer's hard disk drive, and preferably also to an external backup. This all needs an operator who is familiar with that particular console and its software quirks. Digital consoles necessarily have many common features, but manufacturers have their own proprietary ways of doing things. The typical analog console, in contrast, will be fairly familiar to a user after ten or 15 minutes.

Digitally controlled analog mixers

A digitally controlled analog console will be looked at briefly next. The Midas Heritage 3000 shown in Figure 5.26 is a good example of such a mixer. Its control surface is analog, and the signals remain in the analog domain throughout. Digital control gives such things as mute and mute group automation, VCA assign, and virtual fader automation (a row of LEDs adjacent to each fader displays the audio level of the fader regardless of its physical position; moving the fader to the top lit LED gives the operator manual control). Scene memories can thus be programmed into the desk giving the appropriate fader positions and channel mutes, these being of great value in the live mixing situations for which such consoles are designed. Other consoles also provide automation of EQ in/out, insert, aux send enable, group assign and moving fader automation, albeit at a somewhat higher cost, and such consoles undoubtedly prove their worth in the live sound market where visiting and freelance sound engineers need to become quickly familiar with a console which does not have too much automation. Such consoles are likely to be in use until well into the second decade of the present century.

MIXER AND WORKSTATION INTEGRATION

Integrated control of digital audio workstations is now a growing feature of either analog or digital mixing consoles. In the case of some designs the mixing console has become little more than a sophisticated control surface, enabling the functions of a workstation to be adjusted using more conventional controls. This is offered as an alternative to using a computer display and mouse, which can be inconvenient when trying to handle complex mixes. In such cases most of the audio processing is handled by the workstation hardware, either using desktop computer processing power or dedicated signal processing cards. Some audio handling can be included, such as monitoring and studio communication control. The mixer control surface is connected to the workstation using a dedicated interface such as MIDI or Ethernet. Such control surfaces often include remote control facilities for the workstation transport and editing functions. An example from the Digidesign ICON series is shown in Figure 5.27.

FIGURE 5.27   *A typical workstation-integrated mixer control surface from the Digidesign ICON series (D-Control ES). (Courtesy of Digidesign.)*

An alternative is to employ an external analog mixer that has dedicated workstation control facilities, such as the SSL Matrix, pictured in Figure 5.28. In this case the mixer handles more of the audio processing itself and can be used as an adjunct or alternative to the onboard digital processing of the workstation, with comprehensive routing to and from conventional analog outboard equipment. It enables analog mixing of audio channels outside the workstation, which has become a popular way of working for some. The control interface to the workstation is by means of MIDI over Ethernet (ipMIDI), and USB to carry the equivalent of computer keyboard commands. The control protocol can be made to conform either to Mackie's HUI (Human User Interface) or MCU (Mackie Control Universal), or to a special configuration of MIDI controllers (see Chapter 14), and the mixer's digitally controlled attenuators can be remote controlled using MIDI commands. The latter facility enables the mixer to be

FIGURE 5.28   *Analog mixer with integrated workstation control: SSL Matrix. (Courtesy of Solid-State Logic.)*

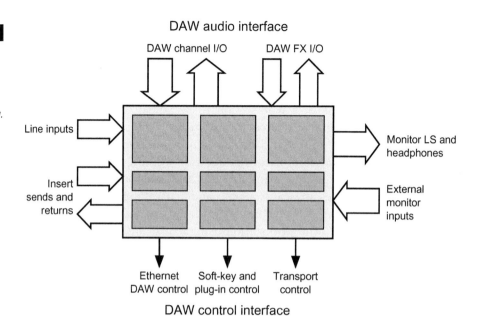

A typical system configuration of the SSL Matrix console showing interconnections to studio equipment and workstation.

automated using MIDI tracks on the workstation. A typical configuration of this system is shown in Figure 5.29.

Euphonix (now part of Avid) developed a remote control protocol known as EUCON to communicate between control surfaces, such as mixers, and computer applications. It uses TCP/IP network communications over Ethernet to carry information about knobs and buttons on the control surface, mapping hardware controls to software objects that control the computer application in question. A variety of 'primitive' control and display types are represented, such as knob, switch, slider, meter and LED. Further information is given in Chapter 14.

INTRODUCTION TO MIXING APPROACHES

This section provides an introduction to basic mixer operation and level setting.

Acoustic sources will be picked up by microphones and fed into the mic inputs of a mixer (which incorporates amplifiers to raise the low-voltage output from microphones), whilst other sources usually produce so-called 'line level' outputs, which can be connected to the mixer without extra amplification. In the mixer, sources are combined in proportions controlled by the engineer and recorded. In 'straight-to-stereo' (or surround) techniques, such as a classical music recording, microphone sources are often mixed 'live' without recording to a multitrack medium, creating a session master which is the collection of original recordings, often consisting of a number of takes of the musical material. The balance between the sources must be correct at this stage, and often only a small number of carefully positioned microphones are used. The session master recordings will then proceed to the editing stage where takes are assembled in an artistically satisfactory manner, under the control of the producer, to create a final master which will be transmitted or made into a commercial release. This final master could be made into a number of production masters which will be used to make different release formats. In the case of 'straight-to-stereo' mixing the console used may be a simpler affair than that used for multitrack recording, since the mixer's job is to take multiple inputs and combine them to a single stereo output, perhaps including processing such as equalization. This method of production is clearly cheaper and less time consuming than multitrack recording, but requires skill to achieve a usable balance quickly. It also limits flexibility in post-production. Occasionally, classical music is recorded in a multitrack form, especially in the case of complex operas or large-force orchestral music with a choir and soloists, where to get a correct balance at the time of the session could be costly and time consuming. In such a case, the production process becomes more similar to the pop recording situation described below.

'Pop' music is rarely recorded live, except at live events such as concerts, but is created in the recording studio. Acoustic and electrical sources are fed into a mixer and recorded on to a multitrack medium, often a few tracks at a time, gradually building up a montage of sounds. The resulting recording then contains a collection of individual sources on multiple tracks which must subsequently be mixed into the final release format. Individual songs or titles are recorded in separate places on the tape, to be compiled later. It is not so common these days to record multitrack pop titles in 'takes' for later editing, as with classical music, since mixer automation allows the engineer to work on a song in sections for automatic execution in sequence by a computer. In any case, multitrack machines have comprehensive 'drop-in' facilities for recording short inserted sections on individual tracks without introducing clicks, and a pop-music master is usually built up by laying down backing tracks for a complete song (drums, keyboards, rhythm guitars, etc.) after which lead lines are overdubbed using drop-in facilities. Occasionally multitrack recordings are edited or compiled ('comped') early on during a recording session to assemble an acceptable backing track from a number of takes, after which further layers are added. Considerable use may be made of computer-sequenced electronic instruments, under MIDI control, often in conjunction with multitrack disk recording. The computer controlling the electronic instruments is synchronized to the recording machine using time code and the outputs of the instruments are fed to the mixer to be combined with the non-sequenced sources.

Once the session is completed, the multitrack recording is mixed down. This is often done somewhere differently from the original session, and involves feeding the outputs of each track into individual inputs of the mixer, treating each track as if it were an original source. The balance between the tracks, and the positioning of the tracks in the stereo image, can then be carried out at leisure (within the budget constraints of the project!), often without all the musicians present, under control of the producer. During the mixdown, further post-production takes place such as the addition of effects from outboard equipment to enhance the mix. An automation system is often used to memorize fader and mute movements on the console, since the large number of channels involved in modern recording makes it difficult if not impossible for the engineer to mix a whole song correctly in one go. Following mixdown, the master that results will be edited very basically, in order to compile titles in the correct order for the production master. The compiled tape will then be mastered for the various distribution media.

BASIC OPERATIONAL TECHNIQUES

Level setting

If one is using a microphone to record speech or classical music then normally a fairly high input gain setting will be required. If the microphone is placed up against a guitar amplifier then the mic's output will be high and a much lower input gain setting can be used. There are essentially three ways of setting the gain control to the optimum position. First, using PFL or prefade listen (see Fact File 5.3).

PFL is pressed, or the fader overpressed (i.e. pressed beyond the bottom of its travel against a sprung microswitch), on the input module concerned and the level read on either a separate PFL meter or with the main meters switched to monitor the PFL bus. The channel input gain should be adjusted to give a meter reading of, say, PPM 5, or 0 VU on older analog desks, and a meter reading of perhaps 6–10 dB below maximum on a digital desk. This gain-setting procedure must be carried out at a realistic input level from the source. It is frequently the case during rehearsals that vocalists and guitarists will produce a level that is rather lower than that which they will use when they actually begin to play.

The pan control should be set next (see Fact File 5.2) to place the source in the stereo image. The main output faders will normally be set to 0 dB on their calibration, which is usually at the top. The channel faders can then be set to give both the desired subjective sound balance and appropriate output meter readings.

The second way of setting the gain is a good way in its own right, and it has to be used if PFL facilities are not provided. First of all both the channel fader and the output faders need to be positioned to the 0 dB point. This will be either at the top of the faders' travels or at a position about a quarter of the way down from the top of their travel. If no 0 dB position is indicated then the latter position should be set. After the pan control and faders have been positioned, the input gain may then be adjusted to give the desired reading on the output level meters. When several incoming signals need to be balanced the gain controls should all be positioned to give both the desired sound balance between them and the appropriate meter readings – normally PPM 6 or just over 0 VU during the loudest passages.

These two gain-setting methods differ in that with the former method the channel fader positions will show a correspondence to the subjective contribution each channel is making towards the overall mix, whereas the latter method places all the channel faders at roughly the same level.

The third way is similar to the second way, but one channel at a time is set up, placing channel and output faders at 0 dB and adjusting the gain for a peak meter reading. That channel fader is then turned completely down and the next channel is set up in a similar way. When all the channels which are to be used have been set up, the channel faders can then be advanced to give both the desired subjective balance and peak meter readings.

Use of the EQ controls often necessitates the resetting of the channel's gain. For example, if a particular instrument requires a bit of bass boost, applying this will also increase the level of signal and so the gain will often need to be reduced a little to compensate. Applying bass or treble cut will sometimes require a small gain increase.

Using auxiliary sends

Aux facilities were described in 'Auxiliary sends', p. 139. The auxiliaries are configured either 'pre-fade' or 'post-fade'. Pre-fade aux sends are useful for providing a monitor mix for musicians, since this balance will be unaffected by movements of the faders which control the main mix. The engineer then retains the freedom to experiment in the control room without disturbing the continuity of feed to the musicians.

Post-fade sends are affected by the channel fader position. These are used to send signals to effects devices and other destinations where it is desirable to have the aux level under the overall control of the channel fader. For example, the engineer may wish to add a little echo to a voice. Aux 2, set to post-fade, is used to send the signal to an echo device, probably positioning the aux 2 control around the number 6 position and the aux 2 master at maximum. The output of the echo device is returned to another input channel or an echo return channel, and this fader can be adjusted to set the amount of echo. The level of echo will then rise and fall with the fader setting for the voice.

The post-fade aux could also be used simply as an additional output to drive separate amplifiers and speakers in another part of a hall, for example.

Using audio groups

The group outputs (see 'Channel grouping', p. 128) or multitrack routing buses (see 'Routing section', above) can be used for overall control of various separate groups of instruments, depending on whether mixing down or track laying. For example, a drum kit may have eight microphones on it. These eight input channels can be routed to groups 1 and 2 with appropriate stereo pan settings. Groups 1 and 2 would then be routed to stereo outputs left and right respectively. Overall control of the drum kit level is now achieved simply by moving group faders 1 and 2.

RECOMMENDED FURTHER READING

See 'General further reading' at the end of this book.

Analog Recording

Successive editions of this book have quite naturally seen the emphasis placed more and more on digital topics and away from analog. Yet even the gramophone record has survived for rather more years than many would have predicted, and the analog open reel tape recorder, particularly in multitrack form in rock recording studios where some engineers and artists value its distortion characteristics, still enjoys a significant amount of use. An analog recording chapter continues to be justified in this edition therefore, albeit in a shortened form.

A SHORT HISTORY OF ANALOG RECORDING

Early recording machines

When Edison and Berliner first developed recording machines in the last years of the nineteenth century they involved little or no electrical apparatus. Certainly the recording and reproduction process itself was completely mechanical or 'acoustic', the system making use of a small horn terminated in a stretched, flexible diaphragm attached to a stylus which cut a groove of varying depth into the malleable tin foil on Edison's 'phonograph' cylinder or of varying lateral deviation in the wax on Berliner's 'gramophone' disc (see Figure 6.1). On replay, the undulations of the groove caused the stylus and diaphragm to vibrate, thus causing the air in the horn to move in sympathy, thus reproducing the sound – albeit with a very limited frequency range and very distorted.

Cylinders for the phonograph could be recorded by the user, but they were difficult to duplicate for mass production, whereas discs for the gramophone were normally replay only, but they could be duplicated readily for mass production. For this reason discs fairly quickly won the day as the mass-market prerecorded music medium. There was no such thing as magnetic recording tape at the time, so recordings were made directly onto a master disc, lasting for the duration of the side of the disc – a maximum of around 4 minutes – with no possibility for editing. Recordings containing errors were either remade or they were passed with mistakes intact. A long item of music would be recorded in short sections with gaps to change the disc, and possibilities arose for discontinuities between the sections as well as variations in pitch and tempo. Owing to the deficiencies of the acoustic recording process, instruments had to be grouped quite tightly around the pickup horn in order for them to be heard on the recording, and often louder instruments were substituted for quieter ones (the double bass was replaced by the tuba, for example) in order to correct for the poor frequency balance. It is perhaps partly because of this that much of the recorded music of the time consisted of vocal soloists and small ensembles, since these were easier to record than large orchestras.

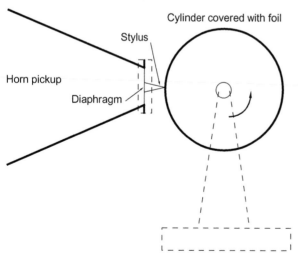

FIGURE 6.1 *The earliest phonograph used a rotating foil-covered cylinder and a stylus attached to a flexible diaphragm. The recordist spoke or sang into the horn causing the stylus to vibrate, thus inscribing a modulated groove into the surface of the soft foil. On replay the modulated groove would cause the stylus and diaphragm to vibrate, resulting in a sound wave being emitted from the horn.*

Electrical recording

During the 1920s, when broadcasting was in its infancy, electrical recording became more widely used, based on the principles of electromagnetic transduction (see Chapter 3). The possibility for a microphone to be connected remotely to a recording machine meant that microphones could be positioned in more suitable places, connected by wires to a complementary transducer at the other end of the wire, which drove the stylus to cut the disc. Even more usefully, the outputs of microphones could be mixed together before being fed to the disc cutter, allowing greater flexibility in the balance. Basic variable resistors could be inserted into the signal chain in order to control the levels from each microphone, and valve amplifiers would be used to increase the electrical level so that it would be suitable to drive the cutting stylus.

The sound quality of electrical recordings shows a marked improvement over acoustic recordings, with a wider frequency range and a greater dynamic range. Experimental work took place both in Europe and the USA on stereo recording and reproduction, but it was not to be until much later that stereo took its place as a common consumer format, nearly all records and broadcasts being in mono at that time.

Later developments

During the 1930s work progressed on the development of magnetic recording equipment, and examples of experimental wire recorders and tape recorders began to appear, based on the principle of using a current flowing through a coil to create a magnetic field which would in turn magnetize a moving metal wire or tape coated with magnetic material. The 1940s, during wartime, saw the introduction of the first AC-biased tape recorders, which brought with them good sound quality and the possibility for editing. Tape itself, though, was first made of paper coated with metal oxide which tended to deteriorate rather quickly, and only later of plastics which proved longer lasting and easier to handle. In the 1950s the microgroove LP record appeared, with markedly lower surface noise and improved frequency response, having a playing time of around 25 minutes per side. This was an ideal medium for distribution of commercial stereo recordings, which began to appear in the late 1950s, although it was not until the 1960s that stereo really took hold. In the early 1960s the first multitrack tape recorders appeared, the Beatles making use of an early four-track recorder for their 'Sergeant Pepper's Lonely Hearts Club Band' album. The machine offered the unprecedented flexibility of allowing sources to be recorded separately, and the results in the stereo mix are panned very crudely to left and right in somewhat 'gimmicky' stereo. Mixing equipment in the 1950s and 1960s was often quite basic, compared with today's sophisticated consoles, and rotary faders were the norm. There simply was not the quantity of tracks involved as exists today.

MAGNETIC TAPE

Structure

Magnetic tape consists of a length of plastic material which is given a surface coating capable of retaining magnetic flux rather in the manner that, say, an iron rod is capable of being magnetized. The earliest recorders actually used a length of iron wire as the recording medium. In practice all modern tape has a polyester base which was chosen, after various trials with other formulations which proved either too brittle (they snapped easily) or too plastic (they stretched), for its good strength and dimensional stability. It is used throughout the tape industry from the dictation microcassette to the 2 inch (5 cm) multitrack variety. The coating is of a metal oxide, or metal alloy particles.

Open-reel tape

Open-reel quarter-inch tape intended for analog recorders has been available in a variety of thicknesses. Standard Play tape has an overall thickness of 50 microns (micrometers), and a playing time (at 15 inches [38 cm] per second) of 33 minutes is obtained from a 10 inch (25 cm) reel. Long Play tape has an overall thickness of 35 microns giving a corresponding 48 minutes of playing time, which is very useful for live recording work. In the past 'Double Play' and even 'Triple Play' thicknesses have been available, these being aimed at the domestic open-reel market. These formulations are prone to snapping or stretching, as well as offering slightly poorer sound quality, and should not really be considered for professional use.

Standard Play tape is almost always 'back coated'. A rough coating is applied to the back of the tape during manufacture which produces neater and more even winding on a tape machine, by providing a certain amount of friction between layers which holds the tape in place. Also, the rough surface helps prevent air being trapped between layers during fast spooling which can contribute to uneven winding. Long Play tape is also available with a back coating, but as often as not it will be absent. It is worth noting that the flanges of a tape spool should only serve to protect the tape from damage. The 'pancake' of tape on the spool should not touch these flanges.

Metal spools are better than plastic spools because they are more rigid and they do not warp. Professional open-reel tape can be purchased either on spools or in 'pancake' form on hubs without flanges. The latter is of course cheaper, but considerable care is needed in its handling so that spillage of the unprotected tape does not occur. Such pancakes are either spooled onto empty reels before use, or they can be placed on top of a special reel with only a lower flange. Professional tape machines are invariably operated with their decks horizontal. Half inch, 1 inch and 2 inch tape intended for multitrack recorders always comes on spools, is always of Standard Play thickness, and is always back coated.

THE MAGNETIC RECORDING PROCESS

Introduction

Since tape is magnetic, the recording process must convert an electrical audio signal into a magnetic form. On replay the recorded magnetic signal must be converted back into electrical form. The process is outlined in Fact File 6.1. Normally a professional tape recorder has three heads, as shown in Figure 6.2, in the order erase–record–replay. This allows for the tape to be first erased, then re-recorded, and then monitored by the third head. The structure of the three heads is similar, but the gap of the replay head is normally smaller than that of the record head. It is possible to use the same head for both purposes, but usually with a compromise in performance. Such a two-head arrangement is often found in cheaper cassette machines which do not allow off-tape monitoring whilst recording. A simplified block diagram of a typical tape recorder is shown in Figure 6.3.

The magnetization characteristics of tape are by no means linear, and therefore a high-frequency signal known as bias is added to the audio signal at the record head, generally a sine wave of between 100 and 200 kHz, which biases the tape towards a more linear part of its operating range. Without bias the tape retains very little magnetization and distortion is

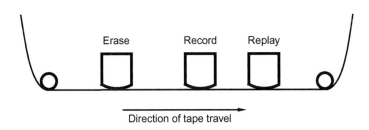

Erase   Record   Replay

Direction of tape travel

FIGURE 6.2

Order of heads on a professional analog tape recorder.

FACT FILE 6.1 A MAGNETIC RECORDING HEAD

When an electrical current flows through a coil of wire a magnetic field is created. If the current only flows in one direction (DC) the electromagnet thus formed will have a north pole at one end and a south pole at the other (see diagram). The audio signal to be recorded onto tape is alternating current (AC), and when this is passed through a similar coil the result is an alternating magnetic field whose direction changes according to the amplitude and phase of the audio signal.

Magnetic flux is rather like the magnetic equivalent of electrical current, in that it flows from one pole of the magnet to the other in invisible 'lines of flux'. For sound recording it is desirable that the tape is magnetized with a pattern of flux representing the sound signal. A recording head is used which is basically an electromagnet with a small gap in it. The tape passes across the gap, as shown in the diagram. The electrical audio signal is applied across the coil and an alternating magnetic field is created across the gap. Since the gap is filled with a non-magnetic material it appears as a very high 'resistance' to magnetic flux, but the tape represents a very low resistance in comparison and thus the flux flows across the gap via the tape, leaving it magnetized.

On replay, the magnetized tape moves across the head gap of a similar or identical head to that used during recording, but this time the magnetic flux on the tape flows through the head and thus induces a current in the coil, providing an electrical output.

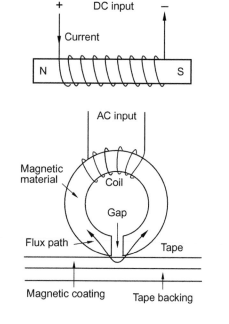

FIGURE 6.3

Simplified block diagram of a typical analog tape recorder. The bias trap is a filter which prevents the HF bias signal feeding back into an earlier stage.

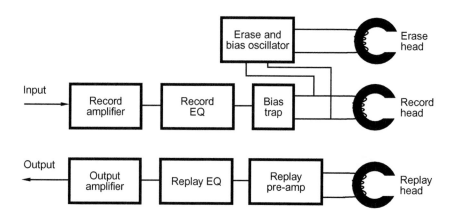

excessive. The bias signal is of too high a frequency to be retained by the tape, so does not appear on the output during replay. Different types of tape require different levels of bias for optimum recording conditions to be achieved, and this will be discussed in bias requirements, below.

Equalization

'Pre-equalization' is applied to the audio signal before recording. This equalization is set in such a way that the replayed short-circuit flux in an ideal head follows a standard frequency response curve (see Figure 6.4). A number of standards exist for different tape speeds, whose time constants are the same as those quoted for replay EQ in Table 6.1. Although the replayed flux level must conform to these curves, the electrical pre-EQ may be very different, since this depends on the individual head and tape characteristics. Replay equalization (see Figure 6.5) is used to ensure that a flat response is available at the tape machine's output. It compensates for losses incurred in the magnetic recording/replay process, the rising output of the replay head with frequency, the recorded flux characteristic, and the fall-off in HF response where the recorded wavelength approaches the head gap width (see Fact File 6.2) Table 6.1 shows the time constants corresponding to the turnover frequencies of replay equalizers at a number of tape speeds. Again a number of standards exist. Time constant (normally

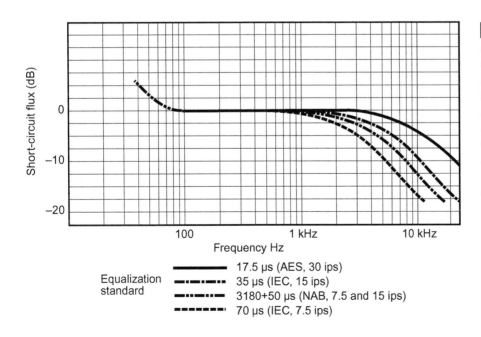

FIGURE 6.4

Examples of standardized recording characteristics for short-circuit flux. (NB: this is not equivalent to the electrical equalization required in the record chain, but represents the resulting flux level replayed from tape, measured using an ideal head).

Table 6.1	Replay equalization time constants		
Tape speed		**Time constants (µs)**	
ips (cm/s)	**Standard**	**HF**	**LF**
30 (76)	AES/IEC	17.5	–
15 (38)	IEC/CCIR	35	–
15 (38)	NAB	50	3180
7.5 (19)	IEC/CCIR	70	–
7.5 (19)	NAB	50	3180
3.75 (9.5)	All	90	3180
1.875 (4.75)	DIN (Type I)	120	3180
1.875 (4.75)	DIN (Type II or IV)	70	3180

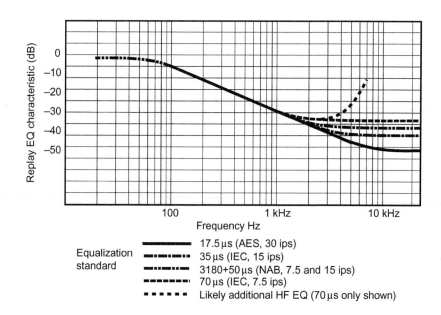

FIGURE 6.5

Examples of replay equalization required to correct for the recording characteristic (see Figure 6.4), replay-head losses, and the rising output of the replay head with frequency.

quoted in microseconds) is the product of resistance and capacitance (RC) in the equivalent equalizing filter, and the turnover frequency corresponding to a particular time constant can be calculated using:

$$f = 1/(2\pi RC)$$

FACT FILE 6.2 REPLAY HEAD EFFECTS

The output level of the replay head coil is proportional to the rate of change of flux, and thus the output level increases by 6 dB per octave as frequency rises (assuming a constant flux recording). Replay equalization is used to correct for this slope.

At high frequencies the recorded wavelength on tape is very short (in other words the distance between magnetic flux reversals is very short). The higher the tape speed, the longer the recorded wavelength. At a certain high frequency the recorded wavelength will equal the replay-head gap width (see diagram) and the net flux in the head will be zero, thus no current will be induced. The result of this is that there is an upper cut-off frequency on replay (the extinction frequency), which is engineered to be as high as possible.

Gap effects are noticeable below the cut-off frequency, resulting in a gradual roll-off in the frequency response as the wavelength approaches the gap length. Clearly, at low tape speeds (in which case the recorded wavelength is short) the cut-off frequency will be lower than at high tape speeds for a given gap width.

At low frequencies, the recorded wavelength approaches the dimensions of the length of tape in contact with the head,

and various additive and cancelation effects occur when not all of the flux from the tape passes through the head, or when flux takes a 'short-circuit' path through the head. This results in low-frequency 'head bumps' or 'woodles' in the frequency response. The diagram below summarizes these effects on the output of the replay head.

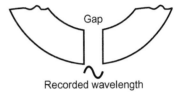

Gap

Recorded wavelength

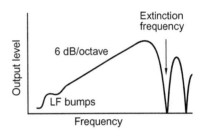

The LF time constant of $3180\,\mu s$ was introduced in the American NAB standard to reduce hum in early tape recorders, and has remained. HF time constants resulting in low turnover frequencies tend to result in greater replay noise, since HF is boosted over a wider band on replay, thus amplifying tape noise considerably. This is mainly why Type I cassette tapes $(120\,\mu s$ EQ) sound noisier than Type II tapes $(70\,\mu s$ EQ). Most professional tape recorders have switchable EQ to allow the replay of NAB- and IEC/CCIR-recorded tapes. EQ switches automatically with tape speed in most machines.

Additional adjustable HF and LF EQ is provided on many tape machines, so that the recorder's frequency response may be optimized for a variety of operational conditions, bias levels and tape types.

THE TAPE RECORDER

Studio recorder

Professional open-reel recorders fall into two categories: console mounted and portable. The stereo console recorder, intended for permanent or semi-permanent installation in a recording studio, outside broadcast truck or whatever generally sports rather few facilities, but has balanced inputs and outputs at line level (no microphone inputs), transport controls, editing modes, possibly a headphone socket, a tape counter (often in real time rather than in arbitrary numbers or revs), tape speed selector, reel size selector, and probably (though not always) a pair of level meters. It is deliberately simple because its job is to accept a signal, store it as faithfully as possible, and then reproduce it on call. It is also robustly built, stays aligned for long periods without the need for frequent adjustment, and will be expected to perform reliably for long periods. A typical example is pictured in Figure 6.6.

The inputs of such a machine will be capable of accepting high electrical levels – up to at least +20 dBu or around 8 volts – so that there is virtually no possibility of electrical input overload. The input impedance will be at least 10 kΩ. The outputs will be capable of driving impedances down to 600 ohms, and will have a source impedance of below 100 ohms. A facility will be provided for connecting a remote control unit so that the transport can be controlled from the mixing console, for instance.

Its semi-professional counterpart will be capable at its best of a performance that is a little inferior, and in addition to being smaller and lighter will sport rather more facilities such as microphone inputs and various alternative input and output options. Headphone outlets will be provided along with record-level meters, source/tape monitor switching, variable output level, and perhaps 'sound on sound'-type facilities for simple overdub work. The semi-professional machine will not usually be as robustly constructed, this being of particular concern for machines which are to be transported since rough treatment can easily send a chassis askew, causing misalignment of the tape transport system which will be virtually impossible to correct. Some chassis are constructed of pressed steel which is not very rigid. A casting is much better.

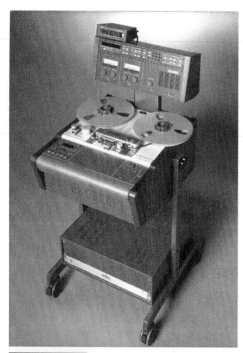

FIGURE 6.6 *A typical professional open-reel two-track analog tape recorder: the Studer A807-TC. (Courtesy of FWO Bauch Ltd.)*

The professional portable tape machine, unlike its console equivalent, needs to offer a wide range of facilities since it will be required to provide such things as balanced outputs and inputs, both at line and microphone level, phantom and A–B mic powering, metering, battery operation which allows usefully long recording times, the facility to record timecode and pilot tone for use in TV and film work, illumination of the important controls and meters, and possibly even basic mixing facilities. It must be robust to stand up to professional field use, and small enough to be carried easily. Nevertheless, it should also be capable of accepting professional

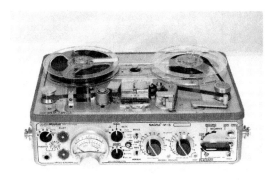

FIGURE 6.7 *A typical professional portable two-track recorder: the Nagra IV-S. (Courtesy of Nagra Kudelski (GB) Ltd.)*

10 inch (25 cm) reels, and adaptors are usually available to facilitate this. A lot has to be provided in a small package, and the miniaturization necessary does not come cheap. The audio performance of such machines is at least as good as that of a studio recorder. A typical commercial example is pictured in Figure 6.7.

The multitrack machine

Multitrack machines come in a variety of track configurations and quality levels. The professional multitrack machine tends to be quite massively engineered and is designed to give consistent, reliable performance on par with the stereo mastering machine. The transport needs to be particularly fine so that consistent performance across the tracks is achieved. A full reel of 2 inch tape is quite heavy, and powerful spooling motors and brakes are required to keep it under control. Apart from the increased number of tracks, multitrack machines are basically the same as their stereo counterparts and manufacturers tend to offer a range of track configurations within a given model type. Alignment of course takes a lot longer, and computer control of this is most welcome when one considers that 24 tracks implies 168 separate adjustments!

A useful feature to have on a multitrack recorder is an automatic repeat function or autolocate. The real-time counter can be programmed so that the machine will repeat a section of the tape over and over again within the specified start and end points to facilitate mixdown rehearsals. Multitrack recorders will be equipped with a number of unique features which are vital during recording sessions. For example, sync replay (see Fact File 6.3), gapless, noiseless punch-in (allowing any track to be dropped into record at

FACT FILE 6.3 SYNC REPLAY

The overdubbing process used widely in multitrack recording requires musicians to listen to existing tracks on the tape whilst recording others. If replay was to come from the replay head and the new recording was to be made onto the record head, a recorded delay would arise between old and new material due to the distance between the heads. Sync replay allows the record head to be used as a replay head on the tracks which are not currently recording, thus maintaining synchronization. The sound quality coming off the record head (called the sync head in this mode) is not always as good as that coming off the replay head, because the gap is larger, but it is adequate for a cue feed. Often separate EQ is provided for sync replay to optimize this. Mixdown should always be performed from the replay head.

Some manufacturers have optimized their head technology such that record and replay heads are exactly the same, and thus there is no difference between true replay and sync replay.

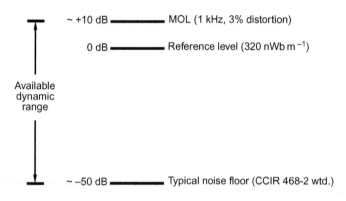

FIGURE 6.8 *The available dynamic range on an analogue tape lies between the noise floor and the MOL. Precise figures depend on tape and tape machine.*

any point without introducing a gap or a click) and spot erasure (allowing a track to be erased manually over a very small portion of tape).

MAGNETIC RECORDING LEVELS

It has already been said that the equivalent of electrical current in magnetic terms is magnetic flux, and it is necessary to understand the relationship between electrical levels and magnetic recording levels on tape. The performance of an analog tape recorder depends very much on the magnetic level recorded on the tape, since at high levels one encounters distortion and saturation, whilst at low levels there is noise (see Figure 6.8). A window exists, between the noise and the distortion, in which the audio signal must be recorded, and the recording level must be controlled to lie optimally within this region. For this reason the relationship between the electrical input level to the tape machine and the flux level on tape must be established so that the engineer knows what meter indication on a mixer corresponds to what magnetic flux level. Once a relationship has been set up it is possible largely to forget about magnetic flux levels and concentrate on the meters. Fact File 6.4 discusses magnetic flux reference levels.

FACT FILE 6.4 MAGNETIC REFERENCE LEVELS

Magnetic flux density on tape is measured in nanowebers per meter ($nWb\,m^{-1}$), the weber being the unit of magnetic flux. Modern tapes have a number of important specifications, probably the most significant being maximum output level (MOL), HF saturation point and noise level. (These parameters are also discussed in Chapter 18.) The MOL is the flux level at which third-harmonic distortion reaches 3% of the fundamental's level, measured at 1 kHz (or 5% and 315 Hz for cassettes), and can be considered as a sensible peak recording level unless excessive distortion is required for some reason. The MOL for a modern high-quality tape lies at a magnetic level of around $1000\,nWb\,m^{-1}$, or even slightly higher in some cases, and thus it is wise to align a tape machine such that this magnetic level corresponds fairly closely to the peak level indication on a mixer's meters.

A common reference level in electrical terms is 0 dBu, which often lines up with PPM 4 or −4 VU on a mixer's meter. This must be aligned to correspond to a recognized magnetic reference level on the tape, such as $320\,nWb\,m^{-1}$. Peak recording level, in this case, would normally be around 8 dBu if the maximum allowed PPM indication was to be 6, as is conventional. This would in turn correspond to a magnetic recording level of $804\,nWb\,m^{-1}$, which is close to the MOL of the tape and would probably result in around 2% distortion.

There are a number of accepted magnetic reference levels in use worldwide, the principal ones being 200, 250 and $320\,nWb\,m^{-1}$. There is 4 dB between 200 and $320\,nWb\,m^{-1}$, and thus a $320\,nWb\,m^{-1}$ test tape should replay 4 dB higher in level on a meter than a $200\,nWb\,m^{-1}$ test tape. American test tapes often use $200\,nWb\,m^{-1}$ (so-called NAB level), whilst German tapes often use $250\,nWb\,m^{-1}$ (sometimes called DIN level). Other European tapes tend to use $320\,nWb\,m^{-1}$ (sometimes called IEC level). Test tapes are discussed further in the main text.

There is currently a likelihood in recording studios that analog tapes are being under-recorded, since the performance characteristics of modern tapes are now good enough to allow higher peak recording levels than before. A studio which aligned PPM 4 to equal 0 dBu, in turn to correspond to only $200\,nWb\,m^{-1}$ on tape, would possibly be leaving 4–6 dB of headroom unused on the tape, sacrificing valuable signal-to-noise ratio.

WHAT ARE TEST TAPES FOR?

A test tape is a reference standard recording containing pre-recorded tones at a guaranteed magnetic flux level. A test tape is the only starting point for aligning a tape machine, since otherwise there is no way of knowing what magnetic level will end up on the tape during recording. During alignment, the test tape is replayed, and a 1 kHz tone at the specified magnetic flux level (say $320\,nWb\,m^{-1}$) produces a certain electrical level at the machine's output. The output level would then be adjusted for the desired electrical level, according to the studio's standard (say 0 dBu), to read at a standard meter indication (say PPM 4). It is then absolutely clear that if the output level of the tape machine is 0 dBu then the magnetic level on tape is $320\,nWb\,m^{-1}$. After this relationship has been set up it is then possible to record a signal on tape at a known magnetic level – for example, a 1 kHz tone at 0 dBu could be fed to the input of the tape machine, and the input level adjusted until the output read 0 dBu also. The 1 kHz tone would then be recording at a flux level $320\,nWb\,m^{-1}$.

Test tapes also contain tones at other frequencies for such purposes as azimuth alignment of heads and for frequency response calibration of replay EQ (see below). A test tape with the required magnetic reference level should be used, and it should also conform to the correct EQ standard (NAB or CCIR, see 'Equalization', p. 179). Tapes are available at all speeds, standards and widths, with most being recorded across the full width of the tape.

TAPE MACHINE ALIGNMENT

Head inspection and demagnetization

Heads and tape guides must be periodically inspected for wear. Flats on guides and head surfaces should be looked for; sometimes it is possible to rotate a guide so that a fresh portion contacts the tape. Badly worn guides and heads cause sharp angles to contact the tape which can damage the oxide layer. Heads have been made of several materials. Mu-metal heads have good electromagnetic properties, but are not particularly hard wearing. Ferrite heads wear extremely slowly and their gaps can be machined to tight tolerances. The gap edges can, however, be rather brittle and require careful handling. Permalloy heads last a long time and give a good overall performance, and are often chosen. Head wear is revealed by the presence of a flat area on the surface which contacts the tape. Slight wear does not necessarily indicate that head replacement is required, and if performance is found to be satisfactory during alignment with a test tape then no action need be taken.

Replay-head wear is often signified by exceptionally good high-frequency response, requiring replay EQ to be reduced to the lower limit of its range. This seems odd but is because the replay gap on many designs gets slightly narrower as the head wears down, and is at its narrowest just before it collapses!

Heads should be cleaned regularly using isopropyl alcohol and a cotton bud. They should also be demagnetized fairly regularly, since heads can gradually become slightly permanently magnetized, especially on older machines, resulting in increased noise and a type of 'bubbling' modulation noise in the background on recordings. A demagnetizer is a strong AC electromagnet which should be switched on well away from the tape machine, keeping it clear of anything else magnetic or metal. This device will erase a tape if placed near one! Once turned on the demagger should be drawn smoothly and slowly along the tape path (without a tape present), across the guides and heads, and drawn away gently on the far side. Only then should it be turned off.

Replay alignment

Replay alignment should be carried out before record alignment, as explained above. The method for setting replay and record levels has

FACT FILE 6.5 BIAS ADJUSTMENT

Bias level affects the performance of the recording pro-cess and the correct level of bias is a compromise between output level, distortion, noise level and other fac-tors. The graph below shows a typical tape's performance with increasing bias, and it can be seen that output level increases up to a point, after which it falls off. Distortion and noise go down as bias increases, but unfortunately the point of minimum noise and distortion is not quite the same as the point of maximum output level. Typically the optimum compromise between all the factors, offering the best dynamic range, is where the bias level is set just slightly higher than the point giving peak output. In order to set bias, a 10kHz tone is recorded at, say, 10dB below reference level, whilst bias is gradually increased from the minimum. The output level from the tape machine gradu-ally rises to a peak and then begins to drop off as bias continues to increase. Optimum bias is set for a number of decibels of fall-off in level after this peak – the so-called 'overbias' amount.

The optimum bias point depends on tape speed and formulation, but is typically around 3dB of overbias

at a speed of 15ips (38cms^{-1}). At 7.5ips the overbias increases to 6dB and at 30ips it is only around 1.5dB. If bias is adjusted at 1kHz there is much less change of output level with variation in bias, and thus only between 0.5 and 0.75dB of overbias is required at 15ips. This is difficult to read on most meters.

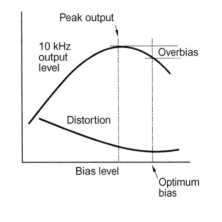

already been covered in the previous section. HF tones for azimuth adjust-ment normally follow (see Fact File 6.5). The test tape will contain a sequence of tones for replay frequency response alignment, often at 10 or 20dB below reference level so that tape saturation is avoided at frequency extremes, starting with a 1kHz reference followed by, say, 31.5Hz, 63Hz, 125Hz, 250Hz, 500Hz, 2kHz, 4kHz, 8kHz and 16kHz. Spoken identifi-cation of each section is provided. As the tape runs, the replay equalization is adjusted so as to achieve the flattest frequency response. Often both LF and HF replay adjustment is provided, sometimes just HF, but normally one should only adjust HF response on replay, since LF can suffer from the head bumps described in Fact File 6.2 and a peak or dip of response may coincide with a frequency on the test tape, leading to potential misalign-ment. Also full-track test tapes can cause 'fringing' at LF, whereby flux from the guard band leaks on to adjacent tracks. (Although it seems strange, replay LF EQ is normally adjusted during recording, to obtain the flattest record–replay response.)

FACT FILE 6.6 AZIMUTH ALIGNMENT

Azimuth

Azimuth describes the orientation of the head gap with respect to the tape. The gap should be exactly perpendicular to the edge of the tape otherwise two consequences follow. First, high frequencies are not efficiently recorded or replayed because the head gap becomes effectively wider as far as the tape is concerned, as shown in the diagram (B is wider than A). Second, the relative phase between tracks is changed.

The high-frequency tone on a test tape (8, 10, or 16 kHz) can be used with the outputs of both channels combined, adjusting replay azimuth so as to give maximum output level which indicates that both channels are in phase. Alternatively, the two channels can be displayed separately on a double-beam oscilloscope, one wave being positioned above the other on the screen, where it can easily be seen if phase errors are present. Azimuth is adjusted until the two sine waves are in step. It is advisable to begin with a lower-frequency tone than 8 kHz if a large azimuth error is suspected, since there is a danger of ending up with tracks a multiple of 360° out of phase otherwise.

In multitrack machines a process of trial and error is required to find a pair of tracks which most closely represents the best phase alignment between all the tracks. Head manufacturing tolerances result in gaps which are not perfectly aligned on all tracks. Cheap multitrack machines display rather wider phase errors between various tracks than do expensive ones.

Azimuth of the replay head is normally adjusted regularly, especially when replaying tapes made on other machines which may have been recorded with a different azimuth. Record-head azimuth is not modified unless there is reason to believe that it may have changed.

Height

Absolute height of the head should be such that the center of the face of the head corresponds with the center of the tape. Height can be adjusted using a test tape that is not recorded across the full width of the tape but with two discrete tracks. The correct height gives both equal output level from both channels and minimum crosstalk between them. It is also possible to buy tapes which are only recorded in the guard band, allowing the user to adjust height for minimum breakthrough onto the audio tracks. It can also sometimes be adjusted visually.

Zenith

Zenith is the vertical orientation of the head with respect to the surface of the tape. The head should neither lean forwards towards the tape, nor lean backwards, otherwise uneven wrap of the tape across the surface of the head results causing inconsistent tape-to-head contact and uneven head wear. Zenith is not normally adjusted unless the head has been changed or there is reason to believe that the zenith has changed.

Wrap

Wrap is the centrality of the head gap in the area of tape in contact with the head. The gap should be exactly in the center of that portion, so that the degree of approach and recede contact of the tape with respect to the gap is exactly equal. Uneven frequency response can be caused if this is not the case. Wrap can be adjusted by painting the head surface with a removable dye and running the tape across it. The tape will remove the dye over the contact area, and adjustments can be made accordingly.

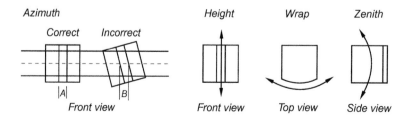

Record alignment

The frequency response of the machine during recording is considerably affected by bias adjustment, and therefore bias is aligned first before record equalization. The effects and alignment of bias are described in Fact File 6.5. It is wise to set a roughly correct input level before adjusting bias, by sending a 1 kHz tone at reference level to the tape machine and adjusting the input gain until it replays at the same level.

After bias levels have been set, record azimuth can be adjusted if necessary (see Fact File 6.5) by recording an HF tone and monitoring the now correctly aligned replay output. It may also be necessary to go back and check the 1 kHz record level if large changes have been made to bias.

Record equalization can now be aligned. Normally only HF EQ is available on record. A 1 kHz tone is recorded at between 10 and 20 dB below reference level and the meter gain adjusted so that this can be seen easily on replay. Spot frequencies are then recorded to check the machine's frequency response, normally only at the extremes of the range. A 5 kHz tone, followed by tones at 10 kHz and 15 kHz can be recorded and monitored off tape. The HF EQ is adjusted for the flattest possible response. The LF replay EQ (see above) can similarly be adjusted, sweeping the oscillator over a range of frequencies from, say, 40 Hz to 150 Hz, and adjusting for the best compromise between the upper and lower limits of the 'head bumps'.

Some machines have a built-in computer which will automatically align it to any tape. The tape is loaded and the command given, and the machine itself runs the tape adjusting bias, level and EQ as it goes. This takes literally seconds. Several settings can be stored in its memory so that a change of tape type can be accompanied simply by telling the machine which type is to be used, and it will automatically set its bias and EQ to the previously stored values. This is of particular value when aligning multi-track machines!

Once the tape machine has been correctly aligned for record and replay, a series of tones should be recorded at the beginning of every tape made on the machine. This allows the replay response of any machine which might subsequently be used for replaying the tape to be adjusted so as to replay the tape with a flat frequency response. The minimum requirement should be a tone at 1 kHz at reference level, followed by tones at HF and LF (say 10 kHz and 63 Hz) at either reference level (if the tape can cope) or at −10 dB. The levels and frequencies of these tones must be marked on the tape box (e.g. 'Tones @ 1 kHz, 320 nWb m^{-1} (=0 dB); 10 kHz and 63 Hz @ −10 dB). Designations on the tape box such as '1 kHz @ 0 VU' mean almost nothing, since 0 VU is not a magnetic level. What the engineer means in this case is that he/she sent a tone from his/her desk to the tape machine, measuring 0 VU on the meters, but this gives no indication of the magnetic level that resulted on the tape. Noted on the box should also be an indication of where peak recording level lies in relation to the 1 kHz reference level (e.g. 'peak recording level @ 8 dB above 320 nWb m^{-1}), in order that the replay chain can be set up to accommodate the likely signal peaks. In broadcasting, for example, it is most important to know where the peak signal level will be, since this must be set to peak at PPM 6 on a program meter, corresponding to maximum transmitter modulation.

When this tape comes to be replayed, the engineer will adjust the replay level and EQ controls of the relevant machine, along with replay azimuth, to ensure that the recorded magnetic reference level replays at his or her studio's electrical reference level, and to ensure a flat response. This is the only way of ensuring that a tape made on one machine replays correctly on another day or on another machine.

MECHANICAL TRANSPORT FUNCTIONS

Properly, mechanical alignment of the tape transport should be looked at before electrical alignment, because the electromagnetic performance is affected by it, but the converse is not the case. Mechanical alignment should be required far less frequently than electrical adjustments, and sometimes it also requires rather specialized tools. Because most mechanical alignments are fairly specialized, and because they differ with each tape machine, detailed techniques will not be covered further here. The manual for a machine normally details the necessary procedures. Looking at the diagram in Figure 6.9, it can be seen that the tape unwinds from the reel on the left, passes through various guides on its way to the head block, and then through various further guides and onto the take-up reel on the right. Some tape guides may be loaded with floppy springs which give on the instant of start-up, then slowly swing back in order to control the tension of the tape as the machine starts. The capstan is the shaft of a motor which pokes up through

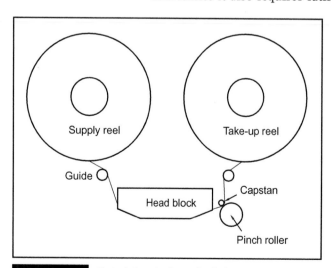

FIGURE 6.9 *Typical layout of mechanical components on the deckplate of an analog open-reel recorder.*

the deck of the machine by a couple of centimeters or so (more of course for multitrack machines with their increased tape widths) and lies fairly close to the tape when the tape is at rest, on the right-hand side of the head block. A large rubber wheel will be located close to the capstan but on the opposite side of the tape. This is called the pinch roller or pinch wheel. The capstan motor rotates at a constant and carefully controlled speed, and its speed of rotation defines the speed at which the tape runs. When record or play is selected the pinch roller rapidly moves towards the capstan, firmly sandwiching the tape in between the two. The rotation of the capstan now controls the speed of tape travel across the heads.

The take-up reel is controlled by a motor which applies a low anti-clockwise torque so that the tape is wound on to it. The supply reel on the left is also controlled by a motor, which now applies a low clockwise torque, attempting to drag the tape back in the opposite direction, and this 'back tension' keeps the tape in firm contact with the heads. Different reel sizes require different degrees of back tension for optimum spooling, and a reel size switch will usually be provided although this is sometimes automatic. One or two transports have been designed without pinch rollers, an enlarged diameter capstan on its own providing speed control. The reel motors need to be rather more finely controlled during record and replay so as to avoid tape slippage across the capstan. Even capstan-less transports have appeared, the tape speed being governed entirely by the reel motors.

When fast wind or rewind is selected the tape is lifted away from the heads by tape lifters, whilst spooling motors apply an appropriately high torque to the reel which is to take up the tape and a low reverse torque to the supply reel to control back tension. The tape is kept away from the heads so that its rapid movement does not cause excessive heating and wear of the tape heads. Also, very high-level, high-frequency energy is induced into the playback head if the tape is in contact with it which can easily damage speakers, particularly tweeters and HF horns. Nevertheless, a facility for moving the tape into contact with the heads during fast spooling is provided so that a particular point in the tape can be listened for.

Motion sensing and logic control is an important feature of a modern open-reel machine. Because the transport controls are electronically governed on modern machines, one can go straight from, say, rewind to play, leaving the machine itself to store the command and bring the tape safely to a halt before allowing the pinch wheel to approach the capstan. Motion sensing can be implemented by a number of means, often either by sensing the speed of the reel motors using tachometers, or by counting pulses from a roller guide.

The tape counter is usually driven by a rotating roller between the head block and a reel. Slight slippage can be expected, this being cumulative over a complete reel of tape, but remarkably accurate real-time counters are nevertheless to be found.

RECOMMENDED FURTHER READING

Jorgensen, F., 1995. The Complete Handbook of Magnetic Recording, fourth edition. McGraw-Hill.

See also 'General further reading' at the end of this book

USEFUL WEBSITES

www.taperecorder.co.uk: A UK source for spare parts and other information.
www.servicesound.com: A US source for spare parts and information.
www.reelprosoundguys.com: A US source for spare parts and information.

Noise Reduction

Much of the material contained in this chapter could be regarded as historical given the widespread adoption of digital recording and processing, but the justification given at the beginning of the last chapter for its continued inclusion requires that its partnering chapter, also in reduced form, should hold a place in this edition. Noise reduction techniques have been applied to analog tape machines of all formats, radio microphones, radio transmission and reception, land lines, satellite relays, gramophone records and even some digital tape machines. The general principles of operation will be outlined, followed by a discussion of particular well-known examples. Detailed descriptions of some individual systems are referred to in the 'Further reading' list at the end of this chapter.

WHY IS NOISE REDUCTION REQUIRED?

A noise reduction system, used correctly, reduces the level of unwanted signals introduced in a recording–replay or transmission–reception process (see Figure. 7.1). Noise such as hiss, hum and interference may be introduced, as well as, say, print-through in analog recording, due to imperfections in the storage or transmission process. In communications, a signal sent along a

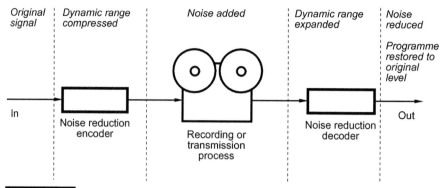

Original signal | Dynamic range compressed | Noise added | Dynamic range expanded | Noise reduced

Programme restored to original level

In

Noise reduction encoder

Recording or transmission process

Noise reduction decoder

Out

FIGURE 7.1   *Graphical representation of a companding noise reduction process.*

land line may be prone to interference from various sources, and will therefore emerge with some of this interference signal mixed with it.

High-quality radio microphone systems routinely incorporate noise reduction in the form of variable ratio compression and expansion, and this can make previously unacceptably noisy reception conditions (weak signal strength coming into the receiver requiring high RF gain with its attendant high noise level) usable.

METHODS OF REDUCING NOISE

Variable pre-emphasis

Pre-emphasis (see Fact File 7.1) is a very straightforward solution to the problem of noise reduction, but is not a panacea. Many sound sources, including music, have a falling energy content at high frequencies, so lower-level HF signals can be boosted to an extent without too much risk of saturating the tape. But tape tends to saturate more easily at HF than at LF (see the previous chapter), so high levels of distortion and compression would result if too much pre-emphasis were applied at the recording stage. What is needed is a circuit which senses the level of the signal on a continuous basis, controlling the degree of pre-emphasis so as to be non-existent at high signal levels but considerable at low signal levels (see Figure. 7.2). This can be achieved by incorporating a filter into a side-chain which passes only high-frequency, low-level signals, adding this component into the unpre-emphasized signal. On replay, a reciprocal de-emphasis circuit could then be used. The lack of noise reduction at high signal levels does not matter, since high-level signals have a masking effect on low-level noise (see Fact File 2.3).

Such a process may be called a compansion process, in other words a process which compresses the dynamic range of a signal during recording

FACT FILE 7.1 PRE-EMPHASIS

One approach to the problem of reducing the apparent level of noise could be to precondition the incoming signal in some way so as to raise it further above the noise. Hiss is most annoying at high frequencies, so one could boost HF on recording. On replay, HF signals would therefore be reproduced with unnatural emphasis, but if the same region is now attenuated to bring the signal down to its original level any hiss in the same band will also be attenuated by a corresponding amount, and so a degree of noise reduction can be achieved without affecting the overall frequency balance of the signal. This is known as pre-emphasis (on record) and de-emphasis (on replay), as shown in the diagram.

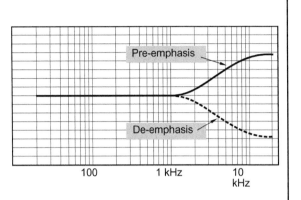

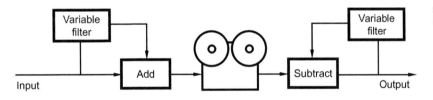

FIGURE 7.2

A simple complementary noise reduction system could boost high frequencies at low signal levels during encoding, and cut them on decoding (encoding characteristic shown).

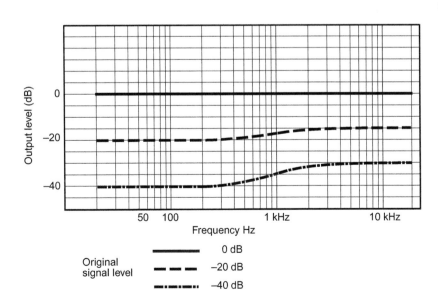

and expands it on replay. The variable HF emphasis described above is an example of selective compansion, acting only on a certain band of frequencies. It is most important to notice that the decoding stage is an exact mirror image of the encoding process, and that it is not possible to use one without the other. Recordings not encoded by a noise reduction system cannot simply be passed through a decoder to reduce their noise. Similarly, encoded tapes sound unusual unless properly decoded, normally sounding overbright and with fluctuations in HF level.

Dolby noise reduction systems

The above process is used as the basis for the Dolby B noise reduction system, found in most cassette decks. Specifically, the threshold below which noise reduction comes into play is around 20 dB below a standard magnetic reference level known as 'Dolby level' (200 nWb m^{-1}). The maximum HF boost of the Dolby B system is 10 dB above 8 kHz, and therefore a maximum of 10 dB of noise reduction is provided. A high-quality cassette deck, without noise reduction, using a good ferric tape, will yield a signal-to-noise ratio of about 50 dB ref. Dolby level. When Dolby B noise reduction is switched in, the 10 dB improvement brings this up to 60 dB (which is more adequate for good-quality music and speech recording). The quoted improvement is seen when noise is measured according to the CCIR 468-2 weighting curve (see 'Dynamic range and signal-to-noise ratio', Appendix 1) and will not be so great when measured unweighted.

Dolby B became widely incorporated into cassette players in the early 1970s, but by the end of the 1970s competition from other companies offering greater levels of noise reduction prompted Dolby to introduce Dolby C, which gives 20 dB of noise reduction. The system acts down to a lower frequency than Dolby B (100 Hz), and incorporates additional circuitry (known as 'anti-saturation') which reduces HF tape squashing when high levels of signal are present. Most of the noise reduction action takes place between 1 kHz and 10 kHz, and less action is taken on frequencies above 10 kHz (where noise is less noticeable) in order to desensitize the system to HF response errors from such factors as azimuth misalignment which would otherwise be exaggerated (this is known as 'spectral skewing'). Dolby C, with its greater compression/expansion ratio compared with Dolby B, will exaggerate tape machine response errors to a correspondingly greater degree, and undecoded Dolby C tapes will sound extremely bright.

Dolby A was introduced in 1965, and is a professional noise reduction system. In essence there is a similarity to the processes described above, but in the Dolby A encoder the noise reduction process is divided into four separate frequency bands, as shown in Figure. 7.3. A low-level 'differential' component is produced for each band, and the differential side-chain output is then recombined with the main signal. The differential component's contribution to the total signal depends on the input level, having maximum effect below −40 dB ref. Dolby level (see Figure. 7.4a and b).

The band splitting means that each band acts independently, such that a high-level signal in one band does not cause a lessening of noise reduction effort in another low-level band, thus maintaining maximum effectiveness with a wide range of program material. The two upper bands are high pass and overlap, offering noise reduction of 10 dB up to around 5 kHz, rising to 15 dB at the upper end of the spectrum.

The decoder is the mirror image of the encoder, except that the differential signal produced by the side-chain is now subtracted from the main signal, restoring the signal to its original state and reducing the noise introduced between encoding and decoding.

The late 1980s saw the introduction of Dolby SR – Spectral Recording – which gives greater noise reduction of around 25 dB. It has been successful

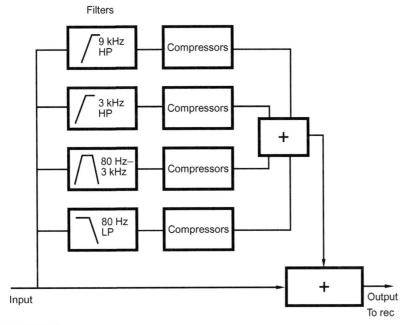

FIGURE 7.3 *In the Dolby A system a low-level 'differential' signal is added to the main signal during encoding. This differential signal is produced in a side-chain which operates independently on four frequency bands. The differential signal is later subtracted during decoding.*

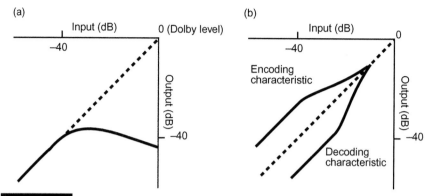

FIGURE 7.4 *(a) Differential signal component produced in a Dolby A side-chain. (b) Input level plotted against output level of Dolby A unit after adding or subtracting differential component.*

in helping to prolong the useful life of analog tape machines, both stereo mastering and multitrack, in the face of the coming of digital tape recorders. Dolby SR differs from Dolby A in that whereas the latter leaves the signal alone until it drops below a certain threshold, the former seeks to maintain full noise reduction (i.e. maximum signal boost during recording) across the whole frequency spectrum until the incoming signal rises above the threshold level. The band of frequencies where this happens is then subject to appropriately less boost. This is rather like looking at the same process from opposite directions, but the SR system attempts to place a comparably high recording level on the tape across the whole frequency spectrum in order that the dynamic range of the tape is always used optimally.

This is achieved by ten fixed and sliding-band filters with gentle slopes. The fixed-band filters can vary in gain. The sliding-band filters can be adjusted to cover different frequency ranges. It is therefore a fairly complex multiband system, requiring analysis of the incoming signal to determine its energy at various frequencies. Spectral skewing and anti-saturation are also incorporated (see 'Dolby C', above). Dolby SR is a particularly inaudible noise reduction system, more tolerant of level mismatches and replay speed changes than previous systems. A simplified 'S'-type version was introduced for the cassette medium, and is also used on some semi-professional multitrack recorders.

The Dolby process, being level dependent, requires that the reproduced signal level on decoding is exactly the same with respect to Dolby level as on encoding, otherwise frequency response errors will result: for instance, a given level of treble boost applied during encoding must be cut by exactly the same amount during replay.

dbx

dbx is another commonly encountered system. It offers around 30 dB of noise reduction and differs from the various Dolby systems as follows. dbx globally compresses the incoming signal across the whole of the frequency spectrum, and in addition gives pre-emphasis at high frequencies (treble boost). It is not level dependent, and seeks to compress an incoming signal with, say, a 90 dB dynamic range into one with a 60 dB dynamic range which will now fit into the dynamic range capabilities of the analog tape recorder. On replay, a recip-rocal amount of expansion is applied together with treble de-emphasis.

Owing to the two factors of high compansion ratios and treble pre- and de-emphasis, frequency response errors can be considerably exaggerated. Therefore, dbx type 1 is offered which may be used with professional equip-ment and type 2 is to be used with domestic equipment such as cassette decks where the noise reduction at high frequencies is relaxed somewhat so as not to exaggerate response errors unduly. The degree of compression/ expansion is fixed, that is, it does not depend on the level of the incoming signal. There is also no division of noise reduction between frequency bands. These factors sometimes produce audible modulation of background hiss with critical program material such as wide dynamic range classical music, and audible 'pumping' noises can sometimes be heard. The system does, however, offer impressive levels of noise reduction, particularly welcome with the cassette medium, and does not require accurate level alignment.

Telcom c4

The ANT telcom c4 noise reduction system arrived somewhat later than did Dolby and dbx, in 1978. Capitalizing on the experience gained by those two systems, the telcom c4 offers a maximum noise reduction of around 30 dB, is level dependent like Dolby, and also splits the frequency spectrum up into four bands which are then treated separately. The makers claim that the c4 system is less affected by record/replay-level errors than is Dolby A. The system works well in operation, and side-effects are minimal.

There is another system offered by the company, called 'hi-com', which is a cheaper, simpler version intended for home studio setups and domestic cassette decks.

LINE-UP OF NOISE REDUCTION SYSTEMS

In order to ensure unity gain through the system on recording and replay, with correct tracking of a Dolby decoder, it is important to align the noise reduction signal chain. Many methods are recommended, some more

FIGURE 7.5 *Dolby level is indicated on Dolby units using either a mechanical meter (shown left), or using red and green LEDs (shown right). The meter is normally aligned to the '18.5 NAB' mark or set such that the two green LEDs are on together.*

rigorous than others, but in a normal studio operation for everyday alignment, the following process should be satisfactory. It should be done after the tape machine has been aligned (this having been done with the NR unit bypassed).

For a Dolby A encoder, a 1 kHz tone should be generated from the mixer at + 4 dBu (usually PPM 5), and fed to the input of the NR unit. The unit should be in 'NR out' mode, and set to 'record'. The input level of the NR unit should normally be adjusted so that this tone reads on the 'NAB' level mark on the meter (see Figure. 7.5). The output of the unit should then be adjusted until its electrical level is also + 4 dBu. (If the tape machine has meters then the level can be read here, provided that these meters are reliable and the line-up is known.)

It is customary to record a passage of 'Dolby tone' (in the case of Dolby A) or Dolby Noise (in the case of Dolby SR) at the beginning of a Dolby-encoded tape, along with the other line-up tones (see 'Record alignment', Chapter 6). During record line-up, the Dolby tone is generated by the Dolby unit itself, and consists of a frequency-modulated 700 Hz tone at the Dolby's internal line-up reference level, which is easily recognized and distinguished from other line-up tones which may be present on a tape. Once the output level of the record Dolby has been set then the Dolby tone button on the relevant unit should be pressed, and the tone recorded at the start of the tape.

To align the replay Dolby (set to 'NR out', 'replay' mode), the recorded Dolby tone should be replayed and the input level adjusted so that the tone reads at the NAB mark on the internal meter. The output level should then be adjusted for + 4 dBu, or so that the mixer's meter reads PPM 5 when switched to monitor the tape machine replay.

For operation, the record and replay units should be switched to 'NR in'.

Dolby SR uses pink noise instead of Dolby tone to distinguish tapes recorded with this system, and it is useful because it allows for line-up of the replay Dolby in cases where accurate level metering is not available. Since level misalignment will result in response errors the effects will be audible on a band of pink noise. A facility is provided for automatic switching between internally generated pink noise and off-tape noise, allowing the user to adjust replay-level alignment until there appears to be no audible difference between the spectra of the two. In normal circumstances Dolby SR systems should be aligned in a similar way to Dolby A, except that a noise band is recorded on the tape instead of a tone. Most systems use LED meters to indicate the correct level, having four LEDs as shown in Figure. 7.5.

SINGLE-ENDED NOISE REDUCTION

General systems

Several companies offer so-called 'single-ended' noise reduction systems, and these are intended to 'clean up' an existing noisy recording or signal. They operate by sensing the level of the incoming signal, and as the level falls below a certain threshold the circuit begins to roll off the treble progressively, thereby reducing the level of hiss. The wanted signal, being low in level, in theory suffers less from this treble reduction than would a high-level signal due to the change in response of the ear with level (see Fact File 2.2). High-level signals are left unprocessed. The system is in fact rather similar to the Dolby B decoding process, but of course the proper reciprocal Dolby B encoding is absent. The input level controls of such systems must be carefully adjusted so as to bring in the effect of the treble roll-off at the appropriate threshold for the particular signal being processed so that a suitable compromise can be achieved between degree of hiss reduction and degree of treble loss during quieter passages. Such single-ended systems should be judiciously used–they are not intended to be left permanently in circuit–and value judgments must always be made as to whether the processed signal is in fact an improvement over the unprocessed one.

If a single-ended system is to be used on a stereo program, units which are capable of being electronically 'ganged' must be employed so that exactly the same degree of treble cut is applied to each channel; otherwise varying frequency balance between channels will cause stereo images to wander.

Noise gates

The noise gate can be looked upon as another single-ended noise reduction system. It operates as follows. A threshold control is provided which can be adjusted such that the output of the unit is muted (the gate is 'closed') when the signal level falls below the threshold. During periods when signal level is very low (possibly consisting of tape or guitar amplifier noise only) or absent the unit shuts down. A very fast attack time is employed so that the sudden appearance of signal opens up the output without audible clipping of the initial transient. The time lapse before the gate closes, after the signal has dropped below the chosen threshold level, can also be varied. The close threshold is engineered to be lower than the open threshold (known as hysteresis) so that a signal level which is on the borderline does not confuse the unit as to whether it should be open or closed, which would cause 'gate flapping'.

Such units are useful when, for instance, a noisy electric guitar setup is being recorded. During passages when the guitarist is not playing the output shuts down so that the noise is removed from the mix. They are sometimes also used in a similar manner during multitrack mixdown where they mute outputs of the tape machine during the times when the tape is unmodulated, thus removing the noise contribution from those tracks.

Noise gates can also be used as effects in themselves, and the 'gated snare drum' is a common effect on pop records. The snare drum is given a heavy degree of gated reverb, and a high threshold level is set on the gate so that around half a second or so after the drum is hit the heavy 'foggy' reverb is abruptly cut off. Drum machines can mimic this effect, as can some effects processors.

Digital noise extraction

Extremely sophisticated single-ended computer-based noise reduction systems have been developed. A given noisy recording will normally have a short period somewhere in which only the noise is present without any program, for instance the run-in groove of an old 78 rpm shellac disc recording provides a sample of that record's characteristic noise. This noise is analyzed by a computer and can subsequently be recognized as an unwanted constituent of the signal, and then extracted electronically from it. Sudden discontinuities in the program caused by scratches and the like can be recognized as such and removed. The gap is filled by new material which is made to be similar to that which exists either side of the gap. Not all of these processes are currently 'real time', and it may take several times longer than the program's duration for the process to be carried out, but as the speed of digital signal processing increases more operations become possible in real time. This is covered in greater detail in Chapter 9.

RECOMMENDED FURTHER READING

Dolby, R., 1967. An audio noise reduction system. J. Audio Eng. Soc. 15, 383–388.

Dolby, R., 1970. A noise reduction system for consumer tape applications. Presented at the 39th AES Convention. J. Audio Eng. Soc. (Abstracts) 18, 704.

Dolby, R., 1983. A 20 dB audio noise reduction system for consumer applications. J. Audio Eng. Soc. 31, 98–113.

Dolby, R. , 1986. The spectral recording process. Presented at the 81st AES Convention. Preprint 2413 (C-6). Audio Engineering Society.

See also 'General further reading' at the end of this book.

Digital Audio Principles

CHAPTER CONTENTS

This chapter contains an introduction to the main principles of digital audio, described in a relatively non-mathematical way. Further reading recommendations at the end of this chapter are given for those who want to study the subject in more depth. Subsequent chapters deal with digital recording and editing systems and with digital audio applications.

DIGITAL AND ANALOG RECORDING CONTRASTED

In analog recording, as described in the previous chapters, sound is recorded by converting continuous variations in sound pressure into continuous variations in electrical voltage, using a microphone. This varying voltage is then converted into a varying pattern of magnetization on a tape, or, alternatively, into a pattern of light and dark areas on an optical-film soundtrack, or a groove of varying deviation on an LP.

Because the physical characteristics of analog recordings relate closely to the sound waveform, replaying them is a relatively simple matter. Variations in the recorded signal can be converted directly into variations in sound pressure using a suitable collection of transducers and amplifiers. The replay system, however, is unable to tell the difference between wanted signals and unwanted signals. Unwanted signals might be distortions, noise and other forms of interference introduced by the recording process. For example, a record player cannot distinguish between the stylus movement it experiences because of a scratch on a record (unwanted) and that caused by a loud transient in the music (wanted). Imperfections in the recording medium are reproduced as clicks, crackles and other noises.

Digital recording, on the other hand, converts the electrical waveform from a microphone into a series of binary numbers, each of which represents the amplitude of the signal at a unique point in time, recording these numbers in a coded form which allows the system to detect whether the replayed signal is correct or not. A reproducing device is then able to distinguish between the wanted and the unwanted signals introduced above, and is thus able to reject all but the wanted original information in most cases. Digital audio can be engineered to be more tolerant of a poor recording channel than analog audio. Distortions and imperfections in the storage or transmission process need not affect the sound quality of the signal provided that they remain within the design limits of the system and that timing and data errors are corrected. These issues are given further coverage in Fact File 8.1.

Digital audio has made it possible for sound engineers to take advantage of developments in the computer industry, and this is particularly beneficial because the size of that industry results in mass production

FACT FILE 8.1 ANALOG AND DIGITAL INFORMATION

Analog information is made up of a continuum of values, which at any instant may have any value between the limits of the system. For example, a rotating knob may have one of an infinite number of positions – it is therefore an analog controller (see the diagram below). A simple switch, on the other hand, can be considered as a digital controller, since it has only two positions – off or on. It cannot take any value in between. The brightness of light that we perceive with our eyes is analog information and as the sun goes down the brightness falls gradually and smoothly, whereas a household light without a dimmer may be either on or off – its state is binary (that is, it has only two possible states).

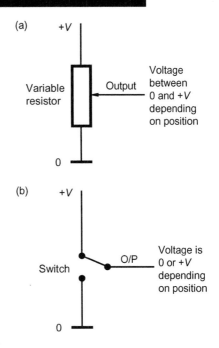

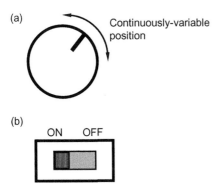

Electrically, analog information may be represented as a varying voltage or current. If a rotary knob is used to control a variable resistor connected to a voltage supply, its position will affect the output voltage as shown below. This, like the knob's position, may occupy any value between the limits – in this case anywhere between zero volts and +V. The switch could be used to control a similar voltage supply and in this case the output voltage could only be either zero volts or +V. In other words the electrical information that resulted would be binary. The high (+V) state could be said to correspond to a binary one and the low state to binary zero (although in many real cases it is actually the other way around).

Binary information is inherently more resilient to noise and interference than analog information, as shown in the diagram below. If noise is added to an analog signal it becomes very difficult to tell what is the wanted signal and what is the unwanted noise, as there is no means of distinguishing between the two. If noise is added to a binary signal it is possible to extract the important information at a later stage. By comparing the signal amplitude with a fixed decision point it is possible for a receiver to treat everything above the decision point as 'high' and everything below it as 'low'. For any noise or interference to influence the state of a digital signal it must be at least large enough in amplitude to cause a high level to be interpreted as 'low', or vice versa.

The timing of digital signals may also be corrected to some extent, giving digital signals another advantage over analog ones. This is because digital information has a discrete time structure in which the intended sample instants are known. If the timing of bits in a digital message becomes unstable, such as after having been passed over a long cable with its associated signal distortions, resulting in timing 'jitter', the signal may be reclocked at a stable rate.

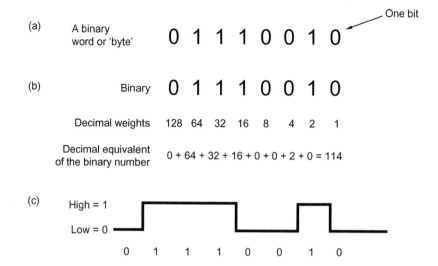

FIGURE 8.1
(a) A binary number (word
or 'byte') consists of bits.
(b) Each bit represents
a power of two.
(c) Binary numbers can
be represented electrically
in pulse code modulation
(PCM) by a string of high
and low voltages.

(and therefore cost savings) on a scale not possible for audio products alone. Today it is common for sound to be recorded, processed and edited on relatively low-cost desktop computer equipment, and this is a trend likely to continue.

BINARY FOR BEGINNERS

First, we introduce the basics of binary number systems, because nearly all digital audio systems are based on this.

In the decimal number system each digit of a number represents a power of ten. In a binary system each digit or bit represents a power of two (see Figure 8.1). It is possible to calculate the decimal equivalent of a binary integer (whole number) by using the method shown. Negative numbers need special treatment, as described in Fact File 8.2. A number made up of more than 1 bit is called a binary 'word', and an 8 bit word is called a 'byte' (from 'by eight'). Four bits is called a 'nibble'. The more bits there are in a word the larger the number of states it can represent, with 8 bits allowing 256 (2^8) states and 16 bits allowing 65536 (2^{16}). The bit with the lowest weight (2^0) is called the least significant bit or LSB and that with the greatest weight is called the most significant bit or MSB. The term kilobyte or Kbyte is used to mean 1024 or 2^{10} bytes and the term megabyte or Mbyte represents 1024 Kbytes.

Electrically it is possible to represent a binary word in either serial or parallel form. In serial communication only one connection need be used

FACT FILE 8.2 NEGATIVE NUMBERS

Negative integers are usually represented´ in a form known as 'two's complement'. Negative values are represented by taking the positive equivalent, inverting all the bits and adding a one. Thus to obtain the 4 bit binary equivalent of decimal minus five (-5^{10}) in binary two's complement form:

$$5^{10} = 0101^2$$
$$-5^{10} = 1010 + 0001 = 1011^2$$

Two's complement numbers have the advantage that the MSB represents the sign (1 = negative, 0 = positive) and that arithmetic may be performed on positive and negative numbers giving the correct result:

e.g. (in decimal): 5
    +(−3)
    = 2
or (in binary): 0101
    +1101
    = 0010

The carry bit that may result from adding the two MSBs is ignored.

An example is shown here of 4 bit, two's complement numbers arranged in a circular fashion. It will be seen that the binary value changes from all zeros to all ones as it crosses the zero point and that the maximum positive value is 0111 whilst the maximum negative value is 1000, so the values wrap around from maximum positive to maximum negative.

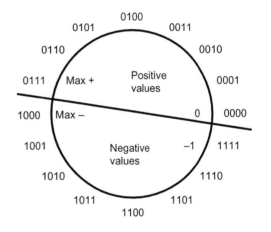

and the word is clocked out one bit at a time using a device known as a shift register. The shift register is previously loaded with the word in parallel form (see Figure 8.2). The rate at which the serial data is transferred depends on the rate of the clock. In parallel communication each bit of the word is transferred over a separate connection.

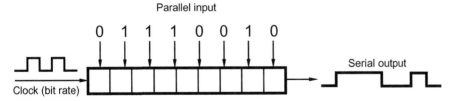

Parallel input

0 1 1 1 0 0 1 0

Clock (bit rate)

Serial output

FIGURE 8.2 *A shift register is used to convert a parallel binary word into a serial format. The clock is used to shift the bits one at a time out of the register, and its frequency determines the bit rate. The data may be clocked out of the register either MSB or LSB first, depending on the device and its configuration.*

Table 8.1	Hexadecimal and decimal equivalents to binary numbers	
Binary	**Hexadecimal**	**Decimal**
0000	0	0
0001	1	1
0010	2	2
0011	3	3
0100	4	4
0101	5	5
0110	6	6
0111	7	7
1000	8	8
1001	9	9
1010	A	10
1011	B	11
1100	C	12
1101	D	13
1110	E	14
1111	F	15

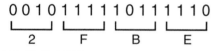

0 0 1 0   1 1 1 1   0 1 1 1   1 1 1 0
2   F   B   E

FIGURE 8.3 *This 16 bit binary number may be represented in hexadecimal as shown, by breaking it up into 4 bit nibbles and representing each nibble as a hex digit.*

Because binary numbers can become fairly unwieldy when they get long, various forms of shorthand are used to make them more manageable. The most common of these is hexadecimal. The hexadecimal system represents decimal values from 0 to 15 using the 16 symbols 0–9 and A–F, according to Table 8.1. Each hexadecimal digit corresponds to 4 bits or one nibble of the binary word. An example showing how a long binary word may be written in hexadecimal (hex) is shown in Figure 8.3 – it is simply a matter of breaking the word up into 4 bit chunks and converting each chunk to hex. Similarly, a hex word can be converted to binary by using the reverse process.

Logical operations can be carried out on binary numbers, which enables various forms of mathematics to be done in binary form, as introduced in Fact File 8.3.

Fixed-point binary numbers are often used in digital audio systems to represent sample values. These are usually integer values represented by a number of bytes (2 bytes for 16 bit samples, 3 bytes for 24 bit samples, etc.). In some applications it is necessary to represent numbers with a very large range, or in a fractional form. Here floating-point representation may

FACT FILE 8.3 LOGICAL OPERATIONS

Most of the apparently complicated processing operations that occur within a computer are actually just a fast sequence of simple logical operations. The apparent power of the computer and its ability to perform complex tasks are really due to the speed with which simple operations are performed.

The basic family of logical operations is shown here in the form of a truth table next to the electrical symbol that represents each 'logic gate'. The AND operation gives an output only when both its inputs are true; the OR operation gives an output when either of its inputs are true; and the XOR (exclusive OR) gives an output only when one of its inputs is true. The inverter or NOT gate gives an output which is the opposite of its input and this is often symbolized using a small circle on inputs or outputs of devices to indicate inversion.

AND

A	B	C
0	0	0
1	0	0
0	1	0
1	1	1

OR

A	B	C
0	0	0
1	0	1
0	1	1
1	1	1

Inverter (NOT)

EXOR

A	B	C
0	0	0
1	0	1
0	1	1
1	1	0

be used. A typical floating-point binary number might consist of 32 bits, arranged as 4 bytes, as shown in Figure 8.4. Three bytes are used to represent the mantissa and 1 byte the exponent (although the choice of number of bits for the exponent and mantissa are open to variance depending on the application). The mantissa is the main part of the numerical value and the exponent determines the power of two to which the mantissa must be raised. The MSB of the exponent is used to represent its sign and the same for the mantissa.

It is normally more straightforward to perform arithmetic processing operations on fixed-point numbers than on floating-point numbers, but signal processing devices are available in both forms.

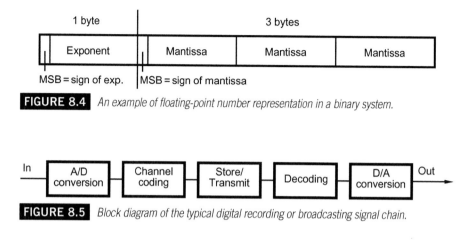

FIGURE 8.4   An example of floating-point number representation in a binary system.

FIGURE 8.5   Block diagram of the typical digital recording or broadcasting signal chain.

THE DIGITAL AUDIO SIGNAL CHAIN

Figure 8.5 shows the signal chain involved in a typical digital recording or broadcasting system. First, the analog audio signal (a time-varying electrical voltage) is passed through an analog-to-digital (A/D) convertor where it is transformed from a continuously varying voltage into a series of 'samples', which are 'snapshots' of the analog signal taken many thousand times per second. Each sample is represented by a number. If the system uses some form of data reduction (see below) this will be carried out here, after A/D conversion and before channel coding. The resulting sequence of audio data is coded into a form that makes it suitable for recording or broadcasting (a process known as coding or channel coding), and the signal is then recorded or transmitted. Upon replay or reception the signal is decoded and subjected to error correction, and it is this latter process which works out what damage has been done to the signal since it was coded. The channel coding and error detection/correction processes are usually integral to the recording or transmission system and modern disk-based recording systems often rely on the built-in processes of generic computer mass storage systems to deal with this. After decoding, any errors in timing or value of the samples are corrected if possible and the result is fed to a digital-to-analog (D/A) convertor, which turns the numerical data back into a time-continuous analog audio signal.

In the following sections each of the main processes involved in this chain will be explained, followed by a discussion of the implementation of this technology in real audio systems.

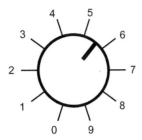

FIGURE 8.6 *A rotary knob's position could be measured against a numbered scale such as the decimal scale shown. Quantizing the knob's position would involve deciding which of the limited number of values (0–9) most closely represented the true position.*

ANALOG-TO-DIGITAL CONVERSION

A basic example

In order to convert analog information into digital information it is necessary to measure its amplitude at specific points in time (called 'sampling') and to assign a binary digital value to each measurement (called 'quantizing'). A simple example of the process can be taken from control technology in which it is wished to convert the position of a rotary knob into a digital control signal that could be used by a computer. This concept can be extended to the conversion of audio signals.

The diagram in Figure 8.6 shows such a rotary knob against a fixed scale running from 0 to 9. The position of the control should be measured or 'sampled' at regular intervals to register changes. The rate at which switches and analog controls are sampled depends on how important it is that they are updated regularly. Some older audio mixing consoles sampled the positions of automated controls once per television frame (40 ms in Europe), whereas some modern digital mixers sample controls as often as once per audio sample period (roughly 20 µs). Clearly the more regularly a control's position is sampled the more data will be produced, since there will be one binary value per sample. A smooth representation of changing control movements is ensured by regular sampling.

To quantize the position of the knob it is necessary to determine which point of the scale it is nearest at each sampling instant and assign a binary number that is equivalent to its position. Unless the pointer is at exactly one of the increments the quantizing process involves a degree of error. The maximum error is plus or minus half of an increment, because once the pointer is more than halfway between one increment and the next it should be quantized to the next.

Introduction to audio A/D conversion

The process of A/D conversion is of paramount importance in determining the inherent sound quality of a digital audio signal. The technical quality of the audio signal, once converted, can never be made any better, only worse. Some applications deal with audio purely in the digital domain, in which case A/D conversion is not an issue, but most operations involve the acquisition of audio material from the analog world at one time or another. The quality of convertors varies very widely in digital audio workstations and their peripherals because the price range of such workstations is also great. Some stand-alone professional convertors can easily cost as much as the complete digital audio hardware and software for a desktop computer. One can find audio A/D convertors built into many multimedia desktop computers now, but these are often rather low performance devices when compared with the best available. As will be seen below, the sampling rate and the number of bits per sample are the main determinants of the quality of a digital audio signal, but the design of the convertors determines how closely the sound quality approaches the theoretical limits.

Despite the above, it must be admitted that to the undiscerning ear one 16 bit convertor sounds very much like another and that there is a law of diminishing returns when one compares the increased cost of good convertors with the perceivable improvement in quality. Convertors are very much like wine in this respect.

Audio sampling

An analog audio signal is a time-continuous electrical waveform and the A/D convertor's task is to turn this signal into a time-discrete sequence of binary numbers. The sampling process employed in an A/D convertor involves the measurement or 'sampling' of the amplitude of the audio waveform at regular intervals in time (see Figure 8.7). From this diagram it will be clear that the sample pulses represent the instantaneous amplitudes of the audio signal at each point in time. The samples can be considered as instantaneous 'still frames' of the audio signal which together and in sequence form a representation of the continuous waveform, rather like the still frames that make up a movie film give the impression of a continuously moving picture when played in quick succession.

In order to represent the fine detail of the signal it is necessary to take a large number of these samples per second. The mathematical sampling theorem proposed by Shannon indicates that at least two samples must be taken per audio cycle if the necessary information about the signal is to be conveyed. This means that the sampling frequency must be at least twice

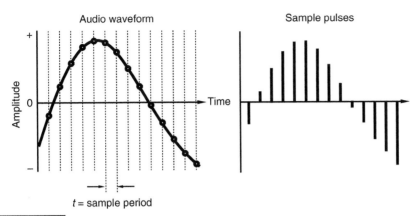

FIGURE 8.7 *An arbitrary audio signal is sampled at regular intervals of time t to create short sample pulses whose amplitudes represent the instantaneous amplitude of the audio signal at each point in time.*

as high as the highest audio frequency to be handled by the system (this is known as the Nyquist criterion).

Another way of visualizing the sampling process is to consider it in terms of modulation, as shown in Figure 8.8. The continuous audio waveform is used to modulate a regular chain of pulses. The frequency of these pulses is the sampling frequency. Before modulation all these pulses have the same amplitude (height), but after modulation the amplitude of the pulses is modified according to the instantaneous amplitude of the audio signal at that point in time. This process is known as pulse amplitude modulation (PAM). Fact File 8.4 describes a frequency domain view of this process.

Filtering and aliasing

It can be seen from Figure 8.9 that if too few samples are taken per cycle of the audio signal then the samples may be interpreted as representing a wave other than that originally sampled. This is one way of understanding the phenomenon known as aliasing. An 'alias' is an unwanted representation of the original signal that arises when the sampled signal is reconstructed during D/A conversion.

It is relatively easy to see why the sampling frequency must be at least twice the highest baseband audio frequency from Figure 8.10. It can be seen that an extension of the baseband above the Nyquist frequency results in the lower sideband of the first spectral repetition overlapping the upper end of the baseband and appearing within the audible range that would be

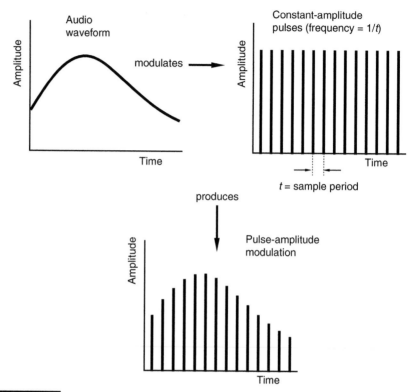

FIGURE 8.8   *In pulse amplitude modulation, the instantaneous amplitude of the sample pulses is modulated by the audio signal amplitude (positive only values shown).*

reconstructed by a D/A convertor. Two further examples are shown to illustrate the point – the first in which a baseband tone has a low enough frequency for the sampled sidebands to lie above the audio frequency range, and the second in which a much higher frequency tone causes the lower sampled sideband to fall well within the baseband, forming an alias of the original tone that would be perceived as an unwanted component in the reconstructed audio signal.

The aliasing phenomenon can be seen in the case of the well-known 'spoked-wheel' effect on films, since moving pictures are also an example of a sampled signal. In film, still pictures (image samples) are normally taken at a rate of 24 per second. If a rotating wheel with a marker on it is filmed it will appear to move round in a forward direction as long as the rate of rotation is much slower than the rate of the still photographs, but as its rotation rate increases it will appear to slow down, stop, and then appear to start moving backwards. The virtual impression of backwards motion

FACT FILE 8.4 SAMPLING – FREQUENCY DOMAIN

Before modulation the audio signal has a frequency spectrum extending over the normal audio range, known as the baseband spectrum (upper diagram). The shape of the waveform and its equivalent spectrum is not significant in this diagram – it is just an artist's impression of a complex audio signal such as music. The sampling pulses, before modulation, have a line spectrum at multiples of the sampling frequency, which is much higher than the highest audio frequency (middle diagram). The frequency spectrum of the pulse-amplitude-modulated (PAM) signal is as shown in the lower diagram. In addition to the 'baseband' audio signal (the original audio spectrum before sampling) there are now a number of additional images of this spectrum, each centered on multiples of the sampling frequency. Sidebands have been produced either side of the sampling frequency and its multiples, as a result of the amplitude modulation, and these extend above and below the sampling frequency and its multiples to the extent of the base bandwidth. In other words these sidebands are pairs of mirror images of the audio baseband.

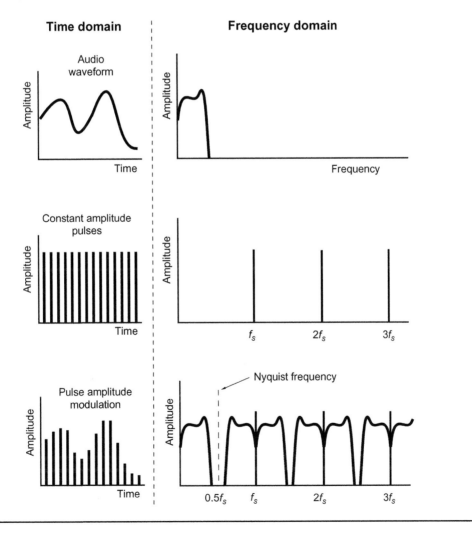

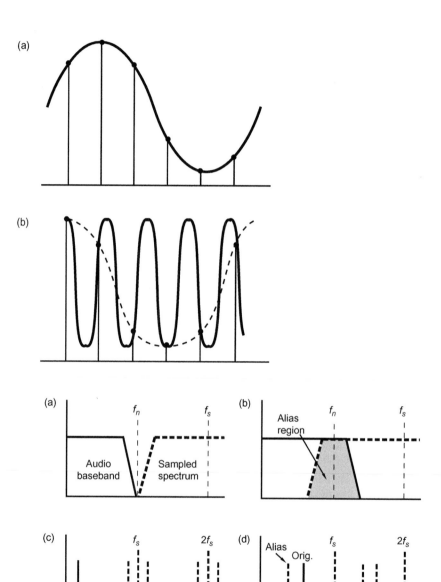

FIGURE 8.9

In example (a) many samples are taken per cycle of the wave. In example (b) less than two samples are taken per cycle, making it possible for another lower-frequency wave to be reconstructed from the samples. This is one way of viewing the problem of aliasing.

FIGURE 8.10   *Aliasing viewed in the frequency domain. In (a) the audio baseband extends up to half the sampling frequency (the Nyquist frequency f_n) and no aliasing occurs. In (b) the audio baseband extends above the Nyquist frequency and consequently overlaps the lower sideband of the first spectral repetition, giving rise to aliased components in the shaded region. In (c) a tone at 1 kHz is sampled at a sampling frequency of 30 kHz, creating sidebands at 29 and 31 kHz (and at 59 and 61 kHz, etc.). These are well above the normal audio frequency range, and will not be audible. In (d) a tone at 17 kHz is sampled at 30 kHz, putting the first lower sideband at 13 kHz – well within the normal audio range. The 13 kHz sideband is said to be an alias of the original wave.*

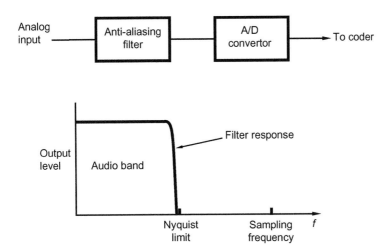

FIGURE 8.11

In simple A/D convertors an analog anti-aliasing filter is used prior to conversion, which removes input signals with a frequency above the Nyquist limit.

gets faster as the rate of rotation of the wheel gets faster and this backwards motion is the aliased result of sampling at too low a rate. Clearly the wheel is not really rotating backwards, it just appears to be. Perhaps ideally one would arrange to filter out moving objects that were rotating faster than half the frame rate of the film, but this is hard to achieve in practice and visible aliasing does not seem to be as annoying subjectively as audible aliasing.

If audio signals are allowed to alias in digital recording one hears the audible equivalent of the backwards-rotating wheel – that is, sound components in the audible spectrum that were not there in the first place, moving downwards in frequency as the original frequency of the signal increases. In basic convertors, therefore, it is necessary to filter the baseband audio signal before the sampling process, as shown in Figure 8.11, so as to remove any components having a frequency higher than half the sampling frequency. It is therefore clear that in practice the choice of sampling frequency governs the high frequency limit of a digital audio system.

In real systems, and because filters are not perfect, the sampling frequency is usually made higher than twice the highest audio frequency to be represented, allowing for the filter to roll off more gently. The filters incorporated into both D/A and A/D convertors have a pronounced effect on sound quality, since they determine the linearity of the frequency response within the audio band, the slope with which it rolls off at high frequency and the phase linearity of the system. In a non-oversampling convertor, the filter must reject all signals above half the sampling frequency with an attenuation of at least 80 dB. Steep filters tend to have an erratic phase response at high frequencies and may exhibit 'ringing' due to the high 'Q' of the filter. Steep filters also have the added disadvantage that they are complicated to produce. Although filter effects are unavoidable to some extent, manufacturers have made considerable improvements to analog anti-aliasing and reconstruction filters and these may be retro-fitted to many existing systems with poor filters. A positive effect is normally noticed on sound quality.

The process of oversampling and the use of higher sampling frequencies (see below) have helped to ease the problems of such filtering. Here the first repetition of the baseband is shifted to a much higher frequency, allowing the use of a shallower anti-aliasing filter and consequently fewer audible side effects.

Sampling frequency and sound quality

The choice of sampling frequency determines the maximum audio bandwidth available. There is a strong argument for choosing a sampling frequency no higher than is strictly necessary, in other words not much higher than twice the highest audio frequency to be represented. This often starts arguments over what is the highest useful audio frequency and this is an area over which heated debates have raged. Conventional wisdom has it that the audio frequency band extends up to 20 kHz, implying the need for a sampling frequency of just over 40 kHz for high quality audio work. There are in fact two standard sampling frequencies between 40 and 50 kHz: the Compact Disc rate of 44.1 kHz and the so-called 'professional' rate of 48 kHz. These are both allowed in the original AES-5 standard of 1984, which sets down preferred sampling frequencies for digital audio equipment. Fact File 8.5 shows commonly encountered sampling frequencies.

The 48 kHz rate was originally specified for professional use because it left a certain amount of leeway for downward varispeed in tape recorders. When many digital recorders are varispeeded, the sampling frequency changes proportionately and the result is a shifting of the first spectral repetition of the audio baseband. If the sampling frequency is reduced too far aliased components may become audible. Most professional digital tape recorders allowed for only around ±12.5% of varispeed for this reason. It is possible now, though, to avoid such problems using digital low-pass filters whose cut-off frequency varies with the sampling frequency, or by using digital signal processing to vary the pitch of audio without varying the output sampling frequency.

The 44.1 kHz frequency had been established earlier on for the consumer Compact Disc and is very widely used in the industry. In fact in many ways it has become the sampling rate of choice for most professional recordings. It allows for full use of the 20 kHz audio band and oversampling

FACT FILE 8.5 AUDIO SAMPLING FREQUENCIES

The table shows commonly encountered sampling frequencies and their applications.

Frequency (kHz)	Application
8	Telephony (speech quality). ITU-T G711 standard.
16	Used in some telephony applications. ITU-T G722 data reduction.
~22.05	Half the CD frequency is 22.05 kHz. Used in some older computer applications. The original Apple Macintosh audio sampling frequency was 22 254.5454 ... Hz.
32	Used in some broadcast coding systems, e.g. NICAM. DAT long play mode. AES-5 secondary rate.
44.056	A slight modification of the 44.1 kHz frequency used in some older equipment to synchronize digital audio with the NTSC television frame rate of 29.97 frames per second. Such 'pull-down' rates are sometimes still encountered in video sync situations.
44.1	CD sampling frequency. AES-5 secondary rate.
47.952	Occasionally encountered when 48 kHz equipment is used in NTSC video operations. Another 'pull-down' rate, ideally to be avoided.
48	AES-5 primary rate for professional applications. Basic rate for Blu-Ray disk (which no longer specifies 44.1 kHz as an option).
88.2	Twice the CD sampling frequency. Optional for DVD-Audio.
96	AES-5-1998 secondary rate for high bandwidth applications. Optional for DVD-Video, DVD-Audio and Blu-Ray disks.
176.4 and 192	Four times the basic standard rates. Optional in DVD-Audio. 192 kHz is the highest sampling frequency allowed on Blu-Ray audio disks.
2.8224 MHz	DSD sampling frequency. A highly oversampled rate used in 1 bit PCM systems such as SuperAudio CD.

convertors allow for the use of shallow analog anti-aliasing filters which avoid phase problems at high audio frequencies. It also generates 10% less data per second than the 48 kHz rate, making it economical from a storage point of view.

A rate of 32 kHz is used in some broadcasting applications, such as NICAM 728 stereo TV transmissions, and in some radio distribution systems. Television and FM radio sound bandwidth is limited to 15 kHz and a considerable economy of transmission bandwidth is achieved by the use of this lower sampling rate. The majority of important audio information lies below 15 kHz in any case and little is lost by removing the top 5 kHz of the audio band. Some professional audio applications offer this rate as an option, but it is not common. It is used for the long play mode of some DAT machines, for example.

Arguments for the standardization of higher sampling rates have become stronger in recent years, quoting evidence from sources claiming that information above 20 kHz is important for higher sound quality, or at least that the avoidance of steep filtering must be a good thing. The DVD standards, for example, incorporate such sampling frequencies as standard features. AES-5-1998 (a revision of the AES standard on sampling frequencies) now allows 96 kHz as an optional rate for applications in which the audio bandwidth exceeds 20 kHz or where relaxation of the anti-alias filtering region is desired. Doubling the sampling frequency leads to a doubling in the overall data rate of a digital audio system and a consequent halving in storage time per megabyte. It also means that any signal processing algorithms need to process twice the amount of data and alter their algorithms accordingly. It follows that these higher sampling rates should be used only after careful consideration of the merits.

Low sampling frequencies such as those below 30 kHz are sometimes encountered for lower quality sound applications such as the storage and transmission of speech, the generation of computer sound effects and so forth. Multimedia applications may need to support these rates because such applications often involve the incorporation of sounds of different qualities. There are also low sampling frequency options for data reduction codecs, as discussed below.

Quantizing

After sampling, the modulated pulse chain is quantized. In quantizing a sampled audio signal the range of sample amplitudes is mapped onto a scale of stepped binary values, as shown in Figure 8.12. The quantizer determines which of a fixed number of quantizing intervals (of size Q) each sample lies within and then assigns it a value that represents the mid-point of that interval. This is done in order that each sample amplitude can be represented by a unique binary number in pulse code modulation (PCM). (PCM is the designation for the form of modulation in which signals are represented as a sequence of sampled and quantized binary data words.) In linear quantizing each quantizing step represents an equal increment of signal voltage and most high-quality audio systems use linear quantizing.

Quantizing error is an inevitable side effect in the process of A/D conversion and the degree of error depends on the quantizing scale used. Considering binary quantization, a 4-bit scale offers 16 possible steps, an 8 bit scale offers 256 steps, and a 16 bit scale 65536. The more bits, the more accurate the process of quantization. The quantizing error magnitude will be a maximum of plus or minus half the amplitude of one quantizing

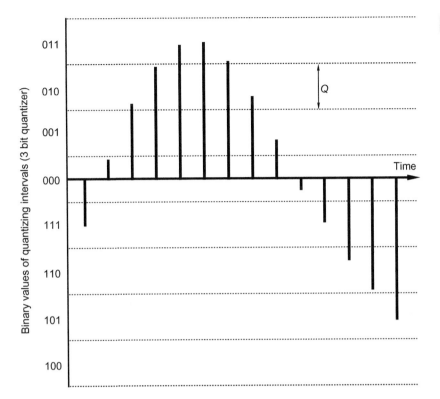

FIGURE 8.12
When a signal is quantized, each sample is mapped to the closest quantizing interval Q, and given the binary value assigned to that interval. (Example of a 3 bit quantizer shown.) On D/A conversion each binary value is assumed to represent the voltage at the mid point of the quantizing interval.

step and a greater number of bits per sample will therefore result in a smaller error (see Figure 8.13), provided that the analog voltage range represented remains the same.

Figure 8.14 shows the binary number range covered by digital audio signals at different resolutions using the usual two's complement hexadecimal representation. It will be seen that the maximum positive sample value of a 16 bit signal is &7FFF, whilst the maximum negative value is &8000. The sample value changes from all zeros (&0000) to all ones (&FFFF) as it crosses the zero point. The maximum digital signal level is normally termed 0 dBFS (FS = full scale).

The quantized output of an A/D convertor can be represented in either serial or parallel form, as shown in Fact File 8.6.

Quantizing resolution and sound quality

The quantizing error may be considered as an unwanted signal added to the wanted signal, as shown in Figure 8.15. Unwanted signals tend to be classified either as distortion or noise, depending on their characteristics,

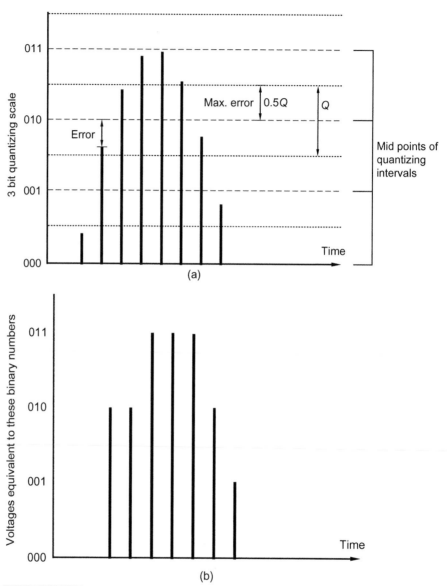

FIGURE 8.13   *In (a) a 3 bit scale is used and only a small number of quantizing intervals covers the analog voltage range, making the maximum quantizing error quite large. The second sample in this picture will be assigned the value 010, for example the corresponding voltage of which is somewhat higher than that of the sample. During D/A conversion the binary sample values from (a) would be turned into pulses with the amplitudes shown in (b), where many samples have been forced to the same level owing to quantizing. In (c) the 4 bit scale means that a larger number of intervals are used to cover the same range and the quantizing error is reduced. (Expanded positive range only shown for clarity.)*

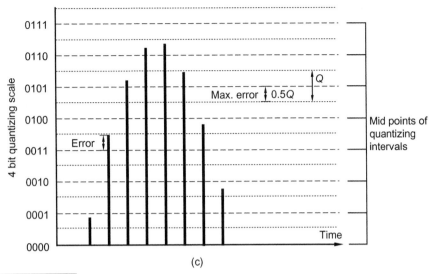

(c)

FIGURE 8.13   (Continued)

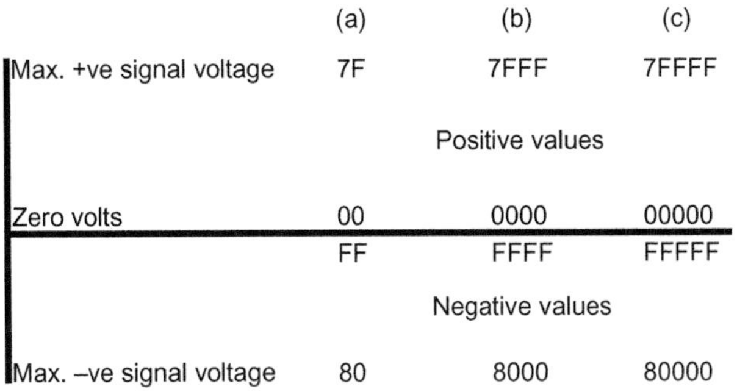

	(a)	(b)	(c)
Max. +ve signal voltage	7F	7FFF	7FFFF
		Positive values	
Zero volts	00	0000	00000
	FF	FFFF	FFFFF
		Negative values	
Max. −ve signal voltage	80	8000	80000

FIGURE 8.14

Binary number ranges (in hexadecimal) related to analog voltage ranges for different convertor resolutions, assuming two's complement representation of negative values. (a) 8 bit quantizer, (b) 16 bit quantizer, (c) 20 bit quantizer.

and the nature of the quantizing error signal depends very much upon the level and nature of the related audio signal. Here are a few examples, the illustrations for which have been prepared in the digital domain for clarity, using 16-bit sample resolution.

First, consider a very low-level sine wave signal, sampled then quantized, having a level only just sufficient to turn the least significant bit of the quantizer on and off at its peak (see Figure 8.16a). Such a signal would have a quantizing error that was periodic, and strongly correlated with the signal, resulting in harmonic distortion. Figure 8.16b shows the frequency spectrum, analyzed in the digital domain of such a signal, showing clearly the distortion products (predominantly odd harmonics) in addition to the original fundamental. Once the signal falls below the level at which it

FACT FILE 8.6 PARALLEL AND SERIAL REPRESENTATION

Electrically it is possible to represent the quantized binary signal in either serial or parallel form. When each bit of the audio sample is carried on a separate wire, the signal is said to be in a parallel format, so a 16 bit convertor would have 16 single bit outputs. If the data is transmitted down a single wire or channel, one bit after the other, the data is said to be in serial format. In serial communication the binary word is clocked out one bit at a time using a device known as a shift register. The shift register is previously loaded with the word in parallel form as shown in the diagram. The rate at which the serial data is transferred depends on the rate of the clock.

Serial form is most useful for transmission over interconnects or transmission links that might cover substantial distances or where the bulk and cost of the interconnect limits the number of paths available. Parallel form tends to be used internally, within high speed digital systems, although serial forms are increasingly used here as well. Most digital audio interfaces are serial, for example, although the Tascam TDIF interface uses a parallel representation of the audio data.

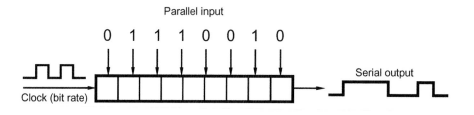

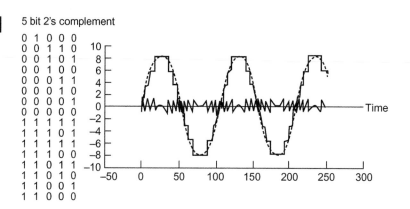

FIGURE 8.15

Quantizing error depicted as an unwanted signal added to the original sample values. Here the error is highly correlated with the signal and will appear as distortion. (Courtesy of Allen Mornington West.)

just turns on the LSB there is no modulation. The audible result, therefore, of fading such a signal down to silence is that of an increasingly distorted signal suddenly disappearing. A higher-level sine wave signal would cross more quantizing intervals and result in more non-zero sample values. As signal level rises the quantizing error, still with a maximum value of $\pm 0.5Q$, becomes increasingly small as a proportion of the total signal level and the error gradually loses its correlation with the signal.

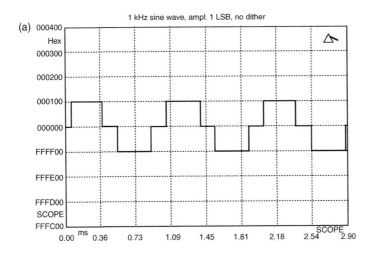

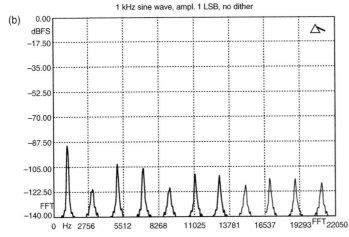

FIGURE 8.16   *(a) A 1 kHz sine wave at very low level (amplitude ±1 LSB) just turns the least significant bit of the quantizer on and off. Analyzed in the digital domain with sample values shown in hex on the vertical axis and time in ms on the horizontal axis. (b) Frequency spectrum of this quantized sine wave, showing distortion products.*

Consider now a music signal of reasonably high level. Such a signal has widely varying amplitude and spectral characteristics and consequently the quantizing error is likely to have a more random nature. In other words it will be more noise-like than distortion-like, hence the term quantizing noise that is often used to describe the audible effect of quantizing error. An analysis of the power of the quantizing error, assuming that it has a noise-like nature, shows that it has an r.m.s. amplitude of $Q/\sqrt{12}$, where Q is the voltage increment represented by one quantizing interval.

Consequently the signal-to-noise ratio of an ideal n bit quantized signal can be shown to be:

$$6.02n + 1.76\,\text{dB}$$

This implies a theoretical S/N ratio that approximates to just over 6 dB per bit. So a 16 bit convertor might be expected to exhibit an S/N ratio of around 98 dB, and an 8 bit convertor around 50 dB. This assumes an undithered convertor, which is not the normal case, as described below. If a convertor is undithered there will only be quantizing noise when a signal is present, but there will be no quiescent noise floor in the absence of a signal. Issues of dynamic range with relation to human hearing are discussed further in Fact File 8.7.

The dynamic range of a digital audio system is limited at high signal levels by the point at which the quantizing range of the convertor has been 'used up' (in other words, when there are no more bits available to represent a higher-level signal). At this point the waveform will be hard clipped (see Figure 8.17) and will become very distorted. This point will normally be set to occur at a certain electrical input voltage, such as +24 dBu in some professional systems. (The effect is very different from that encountered in analog tape recorders which tend to produce gradually

FACT FILE 8.7 DYNAMIC RANGE AND PERCEPTION

It is possible with digital audio to approach the limits of human hearing in terms of sound quality. In other words, the unwanted artefacts of the process can be controlled so as to be close to or below the thresholds of perception. It is also true, though, that badly engineered digital audio can sound poor and that the term 'digital' does not automatically imply high quality. The choice of sampling parameters and noise-shaping methods, as well as more subtle aspects of convertor design, affect the frequency response, distortion and perceived dynamic range of digital audio signals.

The human ear's capabilities should be regarded as the standard against which the quality of digital systems is measured, since it could be argued that the only distortions and noises that matter are those that can be heard. Work carried out by Louis Fielder and Elizabeth Cohen attempted to establish the dynamic range requirements for high-quality digital audio systems by investigating the extremes of sound pressure available from acoustic sources and comparing these with the perceivable noise floors in real acoustic environments. Using psychoacoustic theory, Fielder was able to show what was likely to be heard at different frequencies in terms of noise and distortion, and where the limiting elements might be in a typical recording chain. He determined a dynamic range requirement of 122 dB for natural reproduction. Taking into account microphone performance and the limitations of consumer loudspeakers, this requirement dropped to 115 dB for consumer systems.

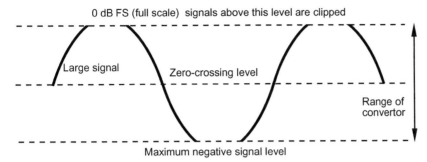

0 dB FS (full scale) signals above this level are clipped

Large signal

Zero-crossing level

Range of convertor

Maximum negative signal level

FIGURE 8.17 *Signals exceeding peak level in a digital system are hard-clipped, since no more digits are available to represent the sample value.*

more distortion as the recording level increases. Digital recorders remain relatively undistorted as the recording level rises until the overload point is reached, at which point very bad distortion occurs.)

The number of bits per sample therefore dictates the signal-to-noise ratio of a linear PCM digital audio system. Fact File 8.8 summarizes the applications for different quantizing resolutions. For many years 16 bit linear PCM was considered the norm for high-quality audio applications. This is the CD standard and is capable of offering a good S/N ratio range of over 90 dB. For most purposes this is adequate, but it fails to reach the psychoacoustic ideal of 122 dB for subjectively noise-free reproduction in professional systems. To achieve such a performance requires a convertor resolution of around 21 bits, which is achievable with today's convertor technology, depending on how the specification is interpreted. So-called 24 bit convertors are indeed available today, but their audio performance is strongly dependent upon the stability of the timing clock, electrical environment, analog stages, grounding and other issues.

For professional recording purposes one may need a certain amount of 'headroom' – in other words some unused dynamic range above the normal peak recording level which can be used in unforeseen circumstances such as when a signal overshoots its expected level. This can be particularly necessary in live recording situations where one is never quite sure what is going to happen with recording levels. This is another reason why many professionals feel that a resolution of greater than 16 bits is desirable for original recording. 20 and 24 bit recording formats are becoming increasingly popular for this reason, with mastering engineers then optimizing the finished recording for 16 bit media (such as CD) using noise-shaped requantizing processes.

FACT FILE 8.8 QUANTIZING RESOLUTIONS

The table shows some commonly encountered quantizing resolutions and their applications.

Bits per sample	Approx. dynamic range with dither (dB)	Application
8	44	Low–moderate quality for older PC internal sound generation. Some older multimedia applications. Usually in the form of unsigned binary numbers.
12	68	Older Akai samplers, e.g. S900.
14	80	Original EIAJ format PCM adaptors, such as Sony PCM-100.
16	92	CD standard. DAT standard. Commonly used high quality resolution for consumer media, some professional recorders and multimedia PCs. Usually two's complement (signed) binary numbers.
20	116	High quality professional audio recording and mastering applications.
24	140	Maximum resolution of most recent professional recording systems, also of AES 3 digital interface. Dynamic range exceeds psychoacoustic requirements. Hard to convert accurately at this resolution.

Use of dither

The use of dither in A/D conversion, as well as in conversion between one sample resolution and another, is now widely accepted as correct. It has the effect of linearizing a normal convertor (in other words it effectively makes each quantizing interval the same size) and turns quantizing distortion into a random, noise-like signal at all times. This is desirable for a number of reasons. First, because white noise at a very low level is less subjectively annoying than distortion; second, because it allows signals to be faded smoothly down without the sudden disappearance noted above; and third, because it often allows signals to be reconstructed even when their level is below the noise floor of the system. Undithered audio signals begin to sound 'grainy' and distorted as the signal level falls. Quiescent hiss will disappear if dither is switched off, making a system seem quieter, but a small amount of continuous hiss is considered preferable to low-level distortion. The resolution of modern high-resolution convertors is such that the noise floor is normally inaudible in any case.

Dithering a convertor involves the addition of a very low-level signal to the audio whose amplitude depends upon the type of dither employed (see Fact File 8.9). The dither signal is usually noise, but may also be a

FACT FILE 8.9 TYPES OF DITHER

Research has shown that certain dither signals are more suitable than others for high quality audio work. Dither noise is often characterized in terms of its probability distribution, which is a statistical method of showing the likelihood of the signal having a certain amplitude. A simple graph is used to indicate the shape of the distribution. The probability is the vertical axis and the amplitude in terms of quantizing steps is the horizontal axis.

Logical probability distributions can be understood simply by thinking of the way in which dice fall when thrown (see the diagram). A single throw has a rectangular probability distribution function (RPDF), as shown in (a), because there is an equal chance of the throw being between 1 and 6. The total value of a pair of dice, on the other hand, has a roughly triangular probability distribution function (TPDF), as shown in (b), with the peak grouped on values from 6 to 8, because there are more combinations that make these totals than there are combinations making 2 or 12. Going back to digital electronics, one could liken the dice to random number generators and see that RPDF dither could be created using a single random number generator, and that TPDF dither could be created by adding the outputs of two RPDF generators.

RPDF dither has equal likelihood that the amplitude of the noise will fall anywhere between zero and maximum, whereas TPDF dither has greater likelihood that the amplitude will be zero than that it will be maximum. Although RPDF and TPDF dither can have the effect of linearizing a digital audio system and removing distortion, RPDF dither tends to result in noise modulation at low signal levels. The most suitable dither noise is found to be TPDF with a peak-to-peak amplitude of $2Q$. If RPDF dither is used it should have a peak-to-peak amplitude of $1Q$. Analog white noise has Gaussian probability, whose shape is like a normal distribution curve. With Gaussian noise, the optimum r.m.s. amplitude for the dither signal is $0.5Q$, at which level noise modulation is minimized but not altogether absent. Dither at this level has the effect of reducing the undithered dynamic range by about 6 dB, making the dithered dynamic range of an ideal 16 bit convertor around 92 dB.

(a)

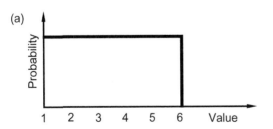

(b)

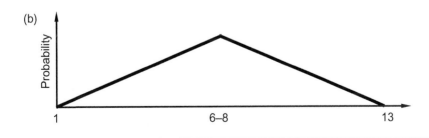

(a) Dither noise added to a sine wave signal prior to quantization. (b) Post-quantization the error signal is now random and noise-like. (Courtesy of Allen Mornington West.)

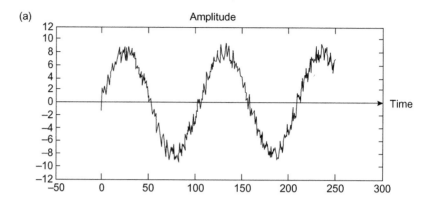

(a)

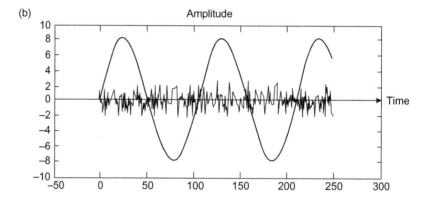

(b)

waveform at half the sampling frequency or a combination of the two. A signal that has not been correctly dithered during the A/D conversion process cannot thereafter be dithered with the same effect, because the signal will have been irrevocably distorted. How then does dither perform the seemingly remarkable task of removing quantizing distortion?

It was stated above that the distortion was a result of the correlation between the signal and the quantizing error, making the error periodic and subjectively annoying. Adding noise, which is a random signal, to the audio has the effect of randomizing the quantizing error and making it noise-like as well (shown in Figure 8.18a and b). If the noise has an amplitude similar in level to the LSB (in other words, one quantizing step) then a signal lying exactly at the decision point between one quantizing interval and the next may be quantized either upwards or downwards, depending on the instantaneous level of the dither noise added to it. Over time this random effect is averaged, leading to a noise-like quantizing error and a fixed noise floor in the system.

Dither is also used in digital processing devices such as mixers, but in such cases it is introduced in the digital domain as a random number sequence (the digital equivalent of white noise). In this context it is used to remove low-level distortion in signals whose gains have been altered and to optimize the conversion from high resolution to lower resolution during post-production.

Oversampling in A/D conversion

Oversampling involves sampling audio at a higher frequency than strictly necessary to satisfy the Nyquist criterion. Normally, though, this high rate is reduced to a lower rate in a subsequent digital filtering process, in order that no more storage space is required than for conventionally sampled audio. It works by trading off quantizing resolution against sampling rate, based on the principle that the information carrying capacity of a channel is related to the product of these two factors. Samples at a high rate with low resolution can be converted into samples at a lower rate with higher resolution, with no overall loss of information. Oversampling has now become so popular that it is the norm in most high-quality audio convertors.

Although oversampling A/D convertors often quote very high sampling rates of up to 128 times the basic rates of 44.1 or 48 kHz, the actual rate at the digital output of the convertor is reduced to a basic rate or a small multiple thereof (e.g. 48, 96 or 192 kHz). Samples acquired at the high rate are quantized to only a few bits' resolution and then digitally filtered to reduce the sampling rate, as shown in Figure 8.19. The digital low-pass filter limits the bandwidth of the signal to half the basic sampling frequency in order to avoid aliasing, and this is coupled with 'decimation'. Decimation reduces the sampling rate by dropping samples from the oversampled stream. A result of the low-pass filtering operation is to increase the word length of the samples very considerably. This is not simply an arbitrary extension of the word length, but an accurate calculation

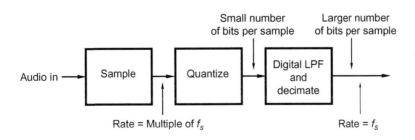

FIGURE 8.19

Block diagram of oversampling A/D conversion process.

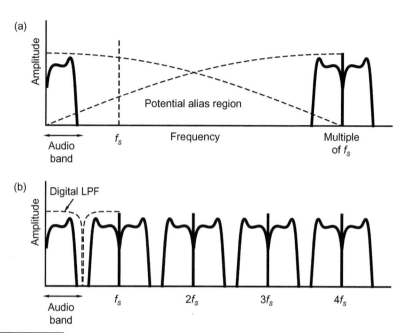

FIGURE 8.20 *(a) Oversampling in A/D conversion initially creates spectral repetitions that lie a long way from the top of the audio baseband. The dotted line shows the theoretical extension of the baseband and the potential for aliasing, but the audio signal only occupies the bottom part of this band. (b) Decimation and digital low-pass filtering limits the baseband to half the sampling frequency, thereby eliminating any aliasing effects, and creates a conventional collection of spectral repetitions at multiples of the sampling frequency.*

of the correct value of each sample, based on the values of surrounding samples. Although oversampling convertors quantize samples initially at a low resolution, the output of the decimator consists of samples at a lower rate with more bits of resolution. The sample resolution can then be shortened as necessary (see 'Requantization', p. 237) to produce the desired word length.

Oversampling brings with it a number of benefits and is the key to improved sound quality at both the A/D and D/A ends of a system. Because the initial sampling rate is well above the audio range (often tens or hundreds of times the nominal rate) the spectral repetitions resulting from PAM are a long way from the upper end of the audio band (see Figure 8.20). The analog anti-aliasing filter used in conventional convertors is replaced by a digital decimation filter. Such filters can be made to have a linear phase response if required, resulting in higher sound quality.

If oversampling is also used in D/A conversion the analog reconstruction filter can have a shallower roll-off. This can have the effect of improving phase linearity within the audio band, which is known to improve audio quality. In oversampled D/A conversion, basic rate audio is up-sampled to a higher rate before conversion and reconstruction filtering. Oversampling also makes it possible to introduce so-called 'noise shaping' into the conversion process, which allows quantizing noise to be shifted out of the most audible parts of the spectrum.

Oversampling without subsequent decimation is a fundamental principle of Sony's Direct Stream Digital system, described below.

Noise shaping in A/D conversion

Noise shaping is a means by which noise within the most audible parts of the audio frequency range is reduced at the expense of increased noise at other frequencies, using a process that 'shapes' the spectral energy of the quantizing noise. It is possible because of the high sampling frequencies used in oversampling convertors. A high sampling frequency extends the frequency range over which quantizing noise is spread, putting much of it outside the audio band.

Quantizing noise energy extends over the whole baseband, up to the Nyquist frequency. Oversampling spreads the quantizing noise energy over a wider spectrum, because in oversampled convertors the Nyquist frequency is well above the upper limit of the audio band. This has the effect of reducing the in-band noise by around 3 dB per octave of oversampling (in other words, a system oversampling at twice the Nyquist rate would see the noise power within the audio band reduced by 3 dB).

In oversampled noise-shaping A/D conversion an integrator (low-pass filter) is introduced before the quantizer, and a D/A convertor is incorporated into a negative feedback loop, as shown in Figure 8.21. This is the so-called 'sigma–delta convertor'. Without going too deeply into the principles of such convertors, the result is that the quantizing noise (introduced after the integrator) is given a rising frequency response at the input to the decimator, whilst the input signal is passed with a flat response. There are clear parallels between such a circuit and analog negative-feedback circuits.

Without noise shaping, the energy spectrum of quantizing noise is flat up to the Nyquist frequency, but with first-order noise shaping this energy spectrum is made non-flat, as shown in Figure 8.22. With second-order noise shaping the in-band reduction in noise is even greater, such that the in-band noise is well below that achieved without noise shaping.

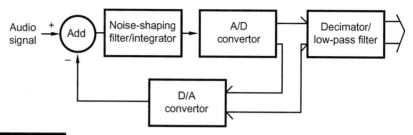

FIGURE 8.21 *Block diagram of a noise shaping delta–sigma A/D convertor.*

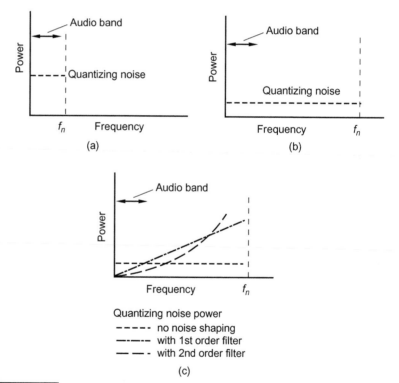

FIGURE 8.22 *Frequency spectra of quantizing noise: In a non-oversampled convertor, as shown in (a), the quantizing noise is constrained to lie within the audio band. In an oversampling convertor, as shown in (b), the quantizing noise power is spread over a much wider range, thus reducing its energy in the audio band. (c) With noise shaping the noise power within the audio band is reduced still further, at the expense of increased noise outside that band.*

D/A CONVERSION

A basic D/A convertor

The basic D/A conversion process is shown in Figure 8.23. Audio sample words are converted back into a staircase-like chain of voltage levels corresponding to the sample values. This is achieved in simple convertors by using the states of bits to turn current sources on or off, making up the required pulse amplitude by the combination of outputs of each of these sources. This staircase is then 'resampled' to reduce the width of the pulses before they are passed through a low-pass reconstruction filter whose cut-off frequency is half the sampling frequency. The effect of the reconstruction filter is to join up the sample points to make a smooth waveform. Resampling is necessary to avoid any discontinuities in signal amplitude at sample boundaries and because otherwise the averaging effect of the filter would result in a reduction in the amplitude of high-frequency audio signals (the so-called 'aperture effect'). Aperture effect may be reduced by limiting the width of the sample pulses to perhaps one-eighth of the sample period. Equalization may be required to correct for aperture effect.

Oversampling in D/A conversion

Oversampling may be used in D/A conversion, as well as in A/D conversion. In the D/A case additional samples must be created in between the Nyquist rate samples in order that conversion can be performed at a higher sampling rate. These are produced by sample rate conversion of the PCM

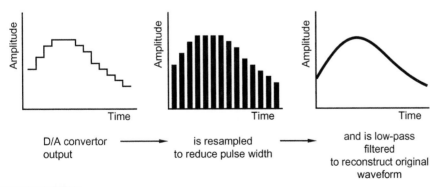

FIGURE 8.23   *Processes involved in D/A conversion (positive sample values only shown).*

data. These samples are then converted back to analog at the higher rate, again avoiding the need for steep analog filters. Noise shaping may also be introduced at the D/A stage, depending on the design of the convertor, to reduce the subjective level of the noise.

A number of advanced D/A convertor designs exist which involve oversampling at a high rate, creating samples with only a few bits of resolution. The extreme version of this approach involves very high rate conversion of single bit samples (so-called 'bit stream conversion'), with noise shaping to optimize the noise spectrum of the signal. The theory of these convertors is outside the scope of this book.

DIRECT STREAM DIGITAL (DSD)

Direct Stream Digital (DSD) is Sony's proprietary name for its 1-bit digital audio coding system that uses a very high sampling frequency (2.8224 MHz as a rule). This system is used for audio representation on the consumer Super Audio CD (SACD) and in various items of professional equipment used for producing SACD material. It is not directly compatible with conventional PCM systems although DSD signals can be down-sampled and converted to multibit PCM if required.

DSD signals are the result of delta–sigma conversion of the analog signal, a technique used at the front end of some oversampling convertors described above. As shown in Figure 8.24, a delta – sigma convertor employs a comparator and a feedback loop containing a low-pass filter that effectively quantizes the difference between the current sample and the accumulated value of previous samples. If it is higher then a '1' results, if it is lower a '0' results. This creates a 1 bit output that simply alternates between one and zero in a pattern that depends on the original signal waveform, as shown in Figure 8.24. Conversion to analog can be as simple a matter as passing the bit stream through a low-pass filter, but is usually somewhat more sophisticated, involving noise shaping and higher order filtering.

Although one would expect 1 bit signals to have an appalling signal-to-noise ratio, the exceptionally high sampling frequency spreads the noise over a very wide frequency range leading to lower noise within the audio band. Additionally, high-order noise shaping is used to reduce the noise in the audio band at the expense of that at much higher (inaudible) frequencies, as discussed earlier. A dynamic range of around 120 dB is therefore claimed, as well as a frequency response extending smoothly to over 100 kHz.

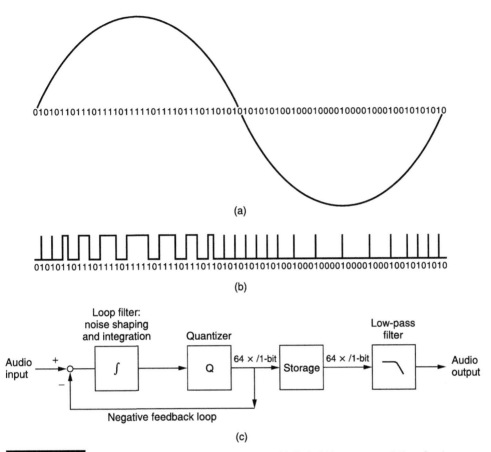

FIGURE 8.24   *Direct Stream Digital bitstream generation. (a) Typical binary representation of a sine wave. (b) Pulse density modulation. (c) DSD signal chain.*

CHANGING THE RESOLUTION OF AN AUDIO SIGNAL (REQUANTIZATION)

There may be points in an audio production when the need arises to change the resolution of a signal. A common example of this in high-quality audio is when mastering 16 bit consumer products from 20 or 24 bit recordings, but it also occurs within signal processors of all types because sample word lengths may vary at different stages. It is important that this operation is performed correctly because incorrect requantization results in unpleasant distortion, just like undithered quantization in A/D conversion. Dynamic range enhancement can also be employed when requantizing for consumer media, as shown in Fact File 8.10.

FACT FILE 8.10 DYNAMIC RANGE ENHANCEMENT

It is possible to maximize the subjective dynamic range of digital audio signals during the process of requantization. This is particularly useful when mastering high resolution recordings for CD because the reduction to 16 bit word lengths would normally result in increased quantizing noise. It is in fact possible to retain most of the dynamic range of a higher resolution recording, even though it is being transferred to a 16 bit medium. This remarkable feat is achieved by a noise-shaping process similar to that described earlier.

During requantization digital filtering is employed to shape the spectrum of the quantizing noise so that as much of it as possible is shifted into the least audible parts of the spectrum. This usually involves moving the noise away from the 4 kHz region where the ear is most sensitive and increasing it at the high-frequency end of the spectrum. The result is often quite high levels of noise at high frequency, but still lying below the audibility threshold. In this way CDs can be made to sound almost as if they had the dynamic range of 20 bit recordings. Some typical weighting curves used in a commercial mastering processor from Meridian are shown in the diagram, although many other shapes are in use. Some approaches allow the mastering

engineer to choose from a number of 'shapes' of noise until one is found that is subjectively the most pleasing for the type of music concerned, whereas others stick to one theoretically derived 'correct' shape.

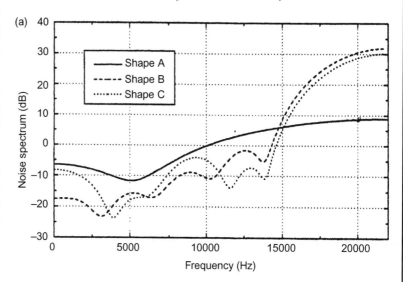

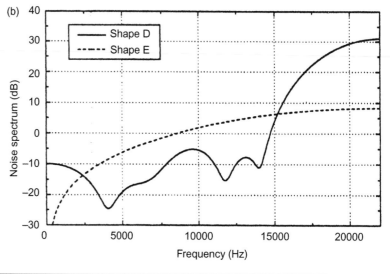

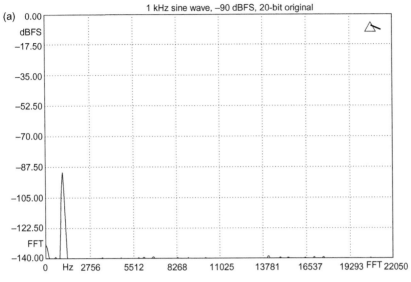

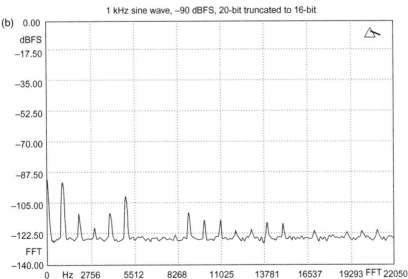

FIGURE 8.25

Truncation of audio samples results in distortion. (a) Shows the spectrum of a 1 kHz signal generated and analyzed at 20 bit resolution. In (b) the signal has been truncated to 16 bit resolution and the distortion products are clearly noticeable.

If the length of audio samples needs to be reduced then the worst possible solution is simply to remove unwanted LSBs. Taking the example of a 20 bit signal being reduced to 16 bits, one should not simply remove the four LSBs and expect everything to be all right. By removing the LSBs one would be creating a similar effect to not using dither in A/D conversion – in other words one would introduce low-level distortion components. Low-level signals would sound grainy and would not fade smoothly into noise. Figure 8.25 shows a

The correct order of events when requantizing an audio signal at a lower resolution is shown here.

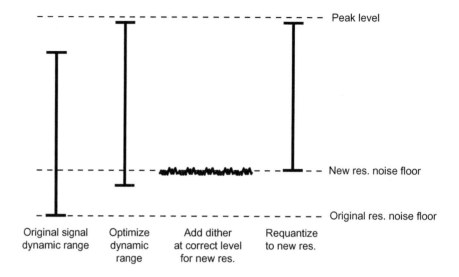

Peak level

New res. noise floor

Original res. noise floor

Original signal dynamic range

Optimize dynamic range

Add dither at correct level for new res.

Requantize to new res.

1 kHz signal at a level of –90 dBFS that originally began life at 20 bit resolution but has been truncated to 16 bits. The harmonic distortion is clearly visible.

The correct approach is to redither the signal for the target resolution by adding dither noise in the digital domain. This digital dither should be at an appropriate level for the new resolution and the LSB of the new sample should then be rounded up or down depending on the total value of the LSBs to be discarded, as shown in Figure 8.26. It is worrying to note how many low-cost digital audio applications fail to perform this operation satisfactorily, leading to complaints about sound quality. Many professional quality audio workstations allow for audio to be stored and output at a variety of resolutions and may make dither user-selectable. They also allow the level of the audio signal to be changed in order that maximum use may be made of the available bits. It is normally important, for example when mastering a CD from a 20 bit recording, to ensure that the highest level signal on the original recording is adjusted during mastering so that it peaks close to the maximum level before requantizing and redithering at 16 bit resolution. In this way as much as possible of the original low-level information is preserved and quantizing noise is minimized. This applies in any requantizing operation, not just CD mastering. A number of applications are available that automatically scale the audio signal so that its level is optimized in this way, allowing the user to set a peak signal value up to which the highest level samples will be scaled. Since some overload detectors on digital meters and CD mastering systems look for repeated samples at maximum level to detect clipping, it is perhaps wise to set peak levels so that they lie just below full modulation. This will ensure that master tapes are not rejected for a suspected recording fault by duplication plants and subsequent users do not complain of 'over' levels.

INTRODUCTION TO DIGITAL SIGNAL PROCESSING

Just as processing operations like equalization, fading and compression can be performed in the analog domain, so they can in the digital domain. Indeed it is often possible to achieve certain operations in the digital domain with fewer side effects such as phase distortion. It is possible to perform operations in the digital domain that are either very difficult or impossible in the analog domain. High-quality, authentic-sounding artificial reverberation is one such example, in which the reflection characteristics of different halls and rooms can be accurately simulated. Digital signal processing (DSP) involves the high-speed manipulation of the binary data representing audio samples. It may involve changing the values and timing order of samples and it may involve the combining of two or more streams of audio data. DSP can affect the sound quality of digital audio in that it can add noise or distortion, although one must assume that the aim of good design is to minimize any such degradation in quality.

In the sections that follow an introduction will be given to some of the main applications of DSP in audio workstations without delving into the mathematical principles involved. In some cases the description is an over-simplification of the process, but the aim has been to illustrate concepts, not to tackle the detailed design considerations involved.

Gain changing (level control)

It is relatively easy to change the level of an audio signal in the digital domain. It is most easy to shift its gain by 6 dB since this involves shifting the whole sample word either one step to the left or right (see Figure 8.27). Effectively the original value has been multiplied or divided by a factor of two. More precise gain control is obtained by multiplying the audio sample value by some other factor representing the increase or decrease in gain. The number of bits in the multiplication factor determines the accuracy of gain adjustment. The result of multiplying two binary numbers together is to create a new sample word which may have many more bits than the original and it is common to find that digital mixers have internal structures capable of handling 32 bit words, even though their inputs and outputs may handle only 20. Because of this, redithering is usually employed

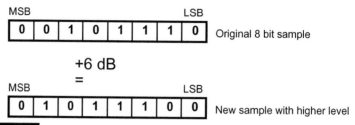

FIGURE 8.27   *The gain of a sample may be changed by 6 dB simply by shifting all the bits one step to the left or right.*

in mixers at points where the sample resolution has to be shortened, such as at any digital outputs or conversion stages, in order to preserve sound quality as described above.

The values used for multiplication in a digital gain control may be derived from any user control such as a fader, rotary knob or on-screen representation, or they may be derived from stored values in an automation system. A simple 'old-fashioned' way of deriving a digital value from an 'analog' fader is to connect the fader to a fixed voltage supply and connect the fader wiper to an A/D convertor, although it is quite common now to find controls capable of providing a direct binary output relating to their position. The 'law' of the fader (the way in which its gain is related to its physical position) can be determined by creating a suitable look-up table of values in memory which are then used as multiplication factors corresponding to each physical fader position.

Mixing

Mixing is the summation of independent data streams representing the different audio channels. Time coincident samples from each input channel are summed to produce a single output channel sample. Clearly it is possible to have many mix 'buses' by having a number of separate summing operations for different output channels. The result of summing a lot of signals may be to increase the overall level considerably and the architecture of the mixer must allow enough headroom for this possibility. In the same way as an analog mixer, the gain structure within a digital mixer must be such that there is an appropriate dynamic range window for the signals at each point in the chain, also allowing for operations such as equalization that change the signal level.

Crossfading is a combination of gain changing and mixing, as described in Fact File 8.11.

FACT FILE 8.11 CROSSFADING

Crossfading is employed widely in audio workstations at points where one section of sound is to be joined to another (edit points). It avoids the abrupt change of waveform that might otherwise result in an audible click and allows one sound to take over smoothly from the other. The process is illustrated conceptually here. It involves two signals each undergoing an automated fade (binary multiplication), one downwards and the other upwards, followed by an addition of the two signals. By controlling the rates and coefficients involved in the fades one can create different styles of crossfade for different purposes.

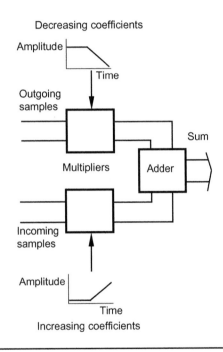

Digital filters and equalization

Digital filtering is something of a 'catch-all' term, and is often used to describe DSP operations that do not at first sight appear to be filtering. A digital filter is essentially a process that involves the time delay, multiplication and recombination of audio samples in all sorts of configurations, from the simplest to the most complex. Using digital filters one can create low- and high-pass filters, peaking and shelving filters, echo and reverberation effects, and even adaptive filters that adjust their characteristics to affect different parts of the signal.

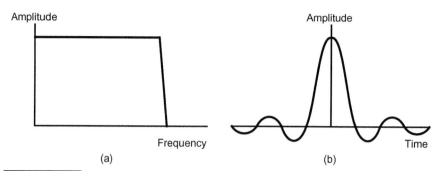

FIGURE 8.28 *Examples of (a) the frequency response of a simple filter, and (b) the equivalent time domain impulse response.*

To understand the basic principle of digital filters it helps to think about how one might emulate a certain analog filtering process digitally. Filter responses can be modeled in two main ways – one by looking at their frequency domain response and the other by looking at their time domain response. (There is another approach involving the so-called z-plane transform, but this is not covered here.) The frequency domain response shows how the amplitude of the filter's output varies with frequency, whereas the time domain response is usually represented in terms of an impulse response (see Figure 8.28). An impulse response shows how the filter's output responds to stimulation at the input by a single short impulse. Every frequency response has a corresponding impulse (time) response because the two are directly related. If you change the way a filter responds in time you also change the way it responds in frequency. A mathematical process known as the Fourier transform is often used as a means of transforming a time domain response into its equivalent frequency domain response. They are simply two ways of looking at the same thing.

Digital audio is time discrete because it is sampled. Each sample represents the amplitude of the sound wave at a certain point in time. It is therefore normal to create certain filtering characteristics digitally by operating on the audio samples in the time domain. In fact if it were desired to emulate a certain analog filter characteristic digitally one would theoretically need only to measure its impulse response and model this in the digital domain. The digital version would then have the same frequency response as the analog version, and one can even envisage the possibility for favorite analog filters to be recreated for the digital workstation. The question, though, is how to create a particular impulse response characteristic digitally, and how to combine this with the audio data.

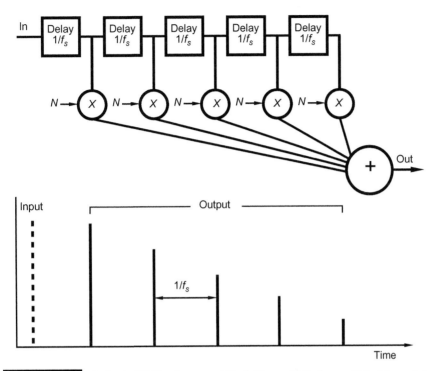

FIGURE 8.29 *A simple FIR filter (transversal filter). N = multiplication coefficient for each tap. Response shown below indicates successive outputs' samples, multiplied by decreasing coefficients.*

As mentioned earlier, all digital filters involve delay, multiplication and recombination of audio samples, and it is the arrangement of these elements that gives a filter its impulse response. A simple filter model is the finite impulse response (FIR) filter, or transversal filter, shown in Figure 8.29. As can be seen, this filter consists of a tapped delay line with each tap being multiplied by a certain coefficient before being summed with the outputs of the other taps. Each delay stage is normally a one sample period delay. An impulse arriving at the input would result in a number of separate versions of the impulse being summed at the output, each with a different amplitude. It is called a finite impulse response filter because a single impulse at the input results in a finite output sequence determined by the number of taps. The more taps there are the more intricate the filter's response can be made, although a simple low-pass filter only requires a few taps.

The other main type is the infinite impulse response (IIR) filter, which is also known as a recursive filter because there is a degree of feedback

A simple IIR filter (recursive filter). The output impulses continue indefinitely but become very small. N in this case is about 0.8. A similar response to the previous FIR filter is achieved but with fewer stages.

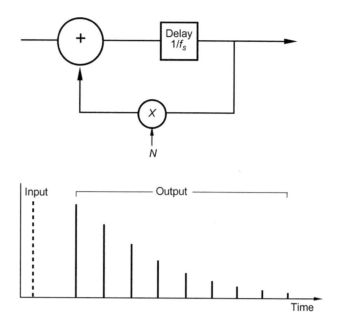

between the output and the input (see Figure 8.30). The response of such a filter to a single impulse is an infinite output sequence, because of the feedback. IIR filters are often used in audio equipment because they involve fewer elements for most variable equalizers than equivalent FIR filters, and they are useful in effects devices. They are unfortunately not phase linear, though, whereas FIR filters can be made phase linear.

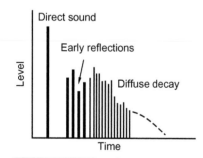

FIGURE 8.31 *The impulse response of a typical reflective room.*

Digital reverberation and other effects

It can probably be seen that the IIR filter described in the previous section forms the basis for certain digital effects, such as reverberation. The impulse response of a typical room looks something like Figure 8.31, that is, an initial direct arrival of sound from the source, followed by a series of early reflections, followed by a diffuse 'tail' of densely packed reflections decaying gradually to almost nothing. Using a number of IIR filters, perhaps together with a few FIR filters, one could create a suitable pattern of delayed and attenuated versions of the original impulse to simulate the decay pattern of a room. By modifying the delays and amplitudes of the early reflections and the nature of the diffuse tail one could simulate different rooms.

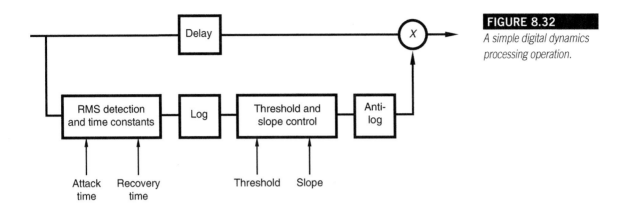

FIGURE 8.32
A simple digital dynamics processing operation.

The design of convincing reverberation algorithms is a skilled task, and the difference between crude approaches and good ones is very noticeable. Some audio workstations offer limited reverberation effects built into the basic software package, but these often sound rather poor because of the limited DSP power available (often processed on the computer's own CPU) and the crude algorithms involved. More convincing reverberation processors are available which exist either as stand-alone devices or as optional plug-ins for the workstation, having access to more DSP capacity and tailor-made software.

Other simple effects can be introduced without much DSP capacity, such as double-tracking and phasing/flanging effects. These often only involve very simple delaying and recombination processes. Pitch shifting can also be implemented digitally, and this involves processes similar to sample rate conversion, as described below. High-quality pitch shifting requires quite considerable horsepower because of the number of calculations required.

Dynamics processing

Digital dynamics processing involves gain control that depends on the instantaneous level of the audio signal. A simple block diagram of such a device is shown in Figure 8.32. A side chain produces coefficients corresponding to the instantaneous gain change required, which are then used to multiply the delayed audio samples. First, the r.m.s. level of the signal must be determined, after which it needs to be converted to a logarithmic value in order to determine the level change in decibels. Only samples above a certain threshold level will be affected, so a constant factor must be added to the values obtained, after which they are multiplied by a factor to represent the compression slope. The coefficient values are then anti-logged to produce linear coefficients by which the audio samples can be multiplied.

Sample rate conversion

Sample rate conversion is necessary whenever audio is to be transferred between systems operating at different rates. The aim is to convert the audio to the new rate without any change in pitch or addition of distortion or noise. These days sample rate conversion can be a very high-quality process, although it is never an entirely transparent process because it involves modifying the sample values and timings. As with requantizing algorithms, it is fairly common to encounter poorly implemented sample rate conversion on low-cost digital audio workstations, often depending very much on the specific software application rather than the hardware involved.

The easiest way to convert from one rate to another is by passing through the analog domain and resampling at the new rate, but this may introduce a small amount of extra noise. The most basic form of digital rate conversion involves the translation of samples at one fixed rate to a new fixed rate, related by a simple fractional ratio. Fractional-ratio conversion involves the mathematical calculation of samples at the new rate based on the values of samples at the old rate. Digital filtering is used to calculate the amplitudes of the new samples such that they are correct based on the impulse response of original samples, after low-pass filtering with an upper limit of the Nyquist frequency of the original sampling rate. A clock rate common to both sample rates is used to control the interpolation process. Using this method, some output samples will coincide with input samples, but only a limited number of possibilities exist for the interval between input and output samples.

If the input and output sampling rates have a variable or non-simple relationship the above does not hold true, since output samples may be required at any interval in between input samples. This requires an interpolator with many more clock phases than for fractional-ratio conversion, the intention being to pick a clock phase that most closely corresponds to the desired output sample instant at which to calculate the necessary coefficient. There will clearly be an error, which may be made smaller by increasing the number of possible interpolator phases. The audible result of the timing error is equivalent to the effects of jitter on an audio signal (see above), and should be minimized in design so that the effects of sample rate conversion are below the noise floor of the signal resolution in hand. If the input sampling rate is continuously varied (as it might be in variable-speed searching or cueing) the position of interpolated samples in relation to original samples must vary also. This requires real-time calculation of the filter phase.

Many workstations now include sample rate conversion as either a standard or optional feature, so that audio material recorded and edited at one rate can be reproduced at another. It is important to ensure that the quality of the sample rate conversion is high enough not to affect the sound quality of your recordings, and it should only be used if it cannot be avoided. Poorly implemented applications sometimes omit to use correct low-pass filtering to avoid aliasing, or incorporate very basic digital filters, resulting in poor sound quality after rate conversion.

Sample rate conversion is also useful as a means of synchronizing an external digital source to a standard sampling frequency reference, when it is outside the range receivable by a workstation.

PITCH SHIFTING AND TIME STRETCHING

Pitch alteration is now a common feature of many digital audio processing systems. It can be used either to modify the musical pitch of notes in order to create alternative versions, including harmony lines, or it can be used to 'correct' the pitch of musical lines such as vocals that were not sung in tune. The most basic way of doing this is to alter the sampling frequency of the audio signal (slowing down the sampling frequency without doing any sophisticated rate conversion will cause the perceived pitch to drop), but this also changes the speed and duration, and the resulting sample data is then at a non-standard sampling frequency. Resampling the original signal in the digital domain at a different frequency, followed by replay at the original sampling frequency, is an alternative, but the speed will be similarly affected. Modern pitch alteration algorithms are usually much more sophisticated than this and can alter the pitch without altering the speed. Time stretching is a related application, and involves altering the duration of a clip without altering its pitch. Pitch correction algorithms attempt to identify the fundamental pitch of individual notes in a phrase and quantize them to a fixed pitch scale in order to force melodies into tune. This can be done with varying degrees of severity and musicality, leading to results anywhere on a range from crude effect to subtle correction of tuning.

Both effects can be achieved by transforming a signal into the frequency domain, modifying it and resynthesizing it in the time domain. Techniques based on phase vocoding or spectral modeling are sometimes used. Both approaches succeed to some extent in enabling pitch and time information to be analyzed and modified independently, although with varying side effects depending on the content of the signal and the parameters of the processing. The signal is transformed to the frequency domain in overlapping blocks using a short time Fourier transform (STFT). It then becomes possible to modify the signal in the frequency domain, for example by scaling certain spectral components, before performing an inverse transform to return it to the time domain with modified pitch. Alternatively the original spectral components can be resynthesized with a new time scale at the stage of the inverse transform in order to change the duration. In the time domain, time stretch processing typically involves identifying the fundamental period of the wave and extracting or adding individual cycles with crossfades, to shorten or lengthen the clip. It may also involve removing or adding samples in silent gaps between notes or phrases. (The latter is particularly used in algorithms that attempt to fit overdubbed speech dynamically to a guide track, for movie sound applications.)

In general, both time and pitch-shifting only work successfully over a limited range of 30% or so either way, although occasionally they can be made to work over wider ranges depending on the nature of the signal and the sophistication of the algorithm. One problem with simple pitch shifting is the well-known 'Pinky and Perky' effect that can be noticed when shifting musical sounds (particularly voices) too far from their original frequency, making them sound unnatural. This is because real musical sounds and human voices have so-called formants, which are peaks and troughs in the spectrum that are due to influences like resonances in the instrument or vocal tract. These give a voice its unique character, for example. When the pitch is shifted the formant structure can become shifted too, so that the peaks and troughs are no longer in the right place in the frequency spectrum for the voice in question. Some sophisticated pitch shifting algorithms therefore employ advanced signal processing methods that can identify the so-called spectral envelope of an instrument or voice (its pattern of peaks and troughs in the frequency spectrum), and attempt to retain this even when the pitch is shifted. In this way a voice can be made to sound like the original singer even when shifted over quite a wide range.

AUDIO DATA REDUCTION

Conventional PCM audio has a high data rate, and there are many applications for which it would be an advantage to have a lower data rate without much (or any) loss of sound quality. 16 bit linear PCM at a sampling rate of 44.1 kHz ('CD quality digital audio') results in a data rate of about 700 kbit/s. For multimedia applications, broadcasting, communications and some consumer purposes (e.g. streaming over the Internet) the data rate may be reduced to a fraction of this with minimal effect on the perceived sound quality. At very low rates the effect on sound quality is traded off with the bit rate required. Simple techniques for reducing the data rate, such as reducing the sampling rate or number of bits per sample would have a very noticeable effect on sound quality, so most modern low bit rate coding works by exploiting the phenomenon of auditory masking to 'hide' the increased noise resulting from bit rate reduction in parts of the audio spectrum where it will hopefully be inaudible. There are a number of types of low bit rate coding used in audio systems, working on similar principles, and used for applications such as consumer disk and tape systems (e.g. Sony ATRAC), digital cinema sound (e.g. Dolby Digital, Sony SDDS, DTS) and multimedia applications (e.g. MPEG).

Why reduce the data rate?

Nothing is inherently wrong with linear PCM from a sound quality point of view, indeed it is probably the best thing to use. The problem is simply that the data rate is too high for a number of applications. Two channels of linear PCM require a rate of around 1.4 Mbit/s, whereas applications such as Digital Audio Broadcasting (DAB) or Digital Radio need it to be more like 128 kbit/s (or perhaps lower for some applications) in order to fit sufficient channels into the radio frequency spectrum – in other words more than ten times less data per second. Some Internet streaming applications need it to be even lower than this, with rates down in the low tens of kilobits per second for mobile communications.

The efficiency of mass storage media and data networks is related to their data transfer rates. The more data can be moved per second, the more audio channels may be handled simultaneously; the faster a disk can be copied, the faster a sound file can be transmitted across the world. In reducing the data rate that each audio channel demands, one also reduces the requirement for such high specifications from storage media and networks, or alternatively one can obtain greater functionality from the same specification. A network connection capable of handling eight channels of linear PCM simultaneously could be made to handle, say, 48 channels of data-reduced audio, without unduly affecting sound quality.

Although this sounds like magic and makes it seem as if there is no point in continuing to use linear PCM, it must be appreciated that the data reduction is achieved by throwing away data from the original audio signal. The more data is thrown away the more likely it is that unwanted audible effects will be noticed. The design aim of most of these systems is to try to retain as much as possible of the sound quality whilst throwing away as much data as possible, so it follows that one should always use the least data reduction necessary, where there is a choice.

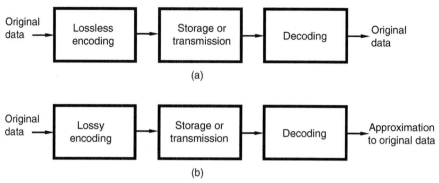

FIGURE 8.33   *(a) In lossless coding the original data is reconstructed perfectly upon decoding, resulting in no loss of information. (b) In lossy coding the decoded information is not the same as that originally coded, but the coder is designed so that the effects of the process are minimal.*

Lossless and lossy coding

There is an important distinction to be made between the type of data reduction used in some computer applications and the approach used in many audio coders. The distinction is really between 'lossless' coding and coding which involves some loss of information (see Figure 8.33). It is quite common to use data compression on computer files in order to fit more information onto a given disk or tape, but such compression is usually lossless in that the original data is reconstructed bit for bit when the file is decompressed. A number of tape backup devices for computers have a compression facility for increasing the apparent capacity of the medium, for example. Methods are used which exploit redundancy in the information, such as coding a string of 80 zeros by replacing them with a short message stating the value of the following data and the number of bytes involved. This is particularly relevant in single-frame bit-mapped picture files where there may be considerable runs of black or white in each line of a scan, where nothing in the image is changing. One may expect files compressed using off-the-shelf PC data compression applications to be reduced to perhaps 25–50% of their original size, but it must be remembered that they are often dealing with static data, and do not have to work in real time. Also, it is not normally acceptable for decompressed computer data to be anything but the original data.

It is possible to use lossless coding on audio signals. Lossless coding allows the original PCM data to be reconstructed perfectly by the decoder and is therefore 'noiseless' since there is no effect on audio quality. The data reduction obtained using these methods ranges from nothing to about 2.5:1

and is variable depending on the program material. This is because audio signals have an unpredictable content, do not make use of a standard limited character set, and do not spend long periods of time in one binary state or the other. Although it is possible to perform this reduction in real time, the coding gains are not sufficient for many applications. Nonetheless, a halving in the average audio data rate is certainly a useful saving. A form of lossless data reduction known as Direct Stream Transfer (DST) can be used for Super Audio CD in order to fit the required multichannel audio data into the space available. A similar system was designed for DVD-Audio, called MLP (Meridian Lossless Packing), which evolved into Dolby TrueHD, used as the lossless coding scheme for the Blu-Ray disc.

'Noisy' or lossy coding methods make possible a far greater degree of data reduction, but require the designer and user to arrive at a compromise between the degree of data reduction and potential effects on sound quality. Here data reduction is achieved by coding the signal less accurately than in the original PCM format (using fewer bits per sample), thereby increasing quantizing noise, but with the intention that increases in noise will be 'masked' (made inaudible) by the signal. The original data is not reconstructed perfectly on decoding. The success of such techniques therefore relies on being able to model the characteristics of the human hearing process in order to predict the masking effect of the signal at any point in time – hence the common term 'perceptual coding' for this approach. Using detailed psychoacoustic models it is possible to code high-quality audio at rates under 100 kbit/s per channel with minimal effects on audio quality. Higher data rates, such as 192 kbit/s, can be used to obtain an audio quality that is demonstrably indistinguishable from the original PCM.

MPEG – an example of lossy coding

The following is a very brief overview of how one approach works, based on the technology involved in the MPEG (Moving Pictures Expert Group) standards.

As shown in Figure 8.34, the incoming digital audio signal is filtered into a number of narrow frequency bands. Parallel to this a computer model of the human hearing process (an auditory model) analyzes a short portion of the audio signal (a few milliseconds). This analysis is used to determine what parts of the audio spectrum will be masked, and to what degree, during that short time period. In bands where there is a strong signal, quantizing noise can be allowed to rise considerably without it being heard, because one signal is very efficient at masking another lower level

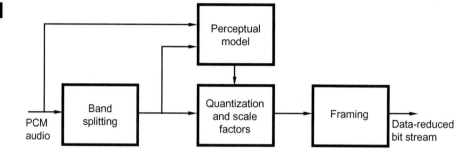

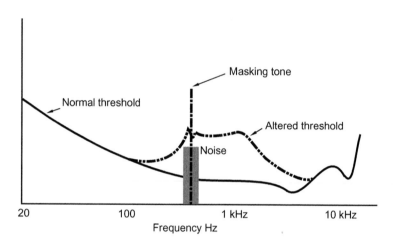

signal in the same band as itself (see Figure 8.35). Provided that the noise is kept below the masking threshold in each band it should be inaudible.

Blocks of audio samples in each narrow band are scaled (low-level signals are amplified so that they use more of the most significant bits of the range) and the scaled samples are then reduced in resolution (requantized) by reducing the number of bits available to represent each sample – a process that results in increased quantizing noise. The output of the auditory model is used to control the requantizing process so that the sound quality remains as high as possible for a given bit rate. The greatest number of bits is allocated to frequency bands where noise would be most audible, and the fewest to those bands where the noise would be effectively masked by the signal. Control information is sent along with the blocks of bit rate-reduced samples to allow them to be reconstructed at the correct level and resolution upon decoding.

The above process is repeated every few milliseconds, so that the masking model is constantly being updated to take account of changes in the

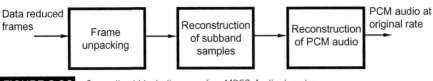

FIGURE 8.36   *Generalized block diagram of an MPEG-Audio decoder.*

Table 8.2	MPEG-1 layers			
Layer	**Complexity**	**Min. delay**	**Bit rate range**	**Target**
1	Low	19 ms	32–448 kbit/s	192 kbit/s
2	Moderate	35 ms	32–384 kbit/s*•	128 kbit/s
3	High	59 ms	32–320 kbit/s	64 kbit/s

•In Layer 2, bit rates of 224 kbit/s and above are for stereo modes only

audio signal. Carefully implemented, such a process can result in a reduction of the original data rate to anything from about one-quarter to less than one-tenth. A decoder uses the control information transmitted with the bit rate-reduced samples to restore the samples to their correct level and can determine how many bits were allocated to each frequency band by the encoder, reconstructing linear PCM samples and then recombining the frequency bands to form a single output (see Figure 8.36). A decoder can be much less complex, and therefore cheaper, than an encoder, because it does not need to contain the auditory model.

A standard known as MPEG-1, published by the International Standards Organization (ISO 11172-3), defines a number of 'layers' of complexity for low bit rate audio coders as shown in Table 8.2. Each of the layers can be operated at any of the bit rates within the ranges shown (although some of the higher rates are intended for stereo modes) and the user must make appropriate decisions about what sound quality is appropriate for each application. The lower the data rate, the lower the sound quality that will be obtained. At high data rates the encoding–decoding process has been judged by many to be audibly 'transparent' – in other words listeners cannot detect that the coded and decoded signal is different from the original input. The target bit rates were for 'transparent' coding.

'MP3' will be for many people the name associated with downloading music files from the Internet. The term MP3 has caused some confusion; it is short for MPEG-1 Layer 3, but MP3 has virtually become a generic term for the system used for receiving compressed audio from the Internet. There is also MPEG-2 which can handle multichannel surround, and further developments in this and later systems will be briefly touched upon.

MPEG-2 BC (Backwards Compatible with MPEG-1) additionally supports sampling frequencies from 16 kHz to 22.05 kHz and 24 kHz at bit rates from 32 to 256 kbit/s for Layer 1. For Layers 2 and 3, bit rates are from 8 to 160 kbit/s. Developments, intended to supersede MPEG-2 BC, have included MPEG-2 AAC (Advanced Audio Coding). This defines a standard for multichannel coding of up to 48 channels, with sampling rates from 8 kHz to 96 kHz. It also incorporates a Modified Discrete Cosine transform system as used in the MiniDisc coding format (ATRAC). MPEG-2AAC was not, however, designed to be backwards compatible with MPEG-1.

MPEG-4 'natural audio coding' is based on the standards outlined for MPEG-2 AAC; it includes further coding techniques for reducing transmission bandwidth and it can scale the bit rate according to the complexity of the decoder. This is used in Apple's iPod, for example. There are also intermediate levels of parametric representation in MPEG-4 such as used in speech coding, whereby speed and pitch of basic signals can be altered over time. One has access to a variety of methods of representing sound at different levels of abstraction and complexity, all the way from natural audio coding (lowest level of abstraction), through parametric coding systems based on speech synthesis and low-level parameter modification (see below), to fully synthetic audio objects.

When audio signals are described in the form of 'objects' and 'scenes', it requires that they be rendered or synthesized by a suitable decoder. Structured Audio (SA) in MPEG-4 enables synthetic sound sources to be represented and controlled at very low bit rates (less than 1 kbit/s). An SA decoder can synthesize music and sound effects. SAOL (Structured Audio Orchestra Language), as used in MPEG-4, was developed at MIT and is an evolution of CSound (a synthesis language used widely in the electroacoustic music and academic communities). It enables 'instruments' and 'scores' to be downloaded. The instruments define the parameters of a number of sound sources that are to be rendered by synthesis (e.g. FM, wavetable, granular, additive) and the 'score' is a list of control information that governs what those instruments play and when (represented in the SASL or Structured Audio Score Language format). This is rather like a more refined version of the established MIDI control protocol, and indeed MIDI can be used if required for basic music performance control. This is discussed further in Chapter 14.

Sound scenes, as distinct from sound objects, are usually made up of two elements – that is, the sound objects and the environment within which they are located. Both elements are integrated within one part of

MPEG-4. This part of MPEG-4 uses so-called BIFS (Binary Format for Scenes) for describing the composition of scenes (both visual and audio). The objects are known as nodes and are based on VRML (virtual reality modeling language). So-called Audio BIFS can be post-processed and represents parametric descriptions of sound objects. Advanced Audio BIFS also enables virtual environments to be described in the form of perceptual room acoustics parameters, including positioning and directivity of sound objects. MPEG-4 audio scene description distinguishes between physical and perceptual representation of scenes, rather like the low- and high-level description information mentioned above.

Parametric audio coding

One variant on lossy low bit rate coding involves the encoding of audio signals in the form of a 'core' signal alongside a sparse stream of 'parameters' that describe one or more features needed to reconstruct an approximation of the original signals. The idea is to code a basic version of the original audio signal, that could be decoded by compatible decoders, and to transmit spatial or spectral enhancements in the form of much lower bit rate 'side' information. For example, parametric stereo coding transmits a basic mono channel plus side parameters that represent the inter-channel intensity, time and correlation differences that enable the spatial impression of the original two-channel stereo to be reconstructed at the decoder. MPEG-Surround transmits a mono or stereo downmix of the original surround, plus side information to enable the surround spatial impression to be approximated upon decoding (see Figure 8.37). The downmix is encoded using a 'legacy' or conventional stereo coder such as MP3. The additional bit rate required for the side information is usually only a few kilobits per second, as opposed to the few hundred that might be needed to transmit the surround information as conventionally coded audio. This enables convincing surround to be transmitted at bit rates as low as 64 kilobits per second.

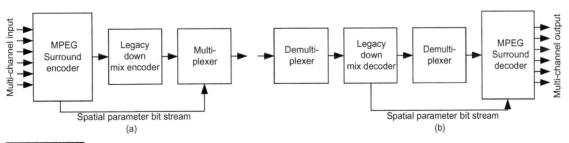

FIGURE 8.37   *Block diagram of MPEG Surround process showing (a) encoder, and (b) decoder. (Courtesy of Jeroen Breebaart)*

Another example is MPEG AAC+ or HE-AAC, which employs a method known as Spectral Band Replication (SBR) to reconstruct the (untransmitted) high-frequency part of the spectrum in the decoder. This achieves greater coding efficiency by transmitting low bit rate side information to describe only the shape and tonality of the high frequency part of the original signal's spectrum. The decoder attempts a synthesis of the missing high frequency content based on lower frequency parts of the spectrum and information in the side parameters. Psychoacoustically the brain does not seem to be as fussy about the precise detail of the very highest frequencies as it is about the lower parts of the spectrum, so this can sound quite convincing while saving a lot of bits.

Surround coding formats

Dolby Digital or AC-3 encoding was developed as a means of delivering 5.1-channel surround to cinemas or the home without the need for analog matrix encoding. The AC-3 coding algorithm can be used for a wide range of different audio signal configurations and bit rates from 32 kbit/s for a single mono channel up to 640 kbit/s for surround signals. It is used widely for the distribution of digital sound tracks on 35 mm movie films, the data being stored optically in the space between the sprocket holes on the film.

It is sufficient to say here that the process involves a number of techniques by which the data representing audio from the source channels is transformed into the frequency domain and requantized to a lower resolution, relying on the masking characteristics of the human hearing process to hide the increased quantizing noise that results from this process. A common bit pool is used so that channels requiring higher data rates than others can trade their bit rate requirements provided that the overall total bit rate does not exceed the constant rate specified.

Aside from the representation of surround sound in a compact digital form, Dolby Digital includes a variety of operational features that enhance system flexibility and help adapt replay to a variety of consumer situations. These include dialog normalization ('dialnorm') and the option to include dynamic range control information alongside the audio data for use in environments where background noise prevents the full dynamic range of the source material being heard. Downmix control information can also be carried alongside the audio data in order that a two-channel version of the surround sound material can be reconstructed in the decoder. As a rule, Dolby Digital data is stored or transmitted with the highest number of channels needed for the end product to be represented and any compatible downmixes are created in the decoder. This differs from some other systems where a two-channel downmix is carried alongside the surround information.

Dialnorm indication can be used on broadcast and other material to ensure that the dialog level remains roughly constant from program to program. It is assumed that dialog level is the main factor governing the listening level used in people's homes, and that they do not want to keep changing this as different programs come on the air (e.g. from advertising to news programs). The dialnorm level is the average dialog level over the duration of the program compared to the maximum level that would be possible, measured using an A-weighted L_{EQ} reading (this averages the level linearly over time). So, for example, if the dialog level averaged at 70 dBA over the program, and the SPL corresponding to peak recording level was 100 dBA, the dialnorm setting would be −30 dB.

The original DTS (Digital Theater Systems) 'Coherent Acoustics' system is another digital signal coding format that can be used to deliver surround sound in consumer or professional applications, using low bit rate coding techniques to reduce the data rate of the audio information. The DTS system can accommodate a wide range of bit rates from 32 kbit/s up to 4.096 Mbit/s (somewhat higher than Dolby Digital), with up to eight source channels and with sampling rates up to 192 kHz. Variable bit rate and lossless coding are also optional. Downmixing and dynamic range control options are provided in the system. Because the maximum data rate is typically somewhat higher than that of Dolby Digital or MPEG, a greater margin can be engineered between the signal and any artefacts of low bit rate coding, leading to potentially higher sound quality. Such judgments, though, are obviously up to the individual and it is impossible to make blanket statements about comparative sound quality between systems.

SDDS stands for Sony Dynamic Digital Sound, and was the third of the main competing formats for digital film sound. Using Sony's ATRAC data reduction system (also used on MiniDiscs), it too encodes audio data with a substantial saving in bit rate compared with the original PCM (about 5:1 compression). It is not as widely encountered as the other two and does not appear to be available in consumer decoders.

Of the MPEG multichannel coding formats, the MPEG-2 BC (backwards compatible) version worked by encoding a matrixed downmix of the surround channels and the center channel into the left and right channels of an MPEG-1 compatible frame structure. Although MPEG-2 BC was originally intended for use with DVD releases in Region 2 countries (primarily Europe), this requirement was dropped in favor of Dolby Digital. MPEG-2 AAC, on the other hand, is a more sophisticated algorithm that codes multichannel audio to create a single bit stream that represents all the channels, in a form that cannot be decoded by an MPEG-1 decoder. Having dropped the requirement for backwards compatibility, the bit rate can now be optimized by coding the channels as a group and taking advantage of interchannel redundancy if required. The MPEG-2 AAC system contained contributions from a wide range of different manufacturers. The parametric MPEG Surround standard was described briefly in the previous section.

Spatial audio object coding

The Spatial Audio Object Coding (SAOC) standard was published as MPEG-D Part 2 - ISO/IEC 23003-2 in 2010 (Part 1 is MPEG Surround, Part 2 is Unified speech and audio coding). It describes a user-controllable rendering of multiple audio objects based on transmission of a mono or stereo downmix of the object signals. SAOC encodes Object Level Differences (OLD), Inter-Object Cross Coherences (IOC) and Downmix Channel Level Differences (DCLD) into a parameter bitstream, and so does not discretely encode input audio signals. An MPEG surround decoder can be manipulated by the user to place the audio objects in the desired positions and at different levels, or attenuated and replaced. An increase in level and/or repositioning of an object can also improve intelligibility with certain speaker layouts and environments. The SAOC bitstream is independent of loudspeaker configuration, and a default downmix option ensures backwards compatibility.

High resolution data-reduced formats

The increased interest in high resolution or 'HD' (high definition) audio distribution to consumers, either on Blu-Ray Disc (BD) or as downloads, has given rise to a number of data-reduced coding formats that are designed specifically for the purpose. Some are lossy (but not very lossy) and some are lossless.

Dolby's TrueHD, based on Meridian Lossless Packing (MLP), is a lossless codec resulting in decoded quality that is identical to the studio master. It enables 7.1-channel playback on BD although it has the capacity to support more than 16 channels of audio. Operating at data rates of up to 18 Mbit/s, it supports the BD standard's requirement for eight full range channels at 96 kHz/24 bits and up to 5.1 channels at 192 kHz/24 bits. An entirely separate artistic stereo mix can be carried if desired. Dolby makes a client–server-based encoder for its HD audio options as well as a stand-alone version.

Dolby Digital Plus is an extension to AC-3 (Dolby Digital), with higher data-rate options and shorter frames if required. It is designed to offer enhanced quality to Dolby Digital, running at data rates up to 6 Mbit/s, although the typical data rate on HD optical disks is said to be between 768 kbit/s and 1.5 Mbit/s. The data stream can be decoded by legacy receivers, which will only decode the Dolby Digital core at up to 640 kbit/s.

DTS offers two codecs that can be used for higher resolution audio on optical disks. Both are backwards compatible with the original DTS Digital Surround decoder because they are based on a lossy core plus extension model. Some other lossless formats take a similar form, for backwards compatibility, whereas others are lossless from the bottom up. DTS-HD High Resolution Audio offers data rates from 2 to 6 Mbit/s, offering quality that is not identical to the studio master but claimed to be close (it's still a lossy coding format). This version allows for a maximum of 7.1 channels at 96 kHz in a CBR (constant bit-rate) stream. DTS-HD Master Audio operates at data rates up to 24.5 Mbit/s in a variable bit rate (VBR) stream, offering 7.1 channels at 96 kHz, or 5.1 at 192 kHz. This version is lossless, and therefore bit-for-bit compatible with the original master. The core coding, which works at up to 1509 kbit/s with 6.1 channels, is at a higher bit rate than typical DVD audio data rates, so non-HD players still get a quality increase. This data stream can be routed to legacy AV receivers using a SPDIF connection. A tool is available (Neural Upmix) that enables one to upmix creatively from 5.1 to surround formats with higher numbers of channels. The encoder enables one to set the downmix coefficients from surround to stereo. There is also a QC control tool that enables one to hear the effect of conversion of 5.1 material to different loudspeaker layouts such as nonstandard 7.1 speaker positions where there are sides and rears.

Free Lossless Audio Encoding (FLAC) is an open-source lossless coding option with data-reduction performance that is very similar to other codecs covered by IP rights. It is claimed to offer fast encoding and decoding with low complexity and is implemented in a lot of software and hardware players used with downloaded audio files. Not all players will decode FLAC files at sampling frequencies above 48 kHz, and only a limited number will handle 192 kHz.

High Definition AAC (HD AAC) has a lossy core accompanied by a lossless extension that enables decoding to provide bit-for-bit compatibility with the original master recording. The AAC core part is compatible with existing decoders in mobile devices such as the iPod and iTunes. It can operate at sampling rates up to 192 kHz and at 24-bit resolution.

RECOMMENDED FURTHER READING

Bosi, M., Goldberg, R., 2003. Introduction to Digital Audio Coding and Standards. Springer.

Pohlmann, K., 2010. Principles of Digital Audio. TAB Electronics.

Watkinson, J., 2001. The Art of Digital Audio, third edition. Focal Press.

Zölzer, U., 2008. Digital Audio Signal Processing. Wiley Blackwell.

Digital Recording, Editing and Mastering Systems

This chapter describes digital audio recording systems and the principles of digital audio editing and mastering.

DIGITAL TAPE RECORDING

Although it is still possible to find examples of dedicated digital tape recording formats in use, they have largely been superseded by recording systems that use computer mass storage media. The economies of scale of the computer industry have made data storage relatively cheap and there is no longer a strong justification for systems dedicated to audio purposes. Tape has a relatively slow access time, because it is a linear storage medium. However, a dedicated tape format can easily be interchanged between recorders, provided that another machine operating to the same standard can be found. Computer mass storage media, on the other hand, come in a very wide variety of sizes and formats, and there are numerous levels at which compatibility must exist between systems before interchange can take place. This matter is discussed in the next chapter.

Background to digital tape recording

When commercial digital audio recording systems were first introduced in the 1970s and early 1980s it was necessary to employ recorders with sufficient bandwidth for the high data rates involved (a machine capable of handling bandwidths of a few megahertz was required). Analog audio tape recorders were out of the question because their bandwidths extended only up to around 35 kHz at best, so video tape recorders (VTRs) were often utilized because of their wide recording bandwidth. PCM adaptors converted digital audio data into a waveform which resembled a television waveform, suitable for recording on to a VTR. The Denon company of Japan developed such a system in partnership with the NHK broadcasting organization and they released the world's first PCM recording onto LP in 1971. In the early 1980s, devices such as Sony's PCM-F1 became available at modest prices, allowing 16 bit, 44.1 kHz digital audio to be recorded on to a consumer VTR, resulting in widespread proliferation of stereo digital recording. Dedicated open-reel digital recorders using stationary heads were also developed (see Fact File 9.1). High-density tape formulations were then

FACT FILE 9.1 ROTARY AND STATIONARY HEADS

There are two fundamental mechanisms for the recording of digital audio on tape, one which uses a relatively low linear tape speed and a quickly rotating head, and one which uses a fast linear tape speed and a stationary head. In the rotary-head system the head either describes tracks almost perpendicular to the direction of tape travel, or it describes tracks which are almost in the same plane as the tape travel. The former is known as transverse scanning, and the latter is known as helical scanning, as shown in (a). Transverse scanning uses more tape when compared with helical scanning. It is not common for digital tape recording to use the transverse scanning method. The reason for using a rotary head is to achieve a high head-to-tape speed, since it is this which governs the available bandwidth. Rotary-head recordings cannot easily be splice-edited because of the track pattern, but they can be electronically edited using at least two machines.

Stationary heads allow the design of tape machines that are very similar in many respects to analog transports. With stationary-head recording it is possible to record a number of narrow tracks in parallel across the width of the tape, as shown in (b). Tape speed can be traded off against the number of parallel tracks used for each audio channel, since the required data rate can be made up by a combination of recordings made on separate tracks. This approach was used in the DASH format, where the tape speed could be 30 ips (76 cm s^{-1}) using one track per channel, 15 ips using two tracks per channel, or 7.5 ips using four tracks per channel.

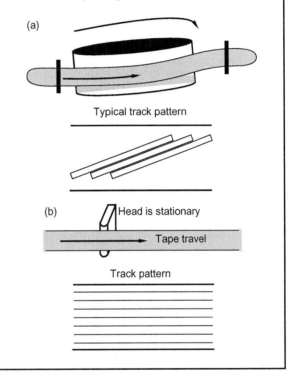

(a)

Typical track pattern

(b) Head is stationary

Tape travel

Track pattern

manufactured for digital use, and this, combined with new channel codes (see below), improvements in error correction and better head design, led to the use of a relatively low number of tracks per channel, or even single-track recording of a given digital signal, combined with playing speeds of 15 or 30 inches per second. Dedicated rotary-head systems, not based on a VTR, were also developed – the R-DAT format being the most well-known.

Digital recording tape is thinner (27.5 microns) than that used for analog recordings; long playing times can be accommodated on a reel, but also thin tape contacts the machine's heads more intimately than does standard 50 micron thickness tape which tends to be stiffer. Intimate contact is essential for reliable recording and replay of such a densely packed and high bandwidth signal.

FACT FILE 9.2 DATA RECOVERY

Channel-coded data must be decoded on replay, but first the audio data must be separated from the clock information which was combined with it before recording. This process is known as data and sync separation, as shown in (a).

It is normal to use a phase-locked loop for the purpose of regenerating the clock signal from the replayed data, as shown in (b), this being based around a voltage-controlled oscillator (VCO) which runs at some multiple of the off-tape clock frequency. A phase comparator compares the relative phases of the divided VCO output and the clock data off tape, producing a voltage proportional to the error which controls the frequency of the VCO. With suitable damping, the phase-locked oscillator will 'flywheel' over short losses or irregularities of the off-tape clock.

Recorded data is usually interspersed with synchronizing patterns in order to give the PLL in the data separator a regular reference in the absence of regular clock data from the encoded audio signal, since many channel codes have long runs without a transition. Even if the off-tape data and clock have timing irregularities, such as might manifest themselves as 'wow' and 'flutter' in analog reproducers (see Chapter 18), these can be removed in digital systems. The erratic data (from tape or disk, for example) is written into a short-term solid state memory (RAM) and read out again a fraction of a second later under control of a crystal clock (which has an exceptionally stable frequency), as shown in (c). Provided that the average rate of input to the buffer is the same as the average rate of output, and the buffer is of sufficient size to soak up short-term irregularities in timing, the buffer will not overflow or become empty.

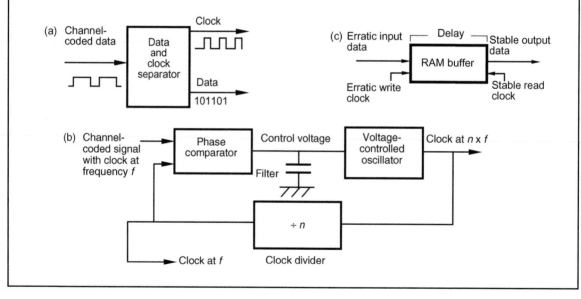

Channel coding for dedicated tape formats

Since 'raw' binary data is normally unsuitable for recording directly by dedicated digital recording systems, a 'channel code' is used which matches the data to the characteristics of the recording system, uses storage space efficiently, and makes the data easy to recover on replay. A wide range of channel codes exists, each with characteristics designed for a specific purpose. The channel code converts a pattern of binary data into a different pattern of transitions in the recording or transmission medium. It is another stage of modulation, in effect. Thus the pattern of bumps in the optical surface of a CD bears little resemblance to the original audio data, and the pattern of magnetic flux transitions on a DAT cassette would be similarly different. Given the correct code book, one could work out what audio data was represented by a given pattern from either of these systems.

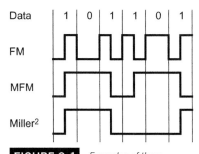

FIGURE 9.1   *Examples of three channel codes used in digital recording. Miller-squared is the most efficient of those shown since it involves the smallest number of transitions for the given data sequence.*

Some examples of channel codes used in audio systems are shown in Figure 9.1. FM is the simplest, being an example of binary frequency modulation. It is otherwise known as 'bi-phase mark', one of the Manchester codes, and is the channel code used by SMPTE/EBU timecode (see Chapter 15). MFM and Miller-squared are more efficient in terms of recording density. MFM is more efficient than FM because it eliminates the transitions between successive ones, only leaving them between successive zeros. Miller-squared eliminates the DC content present in MFM by removing the transition for the last one in an even number of successive ones.

Group codes, such as that used in the Compact Disc and R-DAT, involve the coding of patterns of bits from the original audio data into new codes with more suitable characteristics, using a look-up table or 'code book' to keep track of the relationship between recorded and original codes. This has clear parallels with coding as used in intelligence operations, in which the recipient of a message requires the code book to be able to understand the message. CD uses a method known as 8-to-14 modulation, in which 16 bit audio sample words are each split into two 8 bit words, after which a code book is used to generate a new 14 bit word for each of the 256 possible combinations of 8 bits. Since there are many more words possible with 14 bits than with 8, it is possible to choose those which have appropriate characteristics for the CD recording channel. In this case, it is those words which have no more than 11 consecutive bits in the same state, and no less than three. This limits the bandwidth of the recorded data, and makes it suitable for the optical pickup process, whilst retaining the necessary clock content.

Error correction

There are two stages to the error correction process used in digital tape recording systems. First, the error must be detected, and then it must be corrected. If it cannot be corrected then it must be concealed. In order for the error to be detected it is necessary to build in certain protection mechanisms.

Two principal types of error exist: the burst error and the random error. Burst errors result in the loss of many successive samples and may be due to major momentary signal loss, such as might occur at a tape drop-out or at an instant of impulsive interference such as an electrical spike induced in a cable or piece of dirt on the surface of a CD. Burst error correction capability is usually quoted as the number of consecutive samples which may be corrected perfectly. Random errors result in the loss of single samples in randomly located positions, and are more likely to be the result of noise or poor signal quality. Random error rates are normally quoted as an average rate, for example 1 in 10^6. Error correction systems must be able to cope with the occurrence of both burst and random errors in close proximity.

Audio data is normally interleaved before recording, which means that the order of samples is shuffled (as shown conceptually in Figure 9.2). Samples that had been adjacent in real time are now separated from each other on the tape. The benefit of this is that a burst error, which destroys consecutive samples on tape, will result in a collection of single-sample errors in between good samples when the data is deinterleaved, allowing for the error to be concealed. A common process, associated with interleaving, is the separation of odd and even samples by a delay. The greater the interleave delay, the longer the burst error that can be handled. Redundant data is also added before recording. Redundancy, in simple terms, involves the recording of data in more than one form or place so that if it is damaged in one place it can be retrieved from another.

Cyclic redundancy check (CRC) codes, calculated from the original data and recorded along with that data, are used in many systems to detect the presence and position of errors on replay. Complex mathematical procedures are also used to form codewords from audio data which allow for both burst and random errors to be corrected perfectly up to a given limit. Reed–Solomon encoding is another powerful system which is used to protect digital recordings against errors, but it is beyond the scope of this book to cover these codes in detail.

Up to a certain random error rate or burst error duration an error correction system will be able to reconstitute erroneous samples perfectly. Such corrected samples

Original sample order

| 1 | 2 | 3 | 4 | 5 | 6 | 7 | 8 | 9 | 10 | 11 | 12 | 13 |

Interleaved sample order

| 3 | 7 | 13 | 9 | 4 | 10 | 1 | 5 | 11 | 8 | 2 | 6 | 12 |

Burst error destroys three samples

Consequent random errors in de-interleaved data

| | 2 | 3 | 4 | | 6 | 7 | 8 | 9 | | 11 | 12 | 13 |

FIGURE 9.2 *Interleaving is used in digital recording and broadcasting systems to rearrange the original order of samples for storage or transmission. This can have the effect of converting burst errors into random errors when the samples are deinterleaved.*

are indistinguishable from the originals, and sound quality will not be affected. Such errors are often signaled by green lights showing 'CRC' failure or 'Parity' failure. When the error rate exceeds the limits for perfect correction, interpolation between good samples can be used to arrive at a value for a missing sample. The interpolated value is the mathematical average of the foregoing and succeeding samples. This process is also known as concealment or averaging, and the audible effect is not unpleasant, although it will result in a temporary reduction in audio bandwidth. In extreme cases a system may 'hold'. In other words, it will repeat the last correct sample value. The audible effect of this will not be marked in isolated cases, but is still a severe condition. Most systems will not hold for more than a few samples before muting. Hold is normally indicated by a red light. When an error correction system is completely overwhelmed it will usually mute the audio output of the system. The alternative to muting is to hear the output, regardless of the error. Depending on the severity of the error, it may sound like a small 'spit', click, or even a more severe breakup of the sound.

Digital tape formats

There have been a number of commercial recording formats over the last 20 years, and only a brief summary will be given here of the most common.

Sony's PCM-1610 and PCM-1630 adaptors dominated the CD-mastering market for a number of years, although by today's standards they used a fairly basic recording format and relied on 60 Hz/525 line U-matic cassette VTRs (Figure 9.3). The system operated at a sampling rate of 44.1 kHz and used 16 bit quantization, being designed specifically for the making of tapes to be turned into CDs. Recordings made in this format could be electronically edited using the Sony DAE3000 editing system, and the playing time of tapes ran up to 75 minutes using a tape specially developed for digital audio use.

The R-DAT or DAT format was a small stereo, rotary-head, cassette-based format offering a range of sampling rates and recording times, including the professional rates of 44.1 and 48 kHz. Originally, consumer machines

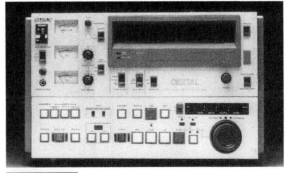

FIGURE 9.3 *Sony DMR-4000 digital master recorder. (Courtesy of Sony Broadcast and Professional Europe.)*

FIGURE 9.4 *Sony PCM-7030 professional DAT machine. (Courtesy of Sony Broadcast and Professional Europe.)*

FIGURE 9.5 *Nagra-D open-reel digital tape recorder. (Courtesy of Sound PR.)*

operated at 48 kHz to avoid the possibility for digital copying of CDs, but professional versions became available which would record at either 44.1 or 48 kHz. Consumer machines would record at 44.1 kHz, but usually only via the analog inputs. DAT was a 16 bit format, but had a non-linearly encoded long-play mode as well, sampled at 32 kHz. Truly professional designs offering editing facilities, external sync and IEC-standard timecode were also developed. The format became exceptionally popular with professionals owing to its low cost, high performance, portability and convenience. Various non-standard modifications were introduced, including a 96 kHz sampling rate machine and adaptors enabling the storage of 20 bit audio on such a high sampling rate machine (sacrificing the high sampling rate for more bits). The IEC timecode standard for R-DAT was devised in 1990. It allowed for SMPTE/EBU timecode of any frame rate to be converted into the internal DAT 'running-time' code, and then converted back into any SMPTE/EBU frame rate on replay. A typical machine is pictured in Figure 9.4.

The Nagra-D recorder (Figure 9.5) was designed as a digital replacement for the world-famous Nagra analog recorders, and as such was intended for professional use in field recording and studios. The format was designed to have considerable commonality with the audio format used in D1- and D2-format digital VTRs, having rotary heads, although it used open reels for operational convenience. Allowing for 20–24 bits of audio resolution, the Nagra-D format was appropriate for use with high-resolution convertors. The error correction and recording density used in this format were designed to make recordings exceptionally robust, and recording time could be up to 6 hours on a 7 inch (18 cm) reel, in two-track mode. The format was also designed for operation in a four-track mode at twice the stereo tape speed, such that in stereo the tape travels at 4.75 cm s^{-1}, and in four track at 9.525 cm s^{-1}.

The DASH (Digital Audio Stationary Head) format consisted of a whole family of open-reel stationary-head recording formats from two tracks up to 48 tracks. DASH-format machines operated at 44.1 kHz or 48 kHz rates (and sometimes optionally at 44.056 kHz), and they allowed varispeed ±12.5%. They were designed to allow gapless punch-in and punch-out, splice editing, electronic editing and easy synchronization. Multitrack DASH machines (an example is shown in Figure 9.6) gained wide acceptance in studios, but the stereo machines did not. Later developments resulted in DASH multitracks capable of storing 24 bit audio instead of the original 16 bits.

Subsequently budget modular multitrack formats were introduced. Most of these were based on eight-track cassettes using rotary head transports borrowed from consumer video technology. The most widely used were the DA-88 format (based on Hi-8 cassettes) and the ADAT format (based on VHS cassettes). These offered most of the features of open-reel machines and a number of them could be synchronized to expand the channel capacity. An example is shown in Figure 9.7.

FIGURE 9.6 *An open-reel digital multitrack recorder: the Sony PCM-3348. (Courtesy of Sony Broadcast and Professional Europe.)*

FIGURE 9.7 *A modular digital multitrack machine, Sony PCM-800. (Courtesy of Sony Broadcast and Professional Europe.)*

Editing digital tape recordings

Razor blade cut-and-splice editing was possible on open-reel digital formats, and the analog cue tracks were monitored during these operations. The thin tape could easily be damaged during the cut-and-splice edit procedure and this method failed to gain an enthusiastic following, despite its having been the norm in the analog world. Electronic editing was far more desirable, and was the usual method.

Electronic editing normally required the use of two machines plus a control unit, as shown in the example in Figure 9.8. A finished master tape was assembled from source takes on player machines and the original source tape was left unaltered. This was a relatively slow process, as it involved real-time copying of audio from one machine to another, and modifications to the finished master were difficult. A crossfade was introduced at edits to smooth the join.

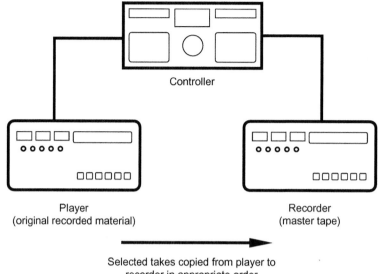

FIGURE 9.8

In electronic tape copy editing selected takes are copied in sequence from player to recorder with appropriate crossfades at joins.

Controller

Player
(original recorded material)

Recorder
(master tape)

Selected takes copied from player to
recorder in appropriate order

MASS STORAGE-BASED SYSTEMS

Once audio is in a digital form it can be handled by a computer, like any other data. The only real difference is that audio requires a high sustained data rate, substantial processing power and large amounts of storage compared with more basic data such as text. The following is an introduction to some of the technology associated with computer-based audio workstations and audio recording using computer mass storage media such as hard disks. The MIDI-based aspects of such systems are covered in Chapter 14, while file formats and interchange are introduced in Chapter 10.

Magnetic hard disks

Magnetic hard disk drives are random-access systems – in other words any data can be accessed at random and with only a short delay. There exist both removable and fixed media disk drives, but in almost all cases the fixed media drives have a higher performance than removable media drives. This is because the design tolerances can be made much finer when the drive does not have to cope with removable media, allowing higher data storage densities to be achieved. Some disk drives have completely removable drive cartridges containing the surfaces and mechanism, enabling hard disk drives to be swapped between systems for easy project management (an example is shown in Figure 9.9).

The general structure of a hard disk drive is shown in Figure 9.10. It consists of a motor connected to a drive mechanism that causes one or more disk surfaces to rotate at anything from a few hundred to many thousands of revolutions per minute. This rotation may either remain constant or may stop and start, and it may either be at a constant rate or a variable rate, depending on the drive. One or more heads are mounted on a positioning mechanism which can move the head across the surface of the disk to access particular points, under the control of hardware and software called a disk controller. The heads read data from and write data to the disk surface.

The disk surface is normally divided up into tracks and sectors, not physically but by means of 'soft' formatting (see Figure 9.11). Low-level formatting places logical markers, which indicate block boundaries, amongst other processes. On most hard disks the tracks are arranged as a series of concentric rings, but with some optical discs there is a continuous spiral track.

Disk drives look after their own channel coding, error detection and correction so there is no need for system designers to devise dedicated audio processes for disk-based recording systems. The formatted capacity of a disk drive is all available for the storage of 'raw' audio data, with no additional overhead required for redundancy and error checking codes. 'Bad blocks' are mapped out during the formatting of a disk, and not used for data storage. If a disk drive detects an error when reading a block of data it will attempt to read it again. If this fails then an error is normally generated and the file cannot be accessed, requiring the user to resort to one of the many file recovery packages on the market. Disk-based audio systems do not resort to error interpolation or

FIGURE 9.9 *A typical removable disk drive system allowing multiple drives to be inserted or removed from the chassis at will. Frame housing multiple removable drives. (Courtesy of Glyph Technologies Inc.)*

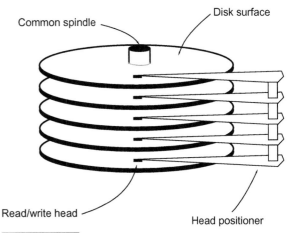

FIGURE 9.10 *The general mechanical structure of a disk drive.*

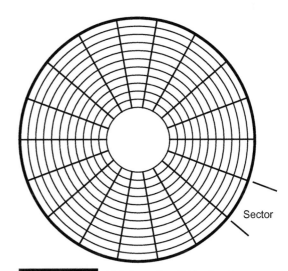

FIGURE 9.11 *Disk formatting divides the storage area into tracks and sectors.*

sample hold operations, unlike tape recorders. Replay is normally either correct or not possible.

RAID arrays enable disk drives to be combined in various ways as described in Fact File 9.3.

Optical discs

There are a number of families of optical disc drive that have differing operational and technical characteristics, although they share the universal benefit of removable media. They are all written and read using a laser, which is a highly focused beam of coherent light, although the method by which the data is actually stored varies from type to type. Optical discs are

---

FACT FILE 9.3 RAID ARRAYS

Hard disk drives can be combined in various ways to improve either data integrity or data throughput. RAID stands for Redundant Array of Independent Disks, and is a means of linking ordinary disk drives under one controller so that they form an array of data storage space. A RAID array can be treated as a single volume by a host computer. Historically there were a number of basic levels of RAID array, each of which was designed for a slightly different purpose, as summarized in the table. However there are also hybrid and non-standard alternatives that combine different features of multi-drive arrays. Recent classifications divide arrays into three categories: 'Failure resistant', 'Failure tolerant' and 'Disaster tolerant', based on increasing ability to withstand various failure and protection criteria. It's now left up to manufacturers how to implement these criteria.

RAID level	Features
0	Data blocks split alternately between a pair of disks, but no redundancy so actually less reliable than a single disk. Transfer rate is higher than a single disk. Can improve access times by intelligent controller positioning of heads so that next block is ready more quickly.
1	Offers disk mirroring. Data from one disk is automatically duplicated on another. A form of real-time backup.
2	Uses bit interleaving to spread the bits of each data word across the disks, so that, say, eight disks each hold one bit of each word, with additional disks carrying error protection data. Non-synchronous head positioning. Slow to read data, and designed for mainframe computers.
3	Similar to level 2, but synchronizes heads on all drives, and ensures that only one drive is used for error protection data. Allows high-speed data transfer, because of multiple disks in parallel. Cannot perform simultaneous read and write operations.
4	Writes whole blocks sequentially to each drive in turn, using one dedicated error protection drive. Allows multiple read operations but only single write operations.
5	As level 4 but splits error protection between drives, avoiding the need for a dedicated check drive. Allows multiple simultaneous reads and writes.
6	As level 5 but incorporates RAM caches for higher performance.

sometimes enclosed in a plastic cartridge that protects the disc from damage, dust and fingerprints, and they have the advantage that the pickup never touches the disc surface making them immune from the 'head crashes' that can affect magnetic hard disks. Drives split between those that handle CD/DVD/BD formats (see Fact File 9.4) and those that handle magneto-optical (M-O) and other cartridge-type ISO standard disc formats. The latter were considered more suitable for 'professional purposes' whereas the former are often encountered in consumer equipment. As mass storage media for computers optical media have declined in importance in recent years, as capacity and speed have failed to keep up with magnetic hard drives, making them less useful for backup and secondary

FACT FILE 9.4 CONSUMER OPTICAL DISK FORMATS

Compact discs and drives

Compact Discs (CDs) are familiar to most people as a consumer read-only optical disc for audio (CD-DA) or data (CD-ROM) storage. Standard audio CDs (CD-DA) conform to the Red Book standard published by Philips. The CD-ROM standard (Yellow Book) divides the CD into a structure with 2048 byte sectors, adds an extra layer of error protection, and makes it useful for general purpose data storage including the distribution of sound and video in the form of computer data files. It is possible to find discs with mixed modes, containing sections in CD-ROM format and sections in CD-Audio format.

CD-R is the recordable CD, and may be used for recording CD-Audio format or other CD formats using a suitable drive and software. The Orange Book, Part 2, contains information on the additional features of CD-R, such as the area in the center of the disc where data specific to CD-R recordings is stored. Audio CDs recorded to the Orange Book standard can be 'fixed' to give them a standard Red Book table of contents (TOC), allowing them to be replayed on any conventional CD player. Once fixed into this form, the CD-R may not subsequently be added to or changed, but prior to this there is a certain amount of flexibility, as discussed below. CD-RW discs are erasable and work on phase-change principles, requiring a drive compatible with this technology, being described in the Orange Book, Part 3.

DVD

DVD was the natural successor to CD, being a higher density optical disc format aimed at the consumer market, having the same diameter as CD and many similar physical features. It uses a different laser wavelength to CD (635–650 nm as opposed to 780 nm) so multi-standard drives need to be able to accommodate both. Data storage capacity depends on the number of sides and layers to the disc, but ranges from 4.7 Gbytes (single-layer, single-sided) up to about 18 Gbytes (double-layer, double-sided). The data transfer rate at 'one times' speed is just over 11 Mbit/s.

DVD can be used as a general purpose data storage medium. Like CD, there are numerous different variants on the recordable DVD, partly owing to competition between the numerous different 'factions' in the DVD consortium. These include DVD-R, DVD-RAM, DVD-RW and DVD + RW, all of which are based on similar principles but have slightly different features, leading to a compatibility minefield. The 'DVD Multi' guidelines produced by the DVD Forum were an attempt to foster greater compatibility between DVD drives and discs, and many drives are now available that will read and write most of the DVD formats.

DVD-Video is the format originally defined for consumer distribution of movies with surround sound, typically incorporating MPEG-2 video encoding and Dolby Digital surround sound encoding. It also allows for up to eight channels of 48 or 96 kHz linear PCM audio, at up to

FACT FILE 9.4 (CONTINUED)

24 bit resolution. DVD-Audio was intended for very high-quality multichannel audio reproduction and allowed for linear PCM sampling rates up to 192 kHz, with numerous configurations of audio channels for different surround modes, and optional lossless data reduction (MLP). However, it has not been widely adopted in the commercial music industry.

Super Audio CD (SACD)

Version 1.0 of the SACD specification is described in the 'Scarlet Book', available from Philips licensing department. SACD uses DSD (Direct Stream Digital) as a means of representing audio signals, as described in Chapter 8, so requires audio to be sourced in or converted to this form. SACD aims to provide a playing time of at least 74 minutes for both two-channel and six-channel balances. The disc is divided into two regions, one for two-channel audio, the other for multichannel. A lossless data packing method known as Direct Stream Transfer (DST) can be used to achieve roughly 2:1 data reduction of the signal stored on disc so as to enable high-quality multichannel audio on the same disc as the two channel mix. SACD has only achieved a relatively modest market penetration compared with formats such as CD and DVD-Video, but is still used by some specialized high-quality record labels. SACDs can be manufactured as single- or dual-layer discs, with the option of the second layer being a Red Book CD layer (the so-called 'hybrid disc' that will also play on a normal CD player).

Blu-Ray disk

The Blu-Ray disk is a higher density optical disk format than DVD, which uses a shorter wavelength blue-violet laser (wavelength 405 nm) to achieve a high packing density of data on the disk surface. Single-layer disks offer 25 Gbytes of storage and dual-layer disks offer 50 Gbytes, and the basic transfer rate is also higher than DVD at around 36 Mbit/s although a higher rate of 54 Mbit/s is required for HD movie replay, which is achieved by using at least 1.5 times playback speed. Like DVD, a range of read-only, writeable and rewriteable formats is possible. There is an audio-only version of the player specification, known as BD-Audio, which does not have to be able to decode video, making possible a high-resolution surround playback format that might offer an alternative to DVD-Audio or SACD. Audio-only transfer rates of the disk vary depending on the format concerned.

As far as audio formats are concerned, Linear PCM, Dolby Digital and DTS Digital Surround are mandatory in Blu-Ray players and recorders, but it is up to individual studios to decide what formats to include on their disk releases. Alternative optional audio formats include higher-resolution versions of Dolby and DTS formats, known as Dolby Digital Plus and DTS-HD respectively, as well as losslessly encoded versions known as Dolby TrueHD and DTS-HD Master Audio. High sampling frequencies (up to 192 kHz) are possible on Blu-Ray, as are audio sample resolutions of 16, 20 or 24 bits. The standard limits audio reproduction to six channels of 192 kHz, 24 bit uncompressed digital audio, which gives rise to a data transfer rate of 27.7 Mbit/s.

Pure Audio Blu-Ray is a variant promoted by msm Studios that has a number of audio-specific features, such as Java code to enable easy remote control, select audio format and download content over a network. These discs are intended to be playable in any standard Profile 2.0 BD player.

storage. However they remain important as consumer data storage and distribution media, particularly in recent formats such as Blu-Ray Disc.

WORM (Write-Once-Read-Many) discs may only be written once by the user, after which the recording is permanent (a CD-R is therefore a type of WORM disc). Other types of optical discs can be written numerous times, either requiring pre-erasure or using direct overwrite methods (where new data is simply written on top of old, erasing it in the process). The read/write process of most current rewritable discs is typically 'phase change' or 'magneto-optical'. The CD-RW is an example of a rewritable disc that now uses direct overwrite principles.

Memory cards

Increasing use is also made in audio systems of small flash memory cards, particularly in portable recorders. These cards are capable of storing many gigabytes of data on a solid state chip with fast access time, and they have no moving parts which makes them relatively robust. Additionally they have the benefit of being removable, which makes them suitable for transfer of some projects between systems, although the capacity and speed limitations still make disks the medium of choice for large professional projects. Such memory cards come in a variety of formats such as Compact Flash (CF), Secure Digital (SD) and Memory Stick, and card readers can be purchased that will read multiple types. There is a limit to the number of times such devices can be rewritten, which is likely to be lower than that for a typical magnetic disk drive.

A number of digital audio recording systems are available that use memory cards as the primary storage medium, having the advantage of minimal mechanical noise pickup by onboard microphones, as well as portability, low power consumption and compactness. One example of a stereo studio recorder using memory cards is shown in Figure 9.12, and these are becoming the natural successor to DAT recorders. Files are typically stored in Broadcast WAVE format (see Chapter 10).

Recording audio on to mass storage media

Mass storage media need to offer at least a minimum level of performance capable of handling the data rates and capacities associated with digital audio, as described in Fact File 9.5. Most standard drives now have no problem meeting this requirement for numerous audio channels.

FIGURE 9.12

The Tascam HS-2 is a solid state memory card recorder, capable of stereo recording at up to 192 kHz, 24 bit resolution. (Courtesy of Tascam.)

FACT FILE 9.5 STORAGE REQUIREMENTS OF DIGITAL AUDIO

The table shows the data rates required to support a single channel of digital audio at various resolutions. Media to be used as primary storage would need to be able to sustain data transfer at a number of times these rates to be useful for multimedia workstations. The table also shows the number of megabytes of storage required per minute of audio, showing that the capacity needed for audio purposes is considerably greater than that required for text or simple graphics applications. Storage requirements increase pro rata with the number of audio channels to be handled.

Sampling rate	Resolution	Bit rate	Capacity/min	Capacity/hour
kHz	bits	kbit/s	Mbytes/min	Mbytes/hour
192	24	4608	33.0	1980
96	24	2304	16.5	989
88.1	16	1410	10.1	605
48	20	960	6.9	412
48	16	768	5.5	330
44.1	16	706	5.0	303
44.1	8	353	2.5	151

Data rates and capacities for linear PCM

The discontinuous 'bursty' nature of recording onto such media usually requires the use of a buffer RAM (Random Access Memory) during replay, which accepts this interrupted data stream and stores it for a short time before releasing it as a continuous stream. It performs the opposite function during recording, as shown in Figure 9.13. Several things cause a delay in the retrieval of information from disks: the time it takes for the head positioner to move across a disk, the time it takes for the required data in a particular track to come around to the pickup head, and the transfer of the data from the disk via the buffer RAM to the outside world, as shown in Figure 9.14. Total delay, or data access time, is usually several milliseconds, depending on the rotational speed. The instantaneous rate at which the system can accept or give out data is called the transfer rate and varies with the storage device.

Sound is stored in named data files on the disk, the files consisting of a number of blocks of data stored either separately or together. A directory stored on the disk keeps track of where the blocks of each file are stored so that they can be retrieved in correct sequence. Each file normally corresponds to a single recording of a single channel of audio, although some stereo file formats exist (see Chapter 10).

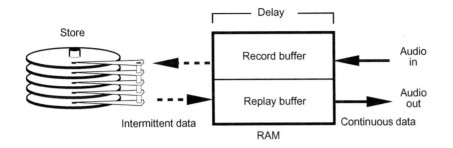

FIGURE 9.13
RAM buffering is used to convert burst data flow to continuous data flow, and vice versa.

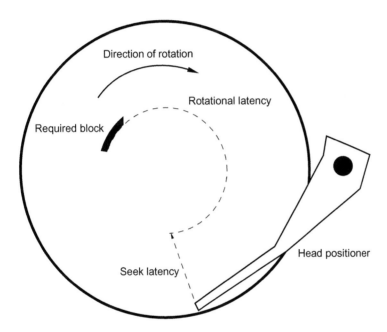

FIGURE 9.14
The delays involved in accessing a block of data stored on a disk.

Multiple channels are handled by accessing multiple files from the disk in a time-shared manner, with synchronization between the tracks being performed subsequently in RAM. The storage capacity of a disk can be divided between channels in whatever proportion is appropriate, and it is not necessary to pre-allocate storage space to particular audio channels. A feature of the disk system is that unused storage capacity is not necessarily 'wasted' as can be the case with a tape system. During recording of a multitrack tape there will often be sections on each track with no information recorded, but that space cannot be allocated elsewhere. On a disk these gaps do not occupy storage space and can be used for additional space on other channels at other times.

The number of audio channels that can be recorded or replayed simultaneously depends on the performance of the storage device, interface, drivers and host computer. Slow systems may only be capable of handling a few channels whereas faster systems with multiple disk drives may be capable of expansion up to a virtually unlimited number of channels. External disks are usually connected using high-speed serial interfaces such as Firewire (IEEE 1394) or Thunderbolt (see Fact File 9.6), and as desktop computers get faster and more capable there is no longer a strong need to have dedicated cards for connecting audio-only disk drives. These days one or more of the host computer's internal or external disks is usually

---

FACT FILE 9.6   PERIPHERAL INTERFACES

A variety of different physical interfaces can be used for interconnecting storage devices and host workstations. Some are internal buses only designed to operate over limited lengths of cable and some are external interfaces that can be connected over several meters. The interfaces can be broadly divided into serial and parallel types, the serial types being by far the most common now. The disk interface can be slower than the drive attached to it in some cases, making it into a bottleneck in some applications. There is no point having a super-fast disk drive if the interface cannot handle data at that rate.

SCSI

For many years the most commonly used interface for connecting mass storage media to host computers was SCSI (the Small Computer Systems Interface), pronounced 'scuzzy'. It is still used quite widely for very high performance applications but other interfaces have taken over for desktop systems. SCSI is a high-speed parallel interface, originally allowing up to seven peripheral devices to be connected to a host on a single bus. SCSI grew through a number of improvements and revisions, resulting in various Ultra SCSI standards. Serial Attached SCSI (SAS) interfaces retain many of the features of SCSI but use a serial format, while iSCSI is designed to be used over Internet connections, usually based on Ethernet.

ATA/IDE

The ATA and IDE family of interfaces evolved through the years as the primary internal interface for connecting disk drives to PC system buses. It is cheap and ubiquitous,

also with various 'Ultra' versions running at high speed. ATAPI (ATA Packet Interface) is a variant used for storage media such as CD drives. Serial ATA (SATA) is designed to enable disk drives to be interfaced serially, thereby reducing the physical complexity of the interface. It is intended primarily for internal connection of disks within host workstations, rather than as an external interface like USB or Firewire, although a version known as eSATA is suitable for external drives.

Firewire and USB

Firewire and USB are both serial interfaces for connecting external peripherals. They both enable disk drives to be connected in a very simple manner, with high transfer rates (many hundreds of megabits per second), although USB 1.0 devices are limited to 12 Mbit/s. A key feature of these interfaces is that they can be 'hot plugged' (in other words devices can be connected and disconnected with the power on). The interfaces also supply basic power that enables some simple devices to be powered from the host device. Interconnection cables can usually be run up to between 5 and 10 meters, depending on the cable and the data rate.

Thunderbolt

Thunderbolt is a very high speed serial interface developed by Apple and Intel that combines PCIe (PCI Express) data and DisplayPort data over the same cable, using a meta-protocol, along with 10 watts of power to peripherals. It is capable of 10 Gbit/s transfer rates. Up to six peripherals can be daisy-chained.

employed, although it is often recommended that this is not the same disk as used for system software in order to avoid conflicts of demand between system housekeeping and audio needs.

Media formatting

The process of formatting a storage device erases all of the information in the volume. (It may not actually do this, but it rewrites the directory and volume map information to make it seem as if the disk is empty again.) Effectively the volume then becomes virgin territory again and data can be written anywhere.

When a disk is formatted at a low level the sector headers are written and the bad blocks mapped out. A map is kept of the locations of bad blocks so that they may be avoided in subsequent storage operations. Low-level formatting can take quite a long time as every block has to be addressed. During a high-level format the disk may be subdivided into a number of 'partitions'. Each of these partitions can behave as an entirely independent 'volume' of information, as if it were a separate disk drive (see Figure 9.15). It may even be possible to format each partition in a different way, such that a different filing system may be used for each partition. Each volume then has a directory created, which is an area of storage set aside to contain information about the contents of the disk. The directory indicates the locations of the files, their sizes and various other vital statistics.

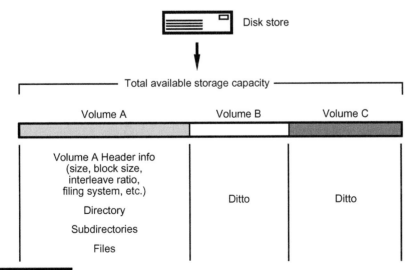

FIGURE 9.15 *A disk may be divided up into a number of different partitions, each acting as an independent volume of information.*

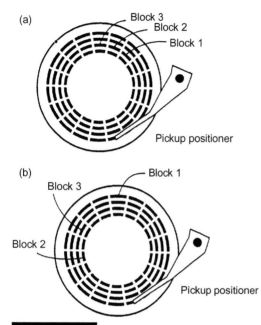

FIGURE 9.16 *At (a) a file is stored in three contiguous blocks and these can be read sequentially without moving the head. At (b) the file is fragmented and is distributed over three remote blocks, involving movement of the head to read it. The latter read operation will take more time.*

The most common general purpose filing systems in audio workstations are HFS (Hierarchical Filing System) or HFS Plus (for Mac OS), FAT 32 (File Allocation Table, for Windows PCs) and NTFS (for Windows NT and 2000). The Unix operating system is used on some multi-user systems and high-powered workstations and also has its own filing system. These were not designed principally with real-time requirements such as audio and video replay in mind but they have the advantage that disks formatted for a widely used filing system will be more easily interchangeable than those using proprietary systems. Further information about audio file formats and interchange is provided in the next chapter.

When an erasable volume like a hard disk has been used for some time there will be a lot of files on the disk, and probably a lot of small spaces where old files have been erased. New files must be stored in the available space and this may involve splitting them up over the remaining smaller areas. This is known as disk fragmentation, and it seriously affects the overall performance of the drive. The reason is clear to see from Figure 9.16. More head seeks are required to access the blocks of a file than if they had been stored contiguously, and this slows down the average transfer rate considerably. It may come to a point where the drive is unable to supply data fast enough for the purpose.

There are only two solutions to this problem: one is to reformat the disk completely (which may be difficult, if one is in the middle of a project), the other is to optimize or consolidate the storage space. Various software utilities exist for this purpose, whose job is to consolidate all the little areas of free space into fewer larger areas. They do this by juggling the blocks of files between disk areas and temporary RAM – a process that often takes a number of hours. Power failure during such an optimization process can result in total corruption of the drive, because the job is not completed and files may be only half moved, so it is advisable to back up the drive before doing this. It has been known for some such utilities to make the files unusable by some audio editing packages, because the software may have relied on certain files being in specific physical places, so it is wise to check first with the manufacturer.

AUDIO PROCESSING FOR COMPUTER WORKSTATIONS

Introduction

Most audio processing now takes place within the workstation, usually relying either on the host computer's processing power (using the CPU to perform signal processing operations) or on one or more DSP (digital signal processing) cards attached to an expansion bus. Professional systems usually use external A/D and D/A convertors, connected to a 'core' card attached to the computer's expansion bus. This is because it is often difficult to obtain the highest technical performance from convertors mounted on internal sound cards, owing to the relatively 'noisy' electrical environment inside most computers. Furthermore, the number of channels required may not fit onto an internal card. As more and more audio work takes place entirely in the digital domain, though, the need for analog convertors decreases. Digital interfaces are also often provided on external 'breakout boxes', partly for convenience and partly because of physical size of the connectors. Compact connectors such as the optical connector used for the ADAT eight-channel interface or the two-channel SPDIF phono connector are accommodated on some cards, but multiple AES/EBU connectors cannot be.

It is also becoming increasingly common for substantial audio processing power to exist on integrated sound cards that contain digital interfaces and possibly A/D and D/A convertors. These cards are typically used for consumer or semi-professional applications on desktop computers, although many now have very impressive features and can be used for advanced operations. Such cards are now available in 'full duplex' configurations that enable audio to be received by the card from the outside world, processed and/or stored, then routed back to an external device. Full duplex operation usually allows recording and replay simultaneously.

Sound cards and DSP cards are commonly connected to the workstation using the PCI (peripheral component interface) expansion bus. Older ISA (PC) buses or NuBus (Mac) slots did not have the same data throughput capabilities and performance was therefore somewhat limited. PCI or the more recent PCI Express (PCIe) bus can be extended to an external expansion chassis that enables a larger number of cards to be connected than allowed for within the host computer. Sufficient processing power can now be installed for the workstation to become the audio processing 'heart' of a larger studio system, as opposed to using an external mixing console and effects units. The higher the sampling frequency, the more DSP operations will be required per second, so it is worth bearing in mind that going up to, say, 96 kHz sampling frequency for a project will require double the processing power and twice the storage space compared with 48 kHz. The same is true of increasing the number of channels to which processing is applied.

The issue of latency is important in the choice of digital audio hardware and software, as discussed in Fact File 9.7.

DSP resources for audio processing

DSP cards can be added to widely used workstation packages such as Avid's Pro Tools. These were termed 'DSP Farms' or 'Mix Farms' in the past, and are expansion cards that connect to the PCI bus of the workstation and take on much of the 'number crunching' work involved in effects processing and mixing. 'Plug-in' processing software is now a popular and cost-effective way of implementing effects processing within the workstation, and this is discussed further in Chapter 13. Such plug-ins rely either on DSP cards or on host-based processing (see below) to handle this load.

Avid's various Pro Tools systems are a useful example of the way in which audio processing can be handled within the workstation. Depending on the 'level' of the system, audio processing is either handled 'natively', using the host CPU, or by PCI-connected DSP cards. Of the two possible card solutions, the company's earlier TDM system, now being retired, processed audio using 24-bit fixed point devices, whereas the more recent HDX system uses 32-bit floating point processing. The latter offers the possibility for greater processing speed and capacity (affecting the number of simultaneous operations and audio channels) and lower latency. Floating

FACT FILE 9.7 AUDIO PROCESSING LATENCY

Latency is the delay incurred in executing audio operations between input and output of a system. The lower the better is the rule, particularly when operating systems in 'full duplex' mode, because processed sound may be routed back to musicians (for foldback purposes) or may be combined with undelayed sound at some point. The management of latency is a software issue and some systems have sophisticated approaches to ensuring that all supposedly synchronous audio reaches the output at the same time no matter what processing it has encountered on the way.

Minimum latency achievable is both a hardware and a software issue. The poorest systems can give rise to tens or even hundreds of milliseconds between input

and output whereas the best reduce this to a few milliseconds. Audio I/O that connects directly to an audio processing card can help to reduce latency, otherwise the communication required between host and various cards can add to the delay. Some real-time audio processing software also implements special routines to minimize and manage critical delays and this is often what distinguishes professional systems from cheaper ones. The audio driver software or 'middleware' that communicates between applications and sound cards influences latency considerably. One example of such middleware intended for low latency audio signal routing in computers is Steinberg's ASIO (Audio Stream Input Output).

point processing typically allows for a greater internal dynamic range, because it represents numbers in the form of a mantissa and exponent. (The mantissa is a basic numerical value and the exponent is a power of two to which it is multiplied.) The latest versions of Pro Tools, both native and HDX card-based, claim an impressive 64-bit floating point resolution for internal mix summing. 24-bit fixed point processing, however, does not necessarily mean that the numerical operations are limited to 24-bit resolution. In fact 'double precision' mode can be used in mixing and plug-ins, allowing 48-bit resolution by splitting the samples into two chunks and operating on them separately.

An alternative to using dedicated DSP cards is to use the internal processing capacity of a typical desktop computer. The success of such 'host-based processing' depends on the number of tasks that the workstation is required to undertake and this capacity may vary with time and context. Typical modern CPUs, though, can easily run DSP plug-ins for implementing equalization, mixing and a wide range of effects, for a considerable number of tracks. The 'multi-core' (e.g. quad-core) processor architectures of some modern computers enables the division of processing power between applications, and in some cases one can allocate a specific number of processor cores to an audio application, leaving, say, one or two for system tasks and other applications. This ensures the greatest degree of hardware independence between processing tasks, and avoids conflicts of demand at times of peak processor load.

The software architecture required to run plug-ins on the host CPU may be different to that used on dedicated DSP cards, so it may be necessary to obtain the correct plug-in for the environment in question. A number of applications, however, enable the integration of host-based (or 'native') plug-ins and dedicated DSP cards. With Pro Tools, for example, the same floating point audio processing algorithms are used for AAX plug-ins in both native and HDX forms, so they should sound the same. This is not necessarily true for TDM compared with RTAS plug-ins (see Chapter 13). Audio processing that runs on the host may be subject to greater latency (input to output delay) than when using dedicated signal processing, and it obviously takes up processing power that could be used for running the user interface or other software. It is nonetheless a cost-effective option for many users.

Audio processing architectures

Apple's Core Audio is an example of that manufacturer's internal tools and architecture for handling audio in computers using the Mac OS X operating system. Core Audio provides plug-in facilities for audio signal processing and synthesis, as well as audio-to-MIDI synchronization. Its audio plug-ins are called Audio Units (AUs). A number of standard AUs are provided with the OS X operating system, offering a range of audio processing options to other Core Audio compatible software that runs on the platform. Audio workstation packages such as Logic, for example, work closely with Core Audio to implement aspects of their functionality, including plug-ins. Core Audio normally expects to work with audio represented as 32-bit floating point linear PCM, but there are means to translate between this and other PCM formats, as well as to coded formats such as MP3, AAC or Apple Lossless Audio Coding (ALAC). It supports the main audio interchange file formats described in Chapter 10, such as AIFF and WAVE.

Core Audio functions are written in C code (a widely used software authoring language) and can be 'called' from other compatible applications using Application Programming Interfaces (APIs) designed for the task. It also uses an internal representation of audio hardware known as a Hardware Abstraction Layer (HAL), which can be used to simplify the interface between Core Audio elements and physical audio devices.

Integrated sound cards

Integrated sound cards typically contain all the components necessary to handle audio for gaming and entertainment purposes within a desktop computer and may be able to operate in full duplex mode (in and out at the same time). They typically incorporate convertors, DSP, a digital interface, and one or more sound synthesis engines. Additionally they may offer 3D audio processing and low bit rate audio decoding, such as Dolby Digital and DTS. Optionally, they may also include some sort of I/O daughter board that can be connected to a break-out audio interface, increasing the number of possible connectors and the options for external analog conversion. Such cards also tend to sport MIDI/joystick interfaces. A typical example of this type of card is the 'SoundBlaster' series from Creative Labs. Any analog audio connections are normally unbalanced and the convertor quality may be lower than the best external devices. For professional purposes it is advisable to use high-quality external convertors and balanced analog audio connections.

Mixing 'in the box'

There has been a lot of debate about the relative merits of audio mixing 'in the box' versus using a conventional external mixer. 'In the box' is the familiar term often used to describe mixing carried out using the DSP facilities of the workstation package, whereas 'out of the box' usually means feeding all the workstation's separate audio outputs to a (possibly analog) mixer. Issues of audio quality have sometimes been questioned when using in-the-box signal processing, which may in some cases have been related to the effects of limited resolution numerical processing or improper dithering in poor systems. With good recent software, though, which uses high-precision internal maths and correct dithering, this is unlikely to be a factor.

Some engineers also simply prefer the sound of analog processing, liking the unique phase and amplitude characteristics of particular equalizers and compressors. Although there is no arguing with the preference for specific external effects, carefully controlled listening tests comparing good in-the-box mixes with external mixes of the same material (avoiding EQ and compression) have not shown reliably detectable differences.

The ergonomic advantages of external mixing can be applied to in-the-box mixing by using a remote control surface designed to interface to the workstation in question, having dedicated faders and knobs, as discussed towards the end of Chapter 5. The argument for in-the-box mixing is taken even further by the availability of digital plug-ins that attempt to emulate the performance of classic analog processors by sampling their impulse responses or otherwise modeling their characteristics. These can provide very convincing plug-in alternatives to the analog hardware they model, and may be badged or marketed by the makers of the original hardware in recognition of the increasingly widespread prevalence of entirely in-the-box mixing.

MASS STORAGE-BASED EDITING SYSTEM PRINCIPLES

Introduction

The random access nature of mass storage media led to the coining of the term non-linear editing for the process of audio editing. With non-linear editing the editor may preview a number of possible masters in their entirety before deciding which should be the final one. Even after this, it is a simple matter to modify the edit list to update the master. Edits may also be previewed and experimented with in order to determine the most appropriate location and processing. Crossfades may be modified and adjustments made to equalization and levels, all in the digital domain. Non-linear editing has also come to feature very widely in post-production for video and film.

Non-linear editing is truly non-destructive in that the edited master only exists as a series of instructions to replay certain parts of sound files at certain times, with specified signal processing overlaid, as shown in Figure 9.17. The original sound files remain intact at all times, and a

FIGURE 9.17

Instructions from an edit decision list (EDL) are used to control the replay of sound file segments from disk, which may be subjected to further processing (also under EDL control) before arriving at the audio outputs.

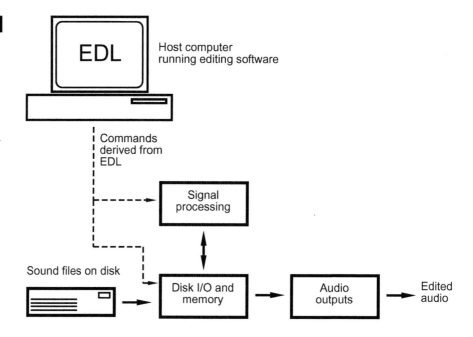

single sound file can be used as many times as desired in different locations and on different tracks without the need for copying the audio data. Editing may involve the simple joining of sections, or it may involve more complex operations such as long crossfades between one album track and the next, or gain offsets between one section and another. All these things are possible without affecting the original source material.

The modern workstation production technique of compiling or 'comping' vocal tracks also relies on this characteristic of random-access storage. Multiple takes of a lead vocal line, for example, can be stored and a final version comped from short sections of different takes arranged in a suitable order with appropriate crossfades or joins.

Sound files and sound segments

In the case of music editing sound files might be session takes, anything from a few bars to a whole movement, while in picture dubbing they might contain a phrase of dialog or a sound effect. In multitrack production, each separately recorded chunk of an individual track is likely to be stored in a separate sound file. Usually such files are mono, but they can be stereo and occasionally multichannel, as discussed in Chapter 10. If the channels of a recording are to be processed or moved around separately then they should be stored in mono. Specific segments of these sound files can be defined while editing, in order to get rid of unwanted material or to select useful extracts. The terminology varies but such identified parts of sound files are usually termed either 'clips' or 'segments'. Rather than creating a copy of the segment or clip and storing it as a separate sound file, it is normal simply to store it as a 'soft' entity – in other words as simply commands in an edit list or project file that identify the start and end addresses of the segment concerned and the sound file to which it relates. It may be given a name by the operator and subsequently used as if it were a sound file in its own right. An almost unlimited number of these segments can be created from original sound files, without the need for any additional audio storage space. Figure 9.18 shows an example of Logic's audio 'bin' containing both entire .wav files, and segments thereof that can be used independently and dragged to appropriate places and tracks in the editing time line.

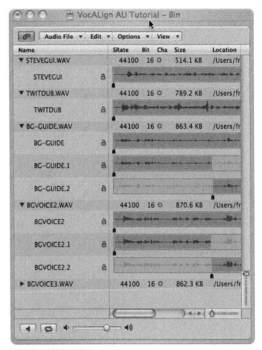

FIGURE 9.18 *Example of Logic's audio bin containing both entire audio files and edited segments thereof.*

Edit point handling

Edit points can be simple butt joins or crossfades. A butt join is very simple because it involves straightforward switching from the replay of one sound segment to another. Since replay involves temporary storage of the sound file blocks in RAM (see above) it is a relatively simple matter to ensure that both outgoing and incoming files in the region of the edit are available in RAM simultaneously (in different address areas). Up until the edit, blocks of the outgoing file are read from the disk into RAM and thence to the audio outputs. As the edit point is reached a switch occurs between outgoing and incoming material by instituting a jump in the memory read address corresponding to the start of the incoming material. Replay then continues by reading subsequent blocks from the incoming sound file. It is normally possible to position edits to single sample accuracy, making the timing resolution as fine as a number of tens of microseconds if required.

The problem with butt joins is that they are quite unsubtle. Audible clicks and bumps may result because of the discontinuity in the waveform that may result, as shown in Figure 9.19. It is normal, therefore, to use at least a short crossfade at edit points to hide the effect of the join. This

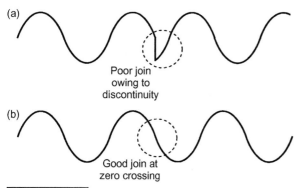

FIGURE 9.19 *(a) A bad butt edit results in a waveform discontinuity. (b) Butt edits can be made to work if there is minimal discontinuity.*

is what happens when analog tape is spliced, because the traditional angled cut has the same effect as a short crossfade (of between 5 and 20 ms depending on the tape speed and angle of cut). Most workstations have considerable flexibility with crossfades and are not limited to short durations. It is now common to use crossfades of many shapes and durations (e.g. linear, root cosine, equal power) for different creative purposes. This, coupled with the ability to preview edits and fine-tune their locations, has made it possible to put edits in places previously considered impossible.

The locations of edit points are kept in an edit decision list (EDL) which contains information about the segments and files to be replayed at each time, the in and the out points of each section and details of the crossfade time and shape at each edit point. It may also contain additional information such as signal processing operations to be performed (gain changes, EQ, etc.). EDL interchange formats for moving projects between systems are discussed in Chapter 10.

Crossfading

Crossfading is similar to butt joining, except that it requires access to data from both incoming and outgoing files for the duration of the crossfade. The crossfade calculation involves simple signal processing, during which the values of outgoing samples are multiplied by gradually decreasing coefficients whilst the values of incoming samples are multiplied by gradually increasing coefficients. Time coincident samples of the two files are then added together to produce output samples, as described in the previous chapter. The duration and shape of the crossfade can be adjusted by altering the coefficients involved and the rate at which the process is executed.

Crossfades are either performed in real time, as the edit point passes, or pre-calculated and written to disk as a file. Real-time crossfades can be varied at any time and are simply stored as commands in the EDL, indicating the nature of the fade to be executed. The process is similar to that for the butt edit, except that as the edit point approaches samples from both incoming and outgoing segments are loaded into RAM in order that there is an overlap in time. During the crossfade samples from both incoming and outgoing segments are loaded into their respective areas of RAM, then routed to the crossfade processor, as shown in Figure 9.20. The resulting samples are then available for routing to the output. Alternatively the crossfade can be calculated in non-real time. This incurs a short delay while the system works out the sums, after which a new sound file is stored which contains only the crossfade. Replay of the edit then involves playing the outgoing segment up to the beginning of the crossfade, then the crossfade file, then the incoming segment from after the crossfade, as shown in Figure 9.21. Load on the disk drive is no higher than normal in this case, except that the crossfade file has to be retrieved.

The shape of the crossfade can usually be changed to suit different operational purposes. Standard linear fades (those where the gain changes uniformly with time) are not always the most suitable for music editing, especially when the crossfade is longer than about ten milliseconds. The

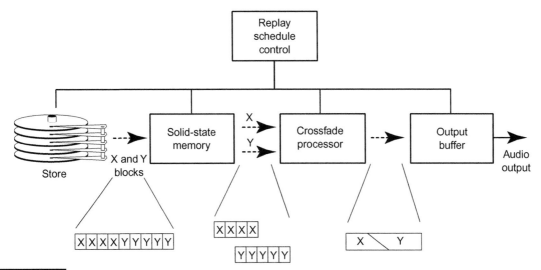

FIGURE 9.20   *Conceptual diagram of the sequence of operations which occur during a crossfade. X and Y are the incoming and outgoing sound segments.*

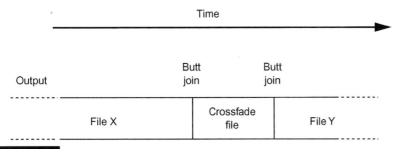

FIGURE 9.21   *Replay of a precalculated crossfade file at an edit point between files X and Y.*

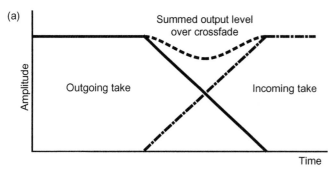

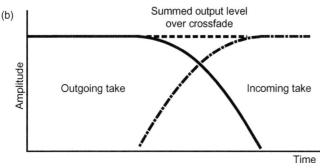

FIGURE 9.22 *Summation of levels at a crossfade. (a) A linear crossfade can result in a level drop if the incoming and outgoing material are non-coherent. (b) An exponential fade, or other similar laws, can help to make the level more constant across the edit.*

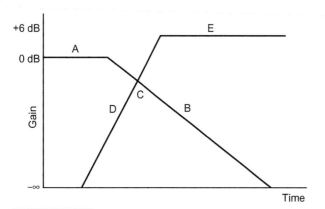

FIGURE 9.23 *The system may allow the user to program a gain profile around an edit point, defining the starting gain (A), the fade-down time (B), the fade-up time (D), the point below unity at which the two files cross over (C) and the final gain (E).*

result may be a momentary drop in the resulting level in the center of the crossfade that is due to the way in which the sound levels from the two files add together. If there is a random phase difference between the signals, as there often is in music, the rise in level resulting from adding the two signals will normally be around 3 dB, but the linear crossfade is 6 dB down in its center resulting in an overall level drop of around 3 dB (see Figure 9.22). Exponential crossfades and other such shapes may be more suitable for these purposes, because they have a smaller level drop in the center. It may even be possible to design customized crossfade laws. It is often possible to alter the offset of the start and end of the fade from the actual edit point and to have a faster fade-up than fade-down.

Many systems also allow automated gain changes to be introduced as well as fades, so that level differences across edit points may be corrected. Figure 9.23 shows a crossfade profile which has a higher level after the edit point than before it, and different slopes for the in- and out-fades. A lot of the difficulties that editors encounter in making edits work can be solved using a combination of these facilities.

Editing modes

During the editing process the operator will load appropriate sound files and audition them, both on their own and in a sequence with other files. The exact method of assembling the edited sequence depends very much on the user interface, but it is common to present the user with a visual analogy of moving tape, allowing files to be 'cut and spliced' or 'copied and pasted' into appropriate locations along the virtual tape. These files, or edited clips of them, are then played out at the time locations corresponding to their positions on this 'virtual tape'. It is also quite common to display a representation of the audio waveform that allows the editor to see as well as hear the signal around the edit point (see Figure 9.24).

In non-linear systems the tape-based approach is often simulated, allowing the user to roughly locate an edit point while playing the virtual tape followed by a fine trim using simulated reel-rocking or a detailed view of the waveform. Some software presents source and destination streams as well, in further simulation of the tape approach. It is also possible to insert or change sections in the middle of a finished master, provided that the EDL and source files are still available. To take an example, assume that an edited opera has been completed and that the producer now wishes to change a take somewhere in the middle (see Figure 9.25). The replacement take is unlikely to be exactly the same length but it is possible simply to shuffle all of the following material along or back slightly to accommodate it, this being only a matter of changing the EDL rather than modifying the stored music in any way. The files are then simply played out at slightly different times than in the first version of the edit.

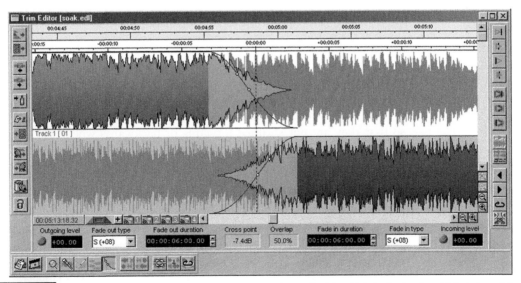

FIGURE 9.24 *Example from SADiE editing system showing the 'trim editor' in which is displayed a detailed view of the audio waveform around the edit point, together with information about the crossfade.*

FIGURE 9.25

Replacing a take in the middle of an edited programme. (a) Tape based copy editing results in a gap of fixed size, which may not match the new take length. (b) Non-linear editing allows the gap size to be adjusted to match the new take.

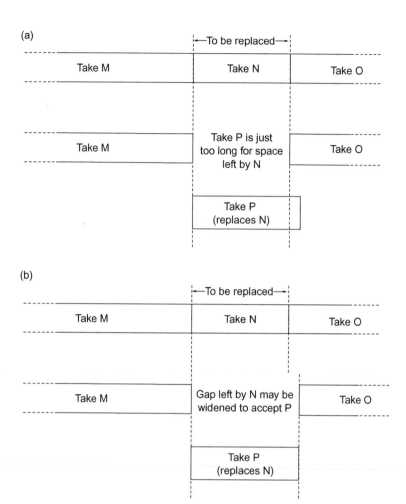

It is also normal to allow edited segments to be fixed in time if desired, so that they are not shuffled forwards or backwards when other segments are inserted. This 'anchoring' of segments is often used in picture dubbing when certain sound effects and dialog have to remain locked to the picture.

Simulation of 'reel-rocking'

It is common to simulate the effect of analog tape 'reel-rocking' in non-linear editors, providing the user with the sonic impression that reels of analog tape are being 'rocked' back and forth. The simulation of variable speed replay in both directions is usually controlled by a wheel or sideways movement of a mouse which moves the 'tape' in either direction around the current play location. The magnitude and direction of this movement is used to control the rate at which samples are read from the disk file, via the buffer. Good simulation requires very fast, responsive action and an ergonomically suitable control. A mouse is rather unsuitable for the purpose. It also requires a certain amount of DSP to filter the signal correctly, in order to avoid the aliasing that can be caused by varying the sampling rate. Systems differ very greatly as to the sound quality achieved in this mode and many current operators, not brought up with tape, prefer to judge edit points accurately 'on the fly', followed by trimming or nudging them either way if they are not successful the first time.

EDITING SOFTWARE

It is increasingly common for MIDI (see Chapter 14) and digital audio editing to be integrated within one software package, particularly for pop music recording and other multitrack productions where control of electronic sound sources is integrated with recorded natural sounds. Such applications used to be called sequencers but this is less common now that MIDI sequencing is only one of many tasks that are possible. Although most sequencers contain some form of audio editing these days, there are some software applications more specifically targeted at high-quality audio editing and production. These have tended to come from a professional audio background rather than a MIDI sequencing background, although it is admitted that the two fields have met in the middle now and it is increasingly hard to distinguish a MIDI sequencer with added audio features from an audio editor with added MIDI features.

Audio applications such as those described here are used in contexts where MIDI is not particularly important and where fine control over editing crossfades, dithering, mixing, mastering and post-production functions are required. Here the editor needs tools for such things as: previewing and trimming edits, such as might be necessary in classical music post-production; PQ editing CD masters; preparing surround sound material for Blu-Ray Disc encoding; MP3 or AAC encoding of audio material; post-production of sound tracks for pictures. The following example, based on the SADiE audio editing system from Prism Sound, demonstrates some of the practical concepts.

SADiE workstations run on the PC platform and most utilize an external audio interface. There are studio versions mounted in rack units as well as location recording systems running on a laptop. There is also a 'native' version that runs using the internal processing and storage of the host computer, and can be used with either proprietary or standard computer audio hardware. Both PCM and DSD signal processing options are available and the system makes provision for various mastering and delivery formats, as well as lossless and lossy audio encoding. A typical user interface for SADiE is shown in Figure 9.26. It is possible to see transport controls, the mixer interface and the playlist display. The main part of the screen is occupied by a horizontal display of recording tracks or 'streams', and these are analogous to the tracks of a multitrack tape recorder. A record icon associated with each stream is used to arm it ready for recording. As recording proceeds, the empty streams are filled from left to right across the screen in real time, led by a vertical moving cursor. These streams can be displayed either as solid continuous blocks or as waveforms, the latter being the usual mode when editing is undertaken. After recording, extra streams can be recorded if required simply by disarming the record icons of the streams already used and arming the record icons of empty streams below them, making it possible to build

FIGURE 9.26 *SADiE editor displays, showing mixer, playlist, transport controls and project elements.*

up a large number of 'virtual' tracks as required. The maximum number that can be replayed simultaneously depends upon the memory and DSP capacity of the system used. A basic two-input/four-output system might allow up to eight streams to be replayed (depending on the amount of DSP being used for other tasks), and a fully equipped system can allow at least 32 simultaneous streams of program material to be recorded and replayed.

Replay involves either using the transport control display or clicking the mouse at a desired position on a time-bar towards the top of the screen, this positioning the moving cursor (which is analogous to a tape head) where one wishes replay to begin. Editing is performed by means of a razor-blade icon, which will make the cut where the moving cursor is positioned. Alternatively, an edit icon can be loaded to the mouse's cursor for positioning anywhere on any individual stream to make a cut.

Audio can be arranged in the playlist by the normal processes of placing, dragging, copying and pasting, and there is a range of options for slipping material left or right in the list to accommodate new material (this ensures that all previous edits remain attached in the right way when the list is slipped backwards or forwards in time). Audio to be edited in detail can be viewed in the trim window (shown earlier in Figure 9.24) which shows a detailed waveform display, allowing edits to be previewed either to or from the edit point, or across the edit, using the play controls in the top right-hand corner (this is particularly useful for music editing). The crossfade region is clearly visible, with different colors and shadings used to indicate the 'live' audio streams before and after the edit. There are many stages of undo and redo so that nothing need be permanent at this stage. When a satisfactory edit is achieved, it can be written back to the main edit list where it will be incorporated. Scrub and jog actions for locating edit points are also possible. A useful 'lock to time' icon is provided which can be activated to prevent horizontal movement of the streams so that they cannot be accidentally moved out of sync with each other during editing.

The mixer section can be thought of in conventional terms, and indeed some systems offer physical mixing controllers with moving fader automation for those who prefer them. As well as mouse control of such things as fader, pan, solo and mute, processing such as EQ, filters, aux send and compression can be selected from an effects 'rack', and each can be dragged across and dropped in above a fader where it will become incorporated into that channel. Third party 'plug-in' software is also available for many systems to enhance the signal processing features, including CEDAR audio restoration software, as described below. The latest software allows for the

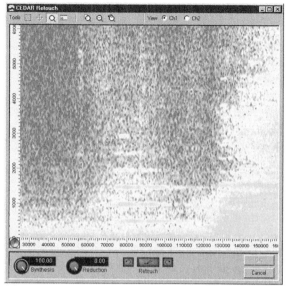

(a)

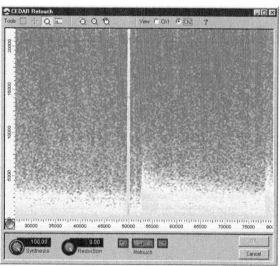

(b)

FIGURE 9.27 *CEDAR Retouch display for SADiE, showing frequency (vertical) against time (horizonal) and amplitude (color/density). Problem areas of the spectrographic display can be highlight and a new signal synthesized using information form the surrounding region. (a) Harmonics of an interfering signal can be clearly seen.(b) A short-term spike crosses most of the frequency range.*

use of DirectX plug-ins for audio processing. Automation of faders and other processing is also possible.

MASTERING AND RESTORATION

Specialized software

Some software applications are designed specifically for the mastering and restoration markets. These products are designed either to enable 'fine tuning' of master recordings prior to commercial release, involving subtle compression, equalization and gain adjustment (mastering), or to enable the 'cleaning up' of old recordings that have hiss, crackle and clicks (restoration).

CEDAR applications or plug-ins are good examples of the restoration group. The company has also introduced advanced visualization tools (known as Retouch) that enable restoration engineers to 'touch up' audio material using an interface not dissimilar to that used for photo editing on computers. Audio anomalies (unwanted content) can be seen in the time and frequency domains, highlighted and interpolated based on information either side of the anomaly. CEDAR's restoration algorithms are typically divided into 'decrackle', 'declick', 'dethump' and 'denoise', each depending on the nature of the anomaly to be corrected. Some typical user interfaces for controlling these processes are shown in Figure 9.27.

Mastering software usually incorporates advanced dynamics control such as the TC Works Master X series, based on its Finalizer products, a user interface of which is pictured in Figure 9.28. Here compressor curves and frequency dependency of dynamics can be adjusted and metered. The display also allows

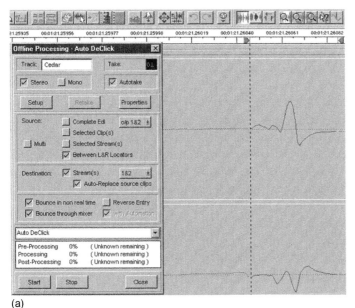

(a)

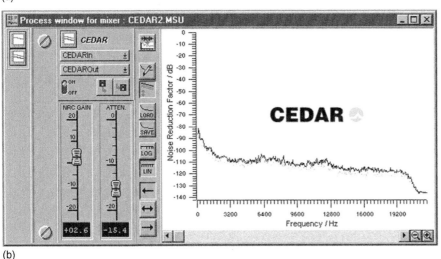

(b)

FIGURE 9.28

CEDAR restoration plug-ins for SADiE, showing (a) Declick and (b) Denoise processes.

the user to view the number of samples at peak level to watch for digital overloads that might be problematic.

Level control and loudness in mastering

Level control, it might be argued, is less crucial than it used to be in the days when a recording engineer struggled to optimize a recording's dynamic range between the noise floor and the distortion ceiling (see Figure 9.29). However, there are still artistic and technical considerations.

The dynamic range of a typical digital audio system can now be well over 100 dB and there is room for the operator to allow a reasonable degree of 'headroom' between the peak audio signal level and the maximum allowable level. Meters are provided to enable the signal level to be observed, and they are usually calibrated in dB, with zero at the top and negative dBs below this. The full dynamic range is not always shown, and there may be a peak bar that can hold the maximum level permanently or temporarily. As explained in Chapter 8, 0 dBFS (full scale) is the point at which all of the bits available to represent the signal have been used. Above this level the signal clips and the effect of this is quite objectionable, except on very short transients where it may not be noticed. It follows that signals should never be allowed to clip (see Figure 9.30).

There is a tendency in modern audio production to want to master everything so that it sounds as loud as possible, and to ensure that the signal peaks as close to 0 dBFS as possible. This level maximizing or normalizing process can be done automatically in most packages, the software searching the audio track for its highest level sample and then adjusting the overall

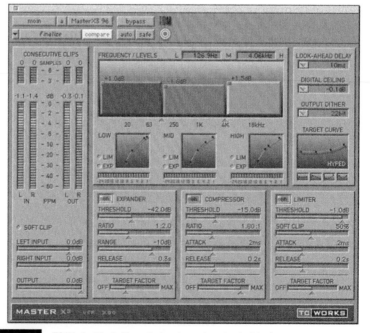

FIGURE 9.29 *TC Works MasterX mastering dynamics plug-in interface.*

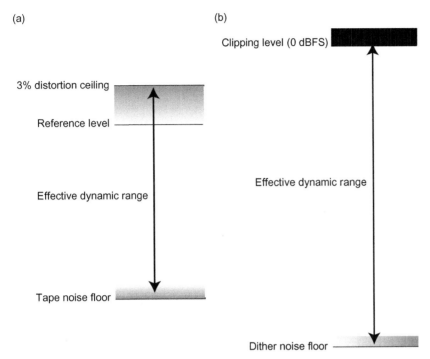

(a)

3% distortion ceiling

Reference level

Effective dynamic range

Tape noise floor

(b)

Clipping level (0 dBFS)

Effective dynamic range

Dither noise floor

FIGURE 9.30 *Comparison of analogue and digital dynamic range. (a) Analogue tape has increasing distortion as the recording level increases, with an effective maximum output level at 3% third harmonic distortion. (b) Modern high resolution digital systems have wider dynamic range with a noise floor fixed by dither noise and a maximum recording level at which clipping occurs. The linearity of digital systems does not normally become poorer as signal level increases, until 0 dBFS is reached. This makes level control a somewhat less important issue at the initial recording stage, provided sufficient headroom is allowed for peaks.*

gain so that this just reaches 0 dBFS. In this way the recording can be made to use all the bits available, which can be useful if it is to be released on a relatively low-resolution consumer medium where noise might be more of a problem. (It is important to make sure that correct redithering is used when altering the level and requantizing, as explained in Chapter 8.) This does not, of course, take into account any production decisions that might be involved in adjusting the overall levels of individual tracks on an album or other compilation, where relative levels should be adjusted according to the nature of the individual items, their loudness and the producer's intent.

However, even if the signal is maximized in the automatic fashion, so that the highest sample value just does not clip, subsequent stages in the signal chain may still do so, such as low bit rate codecs and oversampling convertors. Some equipment, for example, is designed in such a way that the maximum digital signal level is aligned to coincide with the clipping voltage of the analog electronics in a D/A convertor. In fact, owing to the response of the reconstruction filter in an (oversampling) D/A convertor (which reconstructs an analog waveform from the PAM pulse train) intersample signal peaks can be created that slightly exceed the analog level corresponding to 0 dBFS, thereby clipping the analog side of the convertor. For this reason it is recommended that digital-side signals are maximized so that they peak a few dB below 0 dBFS, in order to avoid the possibility of clipping. Some mastering software, such as the Mastered for iTunes suite discussed below, provides detailed analysis of the signal showing exactly how many samples occur in sequence at peak level, which can be a useful warning of potential or previous clipping.

The recent tendency to master recordings as loud as possible has been challenged by the introduction of loudness normalization. Loudness meters are also available that show weighted averages and peaks over a period of time. Loudness normalization is increasingly applied to audio material, particularly in the broadcast domain, and this aims to ensure that program items have a more comparable loudness to each other. The ITU and other organizations now publish detailed standards such as ITU-R BS.1770 and EBU R128, which specify algorithms to measure program loudness, and it is now becoming compulsory in many regions of the world to use these in broadcast delivery. Dolby's Dialnorm (dialogue normalization) is an example of something similar used in movie sound tracks. It is also an optional feature of some music players such as iTunes (Sound Check), and loudness metadata describing the average loudness level of tracks in LUFS (loudness units related to full scale) are increasingly added to recordings either at the mastering stage or by player software. This has the effect of causing material that was mastered at very high levels to be automatically reduced during replay, if the feature is turned on in players. Some online music delivery services do this automatically. The results of replay normalization can also expose the unpleasant sound quality artefacts that may arise from over-compressed and potentially clipped masters. A return to producing masters with more headroom and dynamic range is therefore increasingly recommended by professionals in the field.

PREPARING FOR AND UNDERSTANDING RELEASE MEDIA

Consumer release formats such as CD, DVD and MP3 usually require some form of mastering and pre-release preparation. This can range from subtle tweaks to the sound quality and relative levels on tracks to PQ encoding, data encoding and the addition of graphics, video and text.

Physical media

PQ encoding for CD mastering can often be done in some of the application packages designed for audio editing, such as SADiE and Pyramix. In this case it may involve little more than marking the starts and ends of the tracks in the playlist and allowing the software to work out the relevant frame advances and Red Book requirements for the assembly of the PQ code that will either be written to a CD-R or included in a DDP (Disc Description Protocol) file for sending to the pressing plant. The CD only comes at one resolution and sampling frequency (16 bit, 44.1 kHz) making release preparation a relatively straightforward matter.

DVD mastering is considerably more complicated than CD and requires advanced authoring software that can deal with all the different options possible on this multi-faceted release format. A number of different combinations of players and discs are possible, although the format devised specifically for high resolution surround audio, known as the DVD-Audio format, was not particularly successful commercially. DVD-Video allows for 48 or 96 kHz sampling frequency and 16, 20 or 24 bit PCM encoding. A two-channel downmix must be available on the disc in linear PCM form (for basic compatibility), but most discs also include Dolby Digital or possibly DTS surround audio. Dolby Digital encoding usually involves the preparation of a file or files containing the compressed data, and a range of settings have to be made during this process, such as the bit rate, dialog normalization level, rear channel phase shift and so on. A typical control screen is shown in Figure 9.31. Then of course there are the pictures, but they are not the topic of this book. DVD masters are usually transferred to the pressing plant on DLT tapes, using the Disc Description Protocol, or on DVD-R(A) discs as a disc image with a special CMF (cutting master format) header in the disc lead-in area containing the DDP data.

SACD Authoring software enables the text information to be added, as shown in Figure 9.32. SACD masters are normally submitted to the pressing plant on AIT format data tapes.

Pure Audio Blu-Ray is a format pioneered by Stefan Bock of msm Studios. Pure Audio Blu-Ray can work at up to 192 kHz/24 bit resolution and there are also the losslessly compressed HD formats introduced by Dolby and DTS, as well as FLAC lossless encoding and MP3. It's possible to have all this content, including 7.1 surround, on one Blu-Ray disk without running out of space as one can store around 50 Gbytes of data. mShuttle is an option that was introduced for Pure Audio Blu-Ray disks, which turns the player into a small web server, enabling audio files to be served from the player over the home network to other devices. That way the content can be used on portable media devices or played out of

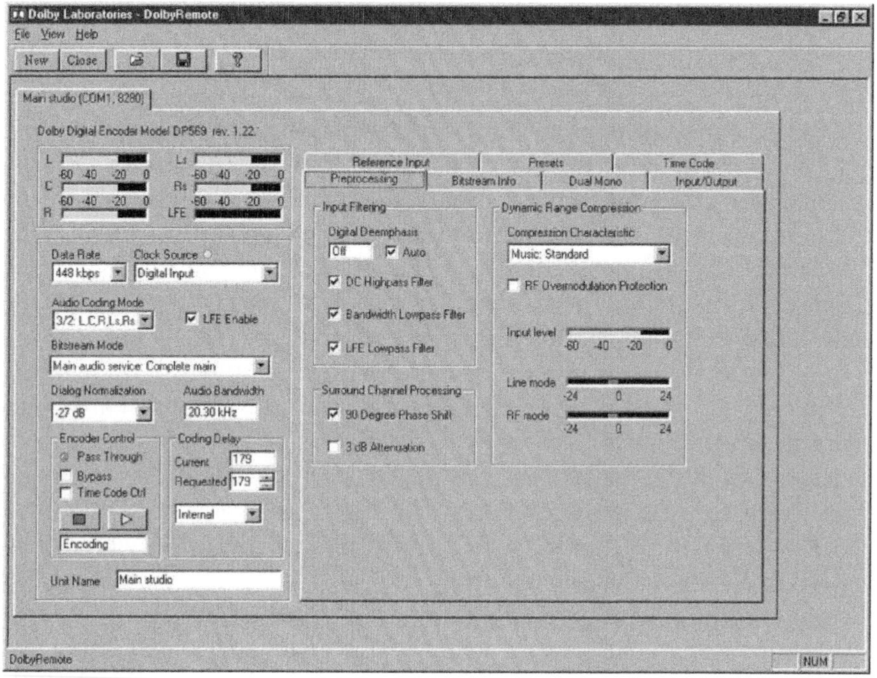

FIGURE 9.31 Screen display of Dolby Digital encoding software options.

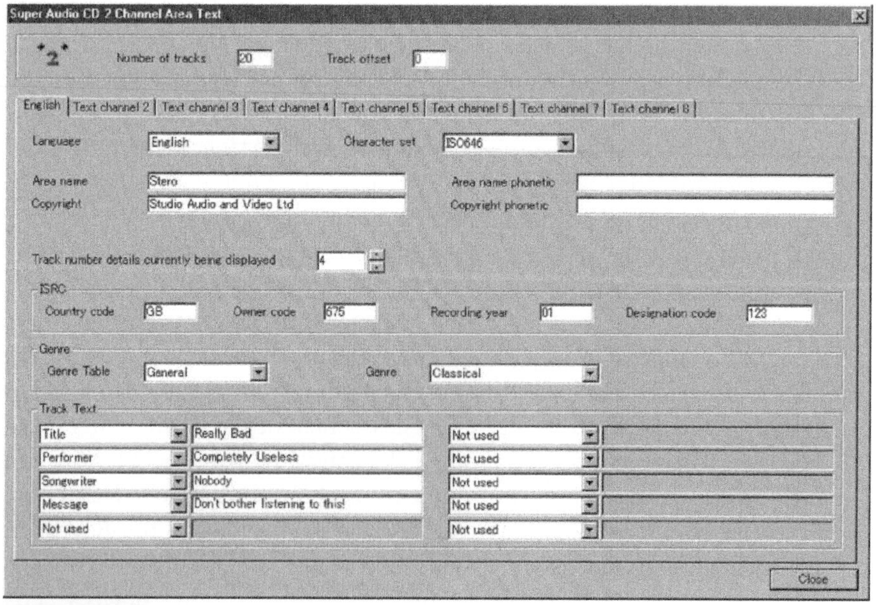

FIGURE 9.32 Example of SACD text authoring screen from SADiE.

alternative file-based audio players. This requires a player working to 'Profile 2.0', which was introduced in 2009 and includes the provision of network functionality. Mastering for these disks is currently relatively specialized, as it involves the inclusion of Java script code to implement the player features specific to the format, such as mShuttle and the ability to select replay mode using the colored buttons on the remote. There are moves to make the format more universally available and standardized.

Downloads and streaming

Mastering and preparation of material for online delivery and downloads is now of at least as much importance as the preparation for physical media, as the Internet has become the dominant mode of delivery in many markets. This has led to the introduction of schemes such as Apple's Mastered for iTunes, which is described in more detail below.

MP3, as already explained in Chapter 8, is actually MPEG-1, Layer 3 encoded audio, stored in a data file, usually for distribution to consumers either on the Internet or on other release media. MP3 mastering requires that the two-channel audio signal is MPEG-encoded, using one of the many MP3 encoders available. Mastering software now usually includes MP3 encoding as an option, as well as other data-reduced formats such as the AAC encoding used for iTunes releases. It is advisable to use a high quality MP3 encoder as the format does not specify how the encoding should be done, only the bit stream and the decoder, so there are definitely good and bad solutions on the market. Fraunhofer and Sonnox joined forces to introduce a plug-in that enables a number of different codecs and bit rates to be compared in real time, so that the mastering engineer can audition the effects of encoding before committing to a final rendering for delivery. This includes blind listening comparison tools for reliable results.

Some of the choices to be made in this process concern the data rate and audio bandwidth to be encoded, as this affects the sound quality. The lowest bit rates (e.g. below 64 kbit/s) will tend to sound noticeably poorer than the higher ones, particularly if full audio bandwidth is retained. For this reason some encoders limit the bandwidth or halve the sampling frequency for very low bit rate encoding, because this tends to minimize the unpleasant side effects of MPEG encoding. It is also possible to select joint stereo coding mode, as this will improve the technical quality somewhat at low bit rates, possibly at the expense of stereo imaging accuracy. As mentioned above, at very low bit rates some audio processing may be required to make sound quality acceptable when squeezed down such a small pipe. For the highest quality it is preferable to use bit rates for MPEG AAC of around 256 kbit/s with constrained variable bit rate, for example, as used in iTunes Plus.

Mastered for iTunes

Apple's Mastered for iTunes program was introduced as a way of trying to give mastering engineers better control over the sound quality of high resolution source material released as iTunes Plus downloads. Before Mastered for iTunes (MfiT), tracks for iTunes were either simply ripped from CDs or taken from the major record company servers and loaded into iTunes Producer (a software package for preparing iTunes tracks). With MfiT, AAC encoding is done from 24 bit masters, often with lowered level to avoid clipping and get a much cleaner result. (The aim is to get the best results out of the 256 kbit/s constrained variable bit-rate [CVBR] of the iTunes Plus format.)

All the encoding for an iTunes release is done by Apple and there is identical free Apple software called 'afconvert' (see Fact File 9.8) that

FACT FILE 9.8 APPLE'S MFIT TOOLS

The tools contained in the free mastering suite that can be downloaded from Apple include the **Master for iTunes Droplet**, which is used to automate the creation of iTunes Plus masters. The suite of tools requires at least the Snow Leopard (10.6) version of OS X to run. The droplet needs either AIFF or WAVE files to be provided as source material and converts them temporarily to Apple's Core Audio Format (CAF) with a Sound Check metadata profile attached that can normalize the relative loudness levels of songs on replay. AAC files are then encoded.

afconvert is a command line utility that enables more direct control over all of the above MfiT encoding operations.

AURoundTripAAC is an Audio Unit (AU) that allows the comparison of encoded audio against the original source file, which also includes clip and peak detection (see screen shots (a)–(c) opposite). There is a listening facility that allows a simple double-blind ABX test to be set up, in order that users can check whether they can reliably tell the difference between source and encoded versions. The plug-in can be used with workstation software that conforms to the AU plugin format, such as Logic, or alternatively the AU Lab application can be used to run the process.

AU Lab is a free standalone digital mixer utility that lets you use AU-type plug-ins without needing an AU-compatible DAW.

afclip is a Unix command line tool that can be used to check a file for on-sample and inter-sample clipping. Inter-sample clipping can arise in oversampling D/A converters used after decoding, for example. (Four-times oversampling is used to estimate sample values in afclip.) When mastering a track for iTunes that peaks very close to digital maximum it's necessary to check it using this tool and reduce the level slightly until an acceptable number of clips is indicated (which may be zero, unless a small number turn out to be inaudible). If there's any on-sample clipping the output of this process is an audio file (.wav) where the left channel data is the original audio and the right channel contains impulses where the audio is clipped, so that clips can be quickly located visually in a digital audio editor. There's also a table that comes up in the Terminal window (see screen shot (d) opposite) to show the timing locations of clips and the amount by which the samples exceed the clipping point. 'Pinned samples' can also be reported – that is, any in a series with a digital level of exactly ± 1.0 (peak level), which suggests on-sample clipping may have occurred.

Finally, the **Audio to WAVE Droplet** converts files that are in other audio file formats (any supported by Mac OS X) to the WAVE format.

FACT FILE 9.8 (CONTINUED)

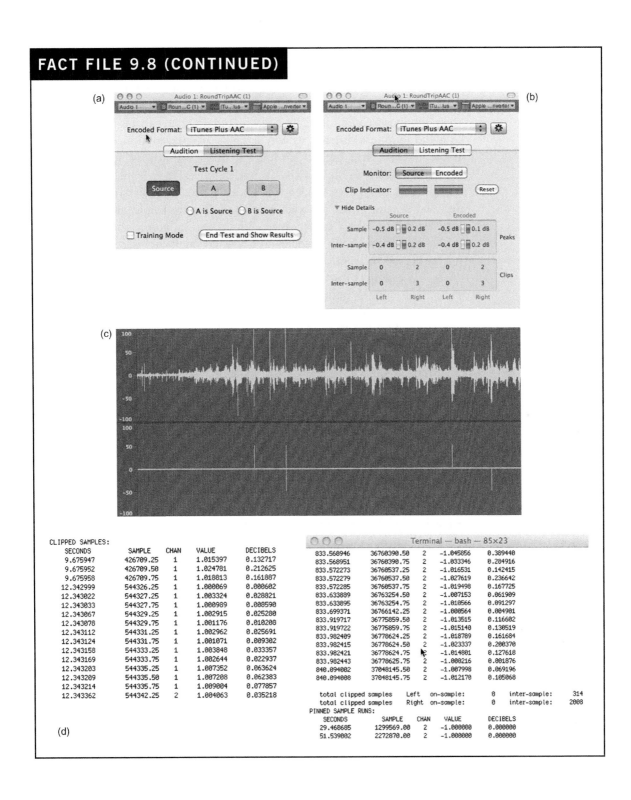

enables users to do the same thing themselves before submitting masters. The first process in this software is Sound Check, which looks at the relative loudness levels of songs to be encoded and attempts to determine how much their levels should be raised or lowered on replay to make their loudness comparable. It adds metadata that can be used by players to avoid loudness differences when tracks are played alongside each other. If the track is at a higher sampling frequency than 44.1 kHz it is downsampled to 44.1 kHz, otherwise it is left alone. There is also a process that will convert the AAC encoded track back to PCM so that you can hear the decoded version. 'afclip' looks at the likely on-sample and inter-sample clips, behaving like a true peak-reading meter, enabling the user to determine the potential for encoder and post-decoder clipping. After the track is transferred to Apple, it is encoded in exactly the same way as the user would have done. 'Test pressings' are then returned to the record company to confirm what is about to be released on iTunes. Usually these turn out to be bit-for-bit the same as the final encoding created by the mastering engineer, which confirms the integrity of the process.

Apple prefers to receive high-resolution masters at sampling frequencies above 44.1 kHz, preferably 96 kHz. That way the encoding process is forced to use its mastering-quality sample rate conversion that generates 32-bit floating point CAF files (see Chapter 10) as the input to AAC encoding. It's claimed that this avoids the need for redithering and preserves all the dynamic range inherent in the original file, avoiding the potential for aliasing or clipping that can otherwise arise in sample rate conversion. (If you supply 44.1 kHz files to Apple the advantages of the above process are bypassed as the sample rate conversion is not initiated.)

RECOMMENDED FURTHER READING

Collins, M., 2013. Pro Tools 11: Music Production, Recording, Editing, and Mixing. Focal Press.

Katz, B., 2007. Mastering Audio: The Art and Science, second edition. Focal Press.

Katz, B., 2012. iTunes Music: Mastering High Resolution Audio Delivery: Produce Great Sounding Music with Mastered for iTunes. Focal Press.

Leider, C., 2004. Digital Audio Workstation: Mixing, Recording and Mastering Your MAC or PC. McGraw-Hill Professional.

Watkinson, J., 2001. The Art of Digital Audio, third edition. Focal Press.

Digital Audio Formats and Interchange

This chapter provides further details about the main formats in which digital audio data is stored and moved between systems. This includes coverage of audio file formats, digital interfaces and networked audio interchange, concentrating mainly on those issues of importance for professional applications.

AUDIO FILE FORMATS FOR DIGITAL WORKSTATIONS

There used to be almost as many file formats for audio as there are days in the year. In the computer games field, for example, this is still true to some extent. For a long time the specific file storage strategy used for disk-based digital audio was the key to success in digital workstation design, because disk drives were relatively slow and needed clever strategies to ensure that they were capable of handling a sufficiently large number of audio channels. Manufacturers also worked in isolation and the size of the market was relatively small, leading to virtually every workstation or piece of software using a different file format for audio and edit list information. Although there may still be theoretical performance advantages in the use of filing structures specially designed for real-time applications such as audio and video editing, interchange of material between systems and applications is now at least as important as ultimate transfer speed. Also the majority of hard disk drives available today are capable of replaying many channels of audio in real time without needing to use a dedicated storage strategy. This has led to the increased use of a few common cross-platform file formats such as WAVE and AIFF, both of which can be used in the IEEE 32-bit floating point data format that is becoming popular on audio workstations, as well as in conventional 16–24 bit PCM mode.

The growth in the importance of metadata (data about data), and the representation of audio, video and metadata as 'objects', has led to the development of interchange methods that are based on object-oriented concepts and project 'packages' as opposed to using simple text files and separate media files. There is increasing integration between audio and other media in multimedia authoring and some of the file formats mentioned below are closely related to international efforts in multimedia file exchange.

It is not proposed to attempt to describe all of the file formats in existence, because that would be a relatively pointless exercise and would not make for interesting reading. It is nonetheless useful to have a look at some examples taken from the most commonly encountered file formats, particularly those used for high-quality audio by desktop and multimedia systems, since these are amongst the most widely used in the world and are often handled by audio workstations even if not their native format. It is not proposed to investigate the large number of specialized file formats developed principally for computer music on various platforms, nor the files used for internal sounds and games on many computers.

File formats in general

A data file is simply a series of data bytes formed into blocks and stored either contiguously or in fragmented form. Files themselves are largely independent of the operating system and filing structure of the host computer, because a file can be transferred to another platform and still exist as an identical series of data blocks. It is the filing system that is often the platform- or operating-system-dependent entity (e.g. FAT32 or HFS).

There are sometimes features of data files that relate directly to the operating system and filing system that created them, but they do not normally prevent such files being translated by other platforms. For example, there are two approaches to byte ordering: the so-called little-endian order in which the least significant byte comes first or at the lowest memory address, and the big-endian format in which the most significant byte comes first or at the highest memory address. These originally related to the byte ordering used in data processing by the two most common microprocessor families and thereby to the two most common operating systems used in desktop audio workstations. Motorola processors, as originally used in the Apple Mac, dealt in big-endian byte ordering, and Intel processors, as used in MS-DOS machines (now in Macs too), deal in little-endian byte ordering.

Second, some older Mac files may have two parts – a resource fork and a data fork – whereas Windows files only have one part. High-level 'resources' were stored in the resource fork (used in some legacy audio files for storing information about the file, such as signal processing to be applied, display information and so forth) whilst the raw data content of the file was stored in the data fork (used in audio applications for audio sample data). The resource fork is not always there, but may be. The resource fork can get lost when transferring such files between machines or to servers, unless Mac-specific protocols are used (e.g. MacBinary or BinHex). Mac OS X files are not supposed to use resource forks, in order that they can be easily transferred between systems.

Some data files include a 'header', that is a number of bytes at the start of the file containing information about the data that follows. In audio systems this may include the sampling rate and resolution of the file. Audio replay would normally be started immediately after the header. On the other hand, some files are simply raw data, usually in cases where the format is fixed. ASCII text files are a well-known example of raw data files – they simply begin with the first character of the text. More recently file structures have been developed that are really 'containers' for lots of smaller files, or data objects, each with its own descriptors and data. The RIFF structure, described below, is an early example of the concept of a 'chunk-based' file structure. Apple's Bento container structure, used in OMFI, and the container structure of AAF are more advanced examples of such an approach.

The audio data in most common high-quality audio formats are stored in two's complement form (see Chapter 8) and the majority of files are used for 16 or 24 bit data, thus employing either 2 or 3 bytes per audio sample. 8 bit files use 1 byte per sample. 32 bit floating point files use 4 bytes per sample, three containing the mantissa and one containing the exponent. This makes the storage required to use this number format 50% greater than for 24 bit fixed point operation.

AIFF and AIFF-C formats

The AIFF format is widely used as an audio interchange standard, because it conforms to the EA IFF 85 standard for interchange format files used for various other types of information such as graphical images. AIFF is an Apple standard format for audio data and is encountered widely on Mac-based audio workstations and some Silicon Graphics systems. Audio information can be stored at a number of resolutions and for any number of channels if required, and the related AIFF-C (file type 'AIFC') format allows also for compressed audio data. It consists only of a data fork, with no resource fork, making it easy to transport to other platforms.

All IFF-type files are typically made up of 'chunks' of data as shown in Figure 10.1. A chunk consists of a header and a number of data bytes to follow. The simplest AIFF files contain a 'common chunk', which is equivalent to the header data in other audio files, and a 'sound data' chunk containing the audio sample data. These are contained overall by a 'form' chunk as shown in Figure 10.2. AIFC files must also contain a 'version chunk' before the common chunk to allow for future changes to AIFC.

RIFF WAVE format

The RIFF WAVE (often called WAV) format is the Microsoft equivalent of Apple's AIFF. It has a similar structure, again conforming to the IFF pattern, but with numbers stored in little-endian rather than big-endian form.

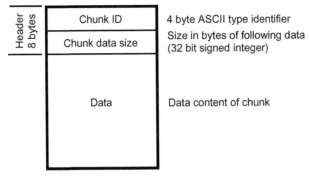

FIGURE 10.1 *General format of an IFF file chunk.*

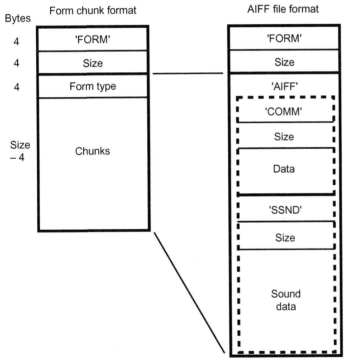

FIGURE 10.2 *General format of an AIFF file.*

It is used widely for sound file storage and interchange on PC workstations, and for multimedia applications involving sound. Within WAVE files it is possible to include information about a number of cue points, and a playlist to indicate the order in which the cues are to be replayed. WAVE files use the file extension 'wav'.

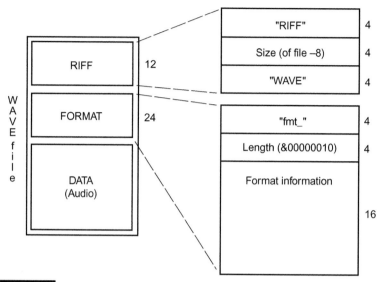

FIGURE 10.3 *Diagrammatic representation of a simple RIFF WAVE file, showing the three principal chunks. Additional chunks may be contained within the overall structure, for example a 'bext' chunk for the Broadcast WAVE file.*

A basic WAV file consists of three principal chunks, as shown in Figure 10.3: the RIFF chunk, the FORMAT chunk and the DATA chunk. The RIFF chunk contains 12 bytes, the first four of which are the ASCII characters 'RIFF', the next four indicating the number of bytes in the remainder of the file (after the first eight) and the last four of which are the ASCII characters 'WAVE'. The format chunk contains information about the format of the sound file, including the number of audio channels, sampling rate and bits per sample, as shown in Table 10.1.

The audio data chunk contains a sequence of bytes of audio sample data, divided as shown in the FORMAT chunk. Unusually, if there are only 8 bits per sample or fewer each value is unsigned and ranges between 0 and 255 (decimal), whereas if the resolution is higher than this the data is signed and ranges both positively and negatively around zero. Audio samples are interleaved by channel in time order, so that if the file contains two channels a sample for the left channel is followed immediately by the associated sample for the right channel. The same is true of multiple channels (one sample for time-coincident sample periods on each channel is inserted at a time, starting with the lowest numbered channel), although basic WAV files were nearly always just mono or two channel.

Table 10.1	Contents of FORMAT Chunk in a Basic WAVE PCM File	
Byte	**ID**	**Contents**
0–3	ckID	'fmt_' (ASCII characters)
4–7	nChunkSize	Length of FORMAT chunk (binary, hex value: &00000010)
8–9	wFormatTag	Audio data format (e.g. &0001 = WAVE format PCM) Other formats are allowed, for example IEEE floating point and MPEG format (&0050 = MPEG 1)
10–11	nChannels	Number of channels (e.g. &0001 = mono, &0002 = stereo)
12–15	nSamplesPerSec	Sample rate (binary, in Hz)
16–19	nAvgBytesPerSec	Bytes per second
20–21	nBlockAlign	Bytes per sample: e.g. &0001 = 8 bit mono; &0002 = 8 bit stereo or 16 bit mono; &0004 = 16 bit stereo
22–23	nBitsPerSample	Bits per sample

The RIFF WAVE format is extensible and can have additional chunks to define enhanced functionality such as surround sound and other forms of coding. This is known as 'WAVE-format extensible'. Chunks can include data relating to cue points, labels and associated data, for example. The Broadcast WAVE format is one example of an enhanced WAVE file (see Fact File 10.1), which is used widely in professional applications for interchange purposes.

DSD-IFF file format

The DSD-IFF file format is based on a similar structure to other IFF-type files, described above, except that it is modified slightly to allow for the large file sizes that may be encountered with the high-resolution Direct Stream Digital format used for SuperAudio CD. Specifically the container FORM chunk is labeled 'FRM8' and this identifies all local chunks that follow as having 'length' indications that are 8 bytes long rather than the normal 4. In other words, rather than a 4 byte chunk ID followed by a 4 byte length indication, these files have a 4 byte ID followed by an 8 byte length indication. This allows for the definition of chunks with a length greater than 2 Gbytes, which may be needed for mastering SuperAudio CDs. There are also various optional chunks that can be used for exchanging more detailed information and comments such as might be used in project interchange. Further details of this file format, and an excellent guide to the use of DSD-IFF in project applications, can be found in the DSD-IFF specification, as described in the 'Recommended further reading' at the end of this chapter.

FACT FILE 10.1 BROADCAST WAVE FORMAT

The Broadcast WAVE format, described in EBU Tech. 3285, was standardized by the European Broadcasting Union (EBU) because of a need to ensure compatibility of sound files and accompanying information when transferred between workstations. It is based on the RIFF WAVE format described above, but contains an additional chunk that is specific to the format (the 'broadcast_audio_extension' chunk, ID 5 'bext') and also limits some aspects of the WAVE format. Version 0 was published in 1997 and Version 1 in 2001, the only difference being the addition of an SMPTE UMID (Unique Material Identifier) in version 1 (this is a form of metadata). Such files currently only contain either PCM or MPEG-format audio data. An optional Extended-BWF file (BWF-E) enables the size to exceed the limits of the basic RIFF WAVE format by extending the address space to 64 bits.

Broadcast WAVE files contain at least three chunks: the broadcast_audio_extension chunk, the format chunk and the audio data chunk. The broadcast extension chunk contains the data shown in the table below. Optionally files may also contain further chunks for specialized purposes and may contain chunks relating to MPEG audio data (the 'fact' and 'mpeg_audio_extension' chunks). MPEG applications of the format are described in EBU Tech. 3285, Supplement 1 and the audio data chunk containing the MPEG data normally conforms to the MP3 frame format.

A multichannel extension chunk defines the channel ordering, surround format, downmix coefficients for creating a two-channel mix, and some descriptive information. There are also chunks defined for metadata describing the audio contained within the file, such as the 'quality chunk' (ckID = 'qlty'), which together with the coding history contained in the 'bext' chunk make up the so-called 'capturing report'. These are described in Supplement 2 to EBU Tech. 3285. Finally there is a chunk describing the peak audio level within a file, which can aid automatic program level setting and program interchange. Recent revisions include the option to add loudness metadata into the file.

BWF files can be either mono, two-channel or multichannel (sometimes called polyfiles, or BWF-P), and utilities exist for separating polyfiles into individual mono files which some applications require.

Broadcast audio extension chunk format

Data	Size (bytes)	Description
ckID	4	Chunk ID = 'bext'
ckSize	4	Size of chunk
Description	256	Description of the sound clip
Originator	32	Name of the originator
OriginatorReference	32	Unique identifier of the originator (issued by the EBU)
OriginationDate	10	'yyyy-mm-dd'
OriginationTime	8	'hh-mm-ss'
TimeReferenceLow	4	Low byte of the first sample count since midnight
TimeReferenceHigh	4	High byte of the first sample count since midnight
Version	2	BWF version number, e.g. &0001 is Version 1
UMID	64	UMID according to SMPTE 330 M; if only a 32 byte UMID then the second half should be padded with zeros
Reserved	190	Reserved for extensions; set to zero in Version 1.
CodingHistory	Unrestricted	A series of ASCII strings, each terminated by CR/LF (carriage return, line feed) describing each stage of the audio coding history, according to EBU R-98

Apple core audio format

Apple's Core Audio Format (CAF) is a chunk-based container structure for storing audio in a way that is highly compatible with the Core Audio architecture of Mac OS X (see Chapter 9). It has a number of advantages over the other standard file types mentioned above in that it can have unlimited size, it can contain audio in a number of data formats and for any number of audio channels, and it can contain a range of metadata types including markers and channel layouts. Recording is also said to be safer because the file header does not have to be rewritten or updated at the end of or during recording, and new data can be appended to the end of existing files in a way that allows applications to determine the length of a file even if the header has not been properly finalized. The Channel Layout chunk describes the way in which channels are allocated to particular surround sound loudspeaker locations in multichannel files, in a similar way to the extended WAVE multichannel formats.

MPEG audio file formats

It is possible to store MPEG-compressed audio in AIFF-C or WAVE files, with the compression type noted in the appropriate header field. There are also older MS-DOS file extensions used to denote MPEG audio files, notably .MPA (MPEG Audio) or .ABS (Audio Bit Stream). However, owing to the ubiquity of the so-called 'MP3' format (MPEG 1, Layer 3) for audio distribution on the Internet, MPEG audio files are increasingly denoted with the extension '.MP3'. Such files are relatively simple, being really no more than MPEG audio frame data in sequence, each frame being preceded by a frame header. MPEG-4 files are containers that can carry multiple streams of audio, video and subtitle information. Two file extensions are commonly used, namely .mp4 and .m4a, having essentially the same structure. The latter was popularized by Apple with its iTunes releases and contains only audio information, including lossy-coded AAC data and Apple Lossless Audio Coding (ALAC) data.

Edit decision list (EDL) files and project interchange

EDL formats were historically proprietary but the need for open interchange of project data has increased the use of standardized EDL structures and 'packaged' project formats to make projects transportable between systems from different manufacturers.

Project interchange can involve the transfer of edit list, mixing, effects and audio data. Many of these are proprietary, such as the Digidesign ProTools session format. Software, such as SSL Pro-Convert, can be obtained for audio and video workstations that translates EDLs or projects between a number of different systems to make interchange easier. The OMFI (Open Media Framework Interchange) structure, originally developed by Avid, was one early attempt at an open project interchange format and contained a format for interchanging edit list data. Other options include XML-tagged formats that identify different items in the edit list in a text-based form. AES-31 is gaining popularity among workstation software manufacturers as a simple means of exchanging audio editing projects between systems, and is described in more detail below.

AES-31 format

AES-31 is an international standard designed to enable straightforward interchange of audio files and projects between systems. Audio editing packages are increasingly offering AES-31 as a simple interchange format for edit lists. In Part 1 the standard specifies a disk format that is compatible with the FAT32 file system, a widely used structure for the formatting of computer hard disks. Part 2 describes the use of the Broadcast WAVE audio file format. Part 3 describes simple project interchange, including a format for the communication of edit lists using ASCII text that can be parsed by a computer as well as read by a human. The basis of this is the edit decision markup language (EDML). It is not necessary to use all the parts of AES-31 to make a satisfactory interchange of elements. For example, one could exchange an edit list according to Part 3 without using a disk based on Part 1. Adherence to all the parts would mean that one could take a removable disk from one system, containing sound files and a project file, and the project would be readable directly by the receiving device.

EDML documents are limited to a 7 bit ASCII character set in which white space delimits fields within records. Standard carriage return (CR) and line-feed (LF) characters can be included to aid the readability of lists but they are ignored by software that might parse the list. An event location is described by a combination of time code value and sample count information. The time code value is represented in ASCII using conventional hours, minutes, seconds and frames (e.g. HH:MM:SS:FF) and the optional sample count is a four-figure number denoting the number of samples after the start of the frame concerned at which the event actually occurs. This enables sample-accurate edit points to be specified. It is slightly more complicated than this because the ASCII delimiters between the time code fields are changed to indicate various parameters:

HH:MM delimiter = Frame count and timebase indicator (see Table 10.2)
MM:SS delimiter = Film frame indicator (if not applicable, use the previous delimiter)
SS:FF delimiter = Video field and timecode type (see Table 10.3)

Table 10.2	Frame Count and Timebase Indicator Coding in AES-31		
		Timebase	
Frame count	**Unknown**	*1.000*	*1.001*
30	?	I	:
25	!	.	/
24	#	=	–

Table 10.3	Video Field and Timecode Type Indicator in AES-31	
	Video Field	
Counting mode	**Field 1**	**Field 2**
PAL	.	:
NTSC non-drop-frame	.	:
NTSC drop-frame	,	;

The delimiter before the sample count value is used to indicate the audio sampling frequency, including all the pull-up and pull-down options (e.g. f_s times 1/1.001). There are too many of these possibilities to list here and the interested reader is referred to the standard for further information. This is an example of a time code and (after the slash denoting 48 kHz sampling frequency) optional sample count value:

14:57:24.03/0175

The Audio Decision List (ADL) is contained between two ASCII keyword tags <ADL> and </ADL>. It in turn contains a number of sections, each contained within other keyword tags such as <VERSION>, <PROJECT>, <SYSTEM> and <SEQUENCE>. The edit points themselves are contained in the <EVENT_LIST> section. Each event begins with the ASCII keyword "(Entry)", which serves to delimit events in the list, followed by an entry number (32 bit integer, incrementing through the list) and an entry type keyword to describe the nature of the event (e.g. "(Cut)"). Each different event te then has a number of bytes following that define the event more specifically. The following is an example of a simple cut edit, as suggested by the standard:

(Entry)0010(Cut)F "FILE://VOL/DIR/FILE"11 03:00:00;00/0000
01:00:00:00/0000 01:00:10:00/0000 _

This sequence essentially describes a cut edit, entry number 0010, the source of which is the file (F) with the path shown, using channel 1 of the source file (or just a mono file), placed on track 1 of the destination timeline, starting at timecode three hours in the source file, placed to begin at one hour in the destination timeline (the 'in point') and to end ten seconds later (the 'out point'). Some workstation software packages store a timecode value along with each sound file to indicate the nominal start time of the original recording (e.g. BWF files contain a timestamp in the 'bext' chunk), otherwise each sound file is assumed to start at time zero.

It is assumed that default crossfades will be handled by the workstation software itself. Most workstations introduce a basic short crossfade at each edit point to avoid clicks, but this can be modified by 'event modifier' information in the ADL. Such modifiers can be used to adjust the shape and duration of a fade in or fade out at an edit point. There is also the option to point at a rendered crossfade file for the edit point, as described in Chapter 9.

MXF – the media exchange format

MXF was developed by the Pro-MPEG forum as a means of exchanging audio, video and metadata between devices, primarily in television operations. It is based on the modern concept of media objects that are split into 'essence' and 'metadata'. Essence files are the raw material (i.e. audio and video) and the metadata describes things about the essence (such as where to put it, where it came from and how to process it).

MXF files attempt to present the material in a 'streaming' format, that is, one that can be played out in real time, but they can also be exchanged in conventional file transfer operations. As such they are normally considered to be finished program material, rather than material that is to be processed somewhere downstream, designed for playout in broadcasting environments. The bit stream is also said to be compatible with recording on digital videotape devices.

AAF – the advanced authoring format

AAF is an authoring format for multimedia data that is supported by numerous vendors, including Avid which adopted it as a migration path from OMFI. Parts of OMFI 2.0 form the basis for parts of AAF and there are also close similarities between AAF and MXF (described in the previous section). Like the formats to which it has similarities, AAF is an object-oriented format that combines essence and metadata within a container structure. Unlike MXF it is designed for project interchange such that elements within the project can be modified, post-processed and resynchronized. It is not, therefore, directly suitable as a streaming format but can easily be converted to MXF for streaming if necessary.

Rather like OMFI it is designed to enable complex relationships to be described between content elements, to map these elements onto a timeline, to describe the processing of effects, synchronize streams of essence, retain historical metadata and refer to external essence (essence not contained within the AAF package itself). It has three essential parts: the AAF Object Specification (which defines a container for essence and metadata, the logical contents of objects and rules for relationships between them); the AAF Low-Level Container Specification (which defines a disk filing structure for the data, based on Microsoft's Structured Storage); and the AAF SDK Reference Implementation (which is a software development kit that enables applications to deal with AAF files). The Object Specification is extensible in that it allows new object classes to be defined for future development purposes.

The basic object hierarchy is illustrated in Figure 10.4, using an example of a typical audio post-production scenario. 'Packages' of metadata are defined that describe either compositions, essence or physical media.

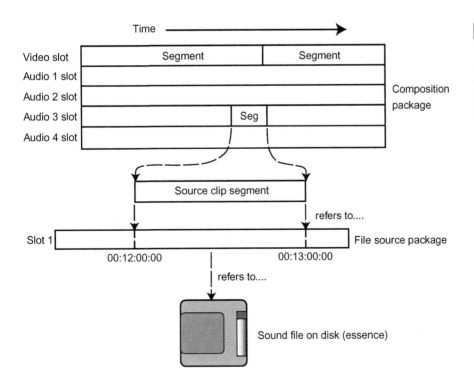

FIGURE 10.4

Graphical conceptualisation of some metadata package relationships in AAF – a simple audio post-production example.

Some package types are very 'close' to the source material (they are at a lower level in the object hierarchy, so to speak) – for example, a 'file source package' might describe a particular sound file stored on disk. The metadata package, however, would not be the file itself, but it would describe its name and where to find it. Higher-level packages would refer to these lower-level packages in order to put together a complex program. A composition package is one that effectively describes how to assemble source clips to make up a finished program. Some composition packages describe effects that require a number of elements of essence to be combined or processed in some way.

Packages can have a number of 'slots'. These are a bit like tracks in more conventional terminology, each slot describing only one kind of essence (e.g. audio, video, graphics). Slots can be static (not time-dependent), timeline (running against a timing reference) or event-based (one-shot, triggered events). Slots have segments that can be source clips, sequences, effects or fillers. A source clip segment can refer to a particular part of a slot in a separate essence package (so it could refer to a short portion of a sound file that is described in an essence package, for example).

Disk pre-mastering formats

The Disk Description Protocol (DDP) developed and licensed by Doug Carson and Associates has been widely adopted for describing consumer optical disk masters, and some optical disk pressing plants require masters to be submitted in this form. Version 1 of the DDP for CD laid down the basic data structure but said little about higher-level issues involved in interchange, making it more than a little complicated for manufacturers to ensure that DDP masters from one system would be readable on another. Version 2 addressed some of these issues. There are versions of DDP for CD, DVD and HD DVD-ROM. The so-called Cutting Master Format (CMF) sanctioned by the DVD Forum is derived from DDP, but the Blu-Ray CMF is not related to DDP.

DDP is a protocol for describing the contents of a disk, which is not medium specific, so a range of tape or disk storage media can be used to transfer files to pressing plants. DDP files can be supplied separately to the audio data if necessary. The protocol consists of a number of 'streams' of data, each of which carries different information to describe the contents of the disk. These streams may be either a series of packets of data transferred over a network, files on a disk or tape, or

raw blocks of data independent of any filing system. The DDP protocol simply maps its data into whatever block or packet size is used by
the medium concerned, provided that the block or packet size is at least
128 bytes. Either a standard computer filing structure can be used, in
which case each stream is contained within a named file, or the storage
medium is used 'raw' with each stream starting at a designated sector or
block address.

The ANSI tape labeling specification is used to label the media used for
DDP transfers. This allows the names and locations of the various streams
to be identified. The principal streams included in a DDP transfer for CD
mastering are as follows:

1. DDP ID stream or 'DDPID' file. 128 bytes long, describing the type
   and level of DDP information, various 'vital statistics' about the
   other DDP files and their location on the medium (in the case of
   physically addressed media), and a user text field (not transferred to
   the CD).

2. DDP Map stream or 'DDPMS' file. This is a stream of 128 byte data
   packets which together give a map of the CD contents, showing
   what types of CD data are to be recorded in each part of the CD,
   how long the streams are, what types of subcode are included, and so
   forth. Pointers are included to the relevant text, subcode and main
   streams (or files) for each part of the CD.

3. Text stream. An optional stream containing text to describe
   the titling information for volumes, tracks or index points
   (not currently stored in CD formats), or for other text comments.
   If stored as a file, its name is indicated in the appropriate map
   packet.

4. Subcode stream. Optionally contains information about the
   subcode data to be included within a part of the disk, particularly for
   CD-DA. If stored as a file, its name is indicated in the appropriate
   map packet.

5. Main stream. Contains the main data to be stored on a part of the
   CD, treated simply as a stream of bytes, irrespective of the block or
   packet size used. More than one of these files can be used in cases
   of mixed-mode disks, but there is normally only one in the case of
   a conventional audio CD. If stored as a file, its name is indicated in
   the appropriate map packet.

INTERCONNECTING DIGITAL AUDIO DEVICES

Introduction

In the case of analog interconnection between devices, replayed digital audio is converted to the analog domain by the replay machine's D/A convertors, routed to the recording machine via a conventional audio cable and then reconverted to the digital domain by the recording machine's A/D convertors. The audio is subject to any gain changes that might be introduced by level differences between output and input, or by the record gain control of the recorder and the replay gain control of the player. Analog domain copying is necessary if any analog processing of the signal is to happen in between one device and another, such as gain correction, equalization, or the addition of effects such as reverberation. Most of these operations, though, are now possible in the digital domain.

An analog domain copy is not a perfect copy or a clone of the original master, because the data values will not be exactly the same (owing to slight differences in recording level, differences between convertors, the addition of noise, and so on). For a clone it is necessary to make a true digital copy. This can either involve a file copying process, perhaps over a network using a workstation, or a digital interface or network may be used for the streamed interconnection of recording systems and other audio devices such as mixers and effects units. The essential differences between digital audio interfaces and networked data exchange are explained in Fact File 10.2.

Digital interface basics

Digital audio interfaces conforming to one of the standard protocols allow for a number of channels of digital audio data to be transferred between devices with no loss of sound quality. Any number of generations of digital copies can be made without affecting the sound quality of the latest generation, provided that errors have been fully corrected. (This assumes that the audio is in a linear PCM format and has not been subject to low bit rate decoding and re-encoding.) This process takes place in real time, requiring the operator to put the receiving device into record mode such that it simply stores the incoming stream of audio data. Any accompanying metadata may or may not be recorded (often most of it is not). Both machines must be operating at the same sampling frequency (unless a sampling frequency convertor is used) and it may require the recorder to be switched to 'external sync' mode, so that it can lock its sampling frequency to that of the player. Alternatively (and preferably) a common reference (e.g. word

FACT FILE 10.2 COMPUTER NETWORKS VS DIGITAL AUDIO INTERFACES

Dedicated digital audio interfaces are the digital equivalent of analog signal cables, down which signals for one or more channels are carried in real time from one point to another, possibly with some auxiliary information (metadata) attached. An example is the AES-3 interface, described in the main text. Such an audio interface uses a data format dedicated to audio purposes, whereas a computer data network can carry numerous types of information.

Dedicated interfaces are normally unidirectional, point-to-point connections, whereas computer data interconnects and networks are often bidirectional and carry data in a packet format for numerous sources and destinations. With dedicated interfaces sources may be connected to destinations using a routing matrix or by patching individual connections. Audio data are transmitted in an unbroken stream, there is no handshaking process involved in the data transfer, and erroneous data are not retransmitted because there is no mechanism for requesting its retransmission. The data rate of a dedicated audio interface is usually directly related to the audio sampling frequency, word length and number of channels of the audio data to be transmitted, ensuring that the interface is always capable of serving the specified number of channels. If a channel is unused for some reason its capacity is not normally available for assigning to other purposes (such as higher-speed transfer of another channel).

An alternative to dedicated digital audio interfaces, standard computer interconnects and networks are now used widely to transfer audio information. Computer networks are typically general purpose data carriers that may have asynchronous features and may not always have the inherent quality-of-service (QoS) features that are required for 'streaming' applications. They also normally use an addressing structure that enables packets of data to be carried from one of a number of sources to one of a number of destinations and such packets will share the connection in a more or less controlled way. Data transport protocols such as TCP/IP are often used as a universal means of managing the transfer of data from place to place, adding overheads in terms of data rate, delay and error handling that may work against the efficient transfer of audio. Such networks may be designed primarily for file transfer applications where the time taken to transfer the file is not a crucial factor – 'as fast as possible' will do. This has required some special techniques to be developed for carrying real-time data such as audio information.

USB (Universal Serial Bus) and Firewire (IEEE 1394) are examples of personal area network (PAN) technology, allowing a number of devices to be interconnected within a limited range around the user. These have a high enough data rate to carry a number of channels of audio data over relatively short distances, either over copper or optical fiber. Audio protocols also exist for these as discussed in the main text.

clock) signal can be used to synchronize all devices that are to be interconnected digitally. If one of these methods of ensuring a common sampling frequency is not used then either audio will not be decoded at all by the receiver, or regular clicks will be audible at a rate corresponding to the difference between the two sampling frequencies (at which point samples are either skipped or repeated owing to the 'sample slippage' that is occurring between the two machines). A receiver should be capable of at least the same quantizing resolution (number of bits per sample) as the source device, otherwise audio resolution will be lost. If there is a difference in resolution between the systems it is advisable to use a processor in between the machines that optimally dithers the signal for the new resolution, or alternatively, to use redithering options on the source machine to prepare the signal for its new resolution (see Chapter 8).

Dedicated audio interface formats

There are a number of types of digital interface, some of which are international standards and others of which are manufacturer specific. They all carry digital audio for one or more channels with at least 16 bit resolution and will operate at the standard sampling rates of 44.1 and 48 kHz, as well as at 32 kHz if necessary, some having a degree of latitude for varispeed. Some interface standards have been adapted to handle higher sampling frequencies such as 88.2 and 96 kHz. The interfaces vary as to how many physical interconnections are required. Some require one link per channel plus a synchronization signal, whilst others carry all the audio information plus synchronization information over one cable.

The most common interfaces are described below in outline. It is common for subtle incompatibilities to arise between devices, even when interconnected with a standard interface, owing to the different ways in which non-audio information is implemented. This can result in anything from minor operational problems to total non-communication and the causes and remedies are unfortunately far too detailed to go into here. The reader is referred to *The Digital Interface Handbook* by Rumsey and Watkinson, as well as to the standards themselves, if a greater understanding of the intricacies of digital audio interfaces is required.

The AES/EBU interface (AES-3)

The AES-3 interface, described almost identically in AES-3-1992, IEC 60958 and EBU Tech. 3250E among others, allows for two channels of digital audio (A and B) to be transferred serially over one balanced interface, using drivers and receivers similar to those used in the RS422 data transmission standard, with an output voltage of between 2 and 7 volts as shown in Figure 10.5. The interface allows two channels of audio to be transferred over distances up to 100 m, but longer distances may be covered using combinations of appropriate cabling, equalization and termination. Standard XLR-3 connectors are used, often labeled DI (for digital in) and DO (for digital out).

Each audio sample is contained within a 'subframe' (see Figure 10.6), and each subframe begins with one of three synchronizing patterns to

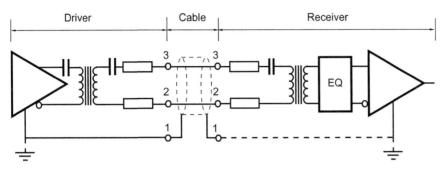

FIGURE 10.5
Recommended electrical circuit for use with the standard two-channel interface.

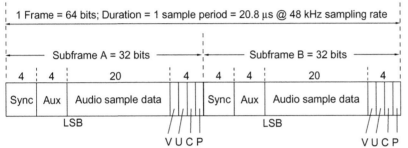

FIGURE 10.6
Format of the standard two-channel interface frame.

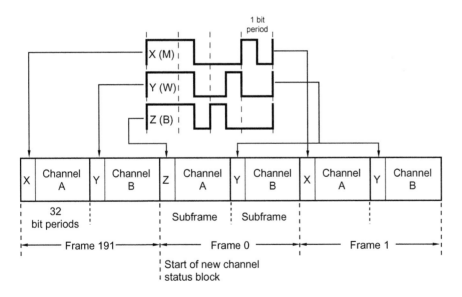

FIGURE 10.7
Three different preambles (X, Y and Z) are used to synchronise a receiver at the starts of subframes.

identify the sample as either the A or B channel, or to mark the start of a new channel status block (see Figure 10.7). These synchronizing patterns violate the rules of bi-phase mark coding (see below) and are easily identified by a decoder. One frame (containing two audio samples) is normally transmitted in the time period of one audio sample, so the data rate varies

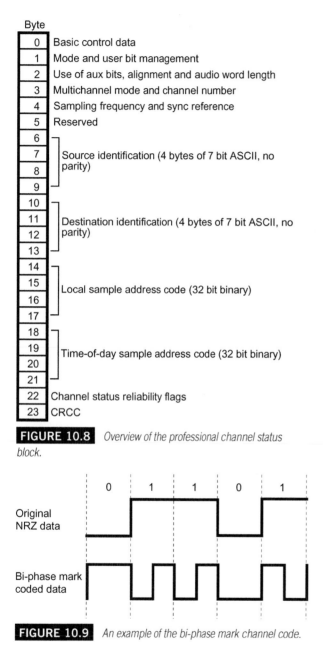

Byte

0	Basic control data
1	Mode and user bit management
2	Use of aux bits, alignment and audio word length
3	Multichannel mode and channel number
4	Sampling frequency and sync reference
5	Reserved
6	
7	Source identification (4 bytes of 7 bit ASCII, no
8	parity)
9	
10	
11	Destination identification (4 bytes of 7 bit ASCII, no
12	parity)
13	
14	
15	Local sample address code (32 bit binary)
16	
17	
18	
19	Time-of-day sample address code (32 bit binary)
20	
21	
22	Channel status reliability flags
23	CRCC

FIGURE 10.8 *Overview of the professional channel status block.*

FIGURE 10.9 *An example of the bi-phase mark channel code.*

with the sampling frequency. (The later 'single-channel-double-sampling-frequency' mode of the interface allows two samples for one channel to be transmitted within a single frame in order to allow the transport of audio at 88.2 or 96 kHz sampling frequency.)

Additional data is carried within the subframe in the form of 4 bits of auxiliary data (which may either be used for additional audio resolution or for other purposes such as low-quality speech), a validity bit (V), a user bit (U), a channel status bit (C) and a parity bit (P), making 32 bits per subframe and 64 bits per frame. Channel status bits are aggregated at the receiver to form a 24 byte word every 192 frames, and each bit of this word has a specific function relating to interface operation, an overview of which is shown in Figure 10.8. Examples of bit usage in this word are the signaling of sampling frequency and pre-emphasis, as well as the carrying of a sample address 'time-code' and labeling of source and destination. Bit 1 of the first byte signifies whether the interface is operating according to the professional (set to 1) or consumer (set to 0) specification.

Bi-phase mark coding, the same channel code as used for SMPTE/EBU timecode, is used in order to ensure that the data is self-clocking, of limited bandwidth, DC free, and polarity independent, as shown in Figure 10.9. The interface has to accommodate a wide range of cable types and a nominal 110 ohm characteristic impedance is recommended. Originally (AES-3-1985) up to four receivers with a nominal input impedance of 250 ohms could be connected across a single professional interface cable, but a later modification to the standard recommended the use of a single receiver per transmitter, having a nominal input impedance of 110 ohms.

Standard consumer interface (IEC 60958-3)

The most common consumer interface (historically related to SPDIF – the Sony/Philips digital interface) is very similar to the AES-3 interface, but uses unbalanced electrical interconnection over a coaxial cable having a characteristic impedance of 75 ohms, as shown in Figure 10.10. It can be found on many items of semi-professional or consumer digital audio equipment, such as CD players, DVD players and DAT machines, and is also widely used on computer sound cards because of the small physical size of the connectors. It usually terminates in an RCA phono connector, although some equipment makes use of optical fiber interconnects (TOS-link) carrying the same data. Format convertors are available for converting consumer format signals to the professional format, and vice versa, and for converting between electrical and optical formats. Both the professional (AES-3 equivalent) and consumer interfaces are capable of carrying data-reduced stereo and surround audio signals such as MPEG and Dolby Digital as described in Fact File 10.3.

The data format of subframes is the same as that used in the professional interface, but the channel status implementation is almost completely different, as shown in Figure 10.11. The second byte of channel status in the consumer interface has been set aside for the indication of 'category codes', these being set to define the type of consumer usage. Examples of defined categories are (00000000) for the General category, (10000000) for Compact Disc and (11000000) for a DAT machine. Once the category has been defined, the receiver is expected to interpret certain bits of the channel status word in a particular way, depending on the category. For example,

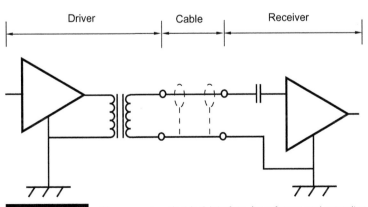

FIGURE 10.10   *The consumer electrical interface (transformer and capacitor are optional but may improve the electrical characteristics of the interface).*

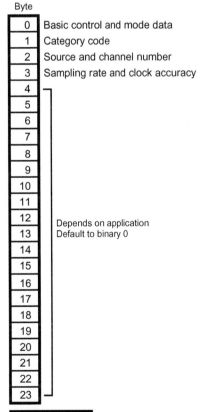

Byte	
0	Basic control and mode data
1	Category code
2	Source and channel number
3	Sampling rate and clock accuracy
4	
5	
6	
7	
8	
9	
10	
11	
12	Depends on application
13	Default to binary 0
14	
15	
16	
17	
18	
19	
20	
21	
22	
23	

FIGURE 10.11   *Overview of the consumer channel status block.*

FACT FILE 10.3 CARRYING DATA-REDUCED AUDIO

The increased use of data-reduced multichannel audio has resulted in methods by which such data can be carried over standard two-channel interfaces, for either professional or consumer purposes. This makes use of the 'non-audio' or 'other uses' mode of the interface, indicated in the second bit of channel status, which tells conventional PCM audio decoders that the information is some other form of data that should not be converted directly to analog audio. Because data-reduced audio has a much lower rate than the PCM audio from which it was derived, a number of audio channels can be carried in a data stream that occupies no more space than two channels of conventional PCM. These applications of the interface are described in SMPTE 337 M (concerned with professional applications) and IEC 61937, although the two are not identical. SMPTE 338 M and 339 M specify data types to be used with this standard. The SMPTE standard packs the compressed audio data into 16, 20 or 24 bits of the audio part of the AES-3 subframe and can use the two subframes independently (e.g. one for PCM audio and the other for data-reduced audio), whereas the IEC standard only uses 16 bits and treats both subframes the same way.

Consumer use of this mode is evident on DVD players, for example, for connecting them to home cinema decoders. Here the Dolby Digital or DTS-encoded surround sound is not decoded in the player but in the attached receiver/decoder. IEC 61937 has parts dealing with a range of different codecs including ATRAC, Dolby AC-3, DTS and MPEG (various flavors). An ordinary PCM convertor trying to decode such a signal would simply reproduce it as a loud, rather unpleasant noise, which is not advised and does not normally happen if the second bit of channel status is correctly observed. Professional applications of the mode vary, but are likely to be increasingly encountered in conjunction with Dolby E data reduction – a relatively recent development involving mild data reduction for professional multichannel applications in which users wish to continue making use of existing AES-3-compatible equipment (e.g. VTRs, switchers and routers). Dolby E enables 5.1-channel surround audio to be carried over conventional two-channel interfaces and through AES-3-transparent equipment at a typical rate of about 1.92 Mbit/s (depending on how many bits of the audio subframe are employed). It is designed so that it can be switched or edited at video frame boundaries without disturbing the audio.

AES55-2012 details ways of transporting MPEG Surround data in an AES-3 bitstream, including the use of Spatial Audio Object Coding (SAOC). The standard specifies how a mono or stereo downmix can be transported in the linear PCM domain, while the MPEG Surround or SAOC data is included in the least significant bits of the PCM audio data.

---

in CD usage, the four control bits from the CD's 'Q' channel subcode are inserted into the first four control bits of the channel status block (bits 1–4). Copy protection can be implemented in consumer-interfaced equipment, according to the Serial Copy Management System (SCMS).

The user bits of the consumer interface are often used to carry information derived from the subcode of recordings, such as track identification and cue point data. This can be used when copying CDs and DAT tapes, for example, to ensure that track start ID markers are copied along with the audio data. This information is not normally carried over AES/EBU interfaces.

MADI

Originally proposed in the UK in 1988 by four manufacturers of professional audio equipment (Sony, Neve, Mitsubishi and Solid State Logic), the so-called 'MADI' interface is an AES and ANSI standard. It was designed to simplify cabling in large installations, especially between multitrack recorders and mixers, and has a lot in common with the format of the AES/EBU interface. The standard concerned is AES10-1991 (ANSI S4.43-1991). This interface was intentionally designed to be transparent to standard two-channel data making the incorporation of two-channel signals into a MADI multiplex a relatively straightforward matter. The original channel status, user and auxiliary data remain intact within the multichannel format.

MADI stands for Multichannel Audio Digital Interface. 32, 56 or 64 channels of audio are transferred serially in asynchronous form and consequently the data rate is much higher than that of the two-channel interface. For this reason the data is transmitted either over a coaxial transmission line with 75 ohm termination (not more than 50 m) or over a fiber optic link. A twisted pair version is also in development. MADI PCIe cards are available for digital audio workstations, which make a convenient way of connecting large numbers of audio channels to and from external systems such as digital mixers.

Proprietary digital interfaces

Tascam's interfaces became popular owing to the widespread use of the company's DA-88 multitrack recorder and derivatives. The primary TDIF-1 interface uses a 25-pin D-sub connector to carry eight channels of audio information in two directions (in and out of the device), sampling frequency and pre-emphasis information (on separate wires, two for f_s and one for emphasis) and a synchronizing signal. The interface is unbalanced and uses CMOS voltage levels. Each data connection carries two channels of audio data, odd channel and MSB first, as shown in Figure 10.12. As can be seen, the audio data can be up to 24 bits long, followed by 2 bits to signal the word length, 1 bit to signal emphasis and 1 bit for parity. There are also 4 user bits per channel that are not usually used.

The Alesis ADAT multichannel optical digital interface, commonly referred to as the 'light pipe' interface or simply 'ADAT Optical', is a serial, self-clocking, optical interface that carries eight channels of audio information. It is described in US Patent 5,297,181: 'Method and apparatus for providing a digital audio interface protocol'. The interface is capable of carrying up to 24 bits of digital audio data for each channel and the eight channels of data are combined into one serial frame that is transmitted at the sampling frequency. The data is encoded in NRZI format for transmission, with forced ones inserted every 5 bits (except during the sync pattern) to provide clock content. This can be used to synchronize the sampling clock of a receiving device if required, although some devices require the use of a separate 9-pin ADAT sync cable for synchronization. The sampling frequency is normally limited to 48 kHz with varispeed up to 50.4 kHz and TOSLINK optical connectors are typically employed (Toshiba TOCP172 or equivalent). In order to operate at 96 kHz sampling frequency some implementations use a 'double-speed' mode in which two channels are used to transmit one channel's audio data (naturally halving the number of channels handled by one serial interface). Although 5 m lengths of optical fiber are the maximum recommended, longer distances may be covered if all the components of the interface are of good quality and clean. Experimentation is required.

As shown in Figure 10.13 the frame consists of an 11 bit sync pattern consisting of 10 zeros followed by a forced one. This is followed by 4 user bits (not normally used and set to zero), the first forced one, then the first audio channel sample (with forced ones every 5 bits), the second audio channel sample, and so on.

SDIF is the original Sony interface for digital audio, most commonly encountered in SDIF-2 format on BNC connectors, along with a word clock signal. However, this is not often used these days. SDIF-3 is Sony's interface for high-resolution DSD data (see Chapter 8), although some early DSD equipment used a data format known as 'DSD-raw', which was simply a stream of DSD samples in non-return-to-zero (NRZ) form, as shown in Figure 10.14. (The latter is essentially the same as SDIF-2.) In SDIF-3 data is carried over 75 ohm unbalanced coaxial cables, terminating

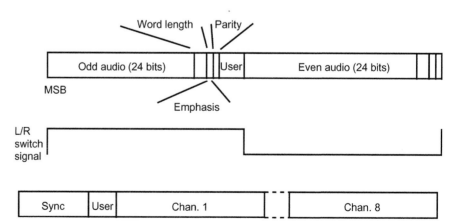

FIGURE 10.12
Format of TDIF data and LRsync signal.

FIGURE 10.13
Basic format of ADAT data.

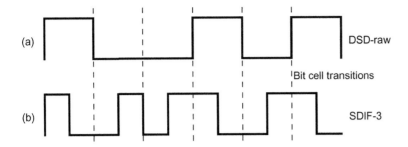

(a)   DSD-raw

                                               Bit cell transitions

(b)   SDIF-3

FIGURE 10.14

*Direct Stream Digital
interface data is either
transmitted 'raw', as shown
at (a) or phase modulated
as in the SDIF-3 format
shown at (b).*

in BNC connectors. The bit rate is twice the DSD sampling frequency (or 5.6448 Mbit/s at the sampling frequency given above) because phase modulation is used for data transmission as shown in Figure 10.14(b). A separate word clock at 44.1 kHz is used for synchronization purposes. It is also possible to encounter a DSD clock signal connection at the 64 times 44.1 kHz (2.8224 MHz).

Sony also developed a multichannel interface for DSD signals, capable of carrying 24 channels over a single physical link. The transmission method is based on the same technology as used for the Ethernet 100BASE-TX (100 Mbit/s) twisted-pair physical layer (PHY), but it is used in this application to create a point-to-point audio interface. Category 5 cabling is used, as for Ethernet, consisting of eight conductors. Two pairs are used for bi-directional audio data and the other two pairs for clock signals, one in each direction.

Twenty-four channels of DSD audio require a total bit rate of 67.7 Mbit/s, leaving an appreciable spare capacity for additional data. In the MAC-DSD interface this is used for error correction (parity) data, frame header and auxiliary information. Data is formed into frames that can contain Ethernet MAC headers and optional network addresses for compatibility with network systems. Audio data within the frame is formed into 352 32 bit blocks, 24 bits of each being individual channel samples, six of which are parity bits and two of which are auxiliary bits.

More recently Sony introduced 'SuperMAC' which is capable of handling either DSD or PCM audio with very low latency (delay), typically less than 50 μs, over Cat-5 Ethernet cables using the 100BASE-TX physical layer. The number of channels carried depends on the sampling frequency. Twenty-four bidirectional DSD channels can be handled, or 48 PCM channels at 44.1/48 kHz, reducing proportionately as the sampling frequency increases. In conventional PCM mode the interface is transparent to AES-3 data including user and channel status information. Up to 5 Mbit/s of Ethernet control information can be carried in addition. A means of interchange based on this was standardized by the AES as AES-50. 'HyperMAC' runs even faster, carrying up to 384 audio channels on gigabit Ethernet Cat-6 cable or optical fiber, together with 100 Mbit/s Ethernet control data. Sony sold this networking technology to Klark Teknik.

The advantage of these interfaces is that audio data thus formatted can be carried over the physical drivers and cables common to Ethernet networks, carrying a lot of audio at high speed. These interfaces bridge the conceptual gap between dedicated audio interfaces and generic computer networks, as they use some of the hardware and the physical layer of a computer network to transfer audio in a convenient form. They do not, however, employ all the higher layers of computer network protocols as mentioned in the next section. This means that the networking protocol overhead is relatively low, minimal buffering is required and latency can be kept to a minimum. Dedicated routing equipment is, however, required. One of the main applications so far has been in a Midas router for live performance mixing.

Data networks and computer interconnects

A network carries data either on wire or optical fiber, and is normally shared between a number of devices and users. The sharing is achieved by containing the data in packets of a limited number of bytes (usually between 64 and 1518), each with an address attached. The packets usually share a common physical link, normally a high-speed serial bus of some kind, being multiplexed in time either using a regular slot structure synchronized to a system clock (isochronous transfer) or in an asynchronous fashion whereby the time interval between packets may be varied or transmission may not be regular, as shown in Figure 10.15. The length of packets may not be constant, depending on the requirements of different protocols sharing the same network. Packets for a

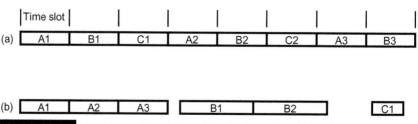

FIGURE 10.15 *Packets for different destinations (A, B and C) multiplexed onto a common serial bus. (a) Time division multiplexed into a regular time slot structure. (b) Asynchronous transfer showing variable time gaps and packet lengths between transfers for different destinations.*

particular file transfer between two devices may not be contiguous and may be transferred erratically, depending on what other traffic is sharing the same physical link.

Figure 10.16 shows some common physical layouts for local area networks (LANs). LANs are networks that operate within a limited area, such as an office building or studio center, within which it is common for every device to 'see' the same data, each picking off that which is addressed to it and ignoring the rest. Routers and bridges can be used to break up complex LANs into subnets. WANs (wide area networks) and MANs (metropolitan area networks) are larger entities that link LANs within communities or regions. PANs (personal area networks) are typically limited to a range of a few tens of meters around the user (e.g. Firewire, USB, Bluetooth). Wireless versions of these network types are increasingly common. Different parts of a network can be interconnected or extended as explained in Fact File 10.4.

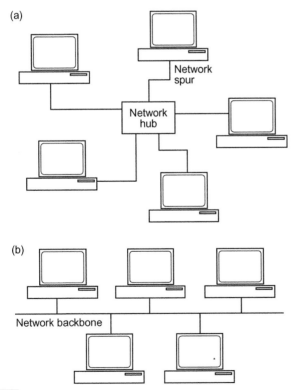

FIGURE 10.16   *Two examples of computer network topologies. (a) Devices connected by spurs to a common hub, and (b) devices connected to a common 'backbone'. The former is now by far the most common, typically using CAT 5 cabling.*

FACT FILE 10.4 EXTENDING A NETWORK

It is common to need to extend a network to a wider area or to more machines. As the number of devices increases so does the traffic, and there comes a point when it is necessary to divide a network into zones, separated by 'repeaters', 'bridges' or 'routers'. Some of these devices allow network traffic to be contained within zones, only communicating between the zones when necessary. This is vital in large interconnected networks because otherwise data placed anywhere on the network would be present at every other point on the network, and overload could quickly occur.

A repeater is a device that links two separate segments of a network so that they can talk to each other, whereas a bridge isolates the two segments in normal use, only transferring data across the bridge when it has a destination address on the other side. A router is very selective in that it examines data packets and decides whether or not to pass them depending on a number of factors. A router can be programmed only to pass certain protocols and only certain source and destination addresses. It therefore acts as something of a network policeman and can be used as a first level of ensuring security of a network from unwanted external access. Routers can also operate between different standards of network, such as between FDDI and Ethernet, and ensure that packets of data are transferred over the most time-/cost-effective route.

One could also use some form of router to link a local network to another that was quite some distance away, forming a wide area network (WAN). Data can be routed either over dialed data links such as ISDN, in which the time is charged according to usage just like a telephone call, or over leased circuits. The choice would depend on the degree of usage and the relative costs. The Internet provides a means by which LANs are easily interconnected, although the data rate available will depend on the route, the service provider and the current traffic.

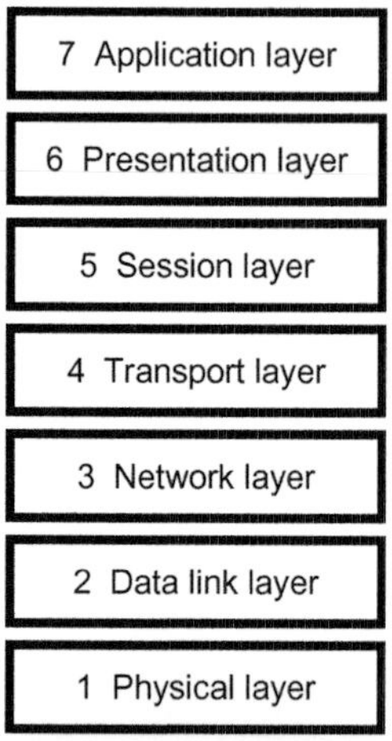

| 7 Application layer |
| 6 Presentation layer |
| 5 Session layer |
| 4 Transport layer |
| 3 Network layer |
| 2 Data link layer |
| 1 Physical layer |

FIGURE 10.17

The ISO model for Open Systems Interconnection is arranged in seven layers, as shown here.

Network communication is divided into a number of conceptual 'layers', each relating to an aspect of the communication protocol and interfacing correctly with the layers either side. The ISO seven-layer model for open systems interconnection (OSI) shows the number of levels at which compatibility between systems needs to exist before seamless interchange of data can be achieved (Figure 10.17). It shows that communication begins when the application is passed down through various stages to the layer most people understand – the physical layer, or the piece of wire over which the information is carried. Layers 3, 4 and 5 can be grouped under the broad heading of 'protocol', determining the way in which data packets are formatted and transferred. There is a strong similarity here with the exchange of data on physical media, as discussed earlier, where a range of compatibility layers from the physical to the application determine whether or not one device can read another's disks.

Audio network requirements

The principal application of computer networks in audio systems is in the transfer of audio data files between workstations, or between workstations and a central 'server' which stores shared files. The device requesting the transfer is known as the 'client' and the device providing the data is known as the 'server'. When a file is transferred in this way a byte-for-byte copy is reconstructed on the client machine, with the file name and any other header data intact. There are considerable advantages in being able to perform this operation at speeds in excess of real time for operations in which real-time feeds of audio are not the aim. For example, in a news editing environment a user might wish to upload a news story file from a remote disk drive in order to incorporate it into a report, this being needed as fast as the system is capable of transferring it. Alternatively, the editor might need access to remotely stored files, such as sound files on another person's system, in order to work on them separately. In audio post-production for films or video there might be a central store of sound effects, accessible by everyone on the network, or it might be desired to pass on a completed portion of a project to the next stage in the post-production process.

Wired Ethernet is fast enough to transfer audio data files faster than real time, depending on network loading and speed. Switched Ethernet architectures allow the bandwidth to be more effectively utilized, by creating switched connections between specific source and destination devices. Approaches using FDDI or ATM are appropriate for handling large numbers of sound file transfers simultaneously at high speed. Unlike a real-time audio interface, the speed of transfer of a sound file over a packet-switched network (when using conventional file transfer protocols) depends on how much traffic is currently using it. If there is a lot of traffic then the file may be transferred more slowly than if the network is quiet (very much like motor traffic on roads). The file might be transferred erratically as traffic volume varies, with the file arriving at its destination in 'spurts'. There therefore arises the need for network communication protocols designed specifically for the transfer of real-time data, which serve the function of reserving a proportion of the network bandwidth for a given period of time. This is known as engineering a certain 'quality of service'.

Without real-time protocols the computer network cannot be relied upon for transferring audio where an unbroken audio output is to be reconstructed at the destination from the data concerned. The faster the network the more likely it is that one would be able to transfer a file fast enough to feed an unbroken audio output, but this should not be taken for granted. Even the highest speed networks can be filled up with traffic! This may seem unnecessarily careful until one considers an application in which a disk drive elsewhere on the network is being used as the source for replay by a local workstation, as illustrated in Figure 10.18. Here it must be possible to ensure guaranteed access to the remote disk at a rate adequate for real-time transfer, otherwise gaps will be heard in the replayed audio.

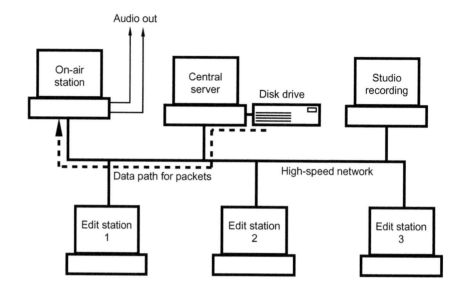

FIGURE 10.18
In this example of a networked system a remote disk is accessed over the network to provide data for real-time audio playout from a workstation used for on-air broadcasting. Continuity of data flow to the on-air workstation is of paramount importance here.

Protocols for the Internet

The common protocol for communication on the Internet is called TCP/IP (Transmission Control Protocol/Internet Protocol). This provides a connection-oriented approach to data transfer, allowing for verification of packet integrity, packet order and retransmission in the case of packet loss. At a more detailed level, as part of the TCP/IP structure, there are high-level protocols for transferring data in different ways. There is a file transfer protocol (FTP) used for downloading files from remote sites, a simple mail transfer protocol (SMTP) and a post office protocol (POP) for transferring email, and a hypertext transfer protocol (HTTP) used for interlinking sites on the world wide web (WWW). The WWW is a collection of file servers connected to the Internet, each with its own unique IP address (the method by which devices connected to the Internet are identified), upon which may be stored text, graphics, sounds and other data.

UDP (user datagram protocol) is a relatively low-level connectionless protocol that is useful for streaming audio over the Internet. Being connectionless, it does not require any handshaking between transmitter and receiver, so the overheads are very low and packets can simply be streamed from a transmitter without worrying about whether or not the receiver gets them. If packets are missed by the receiver, or received in the wrong order, there is little to be done about it except mute or replay distorted audio, but UDP can be efficient when bandwidth is low and quality of service is not the primary issue.

Various real-time protocols have also been developed for use on the Internet, such as RTP (real-time transport protocol). Here packets are time-stamped and may be reassembled in the correct order and synchronized with a receiver clock. RTP does not guarantee quality of service or reserve bandwidth but this can be handled by a protocol known as RSVP (reservation protocol). RTSP is the real-time streaming protocol that manages more sophisticated functionality for streaming media servers and players, such as stream control (play, stop, fast-forward, etc.) and multicast (streaming to numerous receivers).

Audio-specific network standards

A number of proprietary systems have been developed for audio networking in recent years, including Audinate's Dante, which can be licensed to other manufacturers, and Axia's Livewire. Alternatively there are open technology solutions such as RAVENNA, to which a number of parties have signed up. Until recently there have been considerable difficulties in establishing interoperability between the various audio networking systems, even though many of them are based broadly on IP network protocols. Data is almost always transferred using the Real-time Transport Protocol (RTP), for example, although some proprietary solutions use UDP. Synchronization is very often achieved using IEEE 1588 Precision Time Protocol (PTP). Despite this, small differences in implementation often work against interoperability.

The AES X-192 project published a standard for audio network interoperability in 2013. The new standard, known as "AES67-2013, *AES standard for audio applications of networks - High-performance streaming audio-over-IP interoperability*", concentrates on IP (internet protocol) networks for professional audio applications. Professional applications imply low latency and high bandwidth. It uses existing IP network standards and extends interoperability across medium scale networks, such as entire campuses, by including elements of Layer 3 of the internet protocol suite. It does not offer super high performance but occupies the middle space between high performance local networks and the public internet. Latency in AES67-based systems is intended to be 10 ms or below, for example. Synchronization is achieved using the Precision Time Protocol, to which an audio sample clock can be referred, for example.

The European Broadcasting Union (EBU) ACIP recommendations have been developed to standardize broadcast audio contributions over IP networks, as a replacement for ageing and increasingly redundant ISDN connections. The original ACIP interoperability standard itself is described in EBU Tech. Doc. 3326. It recognizes that the quality of service from the public Internet cannot be guaranteed, and a robust codec needed to be developed in order to deliver the material reliably. The ACIP group implemented a very simple solution involving a Voice-over Internet Protocol (VoIP) using the standard session initiation protocol (SIP) that is used for setting up voice calls over the Internet, but with specified codecs. A list of recommended codecs is provided, and the basic G711 and G722 codecs are mandatory because they are specifically designed for VoIP, and so are robust.

The recently developed IEEE 802.1 AVB (Audio Video Bridging) standard can be considered as a set of extensions of the original IEEE 802 network standards that defined Ethernet, designed to facilitate the transfer of real-time media data over IP networks. It currently operates at Layer 2 of the Internet protocol, relying on the physical MAC addresses of devices for packet routing. A number of different components are involved, including Precision Time Protocol to ensure the accurate synchronization of devices, and the possibility to reserve bandwidth for specific time-sensitive data streams. Once a reservation has been made for such a stream the same transport slots cannot be competed for by other types of data. Traffic shaping is a way of attempting to distribute the network load so that requested real-time streams don't overload the resources available.

The AVnu Alliance was formed to advance the Audio/Visual Bridging (AVB) networking standard in practice. There are various options in the AVB standard and not all devices will implement them, or may not implement the standard in the same way, which means that interoperability cannot be guaranteed simply by devices 'conforming' to it. AVnu deals with this problem by introducing a certified base level of interoperability that gives end users the assurance that devices will talk to each other.

Storage area networks

An alternative setup involving the sharing of common storage by a number of workstations is the Storage Area Network (SAN). This employs a networking technology known as Fibre Channel that can run at speeds of 4 Gbit/s, and can also employ fiber optic links to allow long connections between shared storage and remote workstations. RAID arrays (see Chapter 9) are typically employed with SANs, and special software such as Apple's XSAN is needed to enable multiple users to access the files on such common storage. iSCSI is another option for networked storage according to a variant of the SCSI protocol (see Chapter 9).

Wireless networks

Increasing use is made of wireless networks these days, the primary advantage being the lack of need for a physical connection between devices. There are various IEEE 802 standards for wireless networking, including 802.11 which covers wireless Ethernet or 'Wi-Fi'. These typically operate on either the 2.4 GHz or 5 GHz radio frequency bands, at relatively low power, and use various interference reduction and avoidance mechanisms to enable networks to coexist with other services. It should, however, be recognized that wireless networks will never be as reliable as wired networks owing to the differing conditions under which they operate, and that any critical applications in which real-time streaming is required would do well to stick to wired networks where the chances of experiencing dropouts owing to interference or RF fading are almost non-existent. They are, however, extremely convenient for mobile applications and when people move around with computing devices, enabling reasonably high data rates to be achieved with the latest technology.

Bluetooth is one example of a wireless personal area network (WPAN) designed to operate over limited range at data rates of up to 1 Mbit/s. Within this there is the capacity for a number of channels of voice quality audio at data rates of 64 kbit/s and asynchronous channels up to 723 kbit/s. Taking into account the overhead for communication and error protection, the actual data rate achievable for audio communication is usually only sufficient to transfer data-reduced audio for a few channels at a time.

Audio over Firewire (IEEE 1394)

Firewire is an international standard serial data interface specified in IEEE 1394-1995. One of its key applications has been as a replacement for SCSI (Small Computer Systems Interface) for connecting disk drives and other peripherals to computers. It is extremely fast, running at rates of 100, 200 and 400 Mbit/s in its original form, with higher rates appearing all the time up to 3.2 Gbit/s. It is intended for optical fiber or copper interconnection. The S100 version has a maximum realistic data capacity of 65 Mbit/s, a maximum of 16 hops between nodes and no more than 63 nodes on up to 1024 separate buses. On the copper version there are three twisted pairs – data, strobe and power – and the interface operates in half duplex mode, which means that communications in two directions are possible, but only in one direction at a time. Connections are 'hot pluggable' with auto-reconfiguration – in other words one can connect and disconnect devices without turning off the power and the remaining system will reconfigure itself accordingly. It is also relatively cheap to implement. A recent implementation, 1394c, allows the use of gigabit Ethernet connectors, which may improve the reliability and usefulness of the interface in professional applications.

Firewire combines features of network and point-to-point interfaces, offering both asynchronous and isochronous communication modes, so guaranteed latency and bandwidth are available if needed for time-critical applications. Communications are established between logical addresses, and the end point of an isochronous stream is called a 'plug'. Logical connections between devices can be specified as either 'broadcast' or 'point-to-point'. In the broadcast case either the transmitting or receiving plug is defined, but not both, and broadcast connections are unprotected in that any device can start and stop it. A primary advantage for audio applications is that point-to-point connections are protected – only the device that initiated a transfer can interfere with that connection, so once established the data rate is guaranteed for as long as the link remains intact. The interface can be used for real-time multichannel audio interconnections, file transfer, MIDI and machine control, carrying digital video, carrying any other computer data and connecting peripherals (e.g. disk drives).

Originating partly in Yamaha's 'm-LAN' protocol, the 1394 Audio and Music Data Transmission Protocol is now also available as an IEC PAS component of the IEC 61883 standard (a PAS is a publically available specification that is not strictly defined as a standard but is made available for information purposes by organizations operating under given procedures). It offers a versatile means of transporting digital audio and MIDI control data.

Audio over Universal Serial Bus (USB)

The Universal Serial Bus is not the same as IEEE 1394, but it has some similar implications for desktop multimedia systems, including audio peripherals. USB has been jointly supported by a number of manufacturers including Microsoft, Digital, IBM, NEC, Intel and Compaq. Version 1.0 of the copper interface runs at a lower speed than 1394 (typically either 1.5 or 12 Mbit/s) and is designed to act as a low-cost connection for multiple input devices to computers such as joysticks, keyboards, scanners and so on. USB 2.0 runs at a higher rate of up to 480 Mbit/s and is supposed to be backwards-compatible with 1.0. USB 3.0, introduced in 2008, running at up to 5 Gbit/s.

USB 1.0 supports up to 127 devices for both isochronous and asynchronous communication and can carry data over distances of up to 5 m per hop (similar to 1394). A hub structure is required for multiple connections to the host connector. Like 1394 it is hot pluggable and reconfigures the addressing structure automatically, so when new devices are connected to a USB setup the host device assigns a unique address. Limited power is available over the interface and some devices are capable of being powered solely using this source – known as 'bus-powered' devices – which can be useful for field operation of, say, a simple A/D convertor with a laptop computer.

The way in which audio is handled on USB is well defined and somewhat more clearly explained than the 1394 audio/music protocol. It defines three types of communication: audio control, audio streaming and MIDI streaming. Audio data transmissions fall into one of three types. Type 1 transmissions consist of channel-ordered PCM samples in consecutive subframes, whilst Type 2 transmissions typically contain non-PCM audio data that does not preserve a particular channel order in the bitstream, such as certain types of multichannel data-reduced audio stream. Type 3 transmissions are a hybrid of the two such that non-PCM data is packed into pseudo-stereo data words in order that clock recovery can be made easier.

Audio samples are transferred in subframes, each of which can be 1–4 bytes long (up to 24 bits resolution). An audio frame consists of one or more subframes, each of which represents a sample of different channels in the cluster (see below). As with 1394, a USB packet can contain a number of frames in succession, each containing a cluster of subframes. Frames are described by a format descriptor header that contains a number of bytes describing the audio data type, number of channels, subframe size, as well as information about the sampling frequency and the way it is controlled (for Type 1 data). An example of a simple audio frame would be one containing only two subframes of 24 bit resolution for stereo audio.

Audio of a number of different types can be transferred in Type 1 transmissions, including PCM audio (two's complement, fixed point), PCM-8 format (compatible with original 8 bit WAV, unsigned, fixed point), IEEE floating point, A-law and μ-law (companded audio corresponding to relatively old telephony standards). Type 2 transmissions typically contain data-reduced audio signals such as MPEG or AC-3 streams. Here the data stream contains an encoded representation of a number of channels of audio, formed into encoded audio frames that relate to a large number of original audio samples. An MPEG encoded frame, for example, will typically be longer than a USB packet (a typical MPEG frame might be 8 or 24 ms long), so it is broken up into smaller packets for transmission over USB rather like the way it is streamed over the IEC 60958 interface described in Fact File 10.3.

Audio data for closely related synchronous channels can be clustered for USB transmission in Type 1 format. Up to 254 streams can be clustered and there are 12 defined spatial positions for reproduction, to simplify the relationship between channels and the loudspeaker locations to which they relate. The first six defined streams follow the internationally standardized order of surround sound channels for 5.1 surround, that is left, right, center, LFE (low frequency effects), left surround, right surround (see Chapter 17). Subsequent streams are allocated to other loudspeaker locations around a notional listener. Not all the spatial location streams have to be present but they are supposed to be presented in the defined order. Clusters are defined in a descriptor field that includes 'bNrChannels' (specifying how many logical audio channels are present in the cluster) and 'wChannelConfig' (a bit field that indicates which spatial locations are present in the cluster). If the relevant bit is set then the relevant location is present in the cluster. The bit allocations are shown in Table 10.4

AES-47: audio over ATM

AES-47 defines a method by which linear PCM data, either conforming to AES-3 format or not, can be transferred over ATM (Asynchronous Transfer Mode) networks. There are various arguments for doing this, not least being the increasing use of ATM-based networks for data communications

Table 10.4	Channel Identification in USB Audio Cluster Descriptor
Data bit	**Spatial location**
D0	Left Front (L)
D1	Right Front (R)
D2	Center Front (C)
D3	Low Frequency Enhancement (LFE)
D4	Left Surround (LS)
D5	Right Surround (RS)
D6	Left of Center (LC)
D7	Right of Center (RC)
D8	Surround (S)
D9	Side Left (SL)
D10	Side Right (SR)
D11	Top (T)
D12...15	Reserved

within the broadcasting industry and the need to route audio signals over longer distances than possible using standard digital interfaces. There is also a need for low latency, guaranteed bandwidth and switched circuits, all of which are features of ATM. Essentially an ATM connection is established in a similar way to making a telephone call. A SETUP message is sent at the start of a new 'call' that describes the nature of the data to be transmitted and defines its vital statistics. The AES-47 standard describes a specific professional audio implementation of this procedure that includes information about the audio signal and the structure of audio frames in the SETUP at the beginning of the call.

RECOMMENDED FURTHER READING

1394 Trade Association, 2001. TA Document 2001003: Audio and Music Data Transmission Protocol 2.0.

AES, 2002. AES47-2002: Transmission of Digital Audio over Asynchronous Transfer mode Networks.

AES, 2011. AES50-2011: High-resolution multi-channel audio interconnection.

AES, 2013. AES67-2013: High performance streaming audio-over-IP interoperability.

IEC, 1998. IEC/PAS 61883-6. Consumer Audio/Video Equipment – Digital Interface – Part 6: Audio and Music Data Transmission Protocol.

IEEE, 1995. IEEE 1394: Standard for a High Performance Serial Bus.

Page,M., et al., 2002. Multichannel audio connection for direct stream digital. Presented at AES 113th Convention, Los Angeles, 5–8 October.

Rumsey, F., Watkinson, J., 2003. The Digital Interface Handbook, third edition. Focal Press..

USEFUL WEBSITES

Audio Engineering Society standards, www.aes.org

EBU ACIP, http://www.ebu-acip.org

IEEE 1394, www.1394ta.org

IEEE AVB, http://www.ieee802.org/1/pages/avbridges.html

Universal Serial Bus, www.usb.org

Power Amplifiers

Power amplifiers are uneventful devices. They are usually big and heavy, take up a lot of rack space, and feature very little (or sometimes nothing) beyond input and output sockets. Because one tends to ignore them, it is all the more important that they are chosen and used with due care. Coming in a variety of shapes, sizes and 'generations', they are all required to do the ostensibly simple job of providing voltage amplification – converting line levels of up to a volt or so into several tens of volts, with output currents in the ampere range to develop the necessary power across the loudspeaker terminals. Given these few requirements, it is perhaps surprising how many designs there are on the market.

DOMESTIC POWER AMPLIFIERS

The domestic power amplifier, at its best, is designed for maximum fidelity in the true sense of that word, and this will usually mean that other considerations such as long-term overload protection and complete stability into any type of speaker load are not always given the type of priority which is essential in the professional field. A professional power amp may well be asked to drive a pair of 6 ohm speakers in parallel on the other end of 30 meters of cable, at near to maximum output level for hours on end if used in a rock PA rig. This demands large power supplies and heavy transformers, with plenty of heat sink area (the black fins usually found on the outer casing) to keep it from overheating. Cooling fans are frequently employed which will often run at different speeds depending on the temperature of the amplifier.

The domestic amplifier is unlikely to be operated at high output levels for a significant length of time, and the power supplies are often therefore designed to deliver high currents for short periods to take care of short, loud passages. A power supply big enough to supply high currents for lengthy periods is probably wasted in a domestic amplifier. Also, the thermal inertia of the transformer and the heat sinks means that unacceptable rises in temperature are unlikely. Although there are one or two domestic speakers which are notoriously difficult to drive due to various combinations of low impedance, low efficiency (leading to high power demand), and wide phase swings (current and voltage being out of step with each other due to crossover components and driver behavior in a particular speaker enclosure), the majority of domestic hi-fi speakers are a comfortable load for an amplifier, and usually the speaker leads will be less than 10 meters in length.

It is unlikely that the amplifier will be driven into a short-circuit due to faulty speaker lines for any length of time (silence gives an immediate warning), which is not the case with a professional amplifier which may well be one of many, driving a whole array of speakers. A short-circuit developing soon after a show has begun may cause the amplifier to be driven hard into this condition for the whole evening. Protection circuitry needs to be incorporated into the design to allow the professional amplifier to cope with this without overheating or catastrophically failing which can affect other amplifiers in the same part of the rig.

Several 'classes' of amplifier design have appeared over the years, these being labels identifying the type of output stage topology employed to drive the speaker. These are outlined in Fact File 11.1.

PROFESSIONAL AMPLIFIER FACILITIES

The most straightforward power amplifiers have input sockets and output terminals, and nothing else. Single-channel models are frequently encountered, and in the professional field these are often desirable because if one channel of a stereo power amplifier develops a fault then the other channel

FACT FILE 11.1 AMPLIFIER CLASSES

Class A

The output stage draws a constant high current from the power supply regardless of whether there is an audio signal present or not. Low-current class A stages are used widely in audio circuits. The steady bias current as it is known is employed because transistors are non-linear devices, particularly when operated at very low currents. A steady current is therefore passed through them which biases them into the area of their working range at which they are most linear.

The constant bias current makes class A amplification inefficient due to heat generation, but there is the advantage that the output transistors are at a constant steady temperature. Class A is capable of very high sound quality, and several highly specified upmarket domestic class A power amplifiers exist.

Class B

No current flows through the output transistors when no audio signal is present. The driving signal itself biases the transistors into conduction to drive the speakers. The technique is therefore extremely efficient because the current drawn from the power supply is entirely dependent upon the level of drive signal, and so it is particularly attractive in battery-operated equipment. The disadvantage is that at low signal levels the output transistors operate in a non-linear region. It is usual for pairs (or multiples) of transistors to provide the output current of a power amplifier. Each of the pair handles opposite halves of the output waveform (positive and negative with respect to zero) and therefore as the output swings through zero from positive to negative and vice versa the signal suffers so-called 'crossover distortion'. The result is relatively low sound quality, but class B can be used in applications which do not require high sound quality such as telephone systems, hand-held security transceivers, paging systems and the like.

Class A–B

In this design a relatively low constant bias current flows through the output transistors to give a low-power class A amplifier. As the input drive signal is increased, the output transistors are biased into appropriately higher-current conduction in order to deliver higher power to the speakers. This part of the operation is the class B part, i.e. it depends on input drive signal level. But the low-level class A component keeps the transistors biased into a linear part of their operating range so that crossover distortion is largely avoided. The majority of high-quality amplifiers operate on this principle.

Other Classes

Class C drives a narrow band of frequencies into a resonant load, and is appropriate to radio-frequency (RF) work where an amplifier is required to drive a single frequency into an appropriately tuned aerial.

Class D uses 'pulse width modulation'. It has seen increasing use since the late 1980s, although the technique appeared as far back as the 1960s in Sinclair designs. In a conventional output stage, the voltage across the transistors varies in proportion to the input signal voltage, and on average they spend much of their time under the conditions of moderate voltage and moderate output current which together require them to dissipate power in the form of heat via the amplifier's heat sinks. With class D, however, the output transistors are driven by an ultrasonic square wave whose mark-to-space ratio is varied by the audio signal. The transistors are therefore either in a state of minimum voltage across them combined with maximum current output (full power), or maximum voltage across them combined with virtually no current output. Power dissipation in the transistors is therefore minimal, they run cool, and they are therefore much more efficient. Output low-pass filtering is required to remove the ultrasonic square wave component, leaving just the audio waveform. Combined with efficient switched-mode power supplies, the technique can offer high-powered compact power amplifier designs which do not need fan cooling. The D designation simply means that it was the fourth output topology to emerge, but sometimes it is erroneously termed 'Digital'. There are indeed analogies with digital processing in two respects: the output transistors are always either fully on or fully off (a '1' or a '0'), and the output is low-pass filtered for the same reasons as for digital-to-analog conversion, to remove ultrasonic components.

A variation on class D, called class T (from the Tripath company), has recently been seen. Here, the ultrasonic frequency is continuously varied in accordance with the amplitude of the audio signal. The frequency is about 1.2 MHz at low signal levels, falling to around 200 kHz for very high signal levels; a greater overall efficiency is claimed as a result. Classes E and F were concerned with increasing efficiency, and currently no commercial models conform to these particular categories.

Class G incorporates several different voltage rails which progressively come into action as the drive signal voltage is increased. This technique can give very good efficiency because for much of the time only the lower-voltage, low-current supplies are in operation. Such designs can be rather smaller than their conventional class A–B counterparts of comparable output power rating. Class H is a variation on class G in that the power supply voltage rails are made to track the input signal continuously, maintaining just enough headroom to accommodate the amplifier's requirements for the necessary output voltage swing.

Since the early 1980s the MOSFET (Metal Oxide Semiconductor Field-Effect Transistor) has been widely employed for the output stages of power amplifiers. MOSFET techniques claim lower distortion, better thermal tracking (i.e. good linearity over a wide range of operating temperatures), simpler output stage design, and greater tolerance of adverse loudspeaker loads without the need for elaborate protection circuitry.

also has to be shut down, thus losing a perfectly good circuit. The single-channel power amplifier is thus a good idea when multi-speaker arrays are in use such as in rock PA systems and theater sound.

Other facilities found on power amplifiers include input level controls, output level meters, overload indicators, thermal shutdown (the mains feed is automatically disconnected if the amplifier rises above a certain temperature), earth-lift facility to circumvent earth loops, and 'bridging' switch. This last facility, applicable to a stereo power amplifier, is a facility sometimes provided whereby the two channels of the amp can be bridged together to form a single-channel higher-powered one, the speaker(s) now being connected across the two positive output terminals with the negative terminals left unused. Only one of the input sockets is now used to drive it.

Cooling fans are often incorporated into an amplifier design. Such a force-cooled design can be physically smaller than its convection-cooled counterpart, but fans tend to be noisy. Anything other than a genuinely silent fan is unacceptable in a studio or broadcast control room, or indeed in theater work, and such models will need to be housed in a separate well-ventilated room. Ventilation of course needs to be a consideration with all power amplifiers.

SPECIFICATIONS

Power amplifier specifications include sensitivity, maximum output power into a given load, power bandwidth, frequency response, slew rate, distortion, crosstalk between channels, signal-to-noise ratio, input impedance, output impedance, damping factor, phase response and DC offset. Quite surprising differences in sound quality can be heard between certain models, and steady-state measurements do not, unfortunately, always tell a user what he or she can expect to hear.

Sensitivity

Sensitivity is a measurement of how much voltage input is required to produce the amplifier's maximum rated output. For example, a model may be specified '150 watts into 8 ohms, input sensitivity 775 mV = 0 dBu'. This means that an input voltage of 775 mV will cause the amplifier to deliver 150 watts into an 8 ohm load. Speakers exhibit impedances which vary considerably with frequency, so this is always a nominal specification when real speakers are being driven. Consideration of sensitivity is important because the equipment which is to drive the amp must not be allowed to deliver a greater voltage to the amplifier than its specification states, otherwise the amplifier will be overloaded causing 'clipping' of the output waveform (a squaring-off of the tops and bottoms of the waveform resulting in severe distortion). This manifests itself as a 'breaking-up' of the sound on musical peaks, and will often quickly damage tweeters and high-frequency horns.

Many amplifiers have input level controls so that if, for instance, the peak output level of the mixer which drives the amplifier is normally say 'PPM 6' – about 2 volts – then the amp's input levels can be turned down to prevent overload. In the given example, 2 volts is 8 dB higher than 775 mV (PPM 4 = 0 dBu) and so the input level control should be reduced by 8 dB to allow for this. If a dB calibration is not provided on the level control, and many are not particularly accurate anyway, a reasonable guide is that, compared with its maximum position of about '5 o'clock', reducing the level to about 2 o'clock will reduce the sensitivity by about 10 dB, or by a factor of three. In this position, the power amplifier with an input sensitivity of 775 mV will now require 0.775×3, or about 2 volts, to develop its full output.

If input level controls are not provided, one can build a simple resistive attenuator which reduces the voltage being fed to the amplifier's input. Two examples are shown in Figure 11.1. It is best to place such attenuators

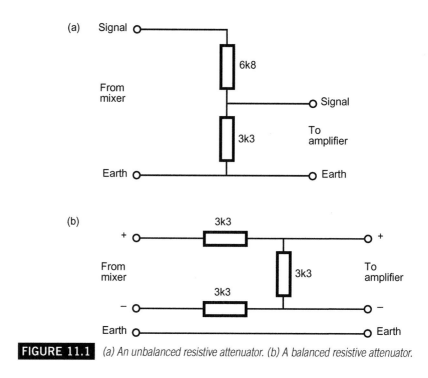

FIGURE 11.1 *(a) An unbalanced resistive attenuator. (b) A balanced resistive attenuator.*

close to the power amp input in order to keep signal levels high while they are traveling down the connecting leads. In both cases the 3k3 resistor which is in parallel with the amplifier's input can be increased in value for less attenuation, and decreased in value for greater attenuation. With care, the resistors can be built into connecting plugs, the latter then needing to be clearly labeled.

Power output

A manufacturer will state the maximum power a particular model can provide into a given load, e.g. '200 watts into 8 ohms', often with 'both channels driven' written after it. This last means that both channels of a stereo amplifier can deliver this simultaneously. When one channel only is being driven, the maximum output is often a bit higher, say 225 watts, because the power supply is less heavily taxed. Thus 200 watts into 8 ohms means that the amplifier is capable of delivering 40 volts into this load, with a current of 5 amps. If the load is now reduced to 4 ohms then the same amplifier should produce 400 watts. A theoretically perfect amplifier should then double its output when the impedance it drives is halved. In practice, this is beyond the great majority of power amplifiers and the 4 ohm specification of the above example may be more like 320 watts, but this is only around 1 dB below the theoretically perfect value. A 2 ohm load is very punishing for an amplifier, and should be avoided even though a manufacturer sometimes claims a model is capable of, say, 800 watts of short-term peaks into 2 ohms. This at least tells us that the amp should be able to drive 4 ohm loads without any trouble.

Because 200 watts is only 3 dB higher than 100 watts, then, other things being equal, the exact wattage of an amplifier is less important than factors such as its ability to drive difficult reactive loads for long periods. Often, 'RMS' will be seen after the wattage rating. This stands for root-mean-square, and defines the raw 'heating' power of an amplifier, rather than its peak output. All amplifiers should be specified RMS so that they can easily be compared. The RMS value is 0.707 times the instantaneous peak capability, and it is unlikely that one would encounter a professional amplifier with just a peak power rating.

Power bandwidth is not the same as power rating, as discussed in Fact File 11.2.

Frequency response

Frequency response, unlike power bandwidth, is simply a measure of the limits within which an amplifier responds equally to all frequencies when delivering a very low power. The frequency response is usually measured with the amplifier delivering 1 watt into 8 ohms. A specification such as

---

FACT FILE 11.2 POWER BANDWIDTH

Power bandwidth is a definition of the frequency response limits within which an amplifier can sustain its specified output. Specifically, a 3 dB drop of output power is allowed in defining a particular amplifier's power bandwidth. For example, a 200 watt amplifier may have a power bandwidth of 10 Hz to 30 kHz, meaning that it can supply 200 watts −3 dB (=100 watts) at 10 Hz and 30 kHz, compared with the full 200 watts at mid frequencies. Such an amplifier would be expected to deliver the full 200 watts at all frequencies between about 30 Hz and 20 kHz, and this should also be looked for in the specification. Often, though, the power rating of an amplifier is much more impressive when measured using single sine-wave tones than with broad-band signals, since the amplifier may be more efficient at a single frequency.

Power bandwidth can indicate whether a given amplifier is capable of driving a subwoofer at high levels in a PA rig, as it will be called upon to deliver much of its power at frequencies below 100 Hz or so. The driving of high-frequency horns also needs good high-frequency power bandwidth so that the amplifier never clips the high frequencies, which easily damages horns as has been said.

'20 Hz – 20 kHz ± 0.5 dB' should be looked for, meaning that the response is virtually flat across the whole of the audible band. Additionally, the −3 dB points are usually also stated, e.g. '−3 dB at 12 Hz and 40 kHz', indicating that the response falls away smoothly below and above the audio range. This is desirable as it gives a degree of protection for the amp and speakers against subsonic disturbances and RF interference.

Distortion

Distortion should be 0.1% THD or less across the audio band, even close to maximum-rated output. It often rises slightly at very high frequencies, but this is of no consequence. Transient distortion, or transient intermodulation distortion (TID), is also a useful specification. It is usually assessed by feeding both a 19 kHz and a 20 kHz sine wave into the amplifier and measuring the relative level of 1 kHz difference tone. The 1 kHz level should be at least 70 dB down, indicating a well-behaved amplifier in this respect. The test should be carried out with the amplifier delivering at least two-thirds of its rated power into 8 ohms. Slew rate distortion is also important (see Fact File 11.3).

Crosstalk

Crosstalk figures of around −70 dB at mid frequencies should be a reasonable minimum, degrading to around −50 dB at 20 kHz, and by perhaps the same amount at 25 Hz or so. 'Dynamic crosstalk' is sometimes specified, this manifesting itself mainly at low frequencies because the power supply works hardest when it is called upon to deliver high currents during high-level, low-frequency drive. Current demand by one channel can modulate the power supply voltage rails, which gets into the other channel. A number of amplifiers have completely separate power supplies for each channel, which eliminates such crosstalk, or at least separate secondary windings on the mains transformer plus two sets of rectifiers and reservoir capacitors which is almost as good.

Signal-to-noise ratio

Signal-to-noise ratio is a measure of the output residual noise voltage expressed as a decibel ratio between that and the maximum output voltage, when the input is short-circuited. Noise should never be a problem with a modern power amplifier and signal-to-noise ratios of at least 100 dB are common. High-powered models (200 watts upwards) should have signal-to-noise ratios correspondingly greater (e.g. 110 dB or so) in order that the output residual noise remains below audibility.

FACT FILE 11.3 SLEW RATE

Slew rate is a measure of the ability of an amplifier to respond accurately to high-level transients. For instance, the leading edge of a transient may demand that the output of an amplifier swings from 0 to 120 watts in a fraction of a millisecond. The slew rate is defined in $V\mu s^{-1}$ (volts per microsecond) and a power amplifier which is capable of 200 watts output will usually have a slew rate of at least $30\,V\mu s^{-1}$. Higher-powered models require a greater slew rate simply because their maximum output voltage swing is greater. A 400 watt model might be required to swing 57 volts into 8 ohms as compared with the 200 watt model's 40, so its slew rate needs to be at least:

$$30 \times 57 \div 40 = 43\,V\mu s^{-1}$$

In practice, modern power amplifiers achieve slew rates comfortably above these figures.

An absolute minimum can be estimated by considering the highest frequency of interest, 20 kHz, then doubling it for safety, 40 kHz, and considering how fast a given amplifier must respond to reproduce this accurately at full output. A sine wave of 40 kHz reaches its positive-going peak in 6.25 µs, as shown in the diagram. A 200 watt model delivers a peak voltage swing of 56.56 volts peak to peak (1.414 times the RMS voltage). It may seem then that it could therefore be required to swing from 0V to 28.28V in 6.25 µs, thus requiring a slew rate of 28.28 ÷ 6.25, or $4.35\,V\mu s^{-1}$. But the actual slew rate requirement is rather higher because the initial portion of the sine wave rises steeply, tailing off towards its maximum level.

Musical waveforms come in all shapes and sizes of course, including near-square waves with their almost vertical leading edges, so a minimum slew rate of around eight times this (i.e. $30\,V\mu s^{-1}$) might be considered as necessary. It should be remembered, though, that the harmonics of an HF square wave are well outside the audible spectrum, and thus slew rate distortion of such waves at HF is unlikely to be audible. Extremely high slew rates of several hundred volts per microsecond are sometimes encountered. These are achieved in part by a wide frequency response and 'fast' output transistors, which are not always as stable into difficult speaker loads as are their 'ordinary' counterparts. Excessive slew rates are therefore to be viewed with skepticism.

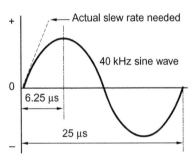

Impedance

The input impedance of an amplifier ought to be at least $10\,k\Omega$, so that if a mixer is required to drive, say, ten amplifiers in parallel, as is often the case with PA rigs, the total load will be 10k ÷ 10, or 1k, which is still a comfortable load for the mixer. Because speakers are of very low impedance, and because their impedance varies greatly with frequency, the amplifier's output impedance must not be greater than a fraction of an ohm, and a value of 0.1 ohms or less is needed. A power amplifier needs to be a virtually perfect 'voltage source', its output voltage remaining substantially constant with different load impedances.

The output impedance does, however, rise a little at frequency extremes. At LF, the output impedance of the power supply rises and therefore so does the amplifier's. It is common practice to place a low-valued inductor of a couple of microhenrys in series with a power amp's output which raises its output impedance a little at HF, this being to protect the amp against particularly reactive speakers or excessively capacitive cables, which can provoke HF oscillation.

Damping factor

Damping factor is a numerical indication of how well an amplifier can 'control' a speaker. There is a tendency for speaker cones and diaphragms to go on vibrating a little after the driving signal has stopped, and a very low output impedance virtually short-circuits the speaker terminals which 'damps' this. Damping factor is the ratio between the amplifier's output impedance and the speaker's rated impedance, so a damping factor of '100 into 8 ohms' means that the output impedance of the amplifier is 8 ÷ 100 ohms, or 0.08 ohms. One hundred is quite a good figure (the higher the better, but a number greater than 200 could imply that the amplifier is insufficiently well protected from reactive loads and the like), but it is better if a frequency is given. Damping factor is most useful at low frequencies because it is the bass cones which vibrate with greatest excursion, requiring the tightest control. A damping factor of '100 at 40 Hz' is therefore a more useful specification than '100 at 1 kHz'.

Phase response

Phase response is a measurement of how well the frequency extremes keep in step with mid frequencies. At very low and very high frequencies, 15° phase leads or phase lags are common, meaning that in the case of phase lag, there is a small delay of the signal compared with mid frequencies, and phase lead means the opposite. At 20 Hz and 20 kHz, the phase lag or phase lead should not be greater than 15°, otherwise this may imply a degree of instability when difficult loads are being driven, particularly if HF phase errors are present.

The absolute phase of a power amplifier is simply a statement of whether the output is in phase with the input. The amplifier should be non-phase-inverting overall. One or two models do phase invert, and this causes difficulties when such models are mixed with non-inverting ones in multi-speaker arrays when phase cancelations between adjacent speakers, and incorrect phase relationships between stereo pairs and the like, crop up. The cause of these problems is not usually apparent and can waste much time.

COUPLING

The vast majority of power amplifier output stages are 'direct coupled', that is the output power transistors are connected to the speakers with nothing in between beyond perhaps a very low-valued resistor and a small inductor. The DC voltage operating points of the circuit must therefore be chosen such that no DC voltage appears across the output terminals of the amplifier. In practice this is achieved by using 'split' voltage rails of opposite polarity (e.g. ±46 volts DC) between which the symmetrical output stage 'hangs', the output being the mid-point of the voltage rails (i.e. 0 V). Small errors are always present, and so 'DC offsets' are produced which means that several millivolts of DC voltage will always be present across the output terminals. This DC flows through the speaker, causing its cone to deflect either forwards or backwards a little from its rest position. As low a DC offset as possible must therefore be achieved, and a value of ±40 mV is an acceptable maximum. Values of 15 mV or less are quite common.

USEFUL WEBSITES

www.lenardaudio.com: Education section, class D amplifiers and links.

Lines and Interconnection

CHAPTER CONTENTS

This chapter is concerned with the interconnection of analog audio signals, and the solving of problems concerned with analog interfacing. It is not intended to cover digital interfacing systems here, since this subject is adequately covered in Chapter 10. The proper interconnection of analog audio signals, and an understanding of the principles of balanced and unbalanced lines, is vital to the maintenance of high quality in an audio system, and will remain important for many years notwithstanding the growing usage of digital systems.

TRANSFORMERS

Mains transformers are widely used throughout the electrical and electronics industries, usually to convert the 240 V AC mains voltage to a rather lower voltage. Audio transformers are widely used in audio equipment for balancing and isolating purposes, and whereas mains transformers are required only to work at 50 Hz, audio transformers must give a satisfactory performance over the complete audio spectrum. Fortunately most audio transformers are only called upon to handle a few volts at negligible power, so they are generally much smaller than their mains counterparts. The principles of transformer operation are outlined in Fact File 12.1.

Transformers and impedances

Consider Figure 12.1a. The turns ratio is 1:2, so the square of the turns ratio (used to calculate the impedance across the secondary) is 1:4, and therefore the impedance across the secondary will be found to be 10 × 4 = 40 k. Another example is shown in Figure 12.1b. The turns ratio is 1:4. 0.7 volts is applied across the primary and gives 2.8 volts across the secondary. The square of the turns ratio is 1:16, so the impedance across the secondary is 2 k × 16 = 32 k. The transformer also works backwards, as shown in Figure 12.1c. A 20 k resistor is now placed across the secondary. The square of the turns ratio is 1:16, and therefore the impedance across the primary is 20 k ÷ 16 = 1 k25.

Consider now a microphone transformer that is loaded with an impedance on both sides, as shown in Figure 12.2. The transformer presents the 2 k impedance of the mixer to the microphone, and the 200 ohm impedance of the microphone to the mixer. With a step-up ratio of 1:4 the square of the turns ratio would be 1:16. The microphone would be presented with an impedance of 2 k ÷ 16 = 125 ohms, whereas the mixer would be presented with an impedance of 200 × 16 = 3200 ohms. In this particular case a 1:4 step-up transformer is unsuitable because microphones like to work

FACT FILE 12.1 THE TRANSFORMER

From the diagrams it can be seen that the transformer consists of a laminated core (i.e. a number of thin sheets of metal 'laminated' together to form a single thick core) around which is wound a 'primary' winding and a 'secondary' winding. If an alternating current is passed through the primary winding, magnetic flux flows in the core (in a similar fashion to the principle of the tape head; see Fact File 6.1), and thus through the secondary winding. Flux changes in the secondary winding cause a current to be induced in it. The voltage across the secondary winding compared with that across the primary is proportional to the ratio between the number of turns on each coil. For example, if the primary and secondary windings each have the same number of turns, then 1 volt across the primary will also appear as 1 volt across the secondary. If the secondary has twice the number of turns as the primary, then twice the voltage will appear across it. The transformer also works in reverse – voltage applied to the secondary will be induced into the primary in proportion to the turns ratio.

The current flowing through the secondary is in inverse proportion to the turns ratio, such that equal power exists on the primary and secondary sides of the transformer (it is not magic – the increased voltage across the secondary of a step-up transformer is traded off against reduced current!).

It is important to remember that the principle of operation of the transformer depends on AC in the windings inducing an alternating field into the core (i.e. it is the change in direction of magnetic flux which induces a current in the secondary, not simply the presence of constant flux). A DC signal, therefore, is not passed by a transformer.

Impedances are proportional to the square of the turns ratio, as discussed in the main text. A transformer will 'reflect' the impedances between which it works. In the case of a 1:1 transformer the impedance across the secondary is equal to the impedance across the primary, but in the case of a 1:2 transformer the impedance seen across the secondary would be four times that across the primary.

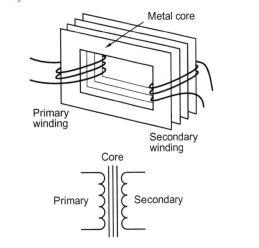

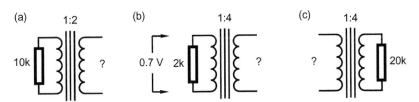

FIGURE 12.1   *Examples of transformer circuits. (a) What is the impedance across the secondary? (b) What are the impedance and voltage across the secondary? (c) What is the impedance across the primary?*

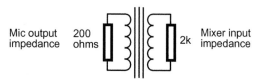

Mic output impedance   200 ohms   2k   Mixer input impedance

FIGURE 12.2   *The input impedance of the mixer is seen by the microphone, modified by the turns ratio of the transformer, and vice versa.*

into an impedance five times or more than their own impedance, so 125 ohms is far too low. Similarly, electronic inputs work best when driven by an impedance considerably lower than their own, so 3200 ohms is far too high.

Limitations of transformers

Earlier it was mentioned that an audio transformer must be able to handle the complete audio range. At very high and very low frequencies this is not easy to achieve, and it is usual to find that distortion rises at low frequencies, and also to a lesser extent at very high frequencies. The frequency response falls away at the frequency extremes, and an average transformer may well be 3 dB down at 20 Hz and 20 kHz compared with mid frequencies. Good (=expensive) transformers have a much better performance than this. All transformers are designed to work within certain limits of voltage and current, and if too high a voltage is applied a rapid increase in distortion results.

The frequency response and distortion performance is affected by the impedances between which the transformer works, and any particular model will be designed to give its optimum performance when used for its intended application. For example, a microphone transformer is designed to handle voltages in the millivolt range up to around 800 mV or so. The primary winding will be terminated with about 200 ohms, and the secondary will be terminated with around 1–2 k (or rather more if a step-up ratio is present). A line level transformer on the other hand must handle voltages up to 8 volts or so, and will probably be driven by a source impedance of below 100 ohms, and will feed an impedance of 10 k or more. Such differing parameters as these require specialized designs. There is no 'universal' transformer.

Transformers are sensitive to electromagnetic fields, and so their siting must be given consideration. Place an audio transformer next to a mains transformer and hum will be induced into it, and thus into the rest of the audio circuit. Most audio transformers are built into metal screening cans which considerably reduce their susceptibility to radio-frequency interference and the like.

UNBALANCED LINES

'Unbalanced' in this context does not mean unstable or faulty. The unbalanced audio line is to be found in virtually all domestic audio equipment, much semi-professional and some professional audio equipment as well. It consists of a 'send' and 'return' path for the audio signal, the return path being an outer screening braid which encloses the send wire and screens it from electromagnetic interference, shown in Figure 12.3. The screening effect considerably reduces interference such as hum, RF and other induction, without eliminating it entirely. If the unbalanced line is used to carry an audio signal over tens of meters, the cumulative effect of interference may be unacceptable. Earth loops can also be formed (see Fact File 12.2). Unbalanced lines are normally terminated in connectors such as phono plugs, DIN plugs and quarter-inch 'A-gauge' jack plugs.

An improved means of unbalanced interconnection is shown in Figure 12.4. The connecting lead now has two wires inside the outer screen. One is used as the signal wire, and instead of the return being provided by the outer screen, it is provided by the second inner wire. The screening braid is connected to earth at one end only, and so it merely provides an interference screen without affecting the audio signal.

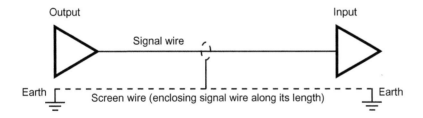

FIGURE 12.3
Simple unbalanced interconnection.

FACT FILE 12.2 EARTH LOOPS

It is possible to wire cables such that the screening braid of a line is connected to earth at both ends. In many pieces of audio equipment the earth side of the audio circuit is connected to the mains earth. When two or more pieces of equipment are connected together this creates multiple paths to the mains earth, and low-level mains currents can circulate around the screening braids of the connecting leads if the earths are at even slightly different potentials. This induces 50 Hz mains hum into the inner conductor. A common remedy for this problem is to disconnect the earth wires in the mains plugs on all the interconnected pieces of equipment except one, the remaining connection providing the earth for all the other pieces of equipment via the audio screening braids. This, though, is potentially dangerous, since if a piece of equipment develops a fault and the mains plug with the earth connection is unplugged, then the rest of the system is now unearthed and the fault could in serious cases place a mains voltage on the metal parts of the equipment. A lot of units are now 'double insulated', so that internal mains wiring cannot place mains voltage on the metal chassis. The mains lead is just two core, live and neutral.

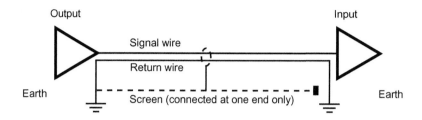

FIGURE 12.4

Alternative unbalanced interconnection.

CABLE EFFECTS WITH UNBALANCED LINES

Cable resistance

'Loop' resistance is the total resistance of both the send and return paths for the signal, and generally, as long as the loop resistance of a cable is a couple of orders of magnitude (i.e. a factor of 100) lower than the input impedance of the equipment it is feeding, it can be ignored. For example, the output impedance of a tape recorder might be 200 ohms. The input impedance of the amplifier it would be connected to would normally be 10 k or more. The DC resistance of a few meters of connecting cable would only be a fraction of an ohm and so would not need to be considered. But what about 100 meters of microphone cable? The input impedance of a microphone amplifier would normally be at least 1000 ohms. Two orders of magnitude lower than this is 10 ohms. Even 100 meters of mic lead will have a lower resistance than this unless very thin cheap wire is used, and so again the DC resistance of microphone cables can be ignored.

Speaker cables do need to be watched, because the input impedance of loudspeakers is of the order of 8 ohms. Wiring manufacturers quote the value of DC resistance per unit length (usually 1 meter) of cable, and a typical cable suitable for speaker use would be of 6 amp rating and about 12 milliohms (0.012 ohms) resistance per meter. Consider a 5 meter length of speaker cable. Its total loop resistance then would be 10 meters multiplied by 0.012 ohms = 0.12 ohms. This is a bit too high to meet the criterion stated above, an 8 ohm speaker requiring a cable of around 0.08 ohm loop resistance. In practice, though, this would probably be perfectly adequate, since there are many other factors which will affect sound quality. Nevertheless, it does illustrate that quite heavy cables are required to feed speakers, otherwise too much power will be wasted in the cable itself before the signal reaches the speaker.

If the same cable as above were used for a 40 meter feed to a remote 8 ohm loudspeaker, the loop resistance would be nearly 1 ohm and nearly one-eighth of the amplifier power would be dissipated in heat in the cable. The moral here is to use the shortest length of cable as is practicable, or if long runs are required use the 100 volt line system (see '100 volt lines', p. 372).

Cable and transformer inductance

The effect of cable inductance (see 'Sound in electrical form', Chapter 1) becomes more serious at high frequencies, but at audio frequencies it is insignificant even over long runs of cable. Conversely, inductance is extremely important in transformers. The coils on the transformer cores consist of a large number of turns of wire, and the electromagnetic field of each turn works against the fields of the other turns. The metallic core greatly enhances this effect. Therefore, the inductance of each transformer coil is very high and presents a high impedance to an audio signal. For a given frequency, the higher the inductance the higher the impedance in ohms.

Cable capacitance

The closer the conductors in a cable are together, the greater the capacitance (see 'Sound in electrical form', Chapter 1). The surface area of the conductors is also important. Capacitance is the opposite of inductance in that, for a given frequency, the greater the capacitance the lower is the impedance in ohms. In a screened cable the screening braid entirely encloses the inner conductor and so the surface area of the braid, as seen by this inner conductor, is quite large. Since large surface area implies high capacitance, screened cable has a much higher capacitance than ordinary mains wiring, for example. So when an audio signal looks into a connecting cable it sees the capacitance between the conductors and therefore a rather less-than-infinite impedance between them, especially at high frequencies. A small amount of the signal can therefore be conducted to earth via the screen.

In the diagram in Figure 12.5 there are two resistors of equal value. A voltage V_1 is applied across the two. Because the value of the resistors is the same, V_1 is divided exactly in half, and V_2 will be found to be exactly half the value of V_1. If the lower resistor were to be increased in value to 400 ohms then twice the voltage would appear across it than across the upper resistor. The ratio of the resistors equals the ratio of the voltages across them.

Consider a 200 ohm microphone looking into a mic lead, as shown in Figure 12.6a. C is the capacitance between the screening braid and

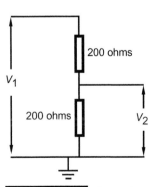

FIGURE 12.5 *The voltage V_2 across the output is half the input voltage (V_1).*

FIGURE 12.6

(a) A microphone with a 200 ohm output impedance is connected to an amplifier. (b) Lead capacitance conducts high frequencies to ground more than low frequencies, and thus the cable introduces HF roll-off. V_2 is lower at HF than at LF.

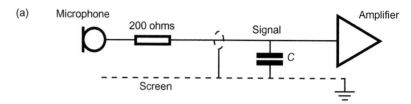

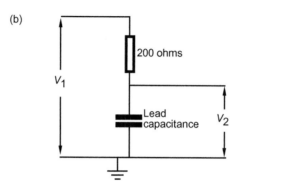

the inner core of the cable. The equivalent of this circuit is shown in Figure 12.6b. Manufacturers quote the capacitance of cables in picofarads (pF) per unit length. A typical value for screened cable is 200 pF (0.0002 µF) per meter. A simple formula exists for determining the frequency at which 3 dB of signal is lost for a given capacitance and source resistance:

$$f = 159155/RC$$

where f = frequency in hertz (Hz), R = resistance in ohms, and C = capacitance in microfarads (µF).

To calculate the capacitance which will cause a 3 dB loss at 40 kHz, putting it safely out of the way of the audio band, the formula must be rearranged to give the maximum value of acceptable capacitance:

$$C = 159155/Rf$$

Thus, if $R=200$ (mic impedance), $f=40000$:

$$C = 159155 \div (200 \times 40000) = 0.02 \,µF$$

So a maximum value of 0.02 µF of lead capacitance is acceptable for a mic lead. Typical lead capacitance was quoted as 0.0002 µF per meter, so 100 meters will give 0.02 µF, which is the calculated acceptable value. Therefore one could safely use up to 100 meters of typical mic cable with a standard 200 ohm microphone without incurring significant signal loss at high frequencies.

The principle applies equally to other audio circuits, and one more example will be worked out. A certain tape recorder has an output impedance of 1 k. How long a cable can it safely drive? From the above formula:

$$C = 159155 \div (200 \times 40000) = 0.004\,\mu F$$

In this case, assuming the same cable capacitance, the maximum safe cable length is $0.004 \div 0.0002 = 20$ meters. In practice, modern audio equipment generally has a low enough source impedance to drive long leads, but it is always wise to check up on this in the manufacturer's specification. Probably of greater concern will be the need to avoid long runs of unbalanced cable due to interference problems.

BALANCED LINES

The balanced line is better at rejecting interference than the unbalanced line, and improvements upon the performance of the unbalanced line in this respect can be 80 dB or more for high-quality microphone lines.

As shown in Figure 12.7, the connecting cable consists of a pair of inner conductors enclosed by a screening braid. At each end of the line is a 'balancing' transformer. The output amplifier feeds the primary of the output transformer and its voltage appears across the secondary. The send and return paths for the audio signal are provided by the two inner conductors, and the screen does not form part of the audio circuit. If an interference signal breaks through the screen it is induced equally into both signal lines. At the input transformer's primary the induced interference current, flowing in the same direction in both legs of the balanced line, cancels out, thus rejecting the interference signal. Two identical signals, flowing in opposite directions, cancel out where they collide.

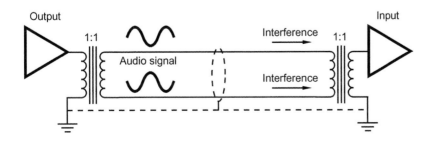

Output   Input

1:1   Interference   1:1

Audio signal

Interference

FIGURE 12.7

A balanced interconnection using transformers.

Such an interfering signal is called a 'common mode' signal because it is equal and common to both audio lines. The rejection of this in the transformer is termed 'common mode rejection' (CMR). A common mode rejection ratio (CMRR) of at least 80 dB may be feasible. Meanwhile, the legitimate audio signal flows through the primary of the transformer as before, because the signal appears at each end of the coil with equal strength but opposite phase. Such a signal is called a 'differential signal', and the balanced input is also termed a 'differential input' because it accepts differential mode signals but rejects common mode signals.

So balanced lines are used for professional audio connections because of their greatly superior rejection of interference, and this is particularly useful when sending just a few precious millivolts from a microphone down many meters of cable to an amplifier.

WORKING WITH BALANCED LINES

In order to avoid earth loops (see Fact File 12.2) with the balanced line, the earth screen is often connected at one end only, as shown in Figure 12.8a, and still acts as a screen for the balanced audio lines. There is now no earth link between the two pieces of equipment, and so both can be safely earthed at the mains without causing an earth loop. The transformers have 'isolated' the two pieces of equipment from each other. The one potential danger with this is that the connecting lead with its earth disconnected in the plug at one end may later be used as a microphone cable. The lack of earth continuity between microphone and amplifier will cause inadequate screening of the microphone, and will also prevent a phantom power circuit being made (see 'Microphone powering options', Chapter 3), so such cables and tie-lines should be marked 'earth off' at the plug without the earth connection.

Unfortunately, not all pieces of audio equipment have balanced inputs and outputs, and one may be faced with the problem of interfacing a balanced output with an unbalanced input, and an unbalanced output with a balanced input. A solution is shown in Figure 12.8b, where the output transformer is connected to the signal and earth of the unbalanced input to give signal continuity. Because the input is unbalanced, there is no common mode rejection and the line is as susceptible to interference as an ordinary unbalanced line is. But notice that the screen is connected at one end only, so at least one can avoid an earth loop.

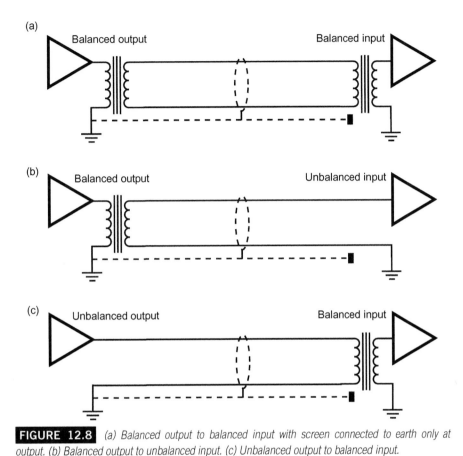

FIGURE 12.8 *(a) Balanced output to balanced input with screen connected to earth only at output. (b) Balanced output to unbalanced input. (c) Unbalanced output to balanced input.*

Figure 12.8c illustrates an unbalanced output feeding a balanced input. The signal and earth from the output feed the primary of the input transformer. Again the screen is not connected at one end so earth loops are avoided. Common mode rejection of interference at the input is again lost, because one side of the transformer primary is connected to earth. A better solution is to use a balancing transformer as close to the unbalanced output as possible, preferably before sending the signal over any length of cable. In the longer term it would be a good idea to fit balancing transformers inside unbalanced equipment with associated three-pin XLR-type sockets (see Fact File 12.3) if space inside the casing will allow. (Wait until the guarantee has elapsed first!)

FACT FILE 12.3 XLR-3 CONNECTORS

The most common balanced connector in professional audio is the XLR-3. This connector has three pins (as shown in the diagram), carrying respectively:

Pin 1 Screen
Pin 2 Signal (Live or 'hot')
Pin 3 Signal (Return or 'cold')

It is easy to remember this configuration, since X–L–R stands for Xternal, Live, Return. Unfortunately, an American convention still hangs on in some equipment which reverses the roles of pins 2 and 3, making pin 2 return and pin 3 live (or 'hot'). The result of this is an apparent absolute phase reversal in signals from devices using this convention when compared with an identical signal leaving a standard device. Modern American equipment mostly uses the European convention, and American manufacturers have now agreed to standardize on this approach.

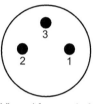

Viewed from end of
male pins

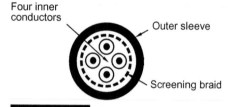

Four inner conductors
Outer sleeve
Screening braid

FIGURE 12.9 *Four conductors are used in star-quad cable.*

STAR-QUAD CABLE

Two audio lines can never occupy exactly the same physical space, and any interference induced into a balanced line may be slightly stronger in one line than in the other. This imbalance is seen by the transformer as a small differential signal which it will pass on, so a small amount of the unwanted signal will still get through. To help combat this, the two audio lines are twisted together during manufacture so as to present, on average, an equal face to the interference along both lines. A further step has been taken in the form of a cable called 'star-quad'. Here, four audio lines are incorporated inside the screen, as shown in Figure 12.9.

It is connected as follows. The screen is connected as usual. The four inner cores are connected in pairs such that two of the opposite wires (top and bottom in the figure) are connected together and used as one line, and the other two opposite wires are used as the other. All four are twisted together along the length of the cable during manufacture. This configuration ensures that for a given length of cable, both audio lines are exposed to an interference signal as equally as possible so that any interference is induced as equally as possible. So the input transformer sees the interference as a virtually perfect common mode signal, and efficiently rejects it.

This may seem like taking things to extremes, but star-quad is in fact quite widely used for microphone cables. When multicore cables are used which contain many separate audio lines in a single thick cable, the balanced system gives good immunity from crosstalk, due to the fact that a signal in a particular wire will be induced equally into the audio pair of the adjacent line, and will therefore be a common mode signal. Star-quad multicores give even lower values of crosstalk.

ELECTRONIC BALANCING

Much audio equipment uses an electronically balanced arrangement instead of a transformer, and it is schematically represented as in Figure 12.10. The transformers have been replaced by a differential amplifier. The differential amplifier is designed to respond only to differential signals, as is the case with the transformer, and has one positive and one negative input. Electronically balanced and transformer-balanced equipment can of course be freely intermixed. Reasons for dispensing with transformers include lower cost (a good transformer is rather more expensive than the electronic components which replace it), smaller size (transformers take up at least a few cubic centimeters, the alternative electronics rather less), less susceptibility to electromagnetic interference, and rather less sensitivity to the impedances between which they work with respect to distortion, frequency response and the like.

Good electronic balancing circuitry is, however, tricky to design, and the use of high-quality transformers in expensive audio equipment may well be a safer bet than electronic balancing of unknown performance. The best electronic balancing is usually capable of equal CMR performance to the transformer. Critics of transformer balancing cite factors such as the low-frequency distortion performance of a transformer, and its inability to pass extremely low frequencies, whereas critics of electronic balancing

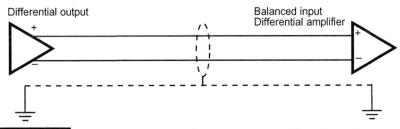

FIGURE 12.10 *An electronically balanced interconnection using differential amplifiers.*

cite the better CMR available from a transformer when compared with a differential amplifier, and the fact that only the transformer provides true isolation between devices. Broadcasters often prefer to use transformer balancing because signals are transferred over very long distances and isolation is required, whereas recording studios often prefer electronic balancing, claiming that the sound quality is better.

100 VOLT LINES

Principles

In 'Cable resistance', p. 364, it was suggested that the resistance of even quite thick cables was still sufficient to cause signal loss in loudspeaker interconnection unless short runs were employed. But long speaker lines are frequently unavoidable, examples being: backstage paging and show relay speakers in theaters; wall-mounted speakers in lecture theaters and halls; paging speakers in supermarkets and factories; and open-air 'tannoy' horns at fairgrounds and fêtes. All these require long speaker runs, or alternatively a separate power amplifier sited close to each speaker, each amplifier being driven from the line output of a mixer or microphone amplifier. The latter solution will in most cases be considered an unnecessarily expensive and complicated solution. So the '100 volt line' was developed so that long speaker cable runs could be employed without too much signal loss along them.

The problem in normal speaker connection is that the speaker cable has a resistance comparable with, or even greater than, the speaker's impedance over longer runs. It was shown in 'limitations of transformers', p. 362, that a transformer reflects impedance according to the square of the turns ratio. Suppose a transformer with a turns ratio of 5:1 is connected to the input of an 8 ohm speaker, as shown in Figure 12.11. The square of the turns ratio is 25:1 so the impedance across the primary of the transformer is $25 \times 8 = 200$ ohms. Now, the effective impedance of the speaker is much greater than the cable resistance, so most of the voltage will now reach the primary of the transformer and thence to the secondary and the speaker itself. But the transformer also transforms voltage, and the voltage across the secondary will only be a fifth of that across the primary. To produce 20 volts across the speaker then, one must apply 100 volts to the primary.

In a 100 volt line system, as shown in Figure 12.12, a 50 W power amplifier drives a transformer with a

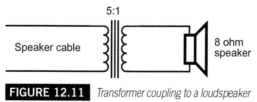

FIGURE 12.11 *Transformer coupling to a loudspeaker as used in 100 volt line systems.*

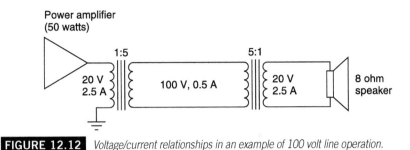

FIGURE 12.12 *Voltage/current relationships in an example of 100 volt line operation.*

step-up ratio of 1:5. Because the output impedance of a power amplifier is designed to be extremely low, the impedance across the secondary is also low enough to be ignored. The 20 volts, 2.5 amp output of the 50 watt amplifier is stepped up to 100 volts. The current is correspondingly stepped down to 0.5 amps (see Fact File 12.1), so that the total power remains the same. Along the speaker line there is a much higher voltage than before, and a much lower current. The voltage drop across the cable resistance is proportional to the current flowing through it, so this reduction in current means that there is a much smaller voltage drop due to the line. At the speaker end, a transformer restores the voltage to 20 volts and the current to 2.5 amps, and so the original 50 watts is delivered to the speaker.

A 50 watt amplifier has been used in the discussion. Any wattage of amplifier can be used, the transformer being chosen so that the step-up ratio gives the standard 100 volts output when the amplifier is delivering its maximum power output. For example, an amplifier rated at 100 watts into 8 ohms produces about 28 V. The step-up ratio of the line transformer would then have to be 28:100, or 1:3.6, to give the standard 100 volt output when the amplifier is being fully driven.

Returning to the loudspeaker end of the circuit. What if the speaker is only rated at 10 watts? The full 100 watts of the above amplifier would burn it out very quickly, and so a step-down ratio of the speaker transformer is chosen so that it receives only 10 watts. As 10 watts across 8 ohms is equivalent to around 9 volts, the required speaker transformer would have a step-down ratio of 100:9, or approximately 11:1.

Working with 100 volt lines

Speaker line transformers usually have a range of terminals labeled such that the primary side has a choice of wattage settings (e.g. 30 W, 20 W, 10 W, 2 W) and the secondary gives a choice of speaker impedance, usually 15 ohms, 8 ohms and 4 ohms. This choice means that a number of speaker systems can be connected along the line (a transformer being required for each speaker enclosure), the wattage setting being appropriate to the speaker's coverage. For example, a paging system in the backstage area of a theater could be required to feed paging to six dressing rooms, a large toilet and a fairly noisy green room. The dressing rooms are small and quiet, and so small speakers rated at 10 watts are employed with line transformers wired for 2 watts output. The large toilet requires greater power from the speaker, so one could use a 10 W speaker with a line transformer wired for 10 W output. The noisy green room could have a rather larger 20 W speaker with a line transformer wired for 20 W output. In this way, each speaker is supplied only with the wattage required to make it loud enough to be clearly heard in that particular room. A 20 W speaker in a small dressing room would be far too loud, and a 2 W speaker in a larger, noisier room would be inadequate.

As a string of loudspeakers is added to the system, one must be careful that the total wattage of the speakers does not exceed the output wattage of the power amplifier, or the latter will be overloaded. In the example, the six dressing rooms were allocated 2 W each, total 12 W. The toilet was allocated 10 W, the green room 20 W. The total is therefore 42 W, and a 50 W amplifier and line transformer would be adequate. In practice, a 100 W amplifier would be chosen to allow for both a good safety margin and plenty of power in hand if extra speakers need to be connected at a later date.

From the foregoing, it might well be asked why the 100 volt line system is not automatically used in all speaker systems. One reason is that 100 volts is high enough to give an electric shock and so is potentially dangerous in the domestic environment and other places where members of the public could interfere with an inadequately installed system. Second, the ultimate sound quality is compromised by the presence of transformers in the speaker lines – they are harder to design than the microphone and line-level transformers already discussed, because they have to handle high voltages as well as several amps – and whereas they still give a perfectly adequate and extremely useful performance in paging and background music applications, they are not therefore used in high-quality PA systems or hi-fi and studio monitor speakers.

600 OHMS

One frequently sees 600 ohms mentioned in the specifications of mixers, microphone amplifiers and other equipment with line-level outputs. Why is 600 ohms so special? The short answer is: it is not.

Principles

As has been shown, the output impedances of audio devices are low, typically 200 ohms for microphones and the same value or rather less for line-level outputs. The input impedances of devices are much higher, at least 1–2 k for microphone inputs and 10 k or more for line inputs. This is to ensure that virtually the whole of the output voltage of a piece of equipment appears across the input it is feeding. Also, lower input impedances draw a higher current for a given voltage, the obvious example of this being an 8 ohm loudspeaker which draws several amps from a power amplifier. So a high input impedance means that only very small currents need to be supplied by the outputs, and one can look upon microphone and line-level signals as being purely voltage without considering current.

This works fine, unless you are a telephone company that needs to send its signals along miles of cable. Over these distances, a hitherto unmentioned parameter comes into play which would cause signal loss if not dealt with, namely the wavelength of the signal in the cable. Now the audio signal is transmitted along a line at close to the speed of light (186000 miles per second or 3×10^8 meters per second). The shortest signal wavelength will occur at the upper limit of the audio spectrum, and will be around 9.3 miles (14.9 km) at 20 kHz.

When a cable is long enough to accommodate a whole wavelength or more, the signal can be reflected back along the line and cause some cancelation of the primary signal. Even when the cable run is somewhat less than a wavelength some reflection and cancelation still occurs. To stop this from happening the cable must be terminated correctly, to form a so-called 'transmission line', and input and output impedances are chosen to be equal. The value of 600 ohms was chosen many decades ago as the standard value for telecommunications, and therefore the '600 ohm balanced line' is used to send audio signals along lines which need to be longer than a mile or so. It was chosen because comparatively little current needs to be supplied to drive this impedance, but it is not high enough to allow much interference, as it is much easier for interference to affect a high-impedance circuit than a low one. Thus, professional equipment began to appear which boasted '600 ohms' to make it compatible with these lines. Unfortunately, many people did not bother to find out, or never understood, why 600 ohms was sometimes needed, and assumed that this was a professional audio standard per se, rather than a telecommunications standard. It was used widely in broadcasting, which has parallels with telecommunications, and may still be found in many cases involving older equipment.

The 600 ohm standard also gave rise to the standard reference level unit of 0 dBm, which corresponds to 1 mW of power dissipated in a resistance of 600 ohms. The corresponding voltage across the 600 ohm resistance at 0 dBm is 0.775 volts, and this leads some people still to confuse dBm with dBu, but 0 dBu refers simply to 0.775 volts with no reference to power or impedance: dBu is much more appropriate in modern equipment, since, as indicated above, the current flowing in most interfaces is negligible and impedances vary; dBm should only correctly be used in 600 ohm systems, unless an alternative impedance is quoted (e.g. dBm [75 ohms] is sometimes used in video equipment where 75 ohm termination is common).

Problems with 600 ohm equipment

A 600 ohm output impedance is too high for normal applications. With 600 ohms, 200 pF per meter cable, and an acceptable 3 dB loss at 40 kHz, the maximum cable length would be only around 33 m, which is inadequate for many installations. (This HF loss does not occur with the properly terminated 600 ohm system because the cable assumes the properties of a transmission line.) Furthermore, consider a situation where a mixer with a 600 ohm output impedance is required to drive five power amplifiers, each with an input impedance of 10 k. Five lots of 10 k in parallel produce an effective impedance of 2 k, as shown in Figure 12.13a. Effectively then 600 ohms is driving 2 k, as shown in Figure 12.13b. If V_1 is 1 volt then V_2 (from Ohm's law) is only 0.77 volts. Almost a quarter of the audio signal has been lost, and only a maximum of 33 meters of cable can be driven anyway.

Despite this, there are still one or two manufacturers who use 600 ohm impedances in order to appear 'professional'. It actually renders their equipment less suitable for professional use, as has been shown. One specification that is frequently encountered for a line output is something like: 'capable of delivering + 20 dBu into 600 ohms'. Here + 20 dBu is 7.75 volts, and 600 ohms is quite a low impedance, thus drawing more current from the source for a given voltage than, say, 10 kΩ does. The above specification is therefore useful, because it tells the user that the equipment can deliver 7.75 volts even into 600 ohms, and can therefore safely drive, say, a stack of power amplifiers, and/or a number of tape recorders, etc., without being overloaded. A domestic cassette recorder, for instance, despite having a low output impedance, could not be expected to do this. Its specification may well state '2 volts into 10 kΩ or greater'. This is fine for domestic applications, but one should do a quick calculation or two before asking it to drive all the power amplifiers in the building at once.

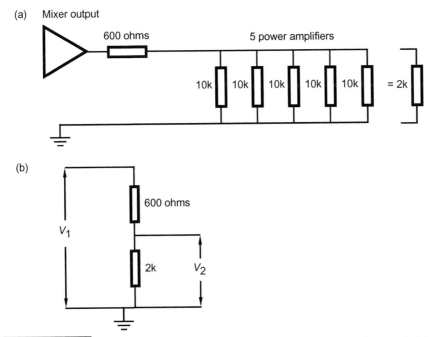

(a) Mixer output

600 ohms

5 power amplifiers

10k   10k   10k   10k   10k   = 2k

(b)

600 ohms

V_1

2k V_2

FIGURE 12.13 *(a) Considerable signal loss can result if a 600 ohm output is connected to a number of 10k inputs in parallel. (b) Electrical equivalent.*

DI BOXES

Overview

A frequent requirement is the need to interface equipment that has basically non-standard unbalanced outputs with the standard balanced inputs of mixers, either at line level or microphone level. An electric guitar, for example, has an unbalanced output of fairly high impedance – around 10 kΩ or so. The standard output socket is the 'mono' quarter-inch jack, and output voltage levels of around a volt or so (with the guitar's volume controls set to maximum) can be expected. Plugging the guitar directly into the mic or line-level input of a mixer is unsatisfactory for several reasons: the input impedance of the mixer will be too low for the guitar, which likes to drive impedances of 500 kΩ or more; the guitar output is unbalanced so the interference-rejecting properties of the mixer's balanced input will be lost; the high output impedance of the guitar renders it incapable of driving long studio tie-lines; and the guitarist will frequently wish to plug the instrument into an amplifier as well as the mixer, and simply using the same guitar output to feed both via a splitter lead electrically connects the amplifier to the studio equipment which causes severe interference and low-frequency hum problems. Similar problems are encountered with other instruments such as synthesizers, electric pianos and pickup systems for acoustic instruments.

To connect such an instrument with the mixer, a special interfacing unit known as a DI box (DI = direct injection) is therefore employed. This unit will convert the instrument's output to a low-impedance balanced signal, and also reduce its output level to the millivolt range suitable for feeding a microphone input. In addition to the input jack socket, it will also have an output jack socket so that the instrument's unprocessed signal can be passed to an amplifier as well. The low-impedance balanced output appears on a standard three-pin XLR panel-mounted plug which can now be looked upon as the output of a microphone. An earth-lift switch is also provided which isolates the earth of the input and output jack sockets from the XLR output, to trap earth loop problems.

Passive DI boxes

The simplest DI boxes contain just a transformer, and are termed 'passive' because they require no power supply. Figure 12.14 shows the circuit. The transformer in this case has a 20:1 step-down ratio, converting the fairly high output of the instrument to a lower output suitable for feeding microphone lines. Impedance is converted according to the square of the turns ratio (400:1), so a typical guitar output impedance of $15\,k\Omega$ will be stepped down to about 40 ohms which is comfortably low enough to drive long microphone lines. But the guitar itself likes to look into a high impedance. If the mixer's microphone input impedance is $2\,k\Omega$, the transformer will step this up to $800\,k\Omega$ which is adequately high for the guitar. The 'link

FIGURE 12.14

A simple passive direct-injection box.

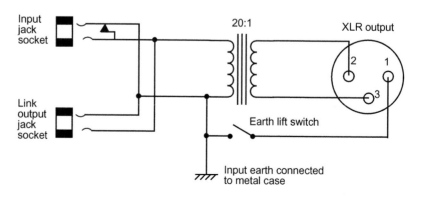

output jack socket' is used to connect the guitar to an amplifier if required. Note the configuration of the input jack socket: the make-and-break contact normally short-circuits the input which gives the box immunity from interference, and also very low noise when an instrument is not plugged in. Insertion of the jack plug opens this contact, removing the short-circuit. The transformer isolates the instrument from phantom power on the microphone line.

This type of DI box design has the advantages of being cheap, simple and requiring no power source – there are no internal batteries to forget to change. On the other hand, its input and output impedances are entirely dependent on the reflected impedances each side of the transformer. Unusually low microphone input impedances will give insufficiently high impedances for many guitars. Also, instruments with passive volume controls can exhibit output impedances as high as $200\,k\Omega$ with the control turned down a few numbers from maximum, and this will cause too high an impedance at the output of the DI box for driving long lines. The fixed turns ratio of the transformer is not equally suited to the wide variety of instruments the DI box will encounter, although several units have additional switches which alter the transformer tapping giving different degrees of attenuation.

Active DI boxes

The active DI box replaces the transformer with an electronic circuit which presents a constant very high impedance to the instrument and provides a constant low-impedance output. Additionally, the presence of electronics provides scope for including other features such as several switched attenuation values (say −20 dB, −40 dB, −60 dB), high and low filters and the like. The box is powered either by internal batteries, or preferably by the phantom power on the microphone line. If batteries are used, the box should include an indication of battery status; a 'test' switch is often included which lights an LED when the battery is good. Alternatively, an LED comes on as a warning when the voltage of the battery drops below a certain level. The make-and-break contacts of the input jack socket are often configured so that insertion of the jack plug automatically switches the unit on. One should be mindful of this because if the jack plug is left plugged into the unit overnight, for instance, this will waste battery power. Usually the current consumption of the DI box is just a few milliamps, so the battery will last for perhaps a hundred hours. Some guitar and keyboard amplifiers offer a separate balanced output on an XLR socket labeled 'DI' or 'studio' which is intended to replace the DI box, and it is often convenient to use this instead.

DI boxes are generally small and light, and they spend much of their time on the floor being kicked around and trodden on by musicians and sound engineers. Therefore, rugged metal (not plastic) boxes should be used and any switches, LEDs, etc. should be mounted such that they are recessed or shrouded for protection. Switches should not be easily moved by trailing guitar leads and feet. The DI box can also be used for interfacing domestic hi-fi equipment such as cassette recorders and radio tuners with balanced microphone inputs.

SPLITTER BOXES

The recording or broadcasting of live events calls for the outputs of microphones and instruments to be fed to at least two destinations, namely the PA mixer and the mixer in the mobile recording or outside broadcast van. The PA engineer can then balance the sound for the live audience, and the recording/broadcast balancer can independently control the mix for these differing requirements. It is possible of course to use two completely separate sets of microphones for this, but when one considers that there may be as many as ten microphones on a drum kit alone, and a vocalist would find the handling of two mics strapped together with two trailing leads rather unacceptable, a single set of microphones plugged into splitter boxes is the obvious way to do it. Ten or 15 years ago recording studios and broadcasting companies would have frowned on this because the quality of some of the microphones then being used for PA was insufficient for their needs; but today PA mics tend to be every bit as good as those found in studios, and indeed many of the same models are common to both environments.

A microphone cannot be plugged into two microphone inputs of two separate mixers directly because on the one hand this will electrically connect one mixer with the other causing ground loop and interference problems, not to mention the fact that one phantom power circuit will be driving directly into the other phantom power circuit; and on the other hand the impedance seen by the microphone will now be the parallel result of the two mixers resulting in impedances of as low as 500 ohms which is too low for many microphones. A splitter box is therefore used which isolates the two mixers from each other and maintains a suitable impedance for the microphone. A splitter box will often contain a transformer with one primary winding for the microphone and two separate secondary windings giving the two outputs, as shown in Figure 12.15.

The diagram requires a bit of explanation. First, phantom power must be conveyed to the microphone. In this case, output 2 provides it via the center tap of its winding which conveys the power to the center tap of the

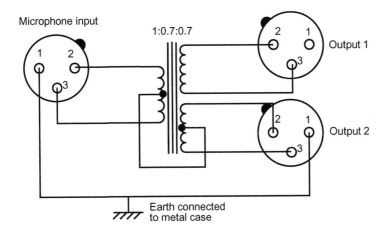

Microphone input

1:0.7:0.7

Output 1

2 1

3

Output 2

2 1

3

Earth connected
to metal case

FIGURE 12.15
A simple passive splitter box.

primary. The earth screen, on pin 1 of the input and output XLR sockets, is connected between the input and output 2 only, to provide screening of the microphone and its lead and also the phantom power return path. Note that pin 1 of output 1 is left unconnected so that earth loops cannot be created between the two outputs.

The turns ratio of the transformer must be considered next. The 1:0.7:0.7 indicates that each secondary coil has only 0.7 times the windings of the primary, and therefore the output voltage of the secondaries will each be 0.7 times the microphone output, which is about 3 dB down. There is therefore a 3 dB insertion loss in the splitter transformer. The reason for this is that the impedance as seen by the microphone must not be allowed to go too low. If there were the same number of turns on each coil, the microphone would be driving the impedance across each output directly in parallel. Two mixers with input impedances of 1 kΩ would therefore together load the microphone with 500 ohms, which is too low. But the illustrated turns ratio means that each 1 kΩ impedance is stepped up by a factor of $1:(0.7)^2 = 1:0.5$, so each mixer input appears as 2 kΩ to the microphone, giving a resultant parallel impedance of 1 kΩ, equal to the value of each mixer on its own which is fine. The 3 dB loss of signal is accompanied by an effective halving of the microphone impedance as seen by each mixer, again due to the transformer's impedance conversion according to the square of the turns ratio, so there need not be a signal-to-noise ratio penalty.

Because of the simple nature of the splitter box, a high-quality transformer and a metal case with the necessary input and output sockets are all that is really needed. Active electronic units are also available which eliminate the insertion loss and can even provide extra gain if required. The advantages of an active splitter box over its transformer counterpart are, however, of far less importance than, say, the advantages that an active DI box has over its passive counterpart.

JACKFIELDS (PATCHBAYS)

Overview

A jackfield (or patchbay) provides a versatile and comprehensive means of interconnecting equipment and tie-lines in a non-permanent manner such that various source and destination configurations can quickly and easily be set up to cater for any requirements that may arise.

For example, a large mixing console may have microphone inputs, line inputs, main outputs, group outputs, auxiliary outputs and insert send and returns for all input channels and all outputs. A jackfield, which usually consists of banks of 19 inch (48 cm) wide rack-mounting modules filled with rows of quarter-inch 'GPO' (Telecom)-type balanced jack sockets, is used as the termination point for all of the above facilities so that any individual input or output can be separately accessed. There are usually 24 jack sockets to a row, but sometimes 20 or 28 are encountered. Multicore cables connect the mixer to the jackfield, multipin connectors normally being employed at the mixer end. At the jackfield end multipin connectors can again be used, but as often as not the multicores will be hard wired to the jack socket terminals themselves. The layout of a mixer jackfield was discussed in 'Patchbay or jackfield', Chapter 5.

In addition to the mixer's jackfield there will be other jackfields either elsewhere in the rack or in adjacent racks which provide connection points for the other equipment and tie-lines. In a recording studio control room there will be such things as multitrack inputs and outputs, mic and line tie-lines linking control room to studio, outboard processor inputs and outputs, and tie-lines to the other control rooms and studios within a studio complex. A broadcasting studio will have similar arrangements, and there may also be tie-lines linking the studio with nearby concert halls or transmitter distribution facilities. A theater jackfield will in addition carry tie-lines leading to various destinations around the auditorium, back stage, in the wings, in the orchestra pit, and almost anywhere else. There is no such thing as too many tie-lines in a theater.

FIGURE 12.16 *A typical busy jackfield.* (Courtesy of the RSC.)

Patch cords

Patch cords (screened leads of around 1 meter in length terminated at each end with a 'B'-gauge jack plug [small tip] giving tip/ring/sleeve connections) are used to link two appropriate sockets. The tip is live (corresponding to pin 2 on an XLR), the ring is return (pin 3 on an XLR), and the sleeve is earth (pin 1 on an XLR), providing balanced interconnection. The wire or 'cord' of a normal patch cord is colored red. Yellow cords should indicate that the patch cord reverses the phase of the signal (i.e. tip at one end is connected to ring of the other end) but this convention is not followed rigidly, leading to potential confusion. Green cords indicate that the earth is left unconnected at one end, and such cords are employed if an earth loop occurs when two separately powered pieces of equipment are connected together via the jackfield.

A large-tipped 'A'-gauge jack plug, found on the end of most stereo headphone leads and often called a stereo jack (although it was originally used as a single-line balanced jack before stereo came into vogue), may sometimes be inserted into a GPO-type socket. This works, but the spring-loaded tip connector is positioned so as to make secure contact with the small 'B'-type tip and the large 'A' tip can sometimes bend the contact. When the correct type of jack plug is later inserted there will be intermittent or no contact between the tip and the socket's tip connector. The insertion of large-tipped jack plugs should therefore be discouraged.

Normaling

Normally, jackfield insertion points will be unused. Therefore the insertion send socket must be connected to the insertion return socket so that signal continuity is achieved. When an outboard unit is to be patched in, the insertion send socket is used to feed the unit's input. The unit's output is fed to the insertion return socket on the jackfield. This means that the send signal must be disconnected from the return socket and replaced by the return from the processor. To effect this requirement, extra make-and-break 'normaling' contacts on the jack socket are employed. Figure 12.17 shows how this is done. The signal is taken from the top jack socket to the bottom jack socket via the black triangle make-and-break contactors on the bottom socket. There is signal continuity. If a jack plug is now inserted into the bottom socket the contacts will be moved away from the make-and-break triangles, disconnecting the upper socket's signal from it. Signal from that jack plug now feeds the return socket. The make-and-break contacts on the upper jack socket are left unused, and so insertion of a jack plug into this socket alone has no effect on signal continuity. The send

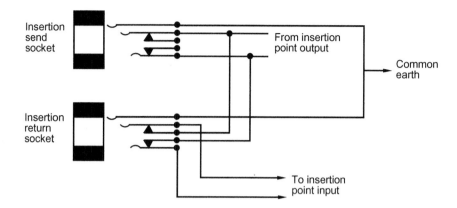

FIGURE 12.17
Normaling at jackfield insertion points.

socket therefore simply provides an output signal to feed the processor. This arrangement is commonly termed 'half normaling' because only the lower socket in Figure 12.17 uses the make-and-break contacts.

Sometimes these make-and-break contacts are wired so that the signal is also interrupted if a jack plug is inserted into the send socket alone. This can be useful if, for instance, an alternative destination is required for, say, a group output. Insertion of a jack plug into the send socket will automatically mute the group's output and allow its signal to be patched in elsewhere, without disturbing the original patching arrangement. Such a wiring scheme is, however, rather less often encountered. It is termed 'full normaling'.

In addition to providing insert points, normaling can also be used to connect the group outputs of a mixer to the inputs of a multitrack recorder or a set of power amplifiers. The recorder or amplifier inputs will have associated sockets on the jackfield, and the mixer's group output sockets will be normaled to these as described. If need be, the input sockets can be overplugged in order to drive the inputs with alternative signals, automatically disconnecting the mixer's outputs in the process. Figure 12.18 illustrates a small section of a mixer's jackfield, and how it could be labeled. The upper row gives access to the mixer's matrix outputs. Patch cords inserted here can convey the signals to other sockets where they are required, for example to processor inputs, or for foldback feeds. The lower row is connected to another jackfield in a room which houses racks of power amplifiers. It will be seen that an N in a circle is written above each of the numbers; this indicates that these are normaled to the sockets above. Thus, with no patch cords inserted, the upper row is by default connected to the lower row, and the matrix outputs automatically drive the amp input tie-lines and thence the power amplifiers. This would be the routine mode

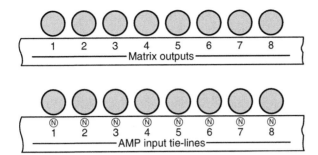

FIGURE 12.18

Part of a mixer's jackfield.

of operation, and the two rows are half normaled together. If, however, it is desired that an amp room tie-line needs to be fed from another device, say, for example, a digital delay unit is required to be inserted between matrix output 1 and tie-line 1, the digital delay's output is plugged into tie-line 1 socket, breaking the normaling to the socket above. Matrix output 1 socket is then patched to drive the digital delay's input.

If an N in a circle also appears beneath the upper sockets, this would indicate full normaling. Inserting a jack plug into an upper socket so labeled would then disconnect that matrix output from the amp input tie-line.

Other jackfield facilities

Other useful facilities in a jackfield include multiple rows of interconnected jacks, or 'mults'. These consist of, say, six or eight or however many adjacent sockets which are wired together in parallel so that a mixer output can be patched into one of the sockets, the signal now appearing on all of the sockets in the chain which can then be used to feed a number of power amplifier inputs in parallel, or several tape machines. The disadvantage of this poor man's distribution amplifier is that if there is a short-circuit or an interference signal on any of the inputs it is feeding, this will affect all of the sockets on the mult. There is no isolation between them.

Outboard equipment will generally be equipped with XLR-type input and output connectors (or sometimes 'mono' jack sockets). It is useful therefore to have a panel of both male and female XLR sockets near to where such outboard gear is usually placed, these being wired to a row of sockets on the jackfield to facilitate connection between the mixer or tie-lines and these units.

Special additional make-and-break contacts can be included in the jack socket which are operated in the usual manner by jack plug insertion but have no contact with any part of the audio signal. These can be used to activate warning indicators to tell operators that a certain piece of equipment is in use, for example.

Since most if not all of the interconnections in a rig pass through the jackfield it is essential that the contacts are of good quality, giving reliable service. Palladium metal plating is employed which is tough, offering good resistance to wear and oxidation. This should always be looked for when jackfield is being ordered. Gold or silver plating is not used because it would quickly wear away in the face of professional use. The latter also tarnishes rather easily.

There is a miniature version known as the Bantam jack. This type is frequently employed in the control surface areas of mixers to give convenient access to the patching. Very high-density jackfields can be assembled, which has implications for the wiring arrangements on the back. Several earlier examples of Bantam-type jackfields were unreliable and unsuited to professional use. Later examples are rather better, and of course palladium contacts should always be specified.

Electronically controlled 'jackfields' dispense with patch cords altogether. Such systems consist of a digitally controlled 'stage box' which carries a number of input and output sockets into which the mixer inputs and outputs and any other tie-lines, processor and tape machine inputs and outputs are plugged. The unit is controlled by a keypad, and a VDU displays the state of the patch. Information can also be entered identifying each input and output by name according to the particular plug-up arrangement of the stage box. Any output can be routed to any input, and an output can be switched to drive any number of inputs as required. Various patches can be stored in the system's memory and recalled; rapid repatching is therefore possible, and this facility can be used in conjunction with timecode to effect automatic repatches at certain chosen points on a tape during mixdown for instance. MIDI control is also a possibility.

DISTRIBUTION AMPLIFIERS

A distribution amplifier is an amplifier used for distributing one input to a number of outputs, with independent level control and isolation for each output. It is used widely in broadcast centers and other locations where signals must be split off and routed to a number of independent locations. This approach is preferable to simple parallel connections, since each output is unaffected by connections made to the others, preventing one from loading down or interfering with the others.

Plug-Ins and Outboard Equipment

PLUG-INS

Plug-ins are now one of the fastest-moving areas of audio development, providing audio signal processing and effects that run either on a workstation's CPU or on dedicated DSP. (The hardware aspects of this were described in Chapter 9.) Audio data can be routed from a sequencer or other audio application, via an API (application programming interface) to another software module called a 'plug-in' that does something to the audio and then returns it to the source application. In this sense it is rather like inserting an effect into an audio signal path, but done in software rather than using physical patch cords and rack-mounted effects units. Plug-ins can be written for the host processor in a language such as C+, using the software development toolkits (SDK) provided by the relevant parties. Plug-in processing introduces a delay that depends on the amount of processing and the type of plug-in architecture used. Clearly low latency architectures are highly desirable for most applications.

Many plug-ins are versions of previously external audio devices that have been modeled in DSP, in order to bring favorite EQs or reverbs into the workstation environment. The sound quality of these depends on the quality of the software modeling that has been done. Some host-based (native) plug-ins do not have as good a quality as dedicated DSP plug-ins as they may have been 'cut to fit' the processing power available, but as hosts become ever more powerful the quality of native plug-ins increases.

A number of proprietary architectures have been developed for plug-ins, including Microsoft's DirectX, Steinberg's VST, Digidesign's TDM, Mark of the Unicorn's MAS, TC Works' PowerCore and EMagic's host-based plug-in format, subsequently bought by Apple and re-named Logic Pro, part of their Logic Studio music application programs. Apple's OS X Audio Units are a feature built into the OS that manages plug-ins without the need for third-party middleware solutions. It is usually necessary to specify for which system any software plug-in is intended, as the architectures are not compatible. As OS-based plug-in architectures for audio become more widely used, the need for proprietary approaches may diminish. Digidesign (now Avid), for example, has had a number of different plug-in approaches that are used variously in its products, as shown in Table 13.1. The oldest of these are being phased out, with the intention that AAX will become the default version.

Chapter 9 deals with audio processing for workstations, but a brief repetition covering some aspects of the Apple system is warranted here. Apple's Core Audio provides plug-in facilities for audio signal processing and synthesis, as well as audio-to-MIDI synchronization. Its audio plug-ins are called Audio Units (AUs). A number of standard AUs are provided with

Table 13.1	Avid/Digidesign Plug-in Formats
Plug-in architecture	**Description**
TDM	Uses dedicated DSP cards for signal processing. Does not affect the host CPU load and processing power can be expanded as required.
HTDM (Host TDM)	Uses the host processor for TDM plug-ins, instead of dedicated DSP.
RTAS (Real Time Audio Suite)	Uses host processor for plug-ins. Not as versatile as HTDM.
AudioSuite	Non-real-time processing that uses the host CPU to perform operations such as time-stretching that require the audio file to be rewritten.
AAX	Avid Audio eXtension. Most recent Avid plug-in format for Pro Tools. Comes in DSP (for Pro Tools HDX systems) and Native (runs on host processor) versions.

the OS X operating system, offering a range of audio processing options to other Core Audio compatible software that runs on the platform. Audio workstation packages such as Logic, for example, work closely with Core Audio to implement aspects of their functionality, including plug-ins. Core Audio normally expects to work with audio represented as 32-bit floating point linear PCM, but there are means to translate between this and other PCM formats, as well as to coded formats such as MP3, AAC or Apple Lossless Audio Coding (ALAC).

DirectX is a suite of multimedia extensions developed by Microsoft for the Windows platform. It includes an element called DirectShow that deals with real-time streaming of media data, together with the insertion of so-called 'filters' at different points. DirectX audio plug-ins work under DirectShow and are compatible with a wide range of Windows-based audio software. They operate at 32 bit resolution, using floating-point arithmetic and can run in real time or can render audio files in non-real time. They do not require dedicated signal processing hardware, running on the host CPU, and the number of concurrent plug-ins depends on CPU power and available memory. DirectX plug-ins are also scalable – in other words they can adapt to the processing resource available. They have the advantage of being compatible with the very wide range of DirectX-compatible software in the general computing marketplace. DXi, for example, is a software synthesizer plug-in architecture developed by Cakewalk, running under DirectX. DirectX 11.1, contained in Windows 8, includes XAudio2, an audio API (Application Programming Interface) for Windows and the Xbox360. XAudio2 succeeds DirectSound on Windows, and operates through the XAudio API on the Xbox 360. It operates through DirectSound on Windows XP, and through the low-level audio mixer WASAPI (Windows Audio Session API) on Windows Vista and higher. Features of XAudio2 include spatial processing and signal processing for high-level audio APIs such as XACT, a Cross-platform Audio Creation Tool which is a part of Microsoft's DirectX SDK.

One example of a proprietary approach used quite widely is VST, Steinberg's Virtual Studio Technology plug-in architecture. It runs on multiple platforms and works in a similar way to DirectX plug-ins. On Windows machines it operates as a DLL (dynamic link library) resource, whereas on Macs it runs as a raw Code resource. It can also run on BeOS and SGI systems, as a Library function. VST incorporates both virtual effects and virtual instruments such as samplers and synthesizers. There is a cross-platform GUI development tool that enables the appearance of the user interface to be ported between platforms without the need to rewrite it each time.

Plug-ins cover all the usual traditional outboard functions including graphic equalization, compressor/limiting, reverb and multi-effects processing, and a variety of 'vintage' examples mimic old guitar amplifiers and analog processors. Examples other than these more traditional types include plug-ins which will essentially transform one instrument into another; and others which are designed to create new and unfamiliar sounds. Fact File 13.1 gives some examples. Outboard equipment, the 'hardware' equivalent of the plug-in, which of course preceded it historically, is still much in use in studios and particularly in the live sound environment. Sometimes a 'half-way house' is encountered where a mixer will incorporate electronics from another manufacturer's effects processor with front panel controls similar to the original unit. This offers useful on-board processing from elsewhere which will normally be a familiar and respected device.

The following describes a selection of devices and their functions, and it applies equally to plug-ins and their more traditional hardware equivalents. Plug-ins by their nature provide a vastly greater range and type of audio processing than can outboard devices, but there are still a number of functions common to both.

The graphic equalizer

The graphic equalizer, pictured in Figure 13.1, consists of a row of faders (or sometimes rotary controls), each of which can cut and boost a relatively narrow band of frequencies. Simple four- or five-band devices exist which are aimed at the electronic music market, these being multiband tone controls. They perform the useful function of expanding existing simple tone controls on guitars and amplifiers, and several of the latter incorporate them.

The professional rack-mounting graphic equalizer will have at least ten frequency bands, spaced at octave or one-third-octave intervals. The ISO (International Standards Organization) center frequencies for octave bands are 31 Hz, 63 Hz, 125 Hz, 250 Hz, 500 Hz, 1 kHz, 2 kHz, 4 kHz, 8 kHz and 16 kHz. Each fader can cut or boost its band by typically 12 dB or more. Figure 13.2 shows two possible types of filter action. The 1 kHz fader is chosen, and three levels of cut and boost are illustrated. Maximum cut and boost of both types produces very similar Q (see Fact File 5.6). A high Q result is obtained by both types when maximum cut or boost is applied. The action of the first type is rather gentler when less cut or boost is applied, and the Q varies according to the degree of deviation from the fader's central position.

FACT FILE 13.1 PLUG-IN EXAMPLES

Three examples of plug-in user interfaces are shown below: a reverb device, a program for creating the sound of another instrument from an existing one and an effects processor. The Applied Acoustic Systems Strum Acoustics GS-1 acoustic guitar synth, taking its input signal from a keyboard, mimics a variety of nylon- and steel-strung acoustic guitars; a voicing module automatically arranges notes of the chords played on the keyboard as a guitar player would play them on a fret board. Strumming and picking actions are also created, helping to give an authentic guitar sound. The Platinum Enigma is an example of a plug-in which processes incoming signals using flanging, phase, delay, filtering and modulation to create a variety of sounds which at the extreme are not recognizably related to the original signal.

The quality of such plug-ins is now getting to the point where it is on a par with the sound quality achievable on external devices, depending primarily on the amount of DSP available.

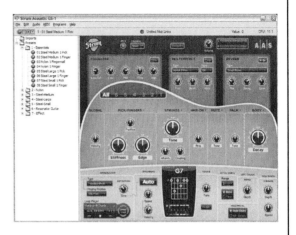

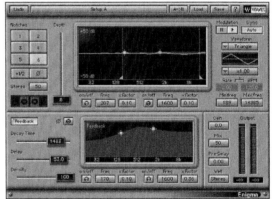

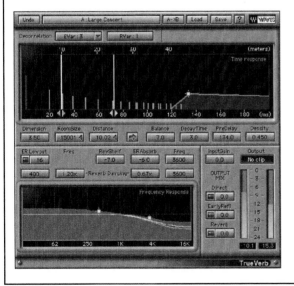

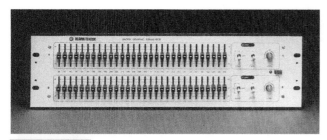

FIGURE 13.1   *A typical two-channel graphic equalizer. (Courtesy of Klark-Teknik Research Ltd.)*

Many graphic equalizers conform to this type of action, and it has the disadvantage that a relatively broad band of frequencies is affected when moderate degrees of boost or cut are applied. The second type maintains a tight control of frequency bandwidth throughout the cut and boost range, and such filters are termed constant Q, the Q remaining virtually the same throughout the fader's travel. This is particularly important in

FIGURE 13.2

Two types of filter action shown with various degrees of boost and cut. (a) Typical graphic equalizer with Q dependent upon degree of boost/cut. (b) Constant Q filter action.

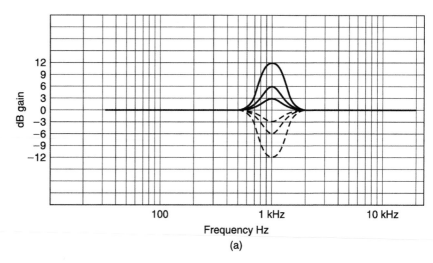

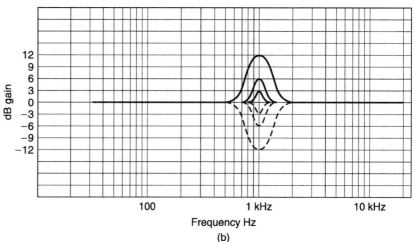

the closely spaced one-third-octave graphic equalizer which has 30 separate bands, so that adjacent bands do not interact with each other too much. The ISO center frequencies for 30 bands are 25 Hz, 31 Hz, 40 Hz, 50 Hz, 63 Hz, 80 Hz, 100 Hz, 125 Hz, 180 Hz, 200 Hz, 250 Hz, 315 Hz, 400 Hz, 500 Hz, 630 Hz, 800 Hz, 1 kHz, 1k25 Hz, 1k8 Hz, 2 kHz, 2k5 Hz, 3k15 Hz, 4 kHz, 5 kHz, 6k3 Hz, 8 kHz, 10 kHz, 12k5 Hz, 16 kHz and 20 kHz. The value of using standard center frequencies is that complementary equipment such as spectrum analyzers which will often be used in conjunction with graphic equalizers have their scales centered on the same frequencies.

Even with tight constant Q filters, the conventional analog graphic equalizer still suffers from adjacent filter interaction. If, say, 12 dB of boost is applied to one frequency and 12 dB of cut applied to the next, the result will be more like a 6 dB boost and cut, the response merging in between to produce an ill-defined Q value. Such extreme settings are, however, unlikely. The digital graphic equalizer applies cut and boost in the digital domain, and extreme settings of adjacent bands can be successfully accomplished without interaction if required.

Some graphic equalizers are single channel, some are stereo. All will have an overall level control, a bypass switch, and many also sport separate steep-cut LF filters. A useful facility is an overload indicator – usually an LED which flashes just before the signal is clipped – which indicates signal clipping anywhere along the circuit path within the unit. Large degrees of boost can sometimes provoke this. Some feature frequency cut only, these being useful as notch filters for getting rid of feedback frequencies in PA/ microphone combinations. Additional dedicated frequency-sweepable notch filters can be incorporated for this purpose. Some can be switched between cut/boost, or cut only. It is quite possible that the graphic equalizer will be asked to drive very long lines, if it is placed between mixer outputs and power amplifiers for example, and so it must be capable of doing this. The '+20 dBu into 600 ohms' specification should be looked for as is the case with mixers. It will be more usual though to patch the graphic equalizer into mixer output inserts, so that the mixer's output level meters display the effect on level the graphic equalizer is having. Signal-to-noise ratio should be at least 100 dB.

The graphic equalizer can be used purely as a creative tool, providing tone control to taste. It will frequently be used to provide overall frequency balance correction for PA rigs. It has formerly been used to equalize control room speakers, but poor results are frequently obtained due to the fact that a spectrum analyzer's microphone samples the complete room frequency response whereas the perceived frequency balance is a complex combination of direct and reflected sound arriving at different times. The graphic equalizer can also change the phase response of signals, and there has been a trend away from their use in the control room for monitor EQ, adjustments being made to the control room acoustics instead.

The parametric equalizer was fully described in 'Equalizer section', Chapter 5.

The compressor/limiter

The compressor/limiter (see Fact File 13.2) is used in applications such as dynamics control and as a guard against signal clipping. Such a device is pictured in Figure 13.3. The three main variable parameters are attack, release and threshold. The attack time, in microseconds (μs) and milliseconds (ms), is the time taken for a limiter to react to a signal. A very fast attack time of 10μs can be used to avoid signal clipping, any high-level

---

FACT FILE 13.2 COMPRESSION AND LIMITING

A compressor is a device whose output level can be made to change at a different rate to input level. For example, a compressor with a ratio of 2:1 will give an output level that changes by only half as much as the input level above a certain threshold (see diagram). For example, if the input level were to change by 6 dB the output level would change by only 3 dB. Other compression ratios are available such as 3:1, 5:1, etc. At the higher ratios, the output level changes only a very small amount with changes in input level, which makes the device useful for reducing the dynamic range of a signal. The threshold of a compressor determines the signal level above which action occurs.

A limiter is a compressor with a very high compression ratio. A limiter is used to ensure that signal level does not rise above a given threshold. A 'soft' limiter has an action which comes in only gently above the threshold, rather than acting as a brick wall, whereas a 'hard' limiter has the effect almost of clipping anything which exceeds the threshold.

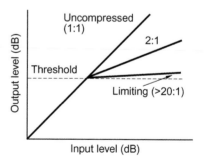

FIGURE 13.3 *A typical compressor/limiter. (Courtesy of Drawmer Distribution Ltd.)*

transients being rapidly brought under control. A fast release time will rapidly restore the gain so that only very short-duration peaks will be truncated. A ducking effect can be produced by using rapid attack plus a release of around 200–300 ms. A threshold level is chosen which causes the limiting to come in at a moderate signal level so that peaks are pushed down before the gain is quickly reinstated. Such a ducking effect is ugly on speech, but is useful for overhead cymbal mics for example.

A slow release time of several seconds, coupled with a moderate threshold, will compress the signal dynamics into a narrower window, allowing a higher mean signal level to be produced. Such a technique is often used in vocals to obtain consistent vocal level from a singer. AM radio is compressed in this way so as to squeeze wide dynamic range material into this narrow dynamic range medium. It is also used on FM radio to a lesser extent, although very bad examples of its application are frequently heard on pop stations. An oppressive, raspy sound is the result, and in the pauses in between items or speech one hears the system gain creeping back up, causing pumping noises. Background noise rapidly ducks back down when the presenter again speaks. Many units offer separate limiting and compressing sections, their attack, release and threshold controls in each section having values appropriate to the two applications. Some also include gates (see 'Noise gates', Chapter 7) with variable threshold and ratio, rather like an upside-down limiter, such that below a certain level the level drops faster than would be expected. This is called 'expansion', described below. 'Gain make-up' is often available to compensate for the overall level-reducing effect of compression. Meters may indicate the amount of level reduction occurring.

One advantage of in-the-box compressor-limiting as opposed to separate outboard processors is that the latter react to incoming signals in real time, and so are unable to anticipate the state of signals before action can be taken. Processors contained within computer workstations can preview a signal, once recorded, on an ongoing basis, anticipating action to be taken beforehand, and by this means processes such as compression and limiting of signal peaks can be handled less abruptly and therefore less obtrusively.

The expander

The expander, or expander/gate as it is sometimes labeled, is the antithesis of the compressor/limiter, offering upward and downward expansion of dynamic range. Upward expansion will increase the maximum level of an existing signal, and is rarely called for, particularly as it can lead to overload difficulties. Downward expansion, reducing the level of the signal when it drops below a certain user-defined threshold, is a far more commonly used facility, and the term 'expander' normally refers to this process. It can be regarded as a more sophisticated form of noise gate; with the latter the gate closes when the signal reaches the user-defined muting threshold, but the expander enables the user to define the rate at which the gain is reduced, tailing the sound gently into silence. This is termed 'slope'. Below the threshold level, a 1:2 slope ratio for instance doubles the attenuation of the signal compared with the un-processed signal. Thus, a signal level that was 3 dB below the threshold would now be reduced to 6 dB below the threshold, and signals 10 dB below would now be reduced a further 10 dB. The release time defines the speed at which the slope activates. The 'hold' control can be used to delay the onset of the expansion for a length of time before it takes action. The attack time defines the time it takes the processor to revert to unity gain once the signal rises above the threshold, and this can be used creatively to alter the sound of the transient attack of short, percussive sounds, for instance.

Echo and reverb devices

Before the advent of electronic reverb and echo processing, somewhat more basic, 'physical' means were used to generate the effects. The echo chamber was literally that, a fairly large reverberant room being equipped with a speaker and at least two spaced microphones. Signal was sent to the speaker and the reverb generated in the room was picked up by the two microphones which constituted the 'stereo return'. The echo plate was a large thin resonant plate of several meters in area suspended in a frame. A driving transducer excited vibrations in the plate, and these were picked up by several transducers placed in various positions on its surface. Some quite high-quality reverb effects were possible. The spring reverb, made popular by the Hammond organ company many decades ago, consists literally of a series of coil springs about the diameter of a pencil and of varying lengths (about 10–30 cm) and tensions depending on the model. A driving transducer excites torsional vibrations in one end of the springs, and these are picked up by transducers at the other end. Quite a pleasing effect can be obtained, and it is still popular for guitar amplifiers. The tape echo consisted of a short tape loop together with a record head followed by several replay heads spaced a few centimeters apart, then lastly an erase head. The output levels from the replay heads could be adjusted as could the speed of the tape, generating a variety of repeat-echo effects. Control of the erase head could generate a huge build-up of multi-echoes.

When one hears real reverb, one hears 'pre-delay' in effects process-ing parlance: a sound from the source travels to the room boundaries and then back to the listener, so there is a delay of several tens of milliseconds between hearing the direct sound and hearing the reverberation. This plays a large part in generating realistic reverb effects, and Fact File 13.3 explains the requirements in more detail.

Digital reverb

The present-day digital reverb and processor, such as that pictured in Figure 13.4, can be quite a sophisticated device. Research into path lengths, boundary and atmospheric absorption, and the physical volume and dimen-sions of real halls, have been taken into account when algorithms have been designed. Typical front panel controls will include selection of internal pre-programmed effects, labeled as 'large hall', 'medium hall', 'cathedral',

---

FACT FILE 13.3 SIMULATING REFLECTIONS

Pre-delay in a reverb device is a means of delaying the first reflection to simulate the effect of a large room with distant surfaces. Early reflections may then be pro-grammed to simulate the first few reflections from the surfaces as the reverberant field builds up, followed by the general decay of reverberant energy in the room as random reflections lose their energy (see diagram).

Pre-delay and early reflections have an important effect on one's perception of the size of a room, and it is these first few milliseconds which provide the brain with one of its main clues as to room size. Reverberation time (RT) alone is not a good guide to room size, since the RT is affected both by room volume and absorption (see Fact Files 1.5 and 1.6); thus the same RT could be obtained

from a certain large room and a smaller, more reflective room. Early reflections, though, are dictated only by the distance of the surfaces.

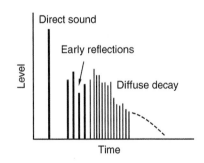

FIGURE 13.4 *The TC Electronic M3000 Digital Reverb and effects processor. (Courtesy of TC Electronic.)*

'church', etc., and parameters such as degree of pre-delay, decay time, frequency balance of delay, dry-to-wet ratio (how much direct untreated sound appears with the effect signal on the output), stereo width, and relative phase between the stereo outputs can often be additionally altered by the user. A small display gives information about the various parameters.

Memory stores generally contain a volatile and a non-volatile section. The non-volatile section contains factory preset effects, and although the parameters can be varied to taste the alterations cannot be stored in that memory. Settings can be stored in the volatile section, and it is usual to adjust an internal preset to taste and then transfer and store this in the volatile section. For example, a unit may contain 300 presets. The first 150 are non-volatile, and cannot be permanently altered. The last 150 can store settings arrived at by the user by transferring existing settings to this section. The method of doing this varies between models, but is usually a simple two- or three-button procedure, for example by pressing 1, 5 and 1 and then 'store'. This means that a setting in the non-volatile memory which has been adjusted to taste will be stored in memory 151. Additional adjustments can then be made later if required.

Several units provide a lock facility so that stored effects can be made safe against accidental overwriting. An internal battery backup protects the memory contents when the unit is switched off. Various unique settings can be stored in the memories, although it is surprising how a particular model will frequently stamp its own character on the sound however it is altered. This can be either good or bad of course, and operators may have a preference for a particular system or house 'sound'. Sometimes the bandwidth of a processor's output reduces with increasing reverberation or echo decay times. Such a shortcoming is sometimes hard to find in a unit's manual.

In some of the above devices it should be noted that the input is mono and the output stereo. In this way 'stereo space' can be added to a mono signal, there being a degree of decorrelation between the outputs. A reverb device may have stereo inputs, so that the source can be assumed to be other than a point.

FIGURE 13.5   *The TC Electronic System 6000 multi-effects processor. (Courtesy of TC Electronic.)*

Multi-effects processors

Digital multi-effects processors such as that shown in Figure 13.5 can offer a great variety of features. Parametric equalization is available, offering variations in Q, frequency and degree of cut and boost. Short memory capacity can store a sample, the unit being able to process this and reproduce it according to the incoming signal's command. MIDI interfacing (see Chapter 14) has become popular for the selection of effects under remote control, as has the RS 232 computer interface, and a USB port or RAM card slot is sometimes encountered for loading and storing information. Some now have Firewire or USB ports for digital audio streaming, or an alternative real-time digital interface (see Chapter 10). Repeat echo, autopan, phase, modulation, flange, high and low filters, straight signal delay, pitch change, gating and added harmony may all be available in the presets, various multifunction nudge buttons being provided for overall control. Many units are only capable of offering one type of effect at a time. Several have software update options so that a basic unit can be purchased and updates later incorporated internally to provide, say, longer delay times, higher sample storage capacity, and new types of effect as funds allow and as the manufacturer develops them. This helps to keep obsolescence at bay in an area which is always rapidly developing.

Frequency shifter

The frequency shifter shifts an incoming signal by a few hertz. It is used for acoustic feedback control in sound reinforcement work, and operates as follows. Feedback is caused by sound from a speaker re-entering a microphone to be reamplified and reproduced again by the speaker, forming a positive feedback loop which builds up to a continuous loud howling noise at a particular frequency. The frequency shifter is placed in the signal path such that the frequencies reproduced by the speakers are displaced by several hertz compared with the sound entering the microphone, preventing additive effects when the sound is recycled, so the positive feedback loop is broken. The very small frequency shift has minimal effect on the perceived pitch of the primary sound.

Pitch shift

Pitch shifting and time stretching, described in Chapter 8, is normally accomplished in computer recording and editing systems and outboard processors using techniques such as altering sampling frequency, re-sampling at a different frequency, or using other more sophisticated pitch alteration algorithms. Another, more basic way of achieving pitch change has been used in effects units and 'stomp boxes' for the music industry to provide effects such as flange and chorus which involve small and varying degrees of pitch shift, as well as for more straightforward pitch change duties in harmonizers.

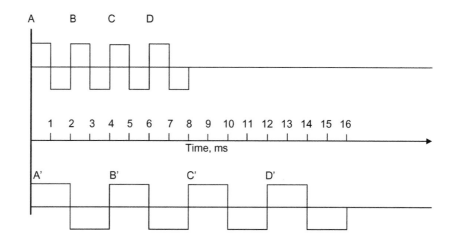

FIGURE 13.6

The process of downward pitch shift involves the application of continuously-increasing delay.

Downward pitch shift can be achieved by introducing a continuously increasing delay to the signal. A continuously *decreasing* delay is applied to achieve upward shift. Figure 13.6 illustrates the process for reducing pitch. A square wave has been chosen as the signal for clarity, and the horizontal time axis is calibrated in milliseconds. The top trace is the original signal, and the bottom trace shows how it has been pitched down by one octave, its total duration being increased from 8 ms to 16 ms. At A and A′, the original and processed signals begin. At B, 2 ms has elapsed. At B′, a delay of 2 ms has been applied to the signal, and 4 ms in all has therefore elapsed since the start of the lower trace. At C, 4 ms after the original signal has begun, a delay of a further 4 ms has been introduced to obtain point C′; and for point D, 6 ms along the time axis, point D′ is obtained by introducing 6 ms of delay. This describes what goes on at several specific points during a continuously increasing delay process which therefore achieves downward pitch shift, the degree of shift being a function of the *rate of increase* of the delay.

A problem with the process as described is that continuously increasing the delay will soon require a prohibitively large amount of digital storage as the program continues, and one way of avoiding this is for the processor to re-start the delay operation after a short period of time, a smooth transition between the two states being managed by the software. The rate of increase of delay determines the degree of pitch shift, this being independent of the length of time the process has continued. Another technique for avoiding prohibitive levels of data storage is for a second processor to begin just before the first processor is due to re-start, and a cross-fade is performed between the two.

Upward pitch shift is the exact opposite of that just described, a continuously *decreasing* delay of the original signal now being applied. Figure 13.6 can now be read from right to left, the bottom trace being regarded as the original signal, the top trace the upward-pitched output. It will immediately be noticed that the processed signal does not start until some time after the original signal has begun. This is because the processor first has to store up a certain amount of the incoming audio before it can begin to apply the continuously decreasing delay. An inherent latency is therefore present in the upward-pitch process, and a delay of up to about 50 ms can be accepted before the ear begins to perceive the emerging signal as having a definite time lag.

Digital delay

The digital delay line is incorporated into any substantial sound reinforcement installation as a matter of course. Consider a typical SR setup which may consist of main speakers each side of the stage (or against the proscenium arch or 'prosc' in a theater); front-fill speakers across the front of the stage; additional speakers covering extreme right- and left-hand sides under galleries; a line of speakers under a gallery covering the rear of the stalls; and a flown cluster covering an upper balcony, these latter speakers being rigged somewhat forward of the stage. Arrival times of the sounds from the various speakers to a particular location in the auditorium will of course vary due to the different physical path lengths, and the sound can be quite blurred as a result. Comb-filtering effects – abrupt attenuation of sound at certain frequencies as sound from one speaker is canceled by anti-phase sound coming from another due to the different path lengths – can also be encountered at some locations, and a slow walk from one side of an auditorium to the other whilst listening to pink noise will quite often make these problems apparent.

Digital delay lines are patched in between the mixer's outputs and the power amplifiers to alleviate the problem (digital mixers often incorporate delays on their outputs for this sort of application) and are set up as follows. First, the sound of sharp clicks, rather like those of a large ticking clock – about one tick per second – is played through the main speakers each side of the stage with the other speakers switched off. Whilst standing close to the front of the stage, the front-fill speakers can then be added in. The clicks from these will reach the listener sooner than those from the main speakers, and the clicks will sound blurred, or perhaps even a double click will be heard each time. Adding a delay of perhaps 10 ms or so to the front-fills will bring the sound back into sharp focus, the exact value of delay needed depending upon the distance between the listener and the two sets of speakers. The front-fills are then switched off, and the listener moves to the side, under the balcony, and the speakers covering these areas are switched on. Again, indistinct clicks will be heard, and a delay to the side speakers of perhaps 25 ms will be required here to bring the sound into focus. For the line of speakers covering the rear stalls, a delay of perhaps 50 ms will be found to be needed. A rule of thumb when setting initial values is that sound travels just over a foot per millisecond, and so if, say, the line of rear stalls speakers is about 50 feet in front of the main stage speakers, an initial setting of 50–55 ms can be set. Many delay devices will also display the delay in terms of feet or meters, which is very useful. Moving up to the balcony, the flown cluster can now be switched on. A delay of perhaps 120 ms will be needed here to time-align the flown cluster with the main speakers each side of the stage. As well as giving a much cleaner sound, the use of delays in this manner also has the effect of making the speakers forward of the stage 'disappear', and the sound appears to come just from the stage itself.

As air temperature rises, sound travels faster, and delay settings obtained in a fairly cold auditorium during the day may well be a little high for the evening concert when the air is somewhat warmer. Some digital delay devices have an input for a temperature sensor, and if this is used the delay settings will automatically adjust to compensate for temperature changes.

Several computer-based systems are available which in conjunction with measuring microphones placed in the auditorium will display required delay settings for the various speaker locations when special test tones are played through the system. Additionally, using pink noise to drive the speakers the EQ curve requirements for flat frequency responses can be displayed, and parametric equalization can be used to mirror the display curves. A word of caution concerning EQ settings – air absorbs high frequencies to a rather greater extent than low frequencies, and sounds coming from a distance naturally sound duller. At a distance of 50 meters, assuming a 20°C temperature and 20% relative humidity, there is about a 10 dB loss at 8 kHz, and about a 35 dB loss at 16 kHz. Measuring microphones placed some distance from flown speakers will therefore register treble loss as a natural consequence of air absorption, and one must add treble boost with caution. An unnaturally bright sound can result, and excessive power can be fed to high-frequency horns.

The cathedral, or large conference hall, is a particularly difficult environment in which to install sound reinforcement. Speech intelligibility is of the highest priority, and a large, reverberant space requires special consideration. The usual approach is to place a number of small loudspeakers discreetly upon pillars along a nave or throughout an auditorium, each reproducing the sound of the person speaking at a low level, providing coverage only for its immediate surroundings. This helps to give a clean, clear sound without exciting much of the building's natural reverberation. One might consider adding progressive delay to the speakers as one approaches the rear of a large venue, and this is usually timed in such a way that the direct sound and that from loudspeakers further toward the front arrive before that from the closest loudspeaker, so that the perceived direction is from the front. (See Fact File 2.4 on the precedence effect.) But the intelligibility of speech is impaired if the sound from a loudspeaker, which may be some distance from the person speaking, is out of synchronization with his lips. Using no delay at all connects the sound one hears intimately with the subject, and this is a special case in which the use of delay would be inappropriate.

Delay can also be a useful creative and corrective tool. If one is recording an electric guitar for instance, one might wish to place a microphone in front of the amplifier's loudspeaker and also DI (direct inject) the signal straight from the guitar, or from its amplifier's pre-amp output. The distance the microphone is from the loudspeaker will cause a slight delay with respect to the DI signal, and this produces frequency-selective reinforcements and cancellations, the 'toothcomb' effect. It can be corrected for by introducing a delay into the DI feed, time-aligning the two signals. Alternatively, different values of delay can be experimented with to modify the tonal quality of the instrument.

Miscellaneous devices

The de-esser cleans up closely miced vocals. Sibilant sounds can produce a rasping quality, and the de-esser dynamically filters the high-level, high-frequency component of the sound to produce a more natural vocal quality.

The Aphex company of America introduced a unit called the Aural Exciter in the 1970s, and for a time the mechanisms by which it achieved its effect were shrouded in a certain amount of mystery. The unit made a signal 'sparkle', enhancing its overall presence and life, and it was usually applied to individual sounds in a mix such as solo instruments and voices, but sometimes also to the complete stereo signal. Such devices succeed entirely by their subjective effect, and several companies later produced similar units. They achieve their psycho-acoustic effect by techniques such as comb filtering, selective boosting of certain frequencies, small increments of pitch change, introducing relatively narrow-band phase shifts between stereo channels, and other such processes which achieve the desired subjective effect.

Effects such as these go back a long way, and in many cases will be long obsolete and unavailable. But the value of such things as old valve (tube) compressors and limiters, distortion devices, tape and plate echoes, room simulators and other vintage-sounding devices such as certain guitar amplifiers is reflected in the market for 'plug-ins' for computer-based digital workstations: computer programs which have been developed to simulate the sounds of such old and well-loved devices. These things succeed purely on their own subjective terms, and are very much part of the creative process. Plug-ins do of course also provide up-to-the-minute effects and processing.

CONNECTION OF OUTBOARD DEVICES

A distinction needs to be made between processors which need to interrupt a signal for treatment (series connection), and those which basically add something to an existing signal (parallel connection). Graphic and parametric equalizers, compressors, de-essers and gates need to be placed in the signal path. One would not normally wish to mix, say, an uncompressed signal with its compressed version, or an ungated with the gated sound. Such processors will generally be patched in via the mixer's channel insertion send and returns (see Figure 13.7), or patched in ahead of the incoming signal or immediately after an output. Devices such as echo, reverb, chorus and flange are generally used to add something to an existing signal and usually a channel aux send will be used to drive them. Their outputs will be brought back to additional input channels and these signals mixed to taste with the existing dry signal (see Figure 13.8).

Sometimes just the effects signal will be required, in which case either the aux send will be switched to pre-fade and that channel's fader closed, or the channel will simply be de-routed from the outputs. The channel is then used merely to send the signal to the effects unit via the aux. The returns will often contain a degree of dry signal anyway, the ratio of dry to effect being adjusted on the processor.

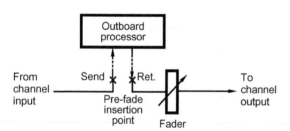

FIGURE 13.7 *Outboard processors such as compressors are normally patched in at an insertion point of the required mixer channel.*

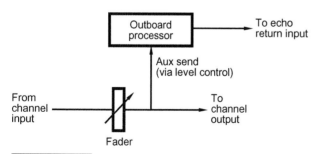

FIGURE 13.8 *Reverberation and echo devices are usually fed from a post-fader auxiliary send, and brought back to a dedicated echo return or another channel input.*

MIDI control for selecting a program has already been mentioned. Additionally, MIDI can be used in a musical way with some devices. For instance, a 'harmonizer' device, designed to add harmony to a vocal or instrumental line, is normally set to add appropriate diatonic harmonies to the incoming line in the appropriate key with the desired number of voices above and/or below it. Results are thereafter in the hands of the machine. Alternatively, a MIDI keyboard can be used to control the device so that the harmonizer adds the notes which are being held down. Composition of the harmonies and voice lines is then under the control of the musician. This can be used in recording for adding harmonies to an existing line, or in a live situation where a keyboard player plays along with a soloist to generate the required harmonies.

RECOMMENDED FURTHER READING

Case, A., 2007. Sound FX: Unlocking the Creative Potential of Recording Studio Effects. Focal Press.

White, P., 2003. Basic Effects and Processors. New Amsterdam Books.

USEFUL WEBSITES

www.gersic.com/plugins: A website listing a large number and variety of plug-ins which are available, including regular updates of the latest releases.

MIDI and Remote Control

MIDI is the Music Instrument Digital Interface, a control protocol and interface standard for electronic musical instruments which has also been used widely in other music and audio products. Although it is relatively dated by modern standards it is still used extensively, which is a testament to its simplicity and success. Even if the MIDI hardware interface is used less these days, either because more synthesis, sampling and processing takes place using software within the workstation, or because other data interfaces such as USB, Firewire and Ethernet are becoming popular, the original protocol for communicating events and other control information is still widely encountered. A lot of software that runs on computers uses MIDI as a basis for controlling the generation of sounds and external devices.

Synthetic audio is used increasingly in audio workstations and mobile devices as a very efficient means of audio representation, because it only requires control information and sound object descriptions to be transmitted. Standards such as MPEG-4 Structured Audio enable synthetic audio to be used as an alternative or an addition to natural audio coding and this can be seen as a natural evolution of the MIDI concept in interactive multimedia applications.

We also include in this chapter coverage of some recent developments in networked remote control of audio systems.

BACKGROUND

Electronic musical instruments existed widely before MIDI was developed in the early 1980s, but no universal means existed of controlling them remotely. Many older musical instruments used analog voltage control, rather than being controlled by a microprocessor, and thus used a variety of analog remote interfaces (if indeed any facility of this kind was provided at all). Such interfaces commonly took the form of one port for timing information, such as might be required by a sequencer or drum machine, and another for pitch and key triggering information, as shown in Figure 14.1. The latter, commonly referred to as 'CV and gate', consisted of a DC (direct current) control line carrying a variable control voltage (CV) which was proportional to the pitch of the note, and a separate line to carry a trigger pulse. A common increment for the CV was 1 volt per octave. Notes on a synthesizer could be triggered remotely by setting the CV to the correct pitch and sending a 'note on' trigger pulse which would initiate a new cycle of the synthesizer's envelope generator. Such an interface would deal with only one note at a time, but many older synths were only monophonic in any case (that is, they were only capable of generating a single voice).

Instruments with onboard sequencers needed a timing reference in order that they could be run in synchronization with other such devices, and this commonly took the form of a square pulse train at a rate related to the current musical tempo, often connected to the device using a DIN-type

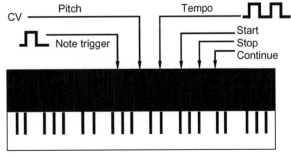

Electronic musical instrument

FIGURE   14.1 *Prior to MIDI control, electronic musical instruments tended to use a DC remote interface for pitch and note triggering. A second interface handled a clock signal to control tempo and trigger pulses to control the execution of a stored sequence.*

connector, along with trigger lines for starting and stopping a sequence's execution. There was no universal agreement over the rate of this external clock, and frequencies measured in pulses per musical quarter note (ppqn), such as 24 ppqn and 48 ppqn, were used by different manufacturers. A number of conversion boxes were available which divided or multiplied clock signals in order that devices from different manufacturers could be made to work together.

As microprocessor control began to be more widely used in musical instruments a number of incompatible digital control interfaces sprang up, promoted by the large synthesizer manufacturers, some serial and some parallel. Needless to say the plethora of non-standardized approaches to remote control made it difficult to construct an integrated system, especially when integrating equipment from different manufacturers. Owing to collaboration between the major parties in America and Japan, the way became clear for agreement over a common hardware interface and command protocol, resulting in the specification of the MIDI standard in late 1982/early 1983. This interface grew out of an amalgamation of a proposed universal interface called USI (the Universal Synthesizer Interface) which was intended mainly for note on and off commands, and a Japanese specification which was rather more complex and which proposed an extensive protocol to cover other operations as well.

The standard has been subject to a number of addenda, extending the functionality of MIDI far beyond the original. The original specification was called the MIDI 1.0 specification, to which has been added such addenda as the MIDI Sample Dump protocol, MIDI Files, General MIDI (1 and 2), MIDI TimeCode, MIDI Show Control, MIDI Machine Control and Downloadable Sounds. A new 'HD' (High Definition) version of the standard was originally planned for release in 2009, which is expected to include support for more channels and controllers, as well as greater controller resolution using single messages. Additional aims include direct control of note pitch and transport of the protocol using UDP over network systems such as Ethernet. It is aimed to make this compatible with existing hardware and software. At the time of writing this seventh edition, in January 2013, the HD Working Group was just about to demonstrate prototype hardware and software, along with a working draft of the new standard, but this was not available for public consumption. The MIDI Manufacturer's Association (MMA) is now the primary association governing formal extensions to the standard, liaising closely with a Japanese association called AMEI (Association of Musical Electronics Industry).

WHAT IS MIDI?

MIDI is a digital remote control interface for music systems, but has come to relate to a wide range of standards and specifications to ensure interoperability between electronic music systems. MIDI-controlled equipment is normally based on microprocessor control, with the MIDI interface forming an I/O port. It is a measure of the popularity of MIDI as a means of control that it has now been adopted in many other audio and visual systems, including the automation of mixing consoles, the control of studio outboard equipment, lighting equipment and other machinery. Although many of its standard commands are music related, it is possible either to adapt music commands to non-musical purposes or to use command sequences designed especially for alternative methods of control.

The adoption of a serial communication standard for MIDI was dictated largely by economic and practical considerations, enabling it to be installed on relatively cheap items of equipment and available to as wide a range of users as possible. The simplicity and ease of installation of MIDI systems was largely responsible for its rapid proliferation as an international standard.

Unlike its analog predecessors, MIDI integrates timing and system control commands with pitch and note triggering commands, such that everything may be carried in the same format over the same piece of wire. MIDI makes it possible to control musical instruments polyphonically in pseudo real time: that is, the speed of transmission is such that delays in the transfer of performance commands are not audible in the majority of cases. It is also possible to address a number of separate receiving devices within a single MIDI data stream, and this allows a controlling device to determine the destination of a command.

MIDI AND DIGITAL AUDIO CONTRASTED

For many the distinction between MIDI and digital audio may be a clear one, but those new to the subject often confuse the two. Any confusion is often due to both MIDI and digital audio equipment appearing to perform the same task – that is, the recording of multiple channels of music using digital equipment – and is not helped by the way in which some manufacturers refer to MIDI sequencing as digital recording.

Digital audio involves a process whereby an audio waveform (such as the line output of a musical instrument) is sampled regularly and then converted into a series of binary words that represent the sound waveform, as described in Chapter 8. A digital audio recorder stores this sequence of data and can replay it by passing the original data through a digital-to-analog convertor that turns the data back into a sound waveform, as shown in Figure 14.2. A multitrack recorder has a number of independent channels that work in the same way, allowing a sound recording to be built up in layers. MIDI, on the other hand, handles digital information that *controls* the generation of sound. MIDI data does not represent the sound waveform itself. When a multitrack music recording is made using a MIDI sequencer (described later) this control data is stored, and can be replayed by transmitting the original data to a collection of MIDI-controlled musical instruments. It is the instruments that actually reproduce the recording.

A digital audio recording, then, allows any sound to be stored and replayed without the need for additional hardware. It is useful for recording acoustic sounds such as voices. A MIDI recording is almost useless without a collection of sound generators. An interesting advantage of the MIDI recording is that, since the stored data represents event information describing a piece of music, it is possible to change the music by changing the event data. MIDI recordings also consume a lot less memory space than digital audio recordings. It is also possible to transmit a MIDI recording to a different collection of instruments from those used during the original recording, thus resulting in a different sound. It is common for MIDI and digital audio recording to be integrated in one software package, allowing the two to be edited and manipulated in parallel. In some cases simple audio information can be converted into MIDI commands (for example a solo melody line converted into the nearest equivalent in terms of MIDI

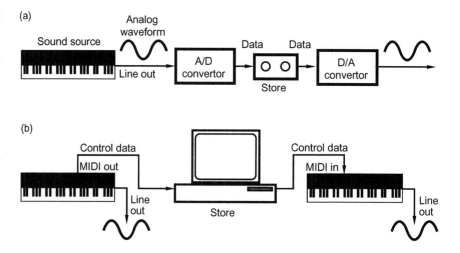

FIGURE 14.2

(a) Digital audio recording and (b) MIDI recording contrasted. In (a) the sound waveform itself is converted into digital data and stored, whereas in (b) only control information is stored, and a MIDI-controlled sound generator is required during replay.

note and controller messages). This preserves essential information about the musical line but not usually anything about the subtleties of acoustics, sound quality or timbre in a digital recording.

BASIC PRINCIPLES

The interface

The MIDI standard specifies a unidirectional serial interface (see Fact File 8.6, Chapter 8) running at 31.25 kbit/s ± 1%. The rate was defined at a time when the clock speeds of microprocessors were typically much slower than they are today, this rate being a convenient division of the typical 1 or 2 MHz master clock rate. The rate had to be slow enough to be carried without excessive losses over simple cables and interface hardware, but fast enough to allow musical information to be transferred from one instrument to another without noticeable delays. Control messages are sent as groups of bytes. Each byte is preceded by one start bit and followed by one stop bit per byte in order to synchronize reception of the data which is transmitted asynchronously, as shown in Figure 14.3. The addition of start and stop bits means that each 8 bit word actually takes ten bit periods to transmit (lasting a total of 320 μs). Standard MIDI messages typically consist of one, two or three bytes, although there are longer messages for some purposes.

The hardware interface is shown in Fact File 14.1. In the MIDI specification, the opto-isolator is defined as having a rise time of no more than 2 μs. The rise time affects the speed with which the device reacts to a change in its input and if slow will tend to distort the leading edge of data bit cells. If a large number of MIDI devices are wired in series (that is from THRU to IN a number of times) the data will be forced to pass through a number of opto-isolators and thus will suffer the combined effects of a

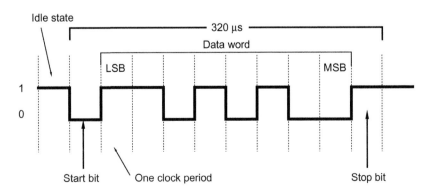

FIGURE 14.3

A MIDI message consists of a number of bytes, each transmitted serially and asynchronously by a UART in this format, with a start and stop bit to synchronize the receiving UART. The total period of a MIDI data byte, including start and stop bits, is 320 μs.

FACT FILE 14.1 MIDI HARDWARE INTERFACE

Most equipment using MIDI has three interface connectors: IN, OUT and THRU. The OUT connector carries data that the device itself has generated. The IN connector receives data from other devices and the THRU connector is a direct throughput of the data that is present at the IN. As can be seen from the hardware interface diagram, it is simply a buffered feed of the input data, and it has not been processed in any way. A few cheaper devices do not have THRU connectors, but it is possible to obtain 'MIDI THRU boxes' which provide a number of 'THRUs' from one input. Occasionally, devices without a THRU socket allow the OUT socket to be switched between OUT and THRU functions.

The interface incorporates an opto-isolator between the MIDI IN (that is the receiving socket) and the device's microprocessor system. This is to ensure that there is no direct electrical link between devices and helps to reduce the effects of any problems which might occur if one instrument in a system were to develop an electrical fault. An opto-isolator is an encapsulated device in which a light-emitting diode (LED) can be turned on or off depending on the voltage applied across its terminals, illuminating a photo-transistor which consequently conducts or not, depending on the state of the LED. Thus the data is transferred optically, rather than electrically.

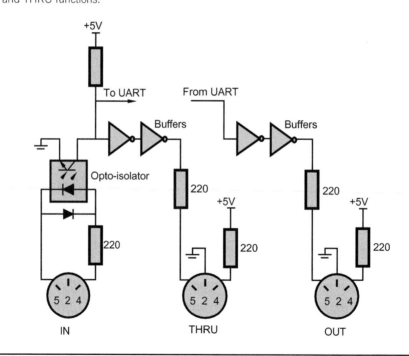

number of stages of rise-time distortion. The better the specification of the opto-isolator, the more stages of device cascading will be possible before unacceptable distortion is introduced. The delay in data passed between IN and THRU is only a matter of microseconds, so this contributes little to any audible delays perceived in the musical outputs of some instruments in a large system. The bulk of any perceived delay will be due to other factors like processing delay, buffer delays and traffic.

The specification of cables and connectors is described in Fact File 14.2. This form of hardware interface is increasingly referred to as 'MIDI-DIN' to distinguish it from other means of transferring MIDI data.

Implementations of MIDI that work over other hardware interfaces such as Ethernet (using Internet Protocol/UDP), USB and Firewire (IEEE 1394) have also been introduced, sometimes in proprietary form. These are described briefly later in the chapter.

FACT FILE 14.2 MIDI CONNECTORS AND CABLES

The connectors used for MIDI interfaces are 5-pin DIN types. Only three of the pins of a 5-pin DIN plug are actually used in most equipment (the three innermost pins). In the cable, pin 5 at one end should be connected to pin 5 at the other, and likewise pin 4 to pin 4, and pin 2 to pin 2. The cable should be a shielded twisted pair. Within the receiver the MIDI IN does not have pin 2 connected to earth. This is to avoid earth loops and makes it possible to use a cable either way round. Professional microphone cable terminated in DIN connectors may be used as a higher-quality solution, because domestic cables will not always be a shielded twisted pair and thus are more susceptible to external interference, as well as radiating more themselves which could interfere with adjacent audio signals. A 5 mA current loop is created between a MIDI OUT or THRU and a MIDI IN, when connected with the appropriate cable, and data bits are signaled by the turning on and off of this current by the sending device. This principle is shown in the diagram.

It is recommended that no more than 15 m of cable is used for a single cable run in a simple MIDI system and investigation of typical cables indicates that corruption of data does indeed ensue after longer distances, although this is gradual and depends on the electromagnetic interference conditions, the quality of cable and the equipment in use. Longer distances may be accommodated with the use of buffer or 'booster' boxes that compensate for some of the cable losses and retransmit the data. It is also possible to extend a MIDI system by using a data network with an appropriate interface.

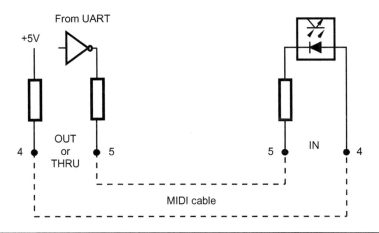

Simple interconnection

In the simplest MIDI system, one instrument could be connected to another as shown in Figure 14.4. Here, instrument 1 sends information relating to actions performed on its own controls (notes pressed, pedals pressed, etc.) to instrument 2, which imitates these actions as far as it is able. This arrangement can be used for 'doubling-up' sounds, 'layering' or 'stacking', such that a composite sound can be made up from two synthesizers' outputs. (The audio outputs of the two instruments would have to be mixed together for this effect to be heard.) Larger MIDI systems could be built up by further 'daisy-chaining' of instruments, such that instruments further down the chain all received information generated by the first (see Figure 14.5), although this is not a very satisfactory way of building a large MIDI system. In large systems some form of central routing helps to avoid MIDI 'traffic jams' and simplifies interconnection.

INTERFACING A COMPUTER TO A MIDI SYSTEM

Adding MIDI ports

In order to use a workstation as a central controller for a MIDI system it must have at least one MIDI interface, consisting of at least an IN and an OUT port. (THRU is not strictly necessary in most cases.) Unless the computer has a built-in interface, as found on the old Atari machines,

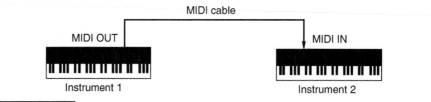

FIGURE 14.4 *The simplest form of MIDI interconnection involves connecting two instruments together as shown.*

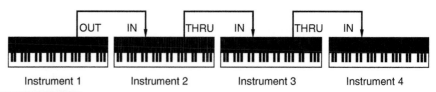

FIGURE 14.5 *Further instruments can be added using THRU ports as shown, in order that messages from instrument 1 may be transmitted to all the other instruments.*

some form of third-party hardware interface must be added and there are many ranging from simple single ports to complex multiple port products.

A typical single port MIDI interface can be connected either to one of the spare I/O ports of the computer (a serial, Firewire or USB port, for example), or can be installed as an expansion slot card (perhaps as part of an integrated sound card). Multiport interfaces have become widely used in MIDI systems where more than 16 MIDI channels are required, and they are also useful as a means of limiting the amount of data sent or received through any one MIDI port. (A single port can become 'overloaded' with MIDI data if serving a large number of devices, resulting in data delays.) Multiport interfaces are normally more than just a parallel distribution of a single MIDI data stream, typically handling a number of independent MIDI data streams that can be separately addressed by the operating system drivers or sequencer software. USB and Firewire MIDI protocols allow a particular stream or 'cable' to be identified so that each stream controlling 16 MIDI channels can be routed to a particular physical port or instrument.

EMagic's Unitor8 interface is pictured in Figure 14.6. It has RS-232 and -422 serial ports as well as a USB port to link with the host workstation. There are eight MIDI ports with two on the front panel for easy connection of 'guest' devices or controllers that are not installed at the back. This device also has VITC and LTC timecode ports in order that synchronization information can be relayed to and from the computer. A multi-device MIDI system is pictured in Figure 14.7, showing a number of multitimbral sound generators connected to separate MIDI ports and a timecode connection to an external video tape recorder for use in synchronized postproduction. As more of these functions are now being provided within the workstation (e.g. synthesis, video, mixing) the number of devices connected in this way will reduce.

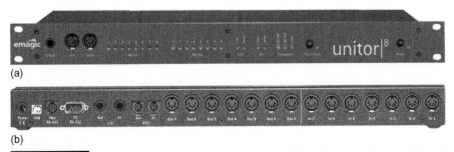

(a)

(b)

FIGURE 14.6   *(a) Front and (b) back panels of the Emagic Unitor 8 interface, showing USB port, RS-422 port, RS-232 port, LTC and VITC ports and multiple MIDI ports.*

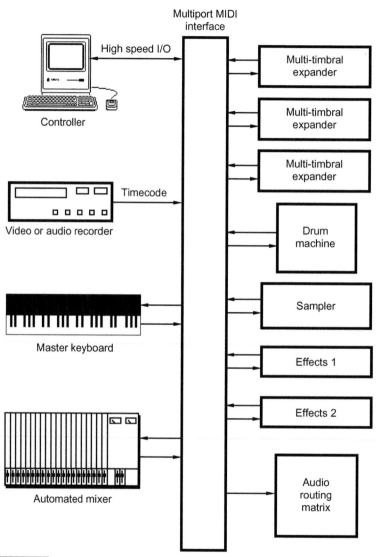

FIGURE 14.7   *A typical multi-machine MIDI system interfaced to a computer via a multiport interface connected by a high-speed link (e.g. USB).*

Drivers and I/O software

Most audio and MIDI hardware requires 'driver' software of some sort to enable the operating system (OS) to 'see' the hardware and use it correctly. These are now designed as 'hardware abstraction layers' (HALs) that enable applications to communicate more effectively with I/O hardware. Whereas previously it would have been necessary to install a third-party MIDI HAL such as OMS (Opcode's Open Music System) or MIDI Manager to route MIDI data to and from multiport interfaces and applications, these features are now included within the Mac and Windows operating system. For example Apple has its Core MIDI specification for OS X, which is a collection of APIs (application programming interfaces) that communicate with MIDI devices. Core Audio is its audio specification (see Chapter 9). Microsoft has also done something similar for Windows systems, with the Windows Driver Model (WDM). DirectSound is the Microsoft equivalent of Apple's OS X Core Audio, while DirectMusic is the equivalent for MIDI data. Steinberg's ASIO (Audio Stream Input Output) is a third-party alternative that handles digital audio and MIDI. It handles a range of audio sampling frequencies and bit depths, as well as multiple channel I/O, and many sound cards and applications are ASIO-compatible.

HOW MIDI CONTROL WORKS

MIDI channels

MIDI messages are made up of a number of bytes as explained in Fact File 14.3. Each part of the message has a specific purpose, and one of these is to define the receiving channel to which the message refers. In this way, a controlling device can make data device specific – in other words it can define which receiving instrument will act on the data sent. If a device is set in software to receive on a specific channel or on a number of channels it will act only on information 'tagged' with its own channel numbers. Everything else it will usually ignore. There are 16 basic MIDI channels and instruments can usually be set to receive on any specific channel or channels (omni off mode), or to receive on all channels (omni on mode). The latter mode is useful as a means of determining whether anything at all is being received by the device.

The limit of 16 MIDI channels can be overcome easily by using multiport MIDI interfaces connected to a computer. In such cases it is important not to confuse the MIDI data channel with the physical port to which a device may be connected, since each physical port will be capable of transmitting on all 16 data channels.

Channel and system messages contrasted

Two primary classes of message exist: those that relate to specific MIDI channels and those that relate to the system as a whole.

FACT FILE 14.3 MIDI MESSAGE FORMAT

There are two basic types of MIDI message byte: the status byte and the data byte. The first byte in a MIDI message is normally a status byte. Standard MIDI messages can be up to 3 bytes long, but not all messages require 3 bytes, and there are some fairly common exceptions to the rule which are described below. The standard has been extended and refined over the years and the following is only an introduction to the basic messages. The prefix '&' will be used to indicate hexadecimal values (see Table 8.1); individual MIDI message bytes will be delineated using square brackets, e.g. [&45], and channel numbers will be denoted using 'n' to indicate that the value may be anything from &0 to &F (channels 1 to 16). The table shows the format and content of MIDI messages under each of the statuses.

Status bytes always begin with a binary one to distinguish them from data bytes, which always begin with a zero. Because the most significant bit (MSB) of each byte is reserved to denote the type (status or data) there are only 7 active bits per byte which allows 2^7 (that is 128)

possible values. As shown in the figure below, the first half of the status byte denotes the message type and the second half denotes the channel number. Because 4 bits of the status byte are set aside to indicate the channel number, this allows for 2^4 (or 16) possible channels. There are only 3 bits to denote the message type, because the first bit must always be a one. This theoretically allows for eight message types, but there are some special cases in the form of system messages (see below).

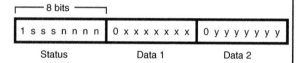

Not all MIDI devices will have all the commands implemented, since it is not mandatory for a device conforming to the MIDI standard to implement every possibility.

Message	Status	Data 1	Data 2
Note off	&8n	Note number	Velocity
Note on	&9n	Note number	Velocity
Polyphonic aftertouch	&An	Note number	Pressure
Control change	&Bn	Controller number	Data
Program change	&Cn	Program number	–
Channel aftertouch	&Dn	Pressure	–
Pitch wheel	&En	LSbyte	MSbyte
System exclusive			
System exclusive start	&F0	Manufacturer ID	Data, (Data), (Data)
End of SysEx	&F7	–	
System common			
Quarter frame	&F1	Data	– Msbyte
Song pointer	&F2	LSbyte	–
Song select	&F3	Song number	
Tune request	&F6	–	
System real time			
Timing clock	&F8	–	–
Start	&FA	–	–
Continue	&FB	–	–
Stop	&FC	–	–
Active sensing	&FE	–	–
Reset	&FF	–	–

Channel messages start with status bytes in the range &8n to &En (they start at hexadecimal eight because the MSB must be a one for a status byte). System messages all begin with &F, and do not contain a channel number. Instead the least significant nibble of the system status byte is used for further identification of the system message, such that there is room for 16 possible system messages running from &F0 to &FF. System messages are themselves split into three groups: system common, system exclusive and system real time. The common messages may apply to any device on the MIDI bus, depending only on the device's ability to handle the message. The exclusive messages apply to whichever manufacturers' devices are specified later in the message (see below) and the real-time messages are for synchronizing devices to be to the prevailing musical tempo. (Some of the so-called real-time messages do not really seem to deserve this appellation, as discussed below.) The status byte &F1 is used for MIDI TimeCode.

MIDI channel numbers are usually referred to as 'channels one to 16', but the binary numbers representing these run from zero to 15 (&0 to &F), as 15 is the largest decimal number which can be represented with 4 bits. Thus the note on message for channel 5 is actually &94 (nine for note on, and four for channel 5).

Note on and note off messages

Much of the musical information sent over a typical MIDI interface will consist of note messages. As indicated by the titles, the note on message turns on a musical note, and the note off message turns it off. Note on takes the general format:

[&8n] [Note number] [Velocity]

and note off takes the form:

[&9n] [Note number] [Velocity]

A MIDI instrument will generate note on messages at its MIDI OUT corresponding to whatever notes are pressed on the keyboard, on whatever channel the instrument is set to transmit. Also, any note which has been turned on must subsequently be turned off in order for it to stop sounding, thus if one instrument receives a note on message from another and then loses the MIDI connection for any reason, the note will continue sounding ad infinitum. This situation can occur if a MIDI cable is pulled out during transmission.

Table 14.1	MIDI Note Numbers Related to the Musical Scale
Musical Note	**MIDI Note Number**
C–2	0
C–1	12
C0	24
C1	36
C2	48
C3 (middle C)	60 (Yamaha convention)
C4	72
C5	84
C6	96
C7	108
C8	120
G8	127

MIDI note numbers relate directly to the western musical chromatic scale and the format of the message allows for 128 note numbers which cover a range of a little over ten octaves – adequate for the full range of most musical material. This quantization of the pitch scale is geared very much towards keyboard instruments, being less suitable for other instruments and cultures where the definition of pitches is not so black and white. Nonetheless, means have been developed of adapting control to situations where unconventional tunings are required. It also seems likely that the upcoming HD MIDI standard will allow for direct control of note pitch. Note numbers normally relate to the musical scale as shown in Table 14.1, although there is a certain degree of confusion here. Yamaha established the use of C3 for middle C, whereas others have used C4. Some software allows the user to decide which convention will be used for display purposes.

Velocity information

Note messages are associated with a velocity byte that is used to represent the speed at which a key was pressed or released. The former will correspond to the force exerted on the key as it is depressed: in other words, 'how hard you hit it' (called 'note on velocity'). It is used to control parameters such as the volume or timbre of the note at the audio output of an instrument and can be applied internally to scale the effect of one or more of the envelope generators in a synthesizer. This velocity value has

128 possible states, but not all MIDI instruments are able to generate or interpret the velocity byte, in which case they will set it to a value half way between the limits, i.e. 64_{10}. Some instruments may act on velocity information even if they are unable to generate it themselves. It is recommended that a logarithmic rather than linear relationship should be established between the velocity value and the parameter which it controls, since this corresponds more closely to the way in which musicians expect an instrument to respond, although some instruments allow customized mapping of velocity values to parameters. The note on, velocity zero value is reserved for the special purpose of turning a note off, for reasons that will become clear under 'Running status', below.

Note off velocity (or 'release velocity') is not widely used, as it relates to the speed at which a note is released, which is not a parameter that affects the sound of many normal keyboard instruments. Nonetheless it is available for special effects if a manufacturer decides to implement it.

Running status

Running status is an accepted method of reducing the amount of data transmitted. It involves the assumption that once a status byte has been asserted by a controller there is no need to reiterate this status for each subsequent message of that status, so long as the status has not changed in between. Thus a string of notes on messages could be sent with the note on status only sent at the start of the series of note data, for example:

[&9n] [Data] [Velocity] [Data] [Velocity] [Data] [Velocity]

For a long string of notes this could reduce the amount of data sent by nearly one-third. But in most music each note on is almost always followed quickly by a note off for the same note number, so note on, velocity zero (see above) allows a string of what appears to be note on messages to act as both note on and note off.

Running status is not used at all times for a string of same-status messages and will often only be called upon by an instrument's software when the rate of data exceeds a certain point. Indeed, an examination of the data from a typical synthesizer indicates that running status is not used during a large amount of ordinary playing.

Polyphonic key pressure (aftertouch)

The key pressure messages are sometimes called 'aftertouch' by keyboard manufacturers. This message refers to the amount of pressure placed on a key at the bottom of its travel, and it is often applied to performance parameters such as vibrato.

The polyphonic key pressure message is not widely used, as it transmits a separate value for every key on the keyboard and thus requires a separate sensor for every key. This can be expensive to implement and is beyond the scope of many keyboards, so most manufacturers have resorted to the use of the channel pressure message (see below). The message takes the general format:

[&An] [Note number] [Pressure]

Implementing polyphonic key pressure messages involves the transmission of a considerable amount of data that might be unnecessary, as the message will be sent for every note in a chord every time the pressure changes. As most people do not maintain a constant pressure on the bottom of a key whilst playing, many redundant messages might be sent per note. A technique known as 'controller thinning' may be used by a device to limit the rate at which such messages are transmitted and this may be implemented either before transmission or at a later stage using a computer. Alternatively this data may be filtered out altogether if it is not required.

Control change

As well as note information, a MIDI device transmits information that corresponds to the various switches, control wheels and pedals associated with it. These come under the control change message group and should be distinguished from program change messages. The controller messages have proliferated enormously since the early days of MIDI and not all devices will implement all of them. The control change message takes the general form:

[&Bn] [Controller number] [Data]

so a number of controllers may be addressed using the same type of status byte by changing the controller number.

Although the original MIDI standard did not lay down any hard and fast rules for the assignment of physical control devices to logical controller numbers, there is now common agreement amongst manufacturers that certain controller numbers will be used for certain purposes. These are assigned by the MMA. There are two distinct kinds of controller: the switch type and the analog type. The analog controller is any continuously

Table 14.2	MIDI Controller Classifications
Controller Number (hex)	**Function**
&00–1F	14 bit controllers, Msbyte
&20–3F	14 bit controllers, Lsbyte
&40–65	7 bit controllers or switches
&66–77	Originally undefined
&78–7F	Channel mode control

variable wheel, lever, slider or pedal that might have any one of a number of positions and these are often known as continuous controllers. There are 128 controller numbers available and these are grouped as shown in Table 14.2. Table 14.3 shows a more detailed breakdown of some of these, as found in the majority of MIDI-controlled musical instruments, although the full list is regularly updated by the MMA.

The first 64 controller numbers (that is up to &3F) relate to only 32 physical controllers (the continuous controllers). This is to allow for greater resolution in the quantization of position than would be feasible with the 7 bits that are offered by a single data byte. The first 32 controllers handle the most significant byte (MSbyte) of the controller data, whilst the second 32 handle the least significant byte (LSbyte). In this way, controller numbers &06 and &38 both represent the data entry slider, for example. Together, the data values can make up a 14 bit number (because the first bit of each data word has to be a zero), which allows the quantization of a control's position to be one part in 2^{14} ($16\,384_{10}$). If a system opts not to use the extra resolution offered by the second byte, it should send only the MSbyte for coarse control. In practice this is all that is transmitted on many devices.

On/off switches can be represented easily in binary form (0 for OFF, 1 for ON), and it would be possible to use just a single bit for this purpose, but, in order to conform to the standard format of the message, switch states are normally represented by data values between &00 and &3F for OFF and &40 and &7F for ON. In other words switches are now considered as 7 bit continuous controllers. In older systems it may be found that only &00 = OFF and &7F = ON.

The data increment and decrement buttons that are present on many devices are assigned to two specific controller numbers (&60 and &61) and an extension to the standard defines four controllers (&62 to &65) that effectively expand the scope of the control change messages. These are the registered and non-registered parameter controllers (RPCs and NRPCs).

426

Table 14.3 MIDI Controller Functions

Controller Number (hex)	Function
00	Bank select
01	Modulation wheel
02	Breath controller
03	Undefined
04	Foot controller
05	Portamento time
06	Data entry slider
07	Main volume
08	Balance
09	Undefined
0A	Pan
0B	Expression controller
0C	Effect control 1
0D	Effect control 2
0E–0F	Undefined
10–13	General purpose controllers 1–4
14–1F	Undefined
20–3F	LSbyte for 14 bit controllers (same function order as 00–1F)
40	Sustain pedal
41	Portamento on/off
42	Sostenuto pedal
43	Soft pedal
44	Legato footswitch
45	Hold 2
46–4F	Sound controllers
50–53	General purpose controllers 5–8
54	Portamento control
55–5A	Undefined
5B–5F	Effects depth 1–5
60	Data increment
61	Data decrement
62	NRPC LSbyte (non-registered parameter controller)
63	NRPC MSbyte
64	RPC LSbyte (registered parameter controller)
65	RPC MSbyte
66–77	Undefined
78	All sounds off
79	Reset all controllers
7A	Local on/off
7B	All notes off
7C	Omni receive mode off
7D	Omni receive mode on
7E	Mono receive mode
7F	Poly receive mode

The 'all notes off' command (frequently abbreviated to 'ANO') was designed as a means of silencing devices, but it does not necessarily have this effect in practice. What actually happens varies between instruments, especially if the sustain pedal is held down or notes are still being pressed manually by a player. All notes off is supposed to put all note generators into the release phase of their envelopes, and clearly the result of this will depend on what a sound is programmed to do at this point. The exception should be notes which are being played whilst the sustain pedal is held down, which should only be released when that pedal is released. 'All sounds off' was designed to overcome the problems with 'all notes off', by turning sounds off as quickly as possible. 'Reset all controllers' is designed to reset all controllers to their default state, in order to return a device to its 'standard' setting.

Channel modes

Although grouped with the controllers, under the same status, the channel mode messages differ somewhat in that they set the mode of operation of the instrument receiving on that particular channel.

'Local on/off' is used to make or break the link between an instrument's keyboard and its own sound generators. Effectively there is a switch between the output of the keyboard and the control input to the sound generators which allows the instrument to play its own sound generators in normal operation when the switch is closed (see Figure 14.8). If the switch is opened, the link is broken and the output from the keyboard feeds the MIDI OUT whilst the sound generators are controlled from the MIDI IN. In this mode the instrument acts as two separate devices: a keyboard without any sound, and a sound generator without a keyboard. This configuration can be useful when the instrument in use is the master keyboard for a large sequencer system, where it may not always be desired that everything played on the master keyboard results in sound from the instrument itself.

'Omni off' ensures that the instrument will only act on data tagged with its own channel number(s), as set by the instrument's controls. 'Omni on' sets the instrument to receive on all of the MIDI channels. In other words, the instrument will ignore the channel number in the status byte and will attempt to act on any data that may arrive, whatever its channel.

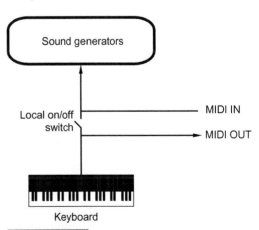

FIGURE 14.8 *The 'local off' switch disconnects a keyboard from its associated sound generators in order that the two parts may be treated independently in a MIDI system.*

Devices should power up in this mode according to the original specification, but more recent devices will tend to power up in the mode in which they were left. Mono mode sets the instrument such that it will only reproduce one note at a time, as opposed to 'Poly' (phonic) in which a number of notes may be sounded together.

Mono mode tends to be used mostly on MIDI guitar synthesizers because each string can then have its own channel and each can control its own set of pitch bend and other parameters. The mode also has the advantage that it is possible to play in a truly legato fashion – that is, with a smooth takeover between the notes of a melody – because the arrival of a second note message acts simply to change the pitch if the first one is still being held down, rather than retriggering the start of a note envelope. The legato switch controller allows a similar type of playing in polyphonic modes by allowing new note messages only to change the pitch.

In poly mode the instrument will sound as many notes as it is able to at the same time. Instruments differ as to the action taken when the number of simultaneous notes is exceeded: some will release the first note played in favor of the new note, whereas others will refuse to play the new note. Some may be able to route excess note messages to their MIDI OUT ports so that they can be played by a chained device. The more intelligent of them may look to see if the same note already exists in the notes currently sounding and only accept a new note if is not already sounding. Even more intelligently, some devices may release the quietest note (that with the lowest velocity value), or the note furthest through its velocity envelope, to make way for a later arrival. It is also common to run a device in poly mode on more than one receive channel, provided that the software can handle the reception of multiple polyphonic channels. A multi-timbral sound generator may well have this facility, commonly referred to as 'multi' mode, making it act as if it were a number of separate instruments each receiving on a separate channel. In multi mode a device may be able to dynamically assign its polyphony between the channels and voices in order that the user does not need to assign a fixed polyphony to each voice.

Program change

The program change message is used most commonly to change the 'patch' of an instrument or other device. A patch is a stored configuration of the device, describing the setup of the tone generators in a synthesizer and the way in which they are interconnected. Program change is channel specific and there is only a single data byte associated with it, specifying to which of 128 possible stored programs the receiving device should switch. On non-musical devices such as effects units, the program change message is often used to switch between different effects and the different effects programs may be mapped to specific program change numbers. The message takes the general form:

&[Cn] [Program number]

If a program change message is sent to a musical device it will usually result in a change of voice, as long as this facility is enabled. Exactly which voice corresponds to which program change number depends on the manufacturer. It is quite common for some manufacturers to implement this function in such a way that a data value of zero gives voice number one. This results in a permanent offset between the program change number and the voice number, which should be taken into account in any software. On some instruments, voices may be split into a number of 'banks' of 8, 16 or 32, and higher banks can be selected over MIDI by setting the program change number to a value which is 8, 16 or 32 higher than the lowest bank number. For example, bank 1, voice 2, might be selected by program change &01, whereas bank 2, voice 2, would probably be selected in this case by program change &11, where there were 16 voices per bank. Where more than 128 voices need to be addressed remotely, the more recent 'bank select' command may be implemented.

Channel aftertouch

Most instruments use a single sensor, often in the form of a pressure-sensitive conductive plastic bar running the length of the keyboard, to detect the pressure applied to keys at the bottom of their travel. In the case of channel aftertouch, one message is sent for the entire instrument and this will correspond to an approximate total of the pressure over the range of the keyboard, the strongest influence being from the key pressed the hardest. (Some manufacturers have split the pressure detector into upper and lower keyboard regions, and some use 'intelligent' zoning.) The message takes the general form:

&[Dn] [Pressure value]

There is only one data byte, so there are 128 possible values and, as with the polyphonic version, many messages may be sent as the pressure is varied at the bottom of a key's travel. Controller 'thinning' may be used to reduce the quantity of these messages, as described above.

Pitch bend wheel

The pitch wheel message has a status byte of its own, and carries information about the movement of the sprung-return control wheel on many keyboards which modifies the pitch of any note(s) played. It uses two data bytes in order to give 14 bits of resolution, in much the same way as the continuous controllers, except that the pitch wheel message carries both bytes together. Fourteen data bits are required so that the pitch appears to change smoothly, rather than in steps (as it might with only 7 bits). The pitch bend message is channel specific so ought to be sent separately for each individual channel. This becomes important when using a single multi-timbral device in mono mode (see above), as one must ensure that a pitch bend message only affects the notes on the intended channel. The message takes the general form:

&[En] [LSbyte] [MSbyte]

The value of the pitch bend controller should be halfway between the lower and upper range limits when it is at rest in its sprung central position, thus allowing bending both down and up. This corresponds to a hex value of &2000, transmitted as &[En] [00] [40]. The range of pitch controlled by the bend message is set on the receiving device itself, or using the RPC designated for this purpose (see 'Control change', p. 424).

System exclusive

A system exclusive message is one that is unique to a particular manufacturer and often a particular instrument. The only thing that is defined about such messages is how they are to start and finish, with the exception of the use of system exclusive messages for universal information, as discussed elsewhere. System exclusive messages generated by a device will naturally be produced at the MIDI OUT, not at the THRU, so a deliberate connection must be made between the transmitting device and the receiving device before data transfer may take place.

Occasionally it is necessary to make a return link from the OUT of the receiver to the IN of the transmitter so that two-way communication is possible and so that the receiver can control the flow of data to some extent by telling the transmitter when it is ready to receive and when it has received correctly (a form of handshaking).

The message takes the general form:

&[F0] [ident.] [data] [data]...[F7]

where [ident.] identifies the relevant manufacturer ID, a number defining which manufacturer's message is to follow. Originally, manufacturer IDs were a single byte but the number of IDs has been extended by setting aside the [00] value of the ID to indicate that two further bytes of ID follow. Manufacturer IDs are therefore either 1 or 3 bytes long. A full list of manufacturer IDs is available from the MMA.

Data of virtually any sort can follow the ID. It can be used for a variety of miscellaneous purposes that have not been defined in the MIDI standard and the message can have virtually any length that the manufacturer requires. It is often split into packets of a manageable size in order not to cause receiver memory buffers to overflow. Exceptions are data bytes that look like other MIDI status bytes (except real-time messages), as they will naturally be interpreted as such by any receiver, which might terminate reception of the system exclusive message. The message should be terminated with &F7, although this is not always observed, in which case the receiving device should 'time-out' after a given period, or terminate the system exclusive message on receipt of the next status byte. It is recommended that some form of error checking (typically a checksum) is employed for long system exclusive data dumps, and many systems employ means of detecting whether the data has been received accurately, asking for retries of sections of the message in the event of failure, via a return link to the transmitter.

Universal system exclusive messages

The three highest numbered IDs within the system exclusive message have been set aside to denote special modes. These are the 'universal non-commercial' messages (ID: &7D), the 'universal non-real-time' messages (ID: &7E) and the 'universal real-time' messages (ID: &7F). Universal sysex messages are often used for controlling device parameters that were not originally specified in the MIDI standard and that now need addressing in most devices. Examples are things like 'chorus modulation depth', 'reverb type' and 'master fine tuning'.

Universal non-commercial messages are set aside for educational and research purposes and should not be used in commercial products. Universal non-real-time messages are used for universal system exclusive events which are not time critical and universal real-time messages deal with time critical events (thus being given a higher priority). The two latter types of message normally take the general form of:

&[F0][ID][dev. ID][sub-ID 1][sub-ID 2][data]......[F7]

Device ID used to be referred to as 'channel number', but this did not really make sense since a whole byte allows for the addressing of 128 channels and this does not correspond to the normal 16 channels of MIDI. The term 'device ID' is now used widely by software as a means of defining one of a number of physical devices in a large MIDI system, rather than defining a MIDI channel number. It should be noted, though, that it is allowable for a device to have more than one ID if this seems appropriate. Modern MIDI devices will normally allow their device ID to be set either over MIDI or from the front panel. The use of &7F in this position signifies that the message applies to all devices as opposed to just one.

The sub-IDs are used to identify first the category or application of the message (sub-ID #1) and second, the type of message within that category (sub-ID #2). For some reason, the original MIDI sample dump messages do not use the sub-ID #2, although some recent additions to the sample dump do.

Tune request

Older analog synthesizers tended to drift somewhat in pitch over the time that they were turned on. The tune request is a request for these synthesizers to retune themselves to a fixed reference. (It is advisable not to transmit pitch bend or note on messages to instruments during a tune-up because of the unpredictable behavior of some products under these conditions.)

Active sensing

Active sensing messages are single status bytes sent roughly three times per second by a controlling device when there is no other activity on the bus. It acts as a means of reassuring the receiving devices that the controller has not disappeared. Not all devices transmit active sensing information, and a receiver's software should be able to detect the presence or lack of it. If a receiver has come to expect active sensing bytes then it will generally act by turning off all notes if these bytes disappear for any reason. This can be a useful function when a MIDI cable has been pulled out during a transmission, as it ensures that notes will not be left sounding for very long. If a receiver has not seen active sensing bytes since last turned on, it should assume that they are not being used.

Reset

This message resets all devices on the bus to their power-on state. The process may take some time and some devices mute their audio outputs, which can result in clicks, therefore the message should be used with care.

MIDI CONTROL OF SOUND GENERATORS

MIDI note assignment in synthesizers and samplers

Many of the replay and signal processing aspects of synthesis and sampling now overlap so that it is more difficult to distinguish between the two. In basic terms a sampler is a device that stores short clips of sound data in RAM, enabling them to be replayed subsequently at different pitches, possibly looped and processed. A synthesizer is a device that enables signals to be artificially generated and modified to create novel sounds. Wavetable synthesis is based on a similar principle to sampling, though, and stored samples can form the basis for synthesis. A sound generator can often generate a number of different sounds at the same time. It is possible that these sounds could be entirely unrelated (perhaps a single drum, an animal noise and a piano note), or that they might have some relationship to each other (perhaps a number of drums in a kit, or a selection of notes from a grand piano). The method by which sounds or samples are assigned to MIDI notes and channels is defined by the replay program.

The most common approach when assigning note numbers to samples is to program the sampler with the range of MIDI note numbers over which a certain sample should be sounded. Akai, one of the most popular sampler manufacturers, called these 'keygroups'. It may be that this 'range' is only one note, in which case the sample in question would be triggered only on receipt of that note number, but in the case of a range of notes the sample would be played on receipt of any note in the range. In the latter case transposition would be required, depending on the relationship between the note number received and the original note number given to the sample (see above). A couple of examples highlight the difference in approach, as shown in Figure 14.9. In the first example, illustrating a possible approach to note assignment for a collection of drum kit sounds, most samples are assigned to only one note number, although it is possible for tuned drum sounds such as tom-toms to be assigned over a range in order to give the impression of 'tuned toms'. Each MIDI note message received would replay the particular percussion sound assigned to that note number in this example.

In the second example, illustrating a suggested approach to note assignment for an organ, notes were originally sampled every musical fifth across the organ's note range. The replay program has been designed so that each of these samples is assigned to a note range of a fifth, centered on the original pitch of each sample, resulting in a maximum transposition of a third up or down. Ideally, of course, every note would have been sampled and assigned to an individual note number on replay, but this

(a) Percussion samples are often assigned to one note per sample, except for tuned percussion which sometimes covers a range of notes. (b) Organ samples could be transposed over a range of notes, centered on the original pitch of the sample.

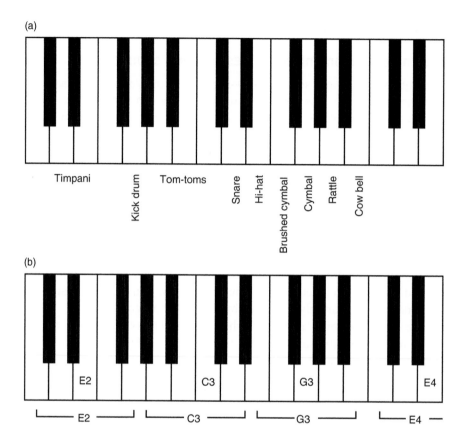

requires very large amounts of memory and painstaking sample acquisition in the first place.

In further pursuit of sonic accuracy, some devices provide the facility for introducing a crossfade between note ranges. This is used where an abrupt change in the sound at the boundary between two note ranges might be undesirable, allowing the takeover from one sample to another to be more gradual. For example, in the organ scenario introduced above, the timbre could change noticeably when playing musical passages that crossed between two note ranges because replay would switch from the upper limit of transposition of one sample to the lower limit of the next (or vice versa). In this case the ranges for the different samples are made to overlap (as illustrated in Figure 14.10). In the overlap range the system mixes a proportion of the two samples together to form the output. The exact proportion depends on the range of overlap and the note's position within this range. Very accurate tuning of the original samples is needed in order to avoid beats when using positional crossfades. Clearly this approach would

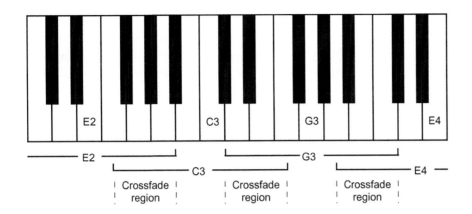

FIGURE 14.10

Overlapped sample ranges can be crossfaded in order that a gradual shift in timbre takes place over the region of takeover between one range and the next.

be of less value when each note was assigned to a completely different sound, as in the drum kit example.

Crossfades based on note velocity allow two or more samples to be assigned to one note or range of notes. This requires at least a 'loud sample' and a 'soft sample' to be stored for each original sound and some systems may accommodate four or more to be assigned over the velocity range. The terminology may vary, but the principle is that a velocity value is set at which the replay switches from one stored sample to another, as many instruments sound quite different when they are loud to when they are soft (it is more than just the volume that changes: it is the timbre also). If a simple switching point is set, then the change from one sample to the other will be abrupt as the velocity crosses either side of the relevant value. This can be illustrated by storing two completely different sounds as the loud and soft samples, in which case the output changes from one to the other at the switching

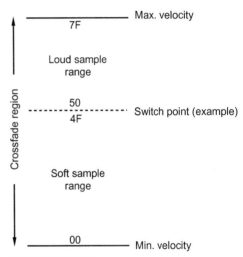

FIGURE 14.11 Illustration of velocity switch and velocity crossfade between two stored samples ('soft' and 'loud') over the range of MIDI note velocity values.

point. A more subtle effect is achieved by using velocity crossfading, in which the proportion of loud and soft samples varies depending on the received note velocity value. At low velocity values the proportion of the soft sample in the output would be greatest and at high values the output content would be almost entirely made up of the loud sample (see Figure 14.11).

MIDI functions of sound generators

The MIDI implementation for a particular sound generator should be described in the manual that accompanies it. A MIDI implementation chart indicates which message types are received and transmitted, together with any comments relating to limitations or unusual features. Functions such as note off velocity and polyphonic aftertouch, for example, are quite rare. It is quite common for a device to be able to accept certain data and act upon it, even if it cannot generate such data from its own controllers. The note range available under MIDI control compared with that available from a device's keyboard is a good example of this, since many devices will respond to note data over a full ten octave range yet still have only a limited (or no) keyboard. This approach can be used by a manufacturer who wishes to make a cheaper synthesizer that omits the expensive physical sensors for such things as velocity and aftertouch, whilst retaining these functions in software for use under MIDI control. Devices conforming to the General MIDI specification described below must conform to certain basic guidelines concerning their MIDI implementation and the structure of their sound generators.

MIDI data buffers and latency

All MIDI-controlled equipment uses some form of data buffering for received MIDI messages. Such buffering acts as a temporary store for messages that have arrived but have not yet been processed and allows for a certain prioritization in the handling of received messages. Cheaper devices tend to have relatively small MIDI input buffers and these can overflow easily unless care is taken in the filtering and distribution of MIDI data around a large system (usually accomplished by a MIDI router or multiport interface). When a buffer overflows it will normally result in an error message displayed on the front panel of the device, indicating that some MIDI data is likely to have been lost. More advanced equipment can store more MIDI data in its input buffer, although this is not necessarily desirable because many messages that are transmitted over MIDI are intended for 'real-time' execution and one would not wish them to be delayed in a temporary buffer. Such buffer delay is one potential cause of latency in MIDI systems. A more useful solution would be to speed up the rate at which incoming messages are processed.

Handling of velocity and aftertouch data

It is common for the user to be able to program a device such that the velocity value affects certain parameters to a greater or lesser extent. For example, it might be decided that the 'brightness' of the sound should increase with greater key velocity, in which case it would be necessary to program the device so that the envelope generator that affected the brightness was subject to control by the velocity value. The exact law of this relationship is up to the manufacturer and may be used to simulate different types of 'keyboard touch'. A device may offer a number of laws or curves relating changes in velocity to changes in the control value, or the received velocity value may be used to scale the preset parameter rather than replace it.

Another common application of velocity value is to control the amplitude envelope of a particular sound, such that the output volume depends on how hard the key is hit. In many synthesizer systems that use multiple interacting digital oscillators, these velocity-sensitive effects can all be achieved by applying velocity control to the envelope generator of one or more of the oscillators, as indicated earlier in this chapter.

Note off velocity is not implemented in many keyboards, and most musicians are not used to thinking about what they do as they release a key, but this parameter can be used to control such factors as the release time of the note or the duration of a reverberation effect. Aftertouch is often used in synthesizers to control the application of low-frequency modulation (tremolo or vibrato) to a note.

Handling of controller messages

The controller messages that begin with a status of &Bn turn up in various forms in sound generator implementations. It should be noted that although there are standard definitions for many of these controller numbers it is often possible to remap them either within sequencer software or within sound modules themselves.

Controllers &07 (Volume) and &0A (Pan) are particularly useful with sound modules as a means of controlling the internal mixing of voices. These controllers work on a per channel basis, and are independent of any velocity control that may be related to note volume. There are two real-time system exclusive controllers that handle similar functions to these, but for the device as a whole rather than for individual voices or channels. The 'master volume' and 'master balance' controls are accessed using:

&[f0][7F][dev. ID][04][01 or 02][data][data][F7]

where the sub-ID #1 of &04 represents a 'device control' message and sub-ID #2s of &01 or &02 select volume or balance respectively. The [data] values allow 14 bit resolution for the parameters concerned, transmitted LSB first. Balance is different to pan because pan sets the stereo positioning (the split in level between left and right) of a mono source, whereas balance

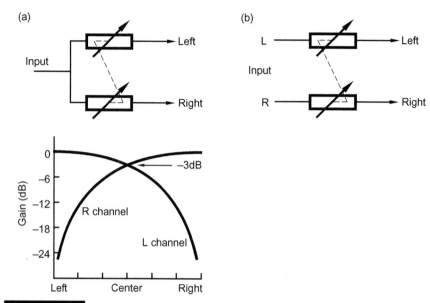

FIGURE 14.12   *(a) A pan control takes a mono input and splits it two ways (left and right), the stereo position depending on the level difference between the two channels. The attenuation law of pan controls is designed to result in a smooth movement of the source across the stereo 'picture' between left and right, with no apparent rise or fall in overall level when the control is altered. A typical pan control gain law is shown below. (b) A balance control simply adjusts the relative level between the two channels of a stereo signal so as to shift the entire stereo image either left or right.*

sets the relative levels of the left and right channels of a stereo source (see Figure 14.12). Since a pan or balance control is used to shift the stereo image either left or right from a center detent position, the MIDI data values representing the setting are ranged either side of a mid-range value that corresponds to the center detent. The channel pan controller is thus normally centered at a data value of 63 (and sometimes over a range of values just below this if the pan has only a limited number of steps), assuming that only a single 7 bit controller value is sent. There may be fewer steps in these controls than there are values of the MIDI controller, depending on the device in question, resulting in a range of controller values that will give rise to the same setting.

Some manufacturers have developed alternative means of expressive control for synthesizers such as the 'breath controller', which is a device which responds to the blowing effort applied by the mouth of the player. It was intended to allow wind players to have more control over expression in performance. Plugged into the synthesizer, it can be applied to various envelope generator or modulator parameters to affect the sound. The breath controller also has its own MIDI controller number. There is also a portamento controller (&54) that defines a note number from which the next note should slide. It is normally transmitted between two note on messages to create an automatic legato portamento effect between two notes.

The 'effects' and 'sound' controllers have been set aside as a form of general purpose control over aspects of the built-in effects and sound quality of a device. How they are applied will depend considerably on the architecture of the sound module and the method of synthesis used, but they give some means by which a manufacturer can provide a more abstracted form of control over the sound without the user needing to know precisely which voice parameters to alter. In this way, a user who is not prepared to get into the increasingly complicated world of voice programming can modify sounds to some extent.

The effects controllers occupy five controller numbers from &5B to &5F and are defined as Effects Depths 1–5. The default names for the effects to be controlled by these messages are respectively 'External Effects Depth', 'Tremolo Depth', 'Chorus Depth', 'Celeste (Detune) Depth' and 'Phaser Depth', although these definitions are open to interpretation and change by manufacturers. There are also ten sound controllers that occupy controller numbers from &46 to &4F. Again these are user or manufacturer definable, but five defaults were originally specified (listed in Table 14.4). They are principally intended as real-time controllers to be used during performance, rather than as a means of editing internal voice patches (the RPCs and NRPCs can be used for this as described in Fact File 14.4).

The sound variation controller is interesting because it is designed to allow the selection of one of a number of variants on a basic sound,

Table 14.4 Sound Controller Functions (byte 2 of status &Bn)

MIDI Controller Number	Function (default)
&46	Sound variation
&47	Timbre/harmonic content
&48	Release time
&49	Attack time
&4A	Brightness
&4B–4F	No default

FACT FILE 14.4 REGISTERED AND NON-REGISTERED PARAMETER NUMBERS

The MIDI standard was extended to allow for the control of individual internal parameters of sound generators by using a specific control change message. This meant, for example, that any aspect of a voice, such as the velocity sensitivity of an envelope generator, could be assigned a parameter number that could then be accessed over MIDI and its setting changed, making external editing of voices much easier. Parameter controllers are a subset of the control change message group, and they are divided into the registered and non-registered numbers (RPNs and NRPNs). RPNs are intended to apply universally and should be registered with the MMA, whilst NRPNs may be manufacturer specific. Only five parameter numbers were originally registered as RPNs, as shown in the table, but more may be added at any time.

Some examples of RPC definitions

RPC number (hex)	Parameter
00 00	Pitch bend sensitivity
00 01	Fine tuning
00 02	Coarse tuning
00 03	Tuning program select
00 04	Tuning bank select
7F 7F	Cancels RPN or NRPN (usually follows Message 3)

Parameter controllers operate by specifying the address of the parameter to be modified, followed by a control change message to increment or decrement the setting concerned. It is also possible to use the data entry slider controller to alter the setting of the parameter. The address of the parameter is set in two stages, with an MSbyte and then an LSbyte message, so as to allow for 16 384 possible parameter addresses. The controller numbers &62 and &63 are used to set the LS- and MSbytes respectively of an NRPN, whilst &64 and &65 are used to address RPNs. The sequence of messages required to modify a parameter is as follows:

Message 1

&[Bn] [62 or 64] [LSB]

Message 2

&[Bn] [63 or 65] [MSB]

Message 3

&[Bn] [60 or 61] [7F] or &[Bn] [06] [DATA] [38] [DATA]

Message 3 represents either data increment (&60) or decrement (&61), or a 14 bit data entry slider control change with MSbyte (&06) and LSbyte (&38) parts (assuming running status). If the control has not moved very far, it is possible that only the MSbyte message need be sent.

depending on the data value that follows the controller number. For example, a piano sound might have variants of 'honky tonk', 'soft pedal', 'lid open' and 'lid closed'. The data value in the message is not intended to act as a continuous controller for certain voice parameters, rather the different data values possible in the message are intended to be used to select certain pre-programmed variations on the voice patch. If there are fewer than the 128 possible variants on the voice then the variants should be spread evenly over the number range so that there is an equal number range between them.

The timbre and brightness controllers can be used to alter the spectral content of the sound. The timbre controller is intended specifically for altering the harmonic content of a sound, whilst the brightness controller is designed to control its high-frequency content. The envelope controllers can be used to modify the attack and release times of certain envelope generators within a synthesizer. Data values less than &40 attached to these messages should result in progressively shorter times, whilst values greater than &40 should result in progressively longer times.

Voice selection

The program change message was adequate for a number of years as a means of selecting one of a number of stored voice patches on a sound generator. Program change on its own allows for up to 128 different voices to be selected and a synthesizer or sound module may allow a program change map to be set up in order that the user may decide which voice is selected on receipt of a particular message. This can be particularly useful when the module has more than 128 voices available, but no other means of selecting voice banks. A number of different program change maps could be stored, perhaps to be selected under system exclusive control.

Modern sound modules tend to have very large patch memories – often too large to be adequately addressed by 128 program change messages. Although some older synthesizers used various odd ways of providing access to further banks of voices, most modern modules have implemented the standard 'bank select' approach. In basic terms, 'bank select' is a means of extending the number of voices that may be addressed by preceding a standard program change message with a message to define the bank from which that program is to be recalled. It uses a 14 bit control change message, with controller numbers &00 and &20, to form a 14 bit bank address, allowing 16 384 banks to be addressed. The bank number is followed directly by a program change message, thus creating the following general message:

&[Bn] [00] [MSbyte (of bank)]

&[Bn] [20] [LSbyte]

&[Cn] [Program number]

GENERAL MIDI

One of the problems with MIDI sound generators is that although voice patches can be selected using MIDI program change commands, there is no guarantee that a particular program change number will recall a particular voice on more than one instrument. General MIDI is an approach to the standardization of a sound generator's behavior, so that MIDI files (see Fact File 14.5) can be exchanged more easily between systems and device behavior can be predicted by controllers. It comes in three flavors: GM 1, GM Lite and GM 2.

FACT FILE 14.5 STANDARD MIDI FILES (SMF)

Sequencers and notation packages typically store data on disk in their own unique file formats. The standard MIDI file was developed in an attempt to make interchange of information between packages more straightforward. MIDI files are most useful for the interchange of performance and control information. They are not so useful for music notation where it is necessary to communicate greater detail about the way music appears on the stave and other notational concepts. For the latter purpose a number of different file formats have been developed, including Music XML which is among the most widely used of the universal interchange formats today. Further information about Music XML resources and other notation formats may be found in the 'Recommended further reading' at the end of this chapter.

Three types of standard MIDI file exist to encourage the interchange of sequencer data between software packages. The MIDI file contains data representing events on individual sequencer tracks, as well as labels such as track names, instrument names and time signatures. File type 0 is the simplest and is used for single-track data, whilst file type 1 supports multiple tracks which are 'vertically' synchronous with each other (such as the parts of a song).

File type 2 contains multiple tracks that have no direct timing relationship and may therefore be asynchronous.

Type 2 could be used for transferring song files made up of a number of discrete sequences, each with a multiple track structure. The basic file format consists of a number of 8 bit words formed into chunk-like parts, very similar to the RIFF and AIFF audio file formats described in Chapter 9. SMFs are not exactly RIFF files though, because they do not contain the highest level FORM chunk. (To encapsulate SMFs in a RIFF structure, use the RMID format.)

The header chunk, which always heads a MIDI file, contains global information relating to the whole file, whilst subsequent track chunks contain event data and labels relating to individual sequencer tracks. Track data should be distinguished from MIDI channel data, since a sequencer track may address more than one MIDI channel. Each chunk is preceded by a preamble of its own, which specifies the type of chunk (header or track) and the length of the chunk in terms of the number of data bytes that are contained in the chunk. There then follow the designated number of data bytes (see the figure below). The chunk preamble contains 4 bytes to identify the chunk type using ASCII representation and 4 bytes to indicate the number of data bytes in the chunk (the length). The number of bytes indicated in the length does not include the preamble (which is always 8 bytes).

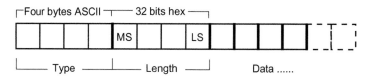

General MIDI Level 1 specifies a standard voice map and a minimum degree of polyphony, requiring that a sound generator should be able to receive MIDI data on all 16 channels simultaneously and polyphonically, with a different voice on each channel. There is also a requirement that the sound generator should support percussion sounds in the form of drum kits, so that a General MIDI sound module is capable of acting as a complete 'band in a box'.

Dynamic voice allocation is the norm in GM sound modules, with a requirement either for at least 24 dynamically allocated voices in total, or 16 for melody and eight for percussion. Voices should all be velocity sensitive and should respond at least to the controller messages 1, 7, 10, 11, 64, 121 and 123 (decimal), RPNs 0, 1 and 2 (see above), pitch bend and channel aftertouch. In order to ensure compatibility between sequences that are replayed on GM modules, percussion sounds are always allocated to MIDI channel 10. Program change numbers are mapped to specific voice names, with ranges of numbers allocated to certain types of sounds, as shown in Table 14.5. Precise voice names may be found in the GM documentation. Channel 10, the percussion channel, has a defined set of note numbers on which particular sounds are to occur, so that the composer may know, for example, that key 39 will always be a 'hand clap'.

Table 14.5 General MIDI Program Number Ranges (Except Channel 10)

Program Change (Decimal)	Sound Type
0–7	Piano
8–15	Chromatic percussion
16–23	Organ
24–31	Guitar
32–39	Bass
40–47	Strings
48–55	Ensemble
56–63	Brass
64–71	Reed
72–79	Pipe
80–87	Synth lead
88–95	Synth pad
96–103	Synth effects
104–111	Ethnic
112–119	Percussive
120–127	Sound effects

General MIDI sound modules may operate in modes other than GM, where voice allocations may be different, and there are two universal non-real-time SysEx messages used to turn GM on or off. These are:

$$\&[F0]\ [7E]\ [dev.\ ID]\ [09]\ [01]\ [F7]$$

to turn GM on, and:

$$\&[F0]\ [7E]\ [dev.\ ID]\ [09]\ [02]\ [F7]$$

to turn it off.

There is some disagreement over the definition of 'voice', as in '24 dynamically allocated voices' – the requirement that dictates the degree of polyphony supplied by a GM module. The spirit of the GM specification suggests that 24 notes should be capable of sounding simultaneously, but some modules combine sound generators to create composite voices, thereby reducing the degree of note polyphony.

General MIDI Lite (GML) is a cut-down GM 1 specification designed mainly for use on mobile devices with limited processing power. It can be used for things like ring tones on mobile phones and for basic music replay from PDAs. It specifies a fixed polyphony of 16 simultaneous notes, with 15 melodic instruments and one percussion kit on channel 10. The voice map is the same as GM Level 1. It also supports basic control change messages and the pitch-bend sensitivity RPN. As a rule, GM Level 1 songs will usually replay on GM Lite devices with acceptable quality, although some information may not be reproduced. An alternative to GM Lite is SPMIDI (see next section) which allows greater flexibility.

GM Level 2 is backwards compatible with Level 1 (GM 1 songs will replay correctly on GM 2 devices) but allows the selection of voice banks and extends polyphony to 32 voices. Percussion kits can run on channel 11 as well as the original channel 10. It adds MIDI tuning, RPN controllers and a range of universal system exclusive messages to the MIDI specification, enabling a wider range of control and greater versatility.

SCALABLE POLYPHONIC MIDI (SPMIDI)

SPMIDI, rather like GM Lite, is designed principally for mobile devices that have issues with battery life and processing power. It has been adopted by the 3GPP wireless standards body for structured audio control of synthetic sounds in ring tones and multimedia messaging. It was developed primarily by Nokia and Beatnik. The SPMIDI basic specification for a device is based on GM Level 2, but a number of selectable profiles are possible, with different levels of sophistication.

The idea is that rather than fixing the polyphony at 16 voices the polyphony should be scalable according to the device profile (a description of the current capabilities of the device). SPMIDI also allows the content creator to decide what should happen when polyphony is limited – for example, what should happen when only four voices are available instead of 16. Conventional 'note stealing' approaches work by stealing notes from sounding voices to supply newly arrived notes, and the outcome of this can be somewhat arbitrary. In SPMIDI this is made more controllable. A process known as channel masking is used, whereby certain channels have a higher priority than others, enabling the content creator to put high priority material on particular channels. The channel priority order and maximum instantaneous polyphony are signaled to the device in a setup message at the initialization stage.

RMID AND XMF FILES

RMID is a version of the RIFF file structure that can be used to combine a standard MIDI file and a downloadable sound file (see Fact File 14.6) within a single structure. In this way all of the data required to replay a song using synthetic sounds can be contained within one file. RMID seems to have been superseded by another file format known as XMF (eXtensible Music Format) that is designed to contain all of the assets required to replay a music file. It is based on Beatnik's RMF (Rich Music Format) which was designed to incorporate standard MIDI files and audio files such as MP3 and WAVE so that a degree of interactivity could be added to audio replay. RMF can also address a Special Bank of MIDI sounds (an extension of GM) in the Beatnik Audio Engine. XMF is now the MMA's recommended way of combining such elements. It is more extensible than RMID and can contain WAVE files and other media elements for streamed or interactive presentations. XMF introduces concepts such as looping and branching into standard MIDI files. RMF included looping but did not incorporate DLS into the file format. In addition to the features just described, XMF can incorporate 40 bit encryption for advanced data security as well as being able to compress standard MIDI files by up to 5:1 and incorporate metadata such as rights information. So far, XMF Type 0 and Type 1 have been defined, both of which contain SMF and DLS data, and which are identical except that Type 0 MIDI data may be streamed.

FACT FILE 14.6 DOWNLOADABLE SOUNDS AND SOUNDFONTS

Downloadable Sounds (DLS) is an MMA specification for synthetic voice description that enables synthesizers to be programmed using voice data downloaded from a variety of sources. In this way a content creator could not only define the musical structure of his or her content in a universally usable way, using standard MIDI files, but could also define the nature of the sounds to be used with downloadable sounds. In these ways content creators can specify more precisely how synthetic audio should be replayed, so that the end result can be more easily predicted across multiple rendering platforms.

The success of these approaches depends on 'wavetable synthesis'. Here basic sound waveforms are stored in wavetables (simply tables of sample values) in RAM, to be read out at different rates and with different sample skip values, for replay at different pitches. Subsequent signal processing and envelope shaping can be used to alter the timbre and temporal characteristics. Such synthesis capabilities exist on the majority of computer sound cards, making it a realistic possibility to implement the standard widely.

DLS Level 1, version 1.1a, was published in 1999 and contains a specification for devices that can deal with DLS as well as a file format for containing the sound descriptions. The basic idea is that a minimal synthesis engine should be able to replay a looped sample from a wavetable, apply two basic envelopes for pitch and volume, use low-frequency oscillator control for tremolo and vibrato, and respond to basic MIDI controls such as pitch bend and modulation wheel. There is no option to implement velocity crossfading or layering of sounds in DLS Level 1, but keyboard splitting into 16 ranges is possible.

DLS Level 2 is somewhat more advanced, requiring two six-segment envelope generators, two LFOs, a low-pass filter with resonance and dynamic cut-off frequency controls. It requires more memory for wavetable storage (2MB), 256 instruments and 1024 regions, amongst other things. DLS Level 2 has been adopted as the MPEG-4 Structured Audio Sample Bank format.

Emu developed so-called SoundFonts for Creative Labs and these have many similar characteristics to downloadable sounds. They have been used widely to define synthetic voices for Sound Blaster and other computer sound cards. In fact the formats have just about been harmonized with the issue of DLS Level 2 which apparently contains many of the advanced features of SoundFonts. SoundFont 2 descriptions are normally stored in RIFF files with the extension '.sf2'.

MIDI OVER USB

The USB Implementers Forum has published a 'USB Device Class Definition for MIDI Devices', version 1.0, which describes how MIDI data may be transported over USB connections. It preserves the protocol of MIDI messages but packages them in such a way as to enable them to be transferred over USB. It also 'virtualizes' the concept of MIDI IN and OUT jacks, enabling USB to MIDI conversion, and vice versa, to take place in software within a synthesizer or other device. Physical MIDI ports can also be created for external connections to conventional MIDI equipment (see Figure 14.13). A so-called 'USB MIDI function' (a device that receives USB MIDI events and transfers) may contain one or more 'elements'.

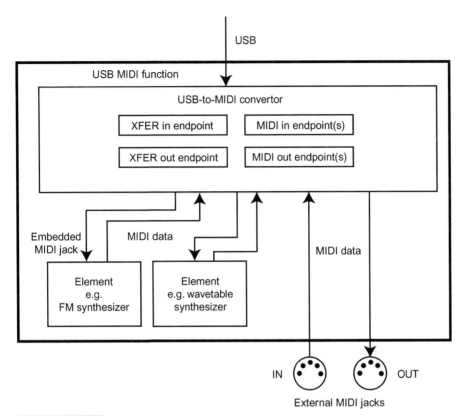

USB

USB MIDI function

USB-to-MIDI convertor

XFER in endpoint

MIDI in endpoint(s)

XFER out endpoint

MIDI out endpoint(s)

Embedded
MIDI jack

MIDI data

MIDI data

Element
e.g.
FM synthesizer

Element
e.g. wavetable
synthesizer

IN   OUT

External MIDI jacks

FIGURE 14.13 *A USB MIDI function contains a USB-to-MIDI convertor that can communicate with both embedded (internal) and external MIDI jacks via MIDI IN and OUT endpoints. Embedded jacks connect to internal elements that may be synthesizers or other MIDI data processors. XFER in and out endpoints are used for bulk dumps such as DLS and can be dynamically connected with elements as required for transfers.*

These elements can be synthesizers, synchronizers, effects processors or other MIDI-controlled objects.

A USB to MIDI convertor within a device will typically have MIDI in and out endpoints as well as what are called 'transfer' (XFER) endpoints. The former are used for streaming MIDI events whereas the latter are used for bulk dumps of data such as those needed for downloadable sounds (DLS). MIDI messages are packaged into 32 bit USB MIDI events, which involve an additional byte at the head of a typical MIDI message. This additional byte contains a cable number address and a code index number (CIN), as shown in Figure 14.14. The cable number enables the MIDI message to be targeted at one of 16 possible 'cables', thereby overcoming the

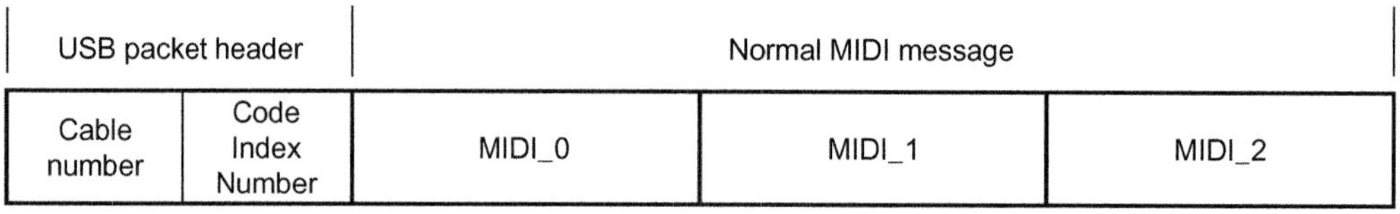

USB packet header		Normal MIDI message		
Cable number	Code Index Number	MIDI_0	MIDI_1	MIDI_2

FIGURE 14.14 *USB MIDI packets have a 1 byte header that contains a cable number to identify the MIDI jack destination and a code index number to identify the contents of the packet and the number of active bytes.*

16 channel limit of conventional MIDI messages, in a similar way to that used in the addressing of multiport MIDI interfaces. The CIN allows the type of MIDI message to be identified (e.g. System Exclusive; Note On), which to some extent duplicates the MIDI status byte. MIDI messages with fewer than 3 bytes should be padded with zeros.

MIDI OVER IEEE 1394

The MMA and AMEI have published a recommended practice, RP-27, 'MIDI Media Adaptation Layer for IEEE 1394', which describes how MIDI data is to be transferred over 1394 (Firewire). This is also referred to in 1394 TA (Trade Association) documents describing the 'Audio and Music Data Transmission Protocol' and IEC standard 61883-6 which deals with the audio part of 1394 interfaces.

The approach is similar to that used with USB, described in the previous section, but has somewhat greater complexity. MIDI 1.0 data streams can be multiplexed into a 1394 'MIDI conformant data channel' which contains eight independent MIDI streams called 'MPX-MIDI data channels'. This way each MIDI conformant data channel can handle $8 \times 16 = 128$ MIDI channels (in the original sense of MIDI channels). The first version of the standard limits the transmission of packets to the MIDI 1.0 data rate of 31.25 kbit/s for compatibility with other MIDI devices; however, provision is made for transmission at substantially faster rates for use in equipment that is capable of it. This includes options for 2X and 3X MIDI 1.0 speed. 1394 cluster events can be defined that contain both audio and MIDI data. This enables the two types of information to be kept together and synchronized.

MIDI OVER ETHERNET

Although not yet standardized by the MMA, the Internet Engineering Task Force (IETF) came up with a method of transporting MIDI over

Ethernet networks using the Real Time Protocol (RTP). This was also the basis for Apple's MIDI over Ethernet. The same format allows for MPEG-4-generic audio object types including General MIDI, Downloadable Sounds and Structured Audio. Each packet of data transmitted over the network includes an RTP header and a MIDI 'payload' that encodes the standard MIDI commands in a suitable form. Each packet header has a baseline timestamp to assist in synchronizing the event to a clock that is defined in the RTP session setup at a particular sampling rate. Each MIDI payload defines the timing of the MIDI event relative to the base-line timestamp.

Because this format uses standard Internet protocols for transporting the RTP packets, such as TCP and UDP, it offers a relatively straightforward way of streaming sound information at very low bit rates over the Internet. Naturally this is limited to the basic music and sound control information provided by MIDI, and the sounds produced by any rendering device will depend on the synthesis engine available.

It is likely that the upcoming HD MIDI standard will also include a method for transporting MIDI over Ethernet.

OPEN SOUND CONTROL

Open Sound Control is an alternative to MIDI that is gradually seeing greater adoption in the computer music and musical instrument control world. Developed by Matt Wright at CNMAT (Center for New Music and Audio Technology) in Berkeley, California, it aims to offer a transport-independent message-based protocol for communication between computers, musical instruments and multimedia devices. It does not specify a particular hardware interface or network for the transport layer, but initial implementations have tended to use UDP (user datagram protocol) over Ethernet or other fast networks as a transport means. It is not proposed to describe this protocol in detail and further details can be found at the website indicated at the end of this chapter. A short summary will be given, however.

OSC uses a form of device addressing that is very similar to an Internet URL (uniform resource locator). In other words a text address with sub-addresses that relate to lower levels in the device hierarchy. For example, '/synthesizer2/voice1/oscillator3/frequency' (not a real address) might refer to a particular device called 'synthesizer2', within which is contained voice 1, within which is oscillator 3, whose frequency value is being addressed. The minimum 'atomic unit' of OSC data is 4 bytes (32 bits) long, so all values are 32 bit aligned, and transmitted packets are made up of multiples of 32 bit information. Packets of OSC data contain either individual messages or so-called 'bundles'. Bundles contain elements that are either messages or further bundles, each having a size designation that precedes it, indicating the length of the element. Bundles have time tags associated with them, indicating that the actions described in the bundle are to take place at a specified time. Individual messages are supposed to be executed immediately. Devices are expected to have access to a representation of the correct current time so that bundle timing can be related to a clock.

SEQUENCING SOFTWARE

Introduction

Sequencers are probably the most ubiquitous of audio and MIDI software applications. Although they used to be available as dedicated devices they are now widely available as sophisticated software packages to run on a desktop computer. A sequencer is capable of storing a number of 'tracks' of MIDI and audio information, editing it and otherwise manipulating it for musical composition purposes. It is also capable of storing MIDI events for non-musical purposes such as studio automation. Some of the more advanced packages are available in cut-down or 'Lite' versions for the new user. Popular packages such as Pro Tools and Logic now combine audio and MIDI manipulation in an almost seamless fashion, and have been developed to the point where they can no longer really be considered as simply sequencers. In fact they are full-blown audio production systems with digital mixers, synchronization, automation, effects and optional video.

The dividing line between sequencer and music notation software is a gray one, since there are features common to both. Music notation software such as Sibelius is designed to allow the user control over the detailed appearance of the printed musical page, rather as page layout packages work for typesetters, and such software often provides facilities for MIDI input and output. MIDI input is used for entering note pitches during setting, whilst output is used for playing the finished score in an audible form. Most major packages will read and write standard MIDI files, and can therefore exchange data with sequencers, allowing sequenced music to be exported to a notation package for fine tuning of printed appearance. It is also common for sequencer packages to offer varying degrees of music notation capability, although the scores that result may not be as professional in appearance as those produced by dedicated notation software.

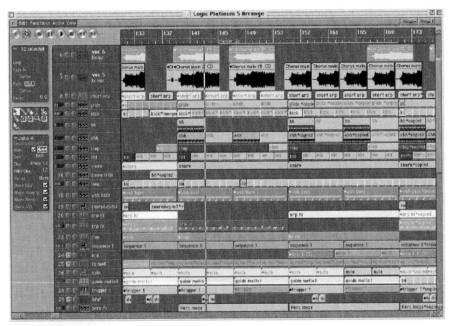

FIGURE 14.15 *Example of a sequencer's primary display, showing tracks and transport controls. (Logic Platinum 5 'arrange' window.)*

Tracks, channels, instruments and environments

A sequencer can be presented to the user so that it emulates a multitrack tape recorder to some extent. The example shown in Figure 14.15 illustrates this point, showing the familiar transport controls as well as a multitrack 'tape-like' display.

A track can be either a MIDI track or an audio track, or it may be a virtual instrument of some sort, perhaps running on the same computer. A project is built up by successively overlaying more and more tracks, all of which may be replayed together. Tracks are not fixed in their time relationship and can be slipped against each other, as they simply consist of data stored in the memory. On older or less advanced sequencers, the replay of each MIDI track was assigned to a particular MIDI channel, but more recent packages offer an almost unlimited number of virtual tracks that can contain data for more than one channel (in order to drive a multi-timbral instrument, for example). Using a multiport MIDI interface it is possible to address a much larger number of instruments than the basic 16 MIDI channels allowed in the past.

In a typical sequencer, instruments are often defined in a separate 'environment' that defines the instruments, the ports to which they are

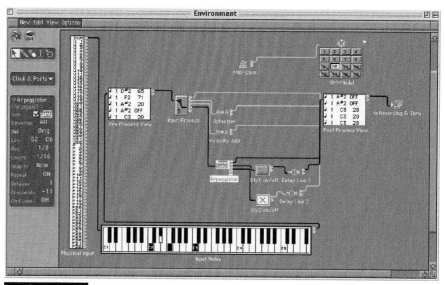

FIGURE 14.16 | *Example of environment window from Logic, showing ways in which various MIDI processes can be inserted between physical input and recording operation.*

connected, any additional MIDI processing to be applied, and so forth. An example is shown in Figure 14.16. When a track is recorded, therefore, the user simply selects the instrument to be used and the environment takes care of managing what that instrument actually means in terms of processing and routing. Now that soft synthesizers are used increasingly, sequencers can often address those directly via plug-in architectures such as DirectX or VST (see Chapter 13), without recourse to MIDI. These are often selected on pull-down menus for individual tracks, with voices selected in a similar way, often using named voice tables.

Input and output filters

After MIDI information is received from the hardware interface it is stored in memory, but it may sometimes be helpful to filter out some information before it can be stored, using an input filter. This will be a subsection of the program that watches out for the presence of certain MIDI status bytes and their associated data as they arrive, so that they can be discarded before storage. The user may be able to select input filters for such data as after-touch, pitch bend, control changes and velocity information, amongst others. Clearly it is only advisable to use input filters if it is envisaged that this data will never be needed, since although filtering saves memory space the information is lost forever. Output filters are often implemented for similar groups of MIDI messages as for the input filters, acting on the replayed rather than recorded information. Filtering may help to reduce MIDI delays, owing to the reduced data flow.

Timing resolution

The timing resolution to which a sequencer can store MIDI events varies between systems. This 'record resolution' may vary with recent systems offering resolution to many thousandths of a note. Audio events are normally stored to sample accuracy. A sequencer with a MIDI resolution of 480 ppqn (pulses per quarter note) can resolve events to 4.1 millisecond steps, for example. The quoted resolution of sequencers, though, tends to be somewhat academic, depending on the operational circumstances, since there are many other factors influencing the time at which MIDI messages arrive and are stored. These include buffer delays and traffic jams. Modern sequencers have sophisticated routines to minimize the latency with which events are routed to MIDI outputs.

The record resolution of a sequencer is really nothing to do with the timing resolution available from MIDI clocks or timecode (see Chapter 15). The sequencer's timing resolution refers to the accuracy with which it time-stamps events and to which it can resolve events internally. Most sequencers attempt to interpolate or 'flywheel' between external timing bytes during replay, in an attempt to maintain a resolution in excess of the 24 ppqn implied by MIDI clocks.

Displaying, manipulating and editing information

A sequencer is the ideal tool for manipulating MIDI and audio information and this may be performed in a number of ways depending on the type of interface provided to the user. The most flexible is the graphical interface employed on many desktop computers which may provide for visual editing of the stored MIDI information either as a musical score, a table or event list of MIDI data, or in the form of a grid of some kind. Figure 14.17 shows a number of examples of different approaches to the display of stored MIDI information. Audio information is manipulated using an audio sample editor display that shows the waveform and allows various changes to be made to the signal, often including sophisticated signal processing, as discussed further below.

Although it might be imagined that a musical score would be the best way of visualizing MIDI data, it is often not the most appropriate. This is partly because unless the input is successfully quantized (see below) the score will represent precisely what was played when the music was

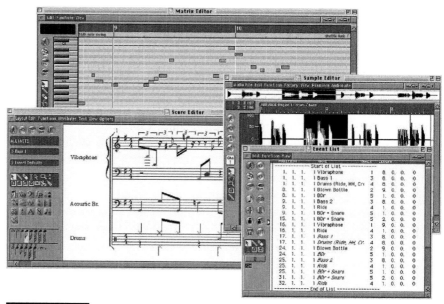

FIGURE 14.17   *Examples of a selection of different editor displays from Logic, showing display of MIDI data as a score, a graphical matrix of events and a list of events. Audio can be shown as an audio waveform display.*

recorded and this is rarely good-looking on a score! The appearance is often messy because some notes were just slightly out of time. Score representation is useful after careful editing and quantization, and can be used to produce a visually satisfactory printed output. Alternatively, the score can be saved as a MIDI file and exported to a music notation package for layout purposes.

In the grid editing (called 'Matrix' in the example shown) display, MIDI notes may be dragged around using a mouse or trackball and audible feedback is often available as the note is dragged up and down, allowing the user to hear the pitch or sound as the position changes. Note lengths can be changed and the timing position may be altered by dragging the note left or right. In the event list form, each MIDI event is listed next to a time value. The information in the list may then be changed by typing in new times or new data values. Also events may be inserted and deleted. In all of these modes the familiar cut and paste techniques used in word processors and other software can be applied, allowing events to be used more than once in different places, repeated so many times over, and other such operations.

A whole range of semi-automatic editing functions are also possible, such as transposition of music, using the computer to operate on the data so as to modify it in a predetermined fashion before sending it out again. Echo effects can be created by duplicating a track and offsetting it by a certain amount, for example. Transposition of MIDI performances is simply a matter of raising or lowering the MIDI note numbers of every stored note by the relevant degree. A number of algorithms have also been developed for converting audio melody lines to MIDI data, or using MIDI data to control the pitch of audio, further blurring the boundary between the two types of information. Silence can also be stripped from audio files, so that individual drum notes or vocal phrases can be turned into events in their own right, allowing them to be manipulated, transposed or time-quantized independently.

A sequencer's ability to search the stored data (both music and control) based on specific criteria, and to perform modifications or transformations to just the data which matches the search criteria, is one of the most powerful features of a modern system. For example, it may be possible to search for the highest-pitched notes of a polyphonic track so that they can be separated off to another track as a melody line. Alternatively it may be possible to apply the rhythm values of one track to the pitch values of another so as to create a new track, or to apply certain algorithmic manipulations to stored durations or pitches for compositional experimentation. The possibilities for searching, altering and transforming stored data are almost endless once musical and control events are stored in the form of unique values, and for those who specialize in advanced composing or experimental music these features will be of particular importance. It is in this field that many of the high-end sequencer packages will continue to develop.

Quantization of rhythm

Rhythmic quantization is a feature of almost all sequencers. In its simplest form it involves the 'pulling-in' of events to the nearest musical time interval at the resolution specified by the user, so that events that were 'out of time' can be played back 'in time'. It is normal to be able to program the quantizing resolution to an accuracy of at least as small as a 32nd note and the choice depends on the audible effect desired. Events can be quantized either permanently or just for replay. Some systems allow 'record quantization' which alters the timing of events as they arrive at the input to the sequencer. This is a form of permanent quantization. It may also be possible to 'quantize' the cursor movement so that it can only drag events to predefined rhythmic divisions.

More complex rhythmic quantization is also possible, in order to maintain the 'natural' feel of rhythm, for example. Simple quantization can result in music that sounds 'mechanical' and electronically produced, whereas the 'human feel' algorithms available in many packages attempt to quantize the rhythm strictly and then reapply some controlled randomness. The parameters of this process may be open to adjustment until the desired effect is achieved.

Automation and non-note MIDI events

In addition to note and audio events, one may either have recorded or may wish to add events for other MIDI control purposes such as program change messages, controller messages or system exclusive messages. Audio automation can also be added to control fades, panning, effects and other mixing features. Such data may be displayed in a number of ways, but again the graphical plot is arguably the most useful. It is common to allow automation data to be plotted as an overlay, such as shown in Figure 14.18.

Some automation data is often stored in a so-called 'segment-based' form. Because automation usually relates to some form of audio processing or control, it usually applies to particular segments on the timeline of the current project. If the segment is moved, often one needs to carry the

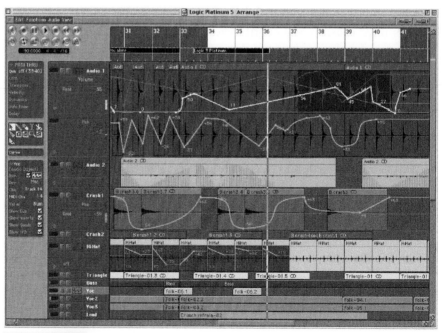

FIGURE 14.18   *Example from Logic of automation data graphically overlaid on sequencer tracks.*

relevant automated audio processing with it. Segment-based processing or automation allows the changes in parameters that take place during a segment to be 'anchored' to that segment so that they can be made to move around with it if required.

It is possible to edit automation or control events in a similar way to note events, by dragging, drawing, adding and deleting points, but there are a number of other possibilities here. For example, a scaling factor may be applied to controller data in order to change the overall effect by so many percent, or a graphical contour may be drawn over the controller information to scale it according to the magnitude of the contour at any point. Such a contour could be used to introduce a gradual increase in MIDI note velocities over a section, or to introduce any other time-varying effect. Program changes can be inserted at any point in a sequence, usually either by inserting the message in the event list, or by drawing it at the appropriate point in the controller chart. This has the effect of switching the receiving device to a new voice or stored program at the point where the message is inserted. It can be used to ensure that all tracks in a sequence use the desired voices from the outset without having to set them up manually each time. Either the name of the program to be selected at that point or its number can be displayed, depending on whether the sequencer is subscribing to a known set of voice names such as General MIDI.

System exclusive data may also be recorded or inserted into sequences in a similar way to the message types described above. Any such data received during recording will normally be stored and may be displayed in a list form. It is also possible to insert SysEx voice dumps into sequences in order that a device may be loaded with new parameters whilst a song is executing if required.

MIDI mixing and external control

Sequencers often combine a facility for mixing audio with one for controlling the volume and panning of MIDI sound generators. Using MIDI volume and pan controller numbers (decimal 7 and 10), a series of graphical faders can be used to control the audio output level of voices on each MIDI channel, and may be able to control the pan position of the source between the left and right outputs of the sound generator if it is a stereo source. On-screen faders may also be available to be assigned to other functions of the software, as a means of continuous graphical control over parameters such as tempo, or to vary certain MIDI continuous controllers in real time.

It is also possible with some packages to control many of the functions of the sequencer using external MIDI controllers. An external MIDI controller with a number of physical faders and buttons could be used as a basic means of mixing, for example, with each fader assigned to a different channel on the sequencer's mixer.

Synchronization

A sequencer's synchronization features are important when locking replay to external timing information such as MIDI clock or timecode. Most sequencers are able to operate in either beat clock or timecode sync modes and some can detect which type of clock data is being received and switch over automatically. To lock the sequencer to another sequencer or to a drum machine beat clock synchronization may be adequate. If you will be using the sequencer for applications involving the timing of events in real rather than musical time, such as the dubbing of sounds to a film, then it is important that the sequencer is able to allow events to be tied to timecode locations, as timecode locations will remain in the same place even if the musical tempo is changed.

Sequencers incorporating audio tracks also need to be able to lock to sources of external audio or video sync information (e.g. word clock or composite video sync), in order that the sampling frequency of the system can be synchronized to that of other equipment in the studio.

Synchronized digital video

Digital video capability is now commonplace in desktop workstations. It is possible to store and replay full motion video on a desktop computer, either using a separate monitor or within a window on an existing monitor, using widely available technology such as QuickTime or Windows Multimedia Extensions. The replay of video from disk can be synchronized to the replay of audio and MIDI, using timecode, and this is particularly useful as an alternative to using video on a separate video tape recorder (which is mechanically much slower, especially in locating distant cues). In some sequencing or editing packages the video can simply be presented as another 'track' alongside audio and MIDI information.

In the applications considered here, compressed digital video is intended principally as a cue picture that can be used for writing music or dubbing sound to picture in post-production environments. In such cases the picture quality must be adequate to be able to see cue points and lip sync, but it does not need to be of professional broadcast quality. What is important is reasonably good slow motion and freeze-frame quality. Good quality digital video (DV), though, can now be transferred to and from workstations using a Firewire interface enabling video editing and audio post-production to be carried out in an integrated fashion, all on the one platform.

AUDIO REMOTE CONTROL USING COMPUTER NETWORKS

MIDI has been around for some decades, and it has already been mentioned how remarkably well it has continued to serve in a variety of applications and environments during that time despite the huge developments and changes the industry has seen. It will continue to perform a useful function, particularly in its original home where electronic musical instruments are required to trigger and control each other, but there has been a growing need for system control standards for computer networks. Options exist for carrying MIDI data over Ethernet, as mentioned above, and the forthcoming HD MIDI standard will most likely be network-oriented. The situation with audio system remote control today is similar to that which existed in about 1980, when several manufacturers were using proprietary signaling systems for the control of musical instruments before MIDI became the industry standard. Several bodies are now working to achieve common technical standards both for the implementation of proprietary formats and with a view to achieving industry consensus, and some formats are in use in the field.

Open control architecture

A number of professional audio manufacturers have formed a group with the aim of securing standardization of the Open Control Architecture (OCA), a media networking system control standard for professional applications. The Alliance has been formed to complete the technical definition of OCA, then to transfer its development to an accredited public standards organization. The latter will render the OCA specification into an open public standard for control of professional media network systems. Developed recently by Bosch Communications Systems, OCA is descended from AES-24, a system control protocol developed by the Audio Engineering Society in the 1990s. OCA defines a flexible and robust control standard that covers the entire range of pro media networking applications, from the smallest to the largest. The OCA Release 1.1 is available on the OCA Alliance website, and is only one of a number of control systems proposed or already in the field. Others will be covered below, but a closer look at the OCA proposals gives an insight into the level of control offered generally by these systems as the industry moves from control via DIN-to-DIN leads towards full computer integration.

The OCA Framework, sometimes abbreviated as OCF, includes the following:

1. Discover devices. This recognizes OCA-compliant devices that are connected to the network.
2. Manage streaming connections. Define and undefine media stream paths between and among devices. Interfaces with features of the media and transport system in order to set up and take down media connections.
3. Control and monitor of operating and configuration parameters of OCA-compliant devices.
4. Define and manage devices that have reconfigurable signal processing and/or control capabilities.
5. Upgrade software/firmware of controlled devices. This to include features for fail-safe upgrades.

OCA operates over industry-standard data network equipment in secure and unsecure mode, coexisting with non-OCA devices. Incorporation of new device types and device upgrades as well as of non-standard devices is allowed, and multiple protocol versions are offered for different kinds of interconnection. The current protocol is OCP.1 for TCP/IP Ethernets. Future specifications, labelled OCP.2, OCP.3 and so on, will be for other kinds of connection such as USB.

The actions of OCA are contained within separate 'classes', each class defining the type of action allowable between devices. This is known as 'object-oriented protocol' and is defined by the combination of four sets:

1. Class definitions: the types of objects existing within devices.
2. Naming and addressing rules: how the objects and their attributes are identified.
3. Protocol Data Unit formats (PDU): specify formats of transmitted and received data.
4. PDU Exchange Rules: defining communication sequences used to effect information exchange.

'Objects' in this context are what actually run in the computer, and are units of code belonging to generic 'classes'. Each OCA device is given a unique object number, ONo, which can be fixed at the time of manufacture or chosen subsequently for configurable devices. The ONos are 32 bits long, so duplication or a later re-use of an ONo would be rare, 2^{32} being a huge number. Allocating blocks of numbers to the various manufacturers however would remove the possibility of duplications entirely. (AES64, described below, addresses this, and EuCon uses a system of 'zones'.) The ONo is sent and received with every command and response, and controllers can allow users to identify devices hierarchically, for instance by using channel numbers.

Actuators, sensors and blocks

Actuators control a device's signal processing and housekeeping. In any device, any actuator class may be instantiated, that is, a specific command pertaining to that class is sent, as many times as required to control that function of a processor. There are 36 actuator classes, and examples that can be specified in the system include control of gain, signal mute, multiposition selection, parametric EQ, delay, compressor/limiting, and temperature setting. The latter can be used to monitor the temperature of power amplifiers, for example.

Sensors detect the value of a parameter and transmit it back to controllers. A sensor's reading may be transmitted either periodically or when it exceeds a defined threshold. There can be up to 20 sensor classes, including sensing of signal level in absolute terms, sensing of level according to VU or PPM ballistics, and the sensing of temperature.

A block is a special type of worker that can contain other objects – units of code. It can contain workers, agents, or certain other blocks, and it has a signal flow topology. An object inside a block is a member of that block, or 'container'. Each block is described by a class, and the base class for all block classes is named OcaBlock. A block class represents a group of workers, agents and nested blocks, the signal flowing within that group; it does not represent specific audio processing.

AES64-2012

The AES64-2012 'standard for audio applications of networks: command, control, and connection management for integrated media' specifies a system of control which shares common ground with OCA. The latter is now a project of the AES Standards Committee, designated X210, and the AES and OCA Alliance are working together to achieve common standards.

The AES64 message structure is as follows:

IP Header: this includes source and destination Internet Protocol (IP) address (unicast, broadcast or multicast).
UDP (User Datagram Protocol) Header: includes source and destination port
(IANA reserved port number 7107).
AES64 Message.

Table 14.6 AES64-2012 Message Structure

Destination Device ID (Source device node ID [total 32 bits]128 bits)			
Destination Device Node ID (32bits)			
Source Device ID (128 bits)			
Source Device Node ID (32 bits)			
Source Parameter ID (32 bits)			
User Level (8 bits)	Message Type (8 bits)	Sequence ID (32 bits)	
Sequence ID, continued		Command Execute (8 bits)	Command Qualifier (8 bits)
Section Block (8 bits)	Section Type (8 bits)	Section Number (24 bits)	
	AES X 170 Message Address Block		
Section No. continued	Parameter Block (8 bits)	Parameter Block Index (24 bits)	
Parameter block index contd	Parameter Type (16 bits)		Parameter Index (16 bits)
Parameter Index contd	Value Format (VALFMT)	Value (variable length)	

The AES message structure has a number of components, as shown in Table 14.6. Each source and destination device has a unique 128-bit ID, comprising:

1. An 8-byte Extended Unique Identifier (EUI64) as registered with IEEE-RA.
2. A 3-byte Organizationally Unique Identifier (OUI) for the manufacturer of the network interface as registered with IEEE-RA.
3. A 2-byte device ID that is unique amongst a manufacturer's products.
4. A 3-byte reserve field.

Every parameter within a device has its own parameter ID comprising a 32-bit value to give unambiguous control of gain, EQ adjustments, routing and other such actions. The User Level informs a parameter of the user's authority to make a change to that parameter. Some Message Types will be requests, some will be responses. Some will contain hierarchical information describing a parameter action, others will contain parameter IDs. The Sequence ID relates each response message with its request message, important when a large number of commands are being transmitted. Command Execute and Command Qualifier involve both the sending of commands to a receiver and the qualifying of the type of command relevant to that action.

AES X170, the area in Table 14.6 enclosed within the heavy lines, is an IP-based protocol for device control, monitoring and configuration over a network, and it forms the address block. It has a seven-level hierarchy:

1. Section block: the highest functional group. A device has a number of these, for example input, output and group sections.
2. Section type: a subgroup within the section block, identifying components within it, such as which input, and which output.
3. Section number: indicates an interface or channel number as appropriate.
4. Parameter block: identifies parameter groups, such as a group of equalizers or a routing section that controls a channel.
5. Parameter block index: differentiates similar components within a parameter block, for example different sections of an eq.
6. Parameter type: indicates the type of parameter, for example eq, gain, or routing.
7. Parameter type index: gives accurate information regarding which of a choice of similar parameters is to be adjusted where there may be ambiguity, such as 'coarse gain' or 'fine gain'.

Where messages are sent to simple devices not requiring or not able to respond to a full set of commands, dummy values are used. Each device has a unique mapping table containing a local index, the device ID and its IP address.

Types of message

Every AES64 message will either be a command message sent from one device to another or a response message once a command has been received. Not all commands require responses. After the Message Type, which can be a full Address Block, an Indexed Message or a Response Message, two other fields indicate the type of command: Command

Executive, and Command Qualifier. The Executive indicates the basic action to be taken, for example 'GET' or 'SET' data values, perform an action – 'ACT' – or create a structure such as a list – 'CREATE'. Command Qualifiers define precisely what action is to be taken: for example the exact number of decibels up or down; the routing; whether the command pertains to single channels or groups of channels; and the appropriate labeling and indication of an action be it a dB scale, an on/off, delay value, or another type of indication.

Table 14.7 gives a brief description of other commands that AES64 is capable of.

EuCon

Avid EuCon is an example of a computer control system which has been operating in the field for several years. Its users include Pro Tools 9.0 and higher, Media Composer, Cubase, Logic Pro, Nuendo, and others. Traditional-looking digital mixing desks are still very much in use, and the need to access a number of faders and knobs rapidly with the fingers rather than control via a mouse and keypad on a computer screen is an important ergonomic consideration in the live sound industry, and also in studios for certain types of recording and mixing sessions. The EuCon system can be used to create control messages from a computer, but an early design aim, however, was also to generate control signals from a digital mixing desk or a control surface designed for operation like one, the operation of faders, EQ and aux knobs, routing and all other functions creating the appropriate control signals to be sent via a computer link. The C++ programming language is used to communicate with other devices using TCP/IP (Transmission Control Protocol/Internet Protocol) via Ethernet, IEEE-1394 or USB connection. The conventional-looking mixer therefore controls what is going on in the computer (the 'box'), the actual audio signals and processing being contained and implemented in the latter. In-the-box/out-of-the-box mixing issues are covered further in Chapter 9. With the EuCon system a number of Control surfaces can be used to control a common Application (the latter in the context of EuCon referring to the audio processing project going on in the computer or workstation) which enables several operators to concentrate on their particular areas of interest, something that has become important in complex audio/visual projects for instance. Also, a single Control surface can be used to control several Applications. Original design aims included high control resolution across thousands of operations with low latency. The following gives a brief overview of some of the details of the EuCon format.

Table 14.7	Some AES64 Features
Feature	**Function**
Snapshot	A set of commands to be saved by a receiving device, the controller sending a 'SAVE SNP' message to instruct it to do so, complete with its unique ID. A future instruction 'SET VAL' with the unique ID attached will recall those commands in the receiver.
Parameter	Each parameter has a number of flags indicating the states the parameters can occupy. A 0 or 1 within a 32-bit flag register indicates 'not set' or 'set'. These flags can include such things as 'select' which will control a certain parameter within a selection group; 'read' which can determine whether a parameter can be read or not; 'write' which determines whether a parameter can be written to or not; and various others such as value protect, isolate, or lock.
'Pushing'	There are situations where the density of information to be handled is unnecessarily high for a requirement, taking up a lot of bandwidth. For example, a large number of meter readings which change frequently can be relayed back to the controlling device, necessitating the sending of much information. The 'push' facility rationalizes the data, allowing a restricted number of meter readings to be used along with a set interval of time between each update of the readings.
Grouping	An application of this can be a VCA-type of grouping parameter where all of the faders controlling the level of instruments in a band need to be moved in relation to each other for overall level control. A master-slave group would operate in this way. A peer-to-peer group however would mean that altering the fader which controls a particular instrument will also alter the faders controlling the other instruments by the same amount.
Modifiers	There will be circumstances where commands sent from a controller will need to be modified to fit the receiving device, or control it in a certain desired way. For instance, a limited number of fader movement commands can be modified to control a large number of movements on the receiving device. The Modifier can be used to assign the original channel numbers to appropriate destinations, and/or control them to a scale and at a rate that is different from the original command. An Event Modifier allows an initial trigger event command to set in motion a series of other events over a period of time, and also to modify such parameters as time intervals between events or whether a particular event is to be triggered or not. Automated mixing is an obvious application of the Event Modifier.
Desk items	As the name applies, these are simply representations on the computer screen, or physical controls on a traditional-style console, of traditional mixer facilities: faders, knobs, switches, buttons. They enable read and write access to the controls to generate the AES64 data to send and receive commands which will be understood by all devices. A large amount of data is needed to describe all aspects of a large console's controls, and these are stored as an Extensible Markup Language (XML) file.

Node objects

Control surfaces and Applications are represented by individual node objects, and all processing actions that a particular Control surface or Application is capable of is registered with its associated node. There is one node per control surface, and one node for each Application; several of the latter can reside in a computer or DAW. All nodes are registered uniquely with a EuCon Discovery distributed database so that any of them on the network can be requested and located to facilitate connection between Control and Application, the database being kept current via the TCP/IP network connection. When a Control surface has located an Application, it proceeds to map its controls to those of the Application in order to coordinate all possible actions, a process called *assignment*. If a new plug-in or other device is added to or removed from the Application at a later stage, an appropriate control knob and label will appear or be deleted on the Control surface.

Protocol

EuCon's protocol is object-orientated, an object in this context being a specific control function such as fader, aux send, routing path or meter; this simplifies programming, and also facilitates object grouping such as all faders, or all routing. These objects have a containment hierarchy as follows.

1. Primitives. Twelve basic controls: knob, fader, switch, meter, LED, bitmap, joystick, graph, wheel, trackball, text display, automation status.

    1.1. Primitive values: 32 bit integer number, 32 bit floating.

2. Controls. A control is an object that contains one or more primitives. For instance, it is desirable to group both a switch and its associated LED into a single control command. A 'knob' control might contain Primitives such as the knob itself, text display, touch switch, automation switch, and function LED.

    2.1. Control Array: a control that contains other controls, in two types. One is a Switch Array which can contain any number of switches across channels. The other is a Knob Cell Array which contains knobs and any associated switches such as pre/post or aux send on/off throughout the control surface.

Processor types

Individual processor objects are grouped together as processor *types* according to a hierarchy rule, a logical procedure which eases programming. Processor types include:

- Channel Strip – groups together traditional input channel objects in a logical order, and arranges channels appropriately on the control surface with number 1 to the left, etc. If there are more channels in the Application than are available on the control surface, channel numbers can be scrolled through, or channel strips on the control surface can be assigned to the channel number in the application.

- Command – Application-defined set of switches and menu choices which appear on the control surface as soft keys on a screen. These include such things as 'file', 'edit', 'undo', 're-do', 'cut', and 'copy'.

- Transport – contains traditional transport control buttons along with text displays for such things as edit locate points and time code. As well as physical button assignment, a series of soft keys can be created on the screen for inclusion of additional Application-specific requirements.

- Edit controller – contains a jog wheel and jog/shuttle mode switches. The latter can assign the jog wheel function to any of the Application's functions, examples including horizontal and vertical zooming, waveform zooming, trimming of clip head and tail, cross fade length, and adjustment of gain for edited and inserted sections.

- Monitor – collects together the traditional control room requirements such as volume, dim, speaker select, and monitor source selection. It will also support any specific Application requirement.

- Project – a slightly confusing term which refers to an array of marked points along an Application which can be located using a row of switches. Operators can thereby locate their specific 'projects' within the Application instantly using one of these buttons.

- System – a collection of global functions such as mute all, clear mute, clear solo or solo in place (SIP), automation mode, and user preferences.

Summary

When hard disk drive editing first appeared in the 1980s, manipulation of a project by a mouse and keypad on a computer screen was a quite adequate way of working, given the basic nature of early systems and the small number of recording tracks available. As the computer formats grew in complexity and the number of recording tracks began to outstrip anything that was available with analog multitrack recorders, ergonomic considerations became more pressing. The digital mixing console is essentially a conventional control surface combined with a computer which carries the audio processing, the two occupying the same box. The use of multi-function knobs and screens facilitate efficient organization and operation whilst maintaining a tactile relationship between machine and operator. Some digital mixers also contain disk drive-based audio recorders for the recording of projects. As computer editing formats have developed, it has been recognized that this relationship continues to be important, not just in the fields of live sound and mixing but also in the recording and audio-visual fields. A logical development is therefore a control surface which provides a conventional and familiar working tool without carrying audio signals, recording or processing. It merely generates and receives control signals which can be transmitted and received via a computer network to and from a computer, workstation, or another control surface. The physical separation of control and processing has provided a very flexible and versatile framework within which the industry can develop, and the next few years should hopefully see the achievement of agreed standards for network control across the industry.

RECOMMENDED FURTHER READING

Hewlett, W., Selfridge-Field, E. (Eds.), 2001. The Virtual Score: Representation, Retrieval, Restoration. MIT Press.

Huber, D., 2007. The MIDI Manual, third edition. Focal Press.

MMA, 1999. Downloadable Sounds Level 1. V1.1a, January. MIDI Manufacturers Association.

MMA, 2000. RP-027: MIDI Media Adaptation Layer for IEEE 1394. MIDI Manufacturers Association.

MMA, 2000. RP-029: Bundling SMF and DLS Data in an RMID file. MIDI Manufacturers Association.

MMA, 2001. XMF Specification Version 1.0. MIDI Manufacturers Association.

MMA, 2002. The Complete MIDI 1.0 Detailed Specification. MIDI Manufacturers Association.

MMA, 2002. Scalable Polyphony MIDI Specification and Device Profiles. MIDI Manufacturers Association.

Rumsey, F., 2004. Desktop Audio Technology. Focal Press.

Scheirer, D., Vercoe, B., 1999. SAOL: the MPEG 4 structured audio orchestra language. Computer Music Journal 23(2), 31–51.

Selfridge-Field, E., Byrd, D., Bainbridge, D., 1997. Beyond MIDI: The Handbook of Musical Codes. MIT Press.

USB Implementers Forum, 1996. USB Device Class Definition for MIDI Devices, version 1.0. Available from http://www.usb.org.

USEFUL WEBSITES

Audio Engineering Society standards: http://www.aes.org/standards.

MIDI Manufacturers Association: http://www.midi.org.

Music XML: http://www.musicxml.org.

Open Control Architecture: http://www.oca-alliance.com.

Open Sound Control: http://opensoundcontrol.org.

Synchronization

In the following chapter the basics of timecode and synchronization are discussed. In the days of analog recording the need for synchronization of audio signals was not obvious, whereas it has always been an issue for video systems. This was because analog recordings were not divided up into samples, blocks or frames that had to happen at specific instances in time – they were time-continuous entities with no explicit time structure. There was nonetheless a requirement to synchronize the speeds of recording and replay machines in some cases, particularly when it became necessary to run them alongside video machines, or to lock two analog recorders together. This was essentially what was meant by machine sychronization, and SMPTE/EBU timecode of some sort, based on video timing structures, was usually used as a timing reference. A form of this timecode is still used as a positional reference in digital audio and video systems, and a MIDI equivalent is also possible, as described below.

With digital audio and video, the use of signal synchronization is unavoidable. For example, in order to handle multiple streams of either type of signal in a mixer or recording system it is usually necessary for them to be running at the same speed, having the same sampling frequency, and often with their frames, blocks or samples aligned in time. If not, all sorts of problems can arise, ranging from complete lack of function to errors, clicks and speed errors. In order to transfer audio from one machine to another the machines generally have to be operating at the same sampling frequency, and may need to be locked to a common reference. At the very least the receiving device needs to be able to lock to the sending device's sample clock. In such cases, timecode is not usually adequate as a reference signal and a more accurate clock signal that relates to digital audio samples is required. In many cases timecode and sample frequency synchronization go hand in hand, the timecode providing a positional reference and the sample or word clock providing a fine-grained reference point for the individual audio samples.

With most production and post-production now being done on computer workstations, and with digital audio, video and MIDI often being run on the same computer, there is perhaps less need for timecode-based machine synchronization. Applications can be kept in sync by means of internal protocols that require less user intervention and understanding, or audio applications like Pro Tools and Logic contain options for handling QuickTime video files and keeping them in sync with audio. However there are still advantages in understanding the principles that underpin video and audio synchronization. It is also still quite common for an audio workstation to be locked to an external video player, in which case timecode and a video sync reference will need to be used, connected by means of a suitable sync interface. Deriving a stable audio sample clock from the video reference is also important if the master clock for an entire system is a video sync reference.

SMPTE/EBU TIMECODE

The American Society of Motion Picture and Television Engineers proposed a system to facilitate the accurate editing of video tape in 1967. This became known as SMPTE ('simpty') code, and it is basically a continuously running eight-digit clock registering time from an arbitrary start point (which may be the time of day) in hours, minutes, seconds and frames, against which the program runs. The clock information was encoded into a signal which could be recorded on the audio track of a tape. Every single frame on a particular video tape had its own unique number called the timecode address and this could be used to pinpoint a precise editing position.

FACT FILE 15.1 DROP-FRAME TIMECODE

When color TV (NTSC standard) was introduced in the USA it proved necessary to change the frame rate of TV broadcasts slightly in order to accommodate the color information within the same spectrum. The 30 fps of monochrome TV, originally chosen so as to lock to the American mains frequency of 60 Hz, was thus changed to 29.97 fps, since there was no longer a need to maintain synchronism with the mains owing to improvements in oscillator stability. In order that 30 fps timecode could be made synchronous with the new frame rate it became necessary to drop two frames every minute, except for every tenth minute, which resulted in minimal long-term drift between timecode and picture (75 ms over 24 hours). The drift in the short-term gradually increased towards the minute boundaries and was then reset.

A flag is set in the timecode word to denote NTSC drop-frame timecode. This type of code should be used for all applications where the recording might be expected to lock to an NTSC video program.

A number of timecode frame rates are used, depending on the television standard to which they relate, the frame rate being the number of still frames per second used to give the impression of continuous motion: 30 frames per second (fps), or true SMPTE, was used for monochrome American television and for CD mastering in the Sony 1630 format (see Chapter 9); 29.97 fps is used for color NTSC television (mainly USA, Japan and parts of the Middle East), and is called 'SMPTE drop-frame' (see Fact File 15.1); 25 fps is used for PAL and SECAM TV and is called 'EBU' (Europe, Australia, etc.); and 24 fps is used for some film work.

Each timecode frame is represented by an 80 bit binary 'word', split principally into groups of 4 bits, with each 4 bits representing a particular parameter such as tens of hours, units of hours, and so forth, in BCD (binary-coded decimal) form (see Figure 15.1). Sometimes, not all 4 bits per group are required – the hours only go up to '23', for example – and in these cases the remaining bits are either used for special control purposes or set to zero (unassigned): 26 bits in total are used for time address information to give each frame its unique hours, minutes, seconds, frame value; 32 are 'user bits' and can be used for encoding information such as reel number, scene number, day of the month and the like; bit 10 can denote drop-frame mode if a binary 1 is encoded there, and bit 11 can denote color frame mode if a binary 1 is encoded. The end of each word consists of 16 bits in a unique sequence, called the 'sync word', and this is used to mark the boundary between one frame and the next. It also allows the reader to tell in which direction the code is being read, since the sync word begins with 11 in one direction and 10 in the other.

If this data is to be recorded as an audio signal it is modulated in a simple scheme known as 'bi-phase mark', or FM, such that a transition from

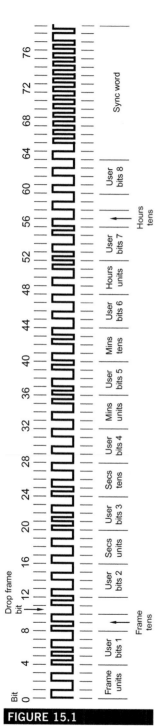

FIGURE 15.1

The data format of an SMPTE/EBU longitudinal timecode frame.

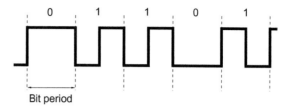

FIGURE 15.2 *Linear timecode data is modulated before recording using a scheme known as 'bi-phase mark' or FM (frequency modulation). A transition from high to low or low to high occurs at every bit-cell boundary, and a binary '1' is represented by an additional transition within a bit cell.*

one state to the other (low to high or high to low) occurs at the edge of each bit period, but an additional transition is forced within the period to denote a binary 1 (see Figure 15.2). The result looks like a square wave with two frequencies, depending on the presence of ones and zeros in the code. Depending on the frame rate, the maximum frequency of square wave contained within the timecode signal is either 2400 Hz (80 bits × 30 fps) or 2000 Hz (80 bits × 25 fps), and the lowest frequency is either 1200 Hz or 1000 Hz, and thus it may easily be recorded on an audio machine. The code can be read forwards or backwards, and phase inverted. Readers are available which will read timecode over a very wide range of speeds, from around 0.1 to 200 times play speed. The rise-time of the signal, that is the time it takes to swing between its two extremes, is specified as $25\mu s \pm 5\mu s$, and this requires an audio bandwidth of about 10 kHz.

There is another form of timecode known as VITC (Vertical Interval Timecode), used widely in VTRs. VITC is recorded not on an audio track, but in the vertical sync period of a video picture, such that it can always be read when video is capable of being read, such as in slow-motion and pause modes.

RECORDING TIMECODE

In the days of tape, timecode was recorded or 'striped' onto tape as an audio signal before, during or after the program material was recorded, depending on the application. In the case of disk-based digital systems the timecode is not usually recorded as an audio signal, but a form of time-stamping can be used to indicate the start times of audio files, from which an ongoing timecode can be synthesized, locked to the sampling frequency. The Broadcast WAVE file format, for example, has an option to store origination time and sample count since midnight, as shown in Chapter 10. The timecode should be locked to the same speed reference as that used to lock the speed of the tape machine or the sampling frequency of a digital system, otherwise a long-term drift can build up between one and the other. Such a reference is usually provided in the form of a video composite sync signal (or black and burst signal), and video sync inputs are often provided on digital tape recorders for this purpose. An alternative is to use a digital audio word clock signal, and this should also be locked to video syncs if they are present.

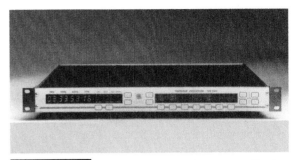

FIGURE 15.3 *A stand-alone timecode generator. (Courtesy of Avitel Electronics Ltd.)*

Timecode generators are available in a number of forms, either as stand-alone devices (such as that pictured in Figure 15.3), as part of a synchronizer or editor, or as part of a recording system. In large centers timecode is sometimes centrally distributed and available on a jackfield point. When generated externally, timecode normally appears as an audio signal on an XLR connector or jack, and this should be routed to the timecode input of any slave systems (slaves are devices expected to lock to the master timecode). Most generators allow the user to preset the start time and the frame-rate standard.

On tape timecode was often recorded onto an outside track of a multitrack machine (usually track 24), or a separate timecode or cue track was provided on digital audio or video machines. The signal was recorded at around 10 dB below reference level, and crosstalk between tracks or cables was often a problem due to the very audible mid-frequency nature of timecode. Some quarter-inch analog machines had a facility for recording timecode in a track which runs down the center of the guard band in the NAB track format (see Chapter 6). This was called 'center-track timecode', and a head arrangement similar to that shown in Figure 15.4 was used for recording and replay. Normally separate heads were used for recording timecode to those for audio, to avoid crosstalk, although some manufacturers circumvented this problem and used the same heads. In the former case a delay line was used to synchronize timecode and audio on the tape.

Professional R-DAT machines were capable of recording timecode, this being converted internally into a DAT running-time code which was recorded in the subcode area of the digital recording. On replay, any frame rate of timecode could be derived, no matter what was used during recording, which was useful in mixed-standard environments.

FIGURE 15.4

The center-track timecode format on quarter-inch tape. (a) Delays are used to record and replay a timecode track in the guard band using separate heads. (Alternatively, specially engineered combination heads may be used.) (b) Physical dimensions of the center-track timecode format.

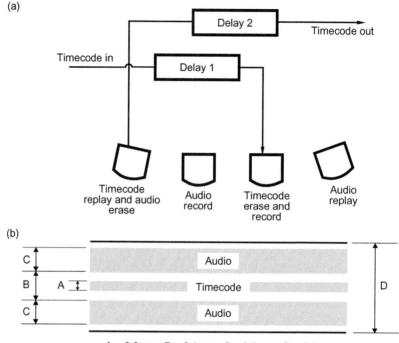

(a)

(b)

A = 0.8 mm, B = 2.1 mm, C = 2.0 mm, D = 6.3 mm

In mobile film and video work which often employs separate machines for recording sound and picture it is necessary to stripe timecode on both the camera's tape or film and on the audio tape. This can be done by using the same timecode generator to feed both machines, but more usually each machine will carry its own generator and the clocks will be synchronized at the beginning of each day's shooting, both reading absolute time of day. Highly stable crystal control ensures that sync between the clocks will be maintained throughout the day, and it does not then matter whether the two (or more) machines are run at different times or for different lengths of time because each frame has a unique time of day address code which enables successful post-production syncing. Clapper boards can still be used to provide a clear audible and visual sync point between the sound and video recordings, so that they can be matched up later.

The code should run for around 20 seconds or more before the program begins in order to give other machines and computers time to lock in. If the program is spread over several reels, the timecode generator should be set and run such that no number repeats itself anywhere throughout the reels, thus avoiding confusion during post-production. Alternatively the reels can be separately numbered.

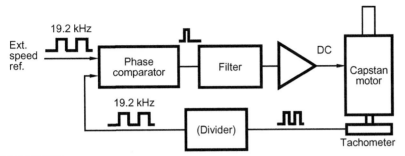

FIGURE 15.5 *Capstan speed control is often effected using a servo circuit similar to this one. The frequency of a square wave pulse generated by the capstan tachometer is compared with an externally generated pulse of nominally the same frequency. A signal based on the difference between the two is used to drive the capstan motor faster or slower.*

MACHINE SYNCHRONIZERS

A machine synchronizer was a device that read timecode from two or more tape machines and controlled the speeds of 'slave' machines so that their timecodes ran at the same rate as the 'master' machine. It did this by modifying the capstan speed of the slave machines, using an externally applied speed reference signal, usually in the form of a 19.2 kHz square wave whose frequency is used as a reference in the capstan servo circuit (see Figure 15.5). Such a synchronizer would be microprocessor controlled, and could incorporate offsets between the master and slave machines, programmed by the user. Often it would store pre-programmed points for such functions as record drop-in, drop-out, looping and autolocation, for use in post-production. Now that dedicated audio tape recorders are not so common these synchronization functions are often found as integral features of digital workstation software or dedicated disk recorders.

DIGITAL AUDIO SYNCHRONIZATION

Requirements for digital audio synchronization

Unlike analog audio, digital audio has a discrete-time structure, because it is a sampled signal in which the samples may be further grouped into frames and blocks having a certain time duration. If digital audio devices are to communicate with each other, or if digital signals are to be combined in any way, then they need to be synchronized to a common reference in order that the sampling frequencies of the devices are identical and do not drift with relation to each other. It is not enough for two devices to be running at nominally the same sampling frequency (say, both at 44.1 kHz). Between the sampling clocks of professional audio equipment it is possible for differences in frequency of up to ±10 parts per million (ppm) to exist and even a very slow drift means that two devices are not truly synchronous. Consumer devices can exhibit an even greater range of sampling frequencies that are nominally the same.

The audible effect resulting from a non-synchronous signal drifting with relation to a sync reference or another signal is usually the occurrence of a glitch or click at the difference frequency between the signal and the reference, typically at an audio level around 50 dB below the signal, due to the repetition or dropping of samples. This will appear when attempting to mix two digital audio signals whose sampling rates differ by a small amount, or when attempting to decode a signal such as an unlocked consumer source by a professional system which is locked to a fixed reference. This said, it is not always easy to detect asynchronous operation by listening, even though sample slippage is occurring, as it depends on the nature of audio signal at the time. Some systems may not operate at all if presented with asynchronous signals.

Furthermore, when digital audio is used with analog or digital video, the sampling rate of the audio needs to be locked to the video reference signal and to any timecode signals which may be used. In single studio operations the problem of ensuring lock to a common clock is not as great as it is in a multi-studio center, or where digital audio signals arrive from remote locations. In distributed system cases either the remote signals must be synchronized to the local sample clock as they arrive, or the remote studio must somehow be fed with the same reference signal as the local studio.

Digital audio signal synchronization

In all-digital systems it is necessary for there to be a fixed sampling frequency, to which all devices in the system lock. This is so that digital audio from one device can be transferred directly to others without conversion to analog or loss of quality, or so that signals from different sources can be processed together. In systems involving video it is often necessary for the digital audio sampling frequency to be locked to the video frame rate and for timecode to be locked to this as well. The relationship between audio sampling rates and video frame rates is discussed in Fact File 15.2.

In very simple digital audio systems it is possible to use one device in the system, such as a mixing console, to act as the sampling frequency reference for the other devices. For example, many digital audio devices will lock to the sample clock contained in AES-3-format signals (see

FACT FILE 15.2 RELATIONSHIPS BETWEEN VIDEO FRAME RATES AND AUDIO SAMPLING RATES

People using the PAL or SECAM television systems are fortunate in that there is a simple integer relationship between the sampling frequency of 48 kHz used in digital audio systems for TV and the video frame rate of 25 Hz (there are 1920 samples per frame). There is also a simple relationship between the other standard sampling frequencies of 44.1 and 32 kHz and the PAL/SECAM frame rate. Users of NTSC TV systems (such as the USA and Japan) are less fortunate because the TV frame rate is 30/1.001 (roughly 29.97) frames per second, resulting in a non-integer relationship with standard audio sampling frequencies. The sampling frequency of 44.056 kHz was introduced in early digital audio recording systems that used NTSC VTRs, as this resulted in an integer relationship with the frame rate. A similar issue can arise with film frame rates, where for example the 24 fps rate of film can be pulled down to 23.98 for a simpler relationship with NTSC video

systems. For a variety of historical reasons it is still quite common to encounter so-called 'pulled-down' sampling frequencies in video environments using the NTSC frame rate, these being 1/1.001 times the standard sampling frequencies. These can cause issues when transferring material to and from workstations running projects at standard sampling frequencies. For example, if doing a telecine transfer from film material to NTSC video, it will probably run the film material at 23.98 fps and it is possible that the corresponding audio sampling rate of imported source material will also be running 0.1% slow. Typical audio editing packages such as ProTools have options for doing audio sample rate conversion during imports, that pull the sampling rate back up to the standard rate if desired, without changing the pitch or length of the material. Thus it can remain in sync with the slowed down 23.98 fps film material while having its sampling rate adjusted.

Chapter 10) arriving at their input. This is sometimes called 'genlock'. This can work if the system primarily involves signal flow in one direction, or is a daisy-chain of devices locked to each other. However, such setups can become problematic when loops are formed and it becomes unclear what is synchronizing what.

Most professional audio equipment has external sync inputs to enable each device to lock to an external reference signal of some kind. Typical sync inputs are word clock (WCLK), which is normally a square-wave TTL-level signal (0–5 V) at the sampling rate, usually available on a BNC-type connector; 'composite video', which is a video reference signal consisting of either normal picture information or just 'black and burst' (a video signal with a blacked-out picture), or a proprietary sync signal such as the optional Alesis sync connection or the LRCK in the Tascam interface (see Chapter 10). HD video systems use a tri-level sync signal that may operate in progressive scan or interlaced modes. WCLK may be 'daisy-chained' (looped through) between devices in cases where the AES/EBU interface is not available. Avid's ProTools system also uses a so-called 'Super Clock' signal at a multiple of 256 times the sampling rate for slaving devices

FIGURE 15.6

In video environments all synchronization signals should be locked to a common clock, as shown here.

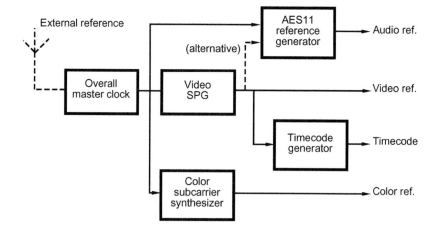

together with low sampling jitter. This is a TTL level (0–5 V) signal on a BNC connector. In all cases one machine or source must be considered to be the 'master', supplying the sync reference to the whole system, and the others as 'slaves'. An alternative to this is a digital audio sync signal such as word clock or an AES-11 standard sync reference (a stable AES-3-format signal, without any audio). Such house sync signals are usually generated by a central sync pulse generator (SPG) that resides in a machine room, whose outputs are widely distributed using digital distribution amplifiers to equipment requiring reference signals. In large systems a central SPG is really the only satisfactory solution. A diagram showing the typical relationship between synchronization signals is shown in Figure 15.6.

For integrating basic audio devices without external clock reference inputs into synchronized digital systems it is possible to employ an external sample frequency convertor that is connected to the digital audio outputs of the device. This convertor can then be locked to the clock reference so that audio from the problematic device can be made to run at the same sampling frequency as the rest of the system.

Sample clock jitter and effects on sound quality

Short-term timing irregularities in sample clocks may affect sound quality in devices such as A/D and D/A convertors and sampling frequency convertors. This is due to modulation in the time domain of the sample instant, resulting in low distortion and noise products within the audio spectrum. This makes it crucial to ensure stable jitter-free clock signals at points in a digital audio system where conversion to and from the analog domain is carried out. In a professional digital audio system, especially in areas where high-quality conversion is required, it is advisable either to re-clock any reference signal or to use a local high-quality reference generator, slaved to the central SPG, with which to clock any A/D or D/A convertors.

TIMECODE AND VIDEO SYNCHRONIZATION FOR AUDIO WORKSTATIONS

Audio applications runing on computer workstations may need to be locked to a source of SMPTE/EBU timecode, because this is used widely as a timing reference in video recording. A number of desktop workstations that have MIDI features lock to MIDI TimeCode (MTC), which is a representation of SMPTE/EBU timecode in the form of MIDI messages (see below). There is also the need to derive a stable audio sample clock from a video sync reference.

External timecode may simply be used as a timing reference against which sound file replay is triggered, or the system may slave to external timecode for the duration of replay. In some cases these modes are switchable because they both have their uses. In the first case replay is simply 'fired off' when a particular timecode is registered, and in such a mode no long-term relationship is maintained between the timecode and the replayed audio. This may be satisfactory for some basic operations but is likely to result in a gradual drift between audio replay and the external reference if files longer than a few seconds are involved. It may be useful though, because replay remains locked to the workstation's internal clock reference, which may be more stable than external references, potentially leading to higher audio quality from the system's convertors. Some cheaper systems do not 'clean up' external clock signals very well before using them as the sample clock for D/A conversion, and this can seriously affect audio quality. In the second mode a continuous relationship is set up between timecode and audio replay, such that long-term lock is achieved and no drift is encountered. This is more difficult to achieve because it involves the continual comparison of timecode to the system's internal timing references and requires that the system follows any drift or jump in the timecode.

Jitter in the external timecode is very common, especially if this timecode derives from a video tape recorder, and this should be minimized in any sample clock signals derived from the external reference. This is normally achieved by the use of a high-quality phase-locked loop, often in two stages. Wow and flutter in the external timecode can be smoothed out using suitable time constants in the software that convert timecode to

FIGURE 15.7 *Avid's Sync HD interface for the Pro Tools system offers a flexible means of locking the computer workstation software to a range of sync sources. (a) Front panel; (b) back panel. (Courtesy of Avid.)*

sample address codes, such that short-term changes in speed are not always reflected in the audio output but longer-term drifts are. Alternatively, since the audio sample/word clock should be locked to the master video sync reference, as should the timecode speed, it is preferable to derive audio conversion clocks from the video sync reference rather than the timecode.

Sample frequency conversion can be employed at the digital audio outputs of a system to ensure that changes in the internal sample rate caused by synchronization action are not reflected in the output sampling rate. This may be required if the system is to be interfaced digitally to other equipment in an all-digital studio.

An example of a common workstation sync interface from Avid is shown in Figure 15.7. It accepts a very wide range of different sync inputs, including LTC, bi-phase and pilot tone (film methods), video, MIDI timecode and AES digital audio, among others. It will accept either standard definition (SD) or high definition (HD) video sync references, the latter being a tri-level sync signal for HD video up to 1080p at 60 frames per second. A highly stable (low jitter) audio sync reference for Pro Tools is derived from the external sync reference.

MIDI AND SYNCHRONIZATION

Introduction to MIDI synchronization

The MIDI interface and protocol can carry synchronization data as well as remote control data for musical instruments and other devices. A sequencer, for example, will need some speed reference to control the rate at which recorded information is replayed and this speed reference could either be internal to the computer or provided by an external device. On the other hand, a normal synthesizer, effects unit or sampler is not usually concerned with timing information, because it has no functions affected by a timing clock. Such devices do not normally store rhythm patterns, although there are some keyboards with onboard sequencers that ought to recognize timing data.

As MIDI equipment has become more integrated with audio and video systems the need has arisen to incorporate timecode handling into the standard and into software. This has allowed sequencers to operate relative either to musical time (e.g. bars and beats) or to 'real' time (e.g. minutes and seconds). Using timecode, MIDI applications can be run in sync with the replay of an external audio or video machine, in order that the long-term speed relationship between the MIDI replay and the machine remains constant. Also relevant to the systems integrator is the MIDI Machine Control standard that specifies a protocol for the remote control of devices such as external recorders using a MIDI interface.

Music-related timing data

This section describes the group of MIDI messages that deals with 'music-related' synchronization – that is, synchronization related to the passing of bars and beats as opposed to 'real' time in hours, minutes and seconds. It is normally possible to choose which type of sync data will be used by a software package or other MIDI receiver when it is set to 'external sync' mode.

A group of system messages called the 'system real-time' messages control the execution of timed sequences in a MIDI system and these are often used in conjunction with the song position pointer (SPP, which is really a system common message) to control autolocation within a stored song. The system real-time messages concerned with synchronization, all of which are single bytes, are:

&F8   Timing clock
&FA   Start
&FB   Continue
&FC   Stop

The timing clock (often referred to as 'MIDI beat clock') is a single status byte (&F8) to be issued by the controlling device six times per MIDI beat. A MIDI beat is equivalent to a musical semiquaver or sixteenth note (see Table 15.1) so the increment of time represented by a MIDI clock byte is related to the duration of a particular musical value, not directly to a unit of real time. Twenty-four MIDI clocks are therefore transmitted per

Table 15.1	Musical durations related to MIDI timing data	
Note value	**Number of MIDI beats**	**Number of MIDI clocks**
Semibreve (whole note)	16	96
Minim (half note)	8	48
Crotchet (quarter note)	4	24
Quaver (eighth note)	2	12
Semiquaver (sixteenth note)	1	6

quarter note, unless the definition is changed. (Some software packages allow the user to redefine the notated musical increment represented by MIDI clocks.) At any one musical tempo, a MIDI beat could be said to represent a fixed increment of time, but this time increment would change if the tempo changed.

The 'start', 'stop' and 'continue' messages are used to remotely control the receiver's replay. A receiver should only begin to increment its internal clock or song pointer after it receives a start or continue message, even though some devices may continue to transmit MIDI clock bytes in the intervening periods. For example, a sequencer may be controlling a number of keyboards, but it may also be linked to a drum machine that is playing back an internally stored sequence. The two need to be locked together, so the sequencer (running in internal sync mode) would send the drum machine (running in external sync mode) a 'start' message at the beginning of the song, followed by MIDI clocks at the correct intervals thereafter to keep the timing between the two devices correctly related. If the sequencer was stopped it would send 'stop' to the drum machine, whereafter 'continue' would carry on playing from the stopped position, and 'start' would restart at the beginning. This method of synchronization appears to be fairly basic, as it allows only for two options: playing the song from the beginning or playing it from where it has been stopped.

SPPs are used when one device needs to tell another where it is in a song. A sequencer or synchronizer should be able to transmit song pointers to other synchronizable devices when a new location is required or detected. For example, one might 'fast-forward' through a song and start again 20 bars later, in which case the other timed devices in the system would have to know where to restart. An SPP would be sent followed by 'continue' and then regular clocks. An SPP represents the position in a stored song in terms of number of MIDI beats (not clocks) from the start of the song. It uses two data bytes so can specify up to 16 384 MIDI beats. SPP is a system common message, not a real-time message. It is often used in conjunction with &F3 (song select), used to define which of a collection of stored song sequences (in a drum machine, say) is to be replayed. SPPs are fine for directing the movements of an entirely musical system, in which every action is related to a particular beat or subdivision of a beat, but not so fine when actions must occur at a particular point in real time. If, for example, one was using a MIDI system to dub music and effects to a picture in which an effect was intended to occur at a particular visual event, that effect would have to maintain its position in time no matter what happened to the music. If the effect was to be triggered by a sequencer at a particular number of beats from the beginning of the song, this point could change in real time if the tempo of the music was altered slightly to fit a particular visual scene. Clearly some means of real-time synchronization is required either instead of, or as well as, the clock and song pointer arrangement, such that certain events in a MIDI-controlled system may be triggered at specific times in hours, minutes and seconds.

Software may also recognize and be able to generate the bar marker and time signature messages. The bar marker message can be used where it is necessary to indicate the point at which the next musical bar begins. It takes effect at the next &F8 clock. Some MIDI synchronizers will also accept an audio input or a tap switch input so that the user can program a tempo track for a sequencer based on the rate of a drum beat or a rate tapped in using a switch. This can be very useful in synchronizing MIDI sequences to recorded music, or fitting music which has been recorded 'rubato' to bar intervals.

MIDI timecode (MTC)

MIDI timecode has two specific functions. First, to provide a means for distributing conventional SMPTE/EBU timecode data (see above) around a MIDI system in a format that is compatible with the MIDI protocol. Second, to provide a means for transmitting 'setup' messages that may be downloaded from a controlling computer to receivers in order to program them with cue points at which certain events are to take place. The intention is that receivers will then read incoming MTC as the program proceeds, executing the pre-programmed events defined in the setup messages. Sequencers and some digital audio systems can use MIDI timecode derived from an external synchronizer or MIDI peripheral when locking to video or to another sequencer. MTC is an alternative to MIDI clocks and song pointers, for use when real-time synchronization is important.

There are two types of MTC synchronizing message: one that updates a receiver regularly with running timecode and another that transmits one-time updates of the timecode position. The latter can be used during high-speed cueing, where regular updating of each single frame would involve too great a rate of transmitted data. The former is known as a quarter-frame message (see Fact File 15.3), denoted by the status byte (&F1), whilst the latter is known as a full-frame message and is transmitted as a universal real-time SysEx message.

FACT FILE 15.3 QUARTER-FRAME MTC MESSAGES

One timecode frame is represented by too much information to be sent in one standard MIDI message, so it is broken down into eight separate messages. Each message of the group of eight represents a part of the timecode frame value, as shown in the figure below, and takes the general form:

&[F1] [DATA]

The data byte begins with zero (as always), and the next 7 bits of the data word are made up of a 3 bit code defining whether the message represents hours, minutes, seconds or frames, MSnibble or LSnibble, followed by the 4 bits representing the binary value of that nibble. In order to reassemble the correct timecode value from the eight quarter-frame messages, the LS and MS nibbles of hours, minutes, seconds and frames are each paired within the receiver to form 8 bit words as follows:

Frames: rrr qqqqq

where 'rrr' is reserved for future use and 'qqqqq' represents the frames value from 0 to 29;

Seconds: rr qqqqqq

where 'rr' is reserved for future use and 'qqqqqq' represents the seconds value from 0 to 59;

Minutes: rr qqqqqq

as for seconds; and

Hours: r qq ppppp

where 'r' is undefined, 'qq' represents the timecode type, and 'ppppp' is the hours value from 0 to 23. The timecode frame rate is denoted as follows in the 'qq' part of the hours value: 00 = 24 fps; 01 = 25 fps; 10 = 30 fps drop-frame; 11 = 30 fps non-drop-frame. Unassigned bits should be set to zero.

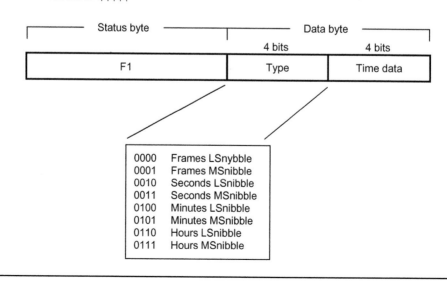

SYNCHRONIZING AUDIO/MIDI/VIDEO COMPUTER APPLICATIONS

It is increasingly common for multiple audio, MIDI and video applications to run simultaneously on the same workstation, and in such cases they may need to run in sync with each other. Furthermore it may be necessary to have audio and MIDI connections between the applications. Some while back a number of these functions were carried out by proprietary 'middleware' such as the Opcode Music System (OMS), but a more recent system that handles many of these functions as well as full audio routing and synchronization between applications is 'Rewire'. This technology, developed by Propellerhead Systems, enables the internal real-time streaming of up to 256 audio channels and 4080 MIDI channels between applications, as well as high precision inter-application synchronization and the communication of common transport commands such as stop, play, record and rewind. One application is designated as the master and others as slaves.

Apple's Core Audio system, for example, includes comprehensive synchronization facilities for applications, handled by the Core Audio Clock API. It can generate clock references in a variety of formats such as SMPTE time, audio sample time and musical bar/beat time, and it can convert clocks between those formats. These clocks can be locked to 'hardware time' such as the computer's system clock or one derived from an

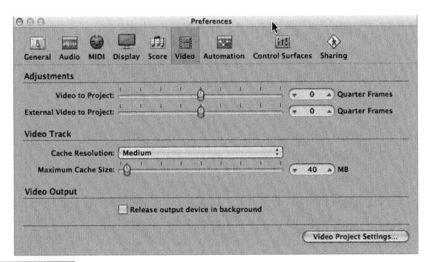

FIGURE 15.8   *Logic's video preferences allow an offset to be adjusted between the start of the video and the audio tracks.*

external interface. Applications such as Logic can choose whether to lock to an internal form of MIDI timecode, to MTC triggers or to external sync sources. Apple's Core Video shares the same timebase references as Core Audio, and similar time stamp data, so application developers can use this commonality to ensure that audio and video applications remain in sync with each other. QuickTime video can be loaded into Logic, for example, and played in sync with the audio using these invisible resources. The timing offset between the start of the video and the audio can be adjusted in a preference window to account for buffering delays or other causes of sync differences between the two, as shown in Figure 15.8. The SMPTE start time of the video may need to be entered into Logic to ensure that it matches up with the appropriate point in the audio track.

RECOMMENDED FURTHER READING

Ratcliff, J., 1999. Timecode: A User's Guide, third edition. Focal Press.

Two-Channel Stereo

This chapter covers the principles and practice of two-channel stereophonic recording and reproduction. Two-channel stereophonic reproduction (in international standard terms '2-0 stereo', meaning two front channels and no surround channels) is often called simply 'stereo' as it is the most common way of conveying some spatial content in sound recording and reproduction. In fact 'stereophony' refers to any sound system that conveys three-dimensional sound images, so it is used more generically in this book and includes surround sound. In international standards describing stereo loudspeaker configurations the nomenclature for the configuration is often in the form 'n–m stereo', where n is the number of front channels and m is the number of rear or side channels (the latter only being encountered in surround systems). This distinction can be helpful as it reinforces the slightly different role of the surround channels as explained in the next chapter. (Readers familiar with earlier editions of this book should note that the broadcasting nomenclature of 'A' and 'B', referring to the left and right signals of a stereo pair, has been replaced in this edition by 'L' and 'R'. This is in order to avoid any possible confusion with the American tradition of referring to spaced microphone pairs as 'AB pairs', as contrasted with 'XY' for coincident microphones.)

It might reasonably be supposed that the best stereo sound system would be that which reproduced the sound signal to the ears as faithfully as possible, with all the original spatial cues intact (see Chapter 2). Possibly that should be the aim, and indeed it is the aim of the so-called 'binaural' techniques discussed later in the chapter, but there are many stereo techniques that rely on loudspeakers for reproduction which only manage to provide some of the spatial cues to the ears. Such techniques are compromises that have varying degrees of success, and they are necessary for the simple reason that they are reasonably straightforward from a recording point of view and result in subjectively high sound quality. The results can be reproduced in anyone's living room and are demonstrably better than mono (single-channel reproduction). Theoretical correctness is one thing, pragmatism and getting a 'commercial sound' is another. The history of stereo could be characterized as being something of a compromise between the two.

Stereo techniques cannot be considered from a purely theoretical point of view, neither can the theory be ignored, the key being in a proper synthesis of theory and subjective assessment. Some techniques which have been judged subjectively to be good do not always stand up to rigorous theoretical analysis, and those which are held up as theoretically 'correct' are sometimes judged subjectively to be poorer than others. Part of the problem is that the mechanisms of spatial perception are not yet entirely understood. Probably more importantly, most commercial stereo reproduction uses only two loudspeakers so the listening situation already represents a serious departure from natural spatial hearing. (Real sonic experience involves sound arriving from all around the head.) The differences between two-channel stereo reproduction and natural listening may lead listeners to prefer 'distorted' sound fields because of other pleasing artefacts such as 'spaciousness'. Most of the stereo techniques used today combine aspects of imaging accuracy with an attempt to give the impression of spaciousness in the sound field, and to some theorists these two are almost mutually exclusive.

It would be reasonable to surmise that in most practical circumstances, for mainstream consumer applications, one is dealing with the business of creating believable illusions. Sound recording is as much an art as a science. In other words, one needs to create the impression of natural spaces, source positions, depth, size and so on, without necessarily being able to replicate the exact sound pressure and velocity vectors that would be needed at each listening position to recreate a sound field accurately. One must remember that listeners rarely sit in the optimum listening position, and often like to move around whilst listening. Whilst it may be possible to achieve greater spatial accuracy using headphone reproduction, head-phones are not always a practical or desirable form of monitoring. Truly accurate soundfield reconstruction covering a wide listening area can only be achieved by using very large numbers of loudspeakers (many thousands) and this is likely to be impractical for most current purposes.

In the following chapters stereo pickup and reproduction is considered from both a theoretical and a practical point of view, recognizing that theoretical rules may have to be bent or broken for operational and subjective reasons. Since the subject is far too large even to be summarized in the short space available, a list of recommended further reading is given at the end of the chapter to allow the reader greater scope for personal study.

PRINCIPLES OF LOUDSPEAKER STEREO

Historical development

We have become used to stereo sound as a two-channel format, although a review of developments during the last century shows that two channels really only became the norm through economic and domestic necessity, and through the practical considerations of encoding directional sound easily for gramophone records and radio. A two-loudspeaker arrangement is practical in the domestic environment, is reasonably cheap to implement, and provides good phantom images for a central listening position.

Early work on directional reproduction undertaken at Bell Labs in the 1930s involved attempts to recreate the 'sound wavefront' which would result from an infinite number of microphone/loudspeaker channels by using a smaller number of channels, as shown in Figure 16.1a and 16.1b. In all cases, spaced pressure response (omnidirectional) microphones were used, each connected via a single amplifier to the appropriate loudspeaker in the listening room. Steinberg and Snow found that when reducing the number of channels from three to two, central sources appeared to recede towards the rear of the sound stage and that the width of the reproduced sound stage appeared to be increased. They attempted to make some calculated rather than measured deductions about the way that loudness differences between the channels affected directional perception, apparently choosing to ignore the effects of time or phase difference between channels.

Some 20 years later Snow made comment on those early results, reconsidering the effects of time difference in a system with a small number

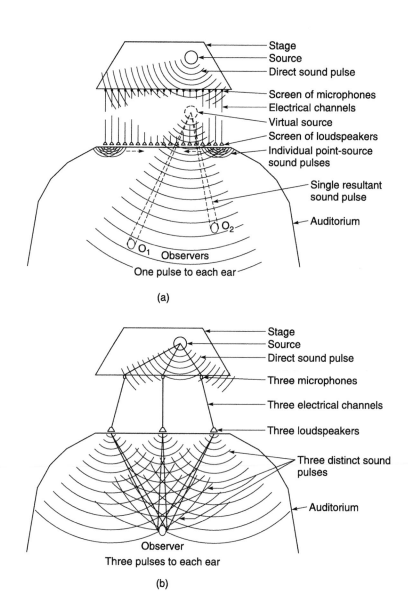

Steinberg and Snow's attempt to reduce the number of channels needed to convey a source wavefront to a reproduction environment with appropriate spatial features intact. (a) 'Ideal' arrangement involving a large number of transducers. (b) Compromise arrangement involving only three channels, relying more on the precedence effect.

of channels, since, as he pointed out, there was in fact a marked difference between the multiple-point-source configuration and the small-number-of-channels configuration. It was suggested that in fact the 'ideal' multi-source system re-created the original wavefront very accurately, allowing the ears to use exactly the same binaural perception mechanisms as used in the real-life sound field. The 'wall' of multiple loudspeakers acted as a source of spherical wavelets, re-creating a new plane wave with its virtual source in the same relative place as the original source, thus resulting in a time-of-arrival difference between the listener's ears in the range $0 - 600\mu s$, depending on source and listener position. (This is the basis of more recent developments in 'wave field synthesis', developed at the University of Delft in the Netherlands, which also relies on large numbers of closely spaced channels to reconstruct sound fields accurately.)

In the two- or three-channel system, far from this simply being a sparse approximation to the 'wavefront' system, the ears are subjected to two or three discrete arrivals of sound, the delays between which are likely to be in excess of those normally experienced in binaural listening. In this case, the effect of directionality relies much more on the precedence effect and on the relative levels of the channels. Snow therefore begs us to remember the fundamental difference between 'binaural' situations and what he calls 'stereophonic' situations (see Fact File 16.1).

This difference was also recognized by Alan Blumlein, whose now-famous patent specification of 1931 (accepted 1933) allows for the conversion of signals from a binaural format suitable for spaced pressure microphones to a format suitable for reproduction on loudspeakers. His patent also covers other formats of pickup which result in an approximation of the original time and phase differences at the ears when reproduced on loudspeakers. This will be discussed in more detail later on, but it is interesting historically to note how much writing on stereo reproduction even in the early 1950s appears unaware of Blumlein's most valuable work, which appears to have been ignored for some time.

A British paper presented by Clark, Dutton and Vanderlyn (of EMI) in 1957 revives the Blumlein theories, and shows in more rigorous mathematical detail how a two-loudspeaker system may be used to create an accurate relationship between the original location of a sound source and its perceived location on reproduction. This is achieved by controlling only the relative signal amplitudes of the two loudspeakers (derived in this case from a pair of coincident figure-eight microphones). The authors discuss the three-channel system of Bell Labs, and suggest that although it produces convincing results in many listening situations it is uneconomical for domestic use. They also conclude that the two-channel simplification (using microphones spaced about 10 feet apart) has a tendency to result in a 'hole-in-the-middle' effect (with which many modern users of spaced microphones may be familiar – sources appearing to bunch towards the left or the right leaving a hole in the center). They concede that the Blumlein method adapted by them does not take advantage of all the mechanisms of binaural hearing, especially the precedence effect, but that they have

FACT FILE 16.1 BINAURAL VERSUS 'STEREOPHONIC' LOCALIZATION

There is a distinct difference between the spatial perception that arises when two ears detect a single wavefront (i.e. from a single source) and that which arises when two arrivals of a similar sound come from different directions and are detected by both ears. The former, shown at (a), gives rise to spatial perceptions based primarily on what is known as the 'binaural delay' (essentially the time-of-arrival difference that arises between the ears for the particular angle of incidence). The latter, shown at (b), gives rise to spatial perceptions based primarily on various forms of 'precedence effect' (or 'law of the first wavefront'). In terms of sound reproduction, the former may be encountered in the headphone presentation context where sound source positions may be implied by using delays between the ear signals within the interaural delay of about 0.65 ms. Headphones enable the two ears to be stimulated independently of each other.

In loudspeaker listening the precedence effect is more relevant, as a rule. The precedence effect is primarily a feature of transient sounds rather than continuous sounds. In this case there are usually at least two sound sources in different places, emitting different versions of the same sound, perhaps with a time or amplitude offset to provide directional information. This is what Snow termed the 'stereophonic' situation. Both ears hear both loudspeakers and the brain tends to localize based on the interaural delay arising from the earliest arriving wavefront, the source appearing to come from a direction towards that of the earliest arriving signal. This effect operates over delays between the sources that are somewhat greater than the interaural delay, of the order of a few milliseconds. Similar sounds arriving within up to 50 ms of each other tend to be perceptually fused together, such that one is not perceived as an echo of the other. The time delay over which this fusing effect obtains depends on the source, with clicks tending to separate before complex sounds like music or speech. The timbre and spatial qualities of this 'fused sound', though, may be affected.

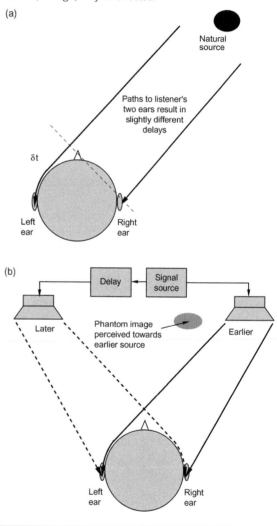

There is therefore a historical basis for both the spaced microphone arrangement which makes use of the time-difference precedence effect (with only moderate level differences between channels), as well as the coincident microphone technique (or any other technique which results in only level differences between channels). There is also some evidence to show that the spaced technique is more effective with three channels than with only two. Later, we shall see that spaced techniques have a fundamental theoretical flaw from a point of view of 'correct' imaging of continuous sounds, which has not always been appreciated, although such techniques may result in subjectively acceptable sounds. Interestingly, three front channels are the norm in cinema sound reproduction, since the central channel has the effect of stabilizing the important central image for off-center listeners, having been used ever since the Disney film *Fantasia* in 1939. (People often misunderstood the intentions of Bell Labs in the 1930s, since it is not generally realized that they were working on a system suitable for auditorium reproduction with wide-screen pictures, as opposed to a domestic system.)

Creating phantom images

Based on a variety of formal research and practical experience, it has become almost universally accepted that the optimum configuration for two-loudspeaker stereo is an equilateral triangle with the listener located just to the rear of the point of the triangle (the loudspeaker forming the baseline). Wider than this, phantom images (the apparent locations of sound sources in between the loudspeakers) become less stable, and the system is more susceptible to the effects of head rotation. This configuration gives rise to an angle subtended by the loudspeakers of ±30° at the listening position, as shown in Figure 16.2. In most cases stereo reproduction from two loudspeakers can only hope to achieve a modest illusion of three-dimensional spatiality, since reproduction is from the front quadrant only.

The so-called 'summing localization' model of stereo reproduction suggests that the best illusion of phantom sources between the loudspeakers will be created when the sound signals present at the two ears are as similar as possible to those perceived in natural listening, or at least that a number of natural localization cues that are non-contradictory are available. It is possible

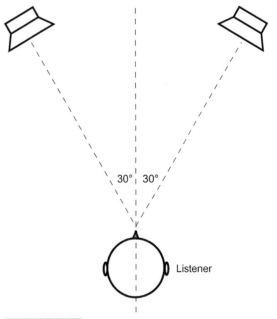

FIGURE 16.2 *Optimum arrangement of two loudspeakers and listener for stereo listening.*

to create this illusion for sources in the angle between the loudspeakers using only amplitude differences between the loudspeakers, where the time difference between the signals is very small (<<1 ms). To reiterate an earlier point, in loudspeaker reproduction both ears receive the signals from both speakers, whereas in headphone listening each ear only receives one signal channel. The result of this is that the loudspeaker listener seated in a center seat (see Figure 16.3) receives at his or her left ear the signal from the left speaker first followed by that from the right speaker, and at his or her right ear the signal from the right speaker first followed by that from the left speaker. The time δt is the time taken for the sound to travel the extra distance from the more distant speaker.

The basis on which 'level difference' or 'Blumlein' stereo works is to use level differences between two loudspeakers to generate low-frequency phase differences between the ears, based on the summation of the loud-speaker signals at the two ears, as described in Fact File 16.2. Depending on which author one believes, an amplitude difference of between 15 and 18 dB between the channels is needed for a source to be panned either fully left or fully right. A useful summary of experimental data on this issue has been drawn by Hugonnet and Walder and is shown in Figure 16.4. A coincident

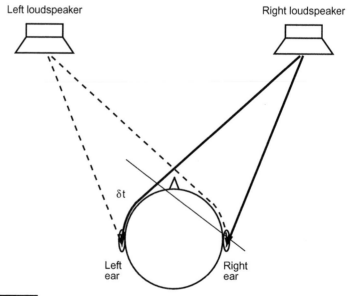

FIGURE 16.3 *An approximation to the situation that arises when listening to sound from two loudspeakers. Both ears hear sound from both loudspeakers, the signal from the right loudspeaker being delayed by δt at the left ear compared with the time it arrives at the right ear (and reversed for the other ear).*

FACT FILE 16.2 STEREO VECTOR SUMMATION

If the outputs of the two speakers differ only in amplitude and not in phase (time) then it can be shown (at least for low frequencies up to around 700 Hz) that the vector summation of the signals from the two speakers at each ear results in two signals that, for a given frequency, differ in phase angle proportional to the relative amplitudes of the two signals (the level difference between the ears being negligible at LF). For a given level difference between the speakers, the phase angle changes approximately linearly with frequency, which is the case when listening to a real point source. At higher frequencies the phase difference cue becomes largely irrelevant but the shadowing effect of the head results in level differences between the ears. If the amplitudes of the two channels are correctly controlled it is possible to produce resultant phase and amplitude differences for continuous sounds that are very close to those experienced with natural sources, thus giving the impression of virtual or 'phantom' images anywhere between the left and right loudspeakers. This is the basis of Blumlein's (1931) stereophonic system 'invention' although the mathematics is quoted by Clark, Dutton and Vanderyn (1957) and further analyzed by others. The result of the mathematical phasor analysis is a simple formula which can be used to determine, for any angle subtended by the loudspeakers at the listener, what the apparent angle of the virtual image will be for a given difference between left and right levels.

First, referring to the diagram, it can be shown that:

$$\sin\alpha = (L - R)/(L + R)\sin\theta_0$$

where α is the apparent angle of offset from the center of the virtual image, and θ_0 is the angle subtended by the speaker at the listener. Second, it can be shown that:

$$(L - R)/(L + R) = \tan\theta_t$$

where θ_t is the true angle of offset of a real source from the center-front of a coincident pair of figure-eight velocity microphones. $(L - R)$ and $(L + R)$ are the well-known difference (S) and sum (M) signals of a stereo pair, defined below.

This is a useful result since it shows that it is possible to use positioning techniques such as 'pan-potting' which rely on the splitting of a mono signal source into two components, with adjustment of the relative proportion fed to the left and right channels without affecting their relative timing. It also makes possible the combining of the two channels into mono without cancelations due to phase difference.

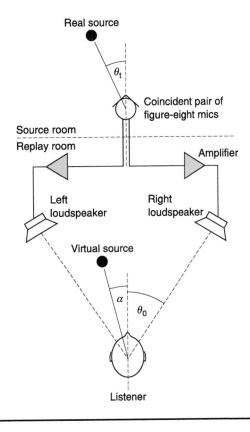

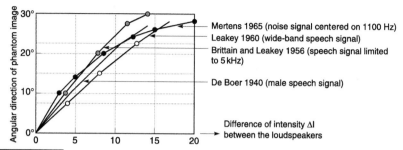

FIGURE 16.4 *A summary of experimental data relating to amplitude differences (here labeled intensity) required between two loudspeaker signals for a particular phantom image location (data compiled by Hugonnet and Walder, 1995). (Courtesy of Christian Hugonnet).*

arrangement of velocity (figure-eight) microphones at 90° to one another produce outputs which differ in amplitude with varying angle over the frontal quadrant by an amount which gives a very close correlation between the true angle of offset of the original source from the center line and the apparent angle on reproduction, assuming loudspeakers which subtend an angle of 120° to the listening position. This angle of loudspeakers is not found to be very satisfactory for practical purposes for reasons such as the tendency to give rise to a 'hole' in the middle of the image. At smaller loudspeaker angles the change in apparent angle is roughly proportionate as a fraction of total loudspeaker spacing, maintaining a correctly proportioned 'sound stage', so the sound stage with loudspeakers at the more typical 60° angle will tend to be narrower than the original sound stage but still in proportion.

If a time difference also exists between the channels, then transient sounds will be 'pulled' towards the advanced speaker because of the precedence effect, the perceived position depending to some extent on the time delay. If the left speaker is advanced in time relative to the right speaker (or more correctly, the right speaker is delayed!) then the sound appears to come more from the left speaker, although this can be corrected by increasing the level to the right speaker. A delay somewhere between 0.5 and 1.5 ms is needed for a signal to appear fully left or fully right at ±30°, depending on the nature of the signal (see Figure 16.5, after Hugonnet and Walder). With time-difference stereo, continuous sounds may give rise to contradictory phantom image positions when compared with the position implied by transients, owing to the phase differences that are created between the channels. Cancelations may also arise at certain frequencies if the channels are summed to mono.

Combinations of time and level difference can also be used to create phantom images, as described in Fact File 16.3.

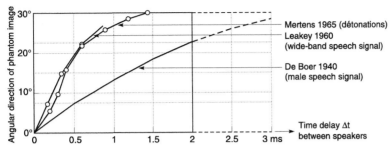

FIGURE 16.5 *A summary of experimental data relating to time differences required between two loudspeaker signals for a particular phantom image location (Hugonnet and Walder, 1995). (Courtsey of Christian Hugonnet).*

FACT FILE 16.3 THE 'WILLIAMS CURVES'

Stereo microphone technique relies on either interchannel level or time difference or a combination of the two. A trade-off is possible between them, although the exact relationship between time and level differences needed to place a source in a certain position is disputed by different authors and seems to depend to some extent on the source characteristics. Michael Williams has based an analysis of microphone arrays on some curves of such

trade-offs which have generally become known as the 'Williams curves', shown below. These curves represent the time and level difference combinations that may be used between two loudspeakers at ±30° in a typical listening room to obtain certain phantom source positions. The data points marked with circles were determined by a Danish researcher, Simonsen, using speech and maracas for signals.

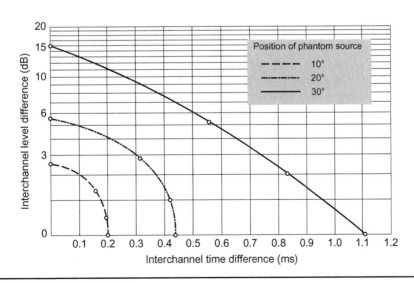

PRINCIPLES OF BINAURAL OR HEADPHONE STEREO

Binaural recording has fascinated researchers for years but it has received very little commercial attention until recently. Part of the problem has been that it is actually very difficult to get it to work properly for a wide range of listeners over a wide range of different headphone types, and partly it is related to the limited compatibility between headphone and loudspeaker listening. Conventional loudspeaker stereo is acceptable on headphones to the majority of people, although it creates a strongly 'in-the-head' effect, but binaural recordings do not sound particularly good on loudspeakers unless some signal processing is used, and the stereo image is dubious.

Recent technical developments have made the signal processing needed to synthesize binaural signals and deal with the conversion between headphone and loudspeaker listening more widely available at reasonable cost. It is now possible to create 3D directional sound cues and to synthesize the acoustics of virtual environments quite accurately using digital signal processors (DSP), and it is this area of virtual environment simulation for computer applications that is receiving the most commercial attention for binaural technology today. Flight simulators, computer games, virtual reality applications and architectural auralization are all areas that are benefiting from these developments.

Basic binaural principles

Binaural approaches to spatial sound representation are based on the premise that the most accurate reproduction of natural spatial listening cues will be achieved if the ears of the listener can be provided with the same signals that they would have experienced in the source environment or during natural listening. In a sense, all stereo reproduction is binaural, but the term is normally taken to mean an approach involving source signals that represent individual ear signals and independent-ear reproduction (such as can be achieved using headphones). Most of the approaches described so far in this chapter have related to loudspeaker reproduction of signals that contain some of the necessary information for the brain to localize phantom images and perceive a sense of spaciousness and depth. Much reproduced sound using loudspeakers relies on a combination of accurate spatial cues and believable illusion. In its purest form, binaural reproduction aims to reproduce all the cues that are needed for accurate spatial perception, but in practice this is something of a tall order and various problems arise.

An obvious and somewhat crude approach to binaural audio is to place two microphones, one at the position of each ear in the source environment, and to reproduce these signals through headphones to the ears of a listener, as shown in Figure 16.6. For binaural reproduction to work well, the HRTFs of sound sources from the source (or synthesized) environment must be accurately re-created at the listener's ears upon reproduction. This means capturing the time and frequency spectrum differences between the two ears accurately. Since each source position results in a unique HRTF, rather like a fingerprint, one might assume that all that is needed is to ensure the listener hears this correctly on reproduction.

Tackling the problems of binaural systems

The primary problems in achieving an accurate reconstruction of spatial cues can be summarized as follows:

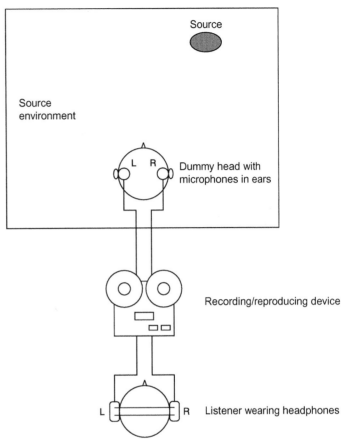

FIGURE 16.6 *Basic binaural recording and reproduction.*

- People's heads and ears are different (to varying degrees), although there are some common features, making it difficult to generalize about the HRTFs that should be used for commercial systems that have to serve lots of people (see above).

- Head movements that help to resolve directional confusion in natural listening are difficult to incorporate in reproduction situations.

- Visual cues are often missing during binaural reproduction and these normally have a strong effect on perception.

- Headphones differ in their equalization and method of mounting, leading to distortions in the perceived HRTFs on reproduction.

- Distortions such as phase and frequency response errors in the signal chain can affect the subtle cues required.

It has been possible to identify the HRTF features that seem to occur in the majority of people and then to create generalized HRTFs that work reasonably well for a wide range of listeners. It has also been found that some people are better at localizing sounds than others, and that the HRTFs of so-called 'good localizers' can be used in preference to those of 'poor localizers'. To summarize, it can be said that although a person's own HRTFs provide them with the most stable and reliable directional cues, generalized functions can be used at the expense of absolute accuracy of reproduction for everyone.

The problem of head movements can be addressed in advanced systems by using head tracking to follow the listener's actions and adapt the signals fed to the ears accordingly. This is generally only possible when using synthesized binaural signals that can be modified in real time. The issue of the lack of visual cues commonly encountered during reproduction can only be resolved in full 'virtual reality' systems that incorporate 3D visual information in addition to sound information. In the absence of visual cues, the listener must rely entirely on the sound cues to resolve things like front–back confusions and elevation/distance estimations.

The issue of headphone equalization is a thorny one as it depends on the design goal for the headphones. Different equalization is required depending on the method of recording, unless the equalization of both ends of the chain is standardized. For a variety of reasons, a diffuse field form of equalization for headphones, dummy heads and synthesized environments has generally been found preferable to free-field equalization. This means that the system is equalized to have a flat response to signals arriving from all angles around the head when averaged in a diffuse sound field. Headphones equalized in this way have been found to be quite suitable for both binaural and loudspeaker stereo signals, provided that the binaural signals are equalized in the same way.

Distortions in the signal chain which can affect the timing and spectral information in binaural signals have been markedly reduced since the introduction of digital audio systems. In the days of analog signal chains and media such as Compact Cassette and LP records, numerous opportunities existed for interchannel phase and frequency response errors to arise, making it difficult to transfer binaural signals with sufficient integrity for success.

LOUDSPEAKER STEREO OVER HEADPHONES AND VICE VERSA

Bauer showed that if stereo signals designed for reproduction on loudspeakers were fed to headphones there would be too great a level difference between the ears compared with the real-life situation, and that the correct interaural delays would not exist. This results in an unnatural stereo image that does not have the expected sense of space and appears to be inside the head. He therefore proposed a network which introduced a measure of delayed crosstalk between the channels to simulate the correct interaural level differences at different frequencies, as well as simulating the interaural time delays which would result from the loudspeaker signals incident at 45° to the listener. He based the characteristics on research done by Weiner which produced graphs for the effects of diffraction around the human head for different angles of incidence. The characteristics of Bauer's circuit are shown in Figure 16.7 (with Weiner's results shown dotted). It may be seen that Bauer chooses to reduce the delay at HF, partially because the circuit design would have been too complicated, and partially because localization relies more on amplitude difference at HF anyway.

Bauer also suggests the reverse process (turning binaural signals into stereo signals for loudspeakers). He points out that crosstalk must be removed between binaural channels for correct loudspeaker reproduction, since the crossfeed between the channels will otherwise occur twice (once between the pair of binaurally spaced microphones, and again at the ears of the listener), resulting in poor separation and a narrow image. He suggests that this may be achieved using the subtraction of an anti-phase component of each channel from the other channel signal, although he does not discuss how the time difference between the binaural channels may be removed. Such processes are the basis of 'transaural stereo' (see Fact File 16.4).

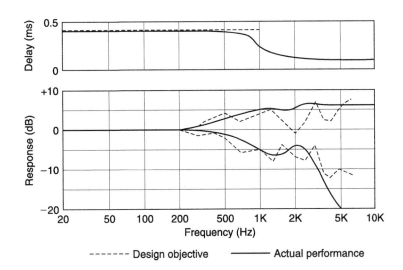

FIGURE 16.7

Bauer's filter for processing loudspeaker signals so that they could be reproduced on headphones. The upper graph shows the delay introduced into the crossfeed between channels. The lower graph shows the left and right channel gains needed to imitate the shadowing effect of the head.

FACT FILE 16.4 TRANSAURAL STEREO

When binaural signals are replayed on loudspeakers there is crosstalk between the signals at the two ears of the listener which does not occur with headphone reproduction. The right ear hears the left channel signal a fraction of a second after it is received by the left ear, with an HRTF corresponding to the location of the left loudspeaker, and vice versa for the other ear. This prevents the correct binaural cues from being established at the listener's ears and eliminates the possibility for full 3D sound reproduction. Binaural stereo tends to sound excessively narrow at low frequencies when replayed on loudspeakers as there is very little difference between the channels which has any effect at a listener's ears. Furthermore the spectral characteristics of binaural recordings can create timbral inaccuracies when reproduced over loudspeakers unless some form of compromise equalization is used.

If the full 3D cues of the original binaural recording are to be conveyed over loudspeakers, some additional processing is required. If the left ear is to be presented only with the left channel signal and the right ear with the right channel signal then some means of removing the interaural crosstalk is required. This is often referred to as crosstalk canceling or 'transaural' processing. Put crudely, transaural crosstalk-canceling systems perform this task by feeding an anti-phase version of the left channel's signal into the right channel and vice versa, filtered and delayed according to the HRTF characteristic representing the crosstalk path, as shown above.

The effect of this technique can be quite striking, and in the best implementations enables fully three-dimensional virtual sources to be perceived, including behind

the listener (from only two loudspeakers located at the front). Crosstalk-canceling filters are usually only valid for a very narrow range of listening positions. Beyond a few tens of centimeters away from the 'hot spot' the effect often disappears almost completely. The effect is sometimes perceived as unnatural, and some listeners find it fatiguing to listen to for extended periods.

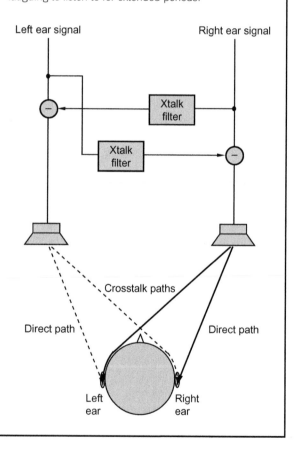

The idea that unprocessed binaural signals are unsuitable for loudspeaker reproduction has been challenged by Theile. He claims that the brain is capable of associating 'head-related' differences between loudspeakers with appropriate spatial cues for stereo reproduction, provided the timbral quality of head-related signals is equalized for a natural-sounding spectrum (e.g. diffuse field equalization, as described above). This theory has led to a variety of companies and recording engineers experimenting with the use of dummy heads such as the Neumann KU100 for generating loudspeaker signals, and created the idea for the Schoeps 'Sphere' microphone described below.

'Spatial equalization' has been proposed by Griesinger to make binaural recordings more suitable for loudspeaker reproduction. He suggested low-frequency difference channel (L–R) boost of about 15 dB at 40 Hz (to increase the LF width of the reproduction) coupled with overall equalization for a flat frequency response in the total energy of the recording to preserve timbral quality. This results in reasonably successful stereo reproduction in front of the listener, but the height and front–back cues are not preserved.

TWO-CHANNEL SIGNAL FORMATS

The two channels of a 'stereo pair' represent the left (L) and the right (R) loudspeaker signals. It is conventional in broadcasting terminology to refer to the left channel of a stereo pair as the 'A' signal and the right channel as the 'B' signal, although this may cause confusion to some who use the term 'AB pair' to refer specifically to a spaced microphone pair. In the case of some stereo microphones or systems the left and right channels are called respectively the 'X' and the 'Y' signals, although some people reserve this convention specifically for coincident microphone pairs. Here we will stick to using L and R for simplicity. In color coding terms (for meters, cables, etc.), particularly in broadcasting, the L signal is colored red and the R signal is colored green. This may be confusing when compared with some domestic hi-fi wiring conventions that use red for the right channel, but it is the same as the convention used for port and starboard on ships. Furthermore there is a German DIN convention which uses yellow for L and red for R.

It is sometimes convenient to work with stereo signals in the so-called 'sum and difference' format, since it allows for the control of image width and ambient signal balance. The sum or main signal is denoted 'M' and is based on the addition of L and R signals. The difference or side signal is denoted 'S' and is based on the subtraction of R from L to obtain a signal which represents the difference between the two channels (see below). The M signal is that which would be heard by someone listening to a stereo program in mono, and thus it is important in situations where the mono listener must be considered, such as in broadcasting. Color-coding convention in broadcasting holds that M is colored white, whilst S is colored yellow, but it is sometimes difficult to distinguish between these two colors on certain meter types leading to the increasing use of orange for S.

Two-channel stereo signals may be derived by many means. Most simply, they may be derived from a pair of coincident directional microphones orientated at a fixed angle to each other. Alternatively they may be derived from a pair of spaced microphones, either directional or non-directional, with an optional third microphone bridged between the left and right channels. Finally stereo signals may be derived by splitting one or more mono signals into two by means of a 'pan-pot'. A pan-pot is simply a dual-ganged variable resistor that controls the relative proportion of the mono signal being fed to the two legs of the stereo pair, such that as the level to the left side is increased that to the right side is decreased.

MS or 'sum and difference' format signals may be derived by conversion from the LR format using a suitable matrix (see Fact File 3.6 in Chapter 3) or by direct pickup in that format. For every stereo pair of signals it is possible to derive an MS equivalent, since M is the sum of L and R, whilst S is the difference between them. Likewise, signals may be converted from MS to LR formats using the reverse process. Misalignment of signals in either format leads to audible effects as described in Fact File 16.5. In order to convert an LR signal into MS format it is necessary to follow some simple rules. First, the M signal is not usually a simple sum of L and R, as this will result in overmodulation of the M channel in the case where a maximum level signal exists on both L and R (representing a central image). A correction factor is normally applied, ranging between $-3\,dB$ and $-6\,dB$ (equivalent to a division of the voltage by between $\sqrt{2}$ and 2 respectively):

$$\text{e.g. } M = (L + R) - 3\,dB \text{ or } (L + R) - 6\,dB$$

The correction factor will depend on the nature of the two signals to be combined. If identical signals exist on the L and R channels (representing 'double mono' in effect), then the level of the uncorrected sum channel (M) will be two times (6 dB) higher than the levels of either L or R. This requires a correction of $-6\,dB$ in the M channel in order for the maximum level of the M signal to be reduced to a satisfactory level. If the L and R signals are non-coherent (random phase relationship), then only a 3 dB rise in the level of M will result when L and R are summed, requiring the $-3\,dB$ correction factor to be applied. This is more likely with stereo music signals. As most stereo material has a degree of coherence between the channels, the actual rise in level of M compared with L and R is likely to be somewhat between the two limits for real program material.

The S signal results from the subtraction of R and L, and is subject to the same correction factor:

$$\text{e.g. } S = (L - R) - 3\,dB \text{ or } (L - R) - 6\,dB$$

FACT FILE 16.5 STEREO MISALIGNMENT EFFECTS

Differences in level, frequency response and phase may arise between signals of a stereo pair, perhaps due to losses in cables, misalignment and performance limitations of equipment. It is important that these are kept to a minimum for stereo work, as inter-channel anomalies result in various audible side-effects. Differences will also result in poor mono compatibility. These differences and their effects are discussed below.

Frequency Response and Level

A difference in level or frequency response between L and R channels will result in a stereo image biased towards the channel with the higher overall level or that with the better HF response. Also, an L channel with excessive HF response compared with that of the R channel will result in the apparent movement of sibilant sounds towards the L loudspeaker. Level and response misalignment on MS signals results in increased crosstalk between the equivalent L and R channels, such that if the S level is too low at any frequency the LR signal will become more monophonic (width narrower),

and if it is too high the apparent stereo width will be increased.

Phase

Inter-channel phase anomalies will affect one's perception of the positioning of sound source, and it will also affect mono compatibility. Phase differences between L and R channels will result in 'comb-filtering' effects in the derived M signal due to cancellation and addition of the two signals at certain frequencies where the signals are either out-of- or in-phase.

Crosstalk

It was stated earlier that an inter-channel level difference of only 18dB was required to give the impression of a signal being either fully left or fully right. Crosstalk between L and R signals is not therefore usually a major problem, since the performance of most audio equipment is far in excess of these requirements. Excessive crosstalk between L and R signals will result in a narrower stereo image, whilst excessive crosstalk between M and S signals will result in a stereo image increasingly biased towards one side.

S can be used to reconstruct L and R when matrixed in the correct way with the M signal (see below), since $(M + S) = 2L$ and $(M - S) = 2R$. It may therefore be appreciated that it is possible at any time to convert a stereo signal from one format to the other and back again.

TWO-CHANNEL MICROPHONE TECHNIQUES

This section contains a review of basic two-channel microphone techniques, upon which many spatial recording techniques are based. Panned spot microphones are often mixed into the basic stereo image created by such techniques.

Coincident-pair principles

The coincident-pair incorporates two directional capsules that may be angled over a range of settings to allow for different configurations and operational requirements. The pair can be operated in either the LR (sometimes known as 'XY') or MS modes (see above), and a matrixing unit is sometimes supplied with microphones which are intended to operate in the MS mode in order to convert the signal to LR format for recording. The directional patterns (polar diagrams) of the two microphones need not necessarily be figure-eight, although if the microphone is used in the MS mode the S capsule must be figure-eight (see below). Directional information is encoded solely in the level differences between the capsule outputs, since the two capsules are mounted physically as close as possible. There are no phase differences between the outputs except at the highest frequencies where inter-capsule spacing may become appreciable in relation to the wavelength of sound.

Coincident pairs are normally mounted vertically in relation to the sound source, so that the two capsules are angled to point symmetrically left and right of the center of the source stage (see Figure 16.8). The choice of angle depends on the polar response of the capsules used. A coincident pair of figure-eight microphones at 90° provides good correspondence between the actual angle of the source and the apparent position of the virtual image when reproduced on loudspeakers, but there are also operational disadvantages to the figure-eight pattern in some cases, such as the amount of reverberation pickup.

Figure 16.9 shows the polar pattern of a coincident pair using figure-eight mics. First, it may be seen that the fully left position corresponds to the null point of the right capsule's pickup. This is the point at which there will be maximum level difference between the two capsules. The fully left position also corresponds to the maximum pickup of the left capsule but it does not always do so in other stereo pairs. As a sound moves across the sound stage from left to right it will result in a gradually decreasing output from the left mic, and an increasing output from the right mic. Since the microphones have cosine responses, the output at 45° off axis is $\sqrt{2}$ times the maximum output, or 3 dB down in level, thus the takeover between left and right microphones is smooth for music signals. Fact File 16.6 goes into greater detail concerning the relationship between capsule angle and stereo width.

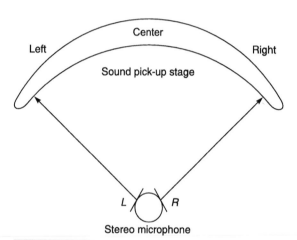

FIGURE 16.8 *A coincident pair's capsules are oriented so as to point left and right of the center of the sound stage.*

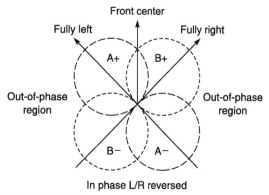

FIGURE 16.9 *Polar pattern of a coincident pair using figure-eight microphones.*

FACT FILE 16.6 STEREO WIDTH ISSUES

With any coincident pair, fully left or fully right corresponds to the null point of pickup of the opposite channel's microphone, although psychoacoustically this point may be reached before the maximum level difference is arrived at. This also corresponds to the point where the M signal equals the S signal (where the sum of the channels is the same as the difference between them). As the angle between the capsules is made larger, the angle between the null points will become smaller, as shown below. Operationally, if one wishes to widen the reproduced sound stage one will widen the angle between the microphones which is intuitively the right thing to do. This results in a narrowing of the angle between fully left and fully right, so sources which had been, say, half left in the original image will now be further towards the left. A narrow angle between fully left and fully right results in a very wide sound stage, since sources have only to move a small

distance to result in large changes in reproduced position. This corresponds to a wide angle between the capsules.

Further coincident pairs are possible using any polar pattern between figure-eight and omni, although the closer one gets to omni, the greater the required angle to achieve adequate separation between the channels. The hypercardioid pattern is often chosen for its smaller rear lobes than the figure-eight, allowing a more distant placement from the source for a given direct-to-reverberant ratio (although in practice hypercardioid pairs tend to be used closer to make the image width similar to that of a figure-eight pair). Since the hypercardioid pattern lies between figure-eight and cardioid, the angle required between the capsules is correctly around 110°.

Psychoacoustic requirements suggest the need for an electrical narrowing of the image at high frequencies in order to preserve the correct angular relationships between low- and high-frequency signals, although this is rarely implemented in practice with coincident pair recording. A further consideration to do with the theoretical versus the practical is that although microphones tend to be referred to as having a particular polar pattern, this pattern is unlikely to be consistent across the frequency range and this will have an effect on the stereo image. Cardioid crossed pairs should theoretically exhibit no out-of-phase region (there should be no negative rear lobes), but in practice most cardioid capsules become more omni at LF and narrower at HF. As a result some out-of-phase components may be noticed in the HF range whilst the width may appear too narrow at LF. Attempts have been made to compensate for this in some stereo microphone designs.

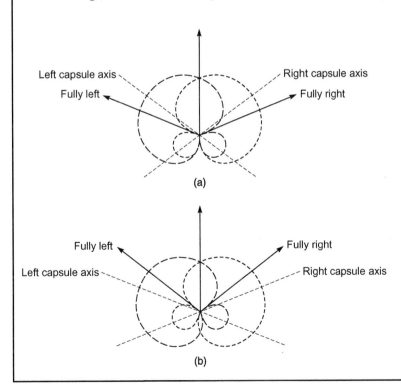

The second point to consider with this pair is that the rear quadrant of pickup suffers a left–right reversal, since the rear lobes of each capsule point in the opposite direction to the front. This is important when considering the use of such a microphone in situations where confusion may arise between sounds picked up on the rear and in front of the mic, such as in television sound where the viewer can also see the positions of sources. The third point is that pickup in both side quadrants results in out-of-phase signals between the channels, since a source further round than 'fully left' results in pickup by both the negative lobe of the right capsule and the positive lobe of the left capsule. There is thus a large region around a crossed pair of figure-eights that results in out-of-phase information, this information often being reflected or reverberant sound. Any sound picked up in this region will suffer cancelation if the channels are summed to mono, with maximum cancelation occurring at 90° and 270°, assuming 0° as the center-front.

The operational advantages of the figure-eight pair are the crisp and accurate phantom imaging of sources, together with a natural blend of ambient sound from the rear. Some cancelation of ambience may occur, especially in mono, if there is a lot of reverberant sound picked up by the side quadrants. Disadvantages lie in the large out-of-phase region, and in the size of the rear pickup which is not desirable in all cases and is left–right reversed. Stereo pairs made up of capsules having less rear pickup may be preferred in cases where a 'drier' or less reverberant balance is required, and where frontal sources are to be favored over rear sources. In such cases the capsule responses may be changed to be nearer the cardioid pattern, and this requires an increased angle between the capsules to maintain good correlation between actual and perceived angle of sources.

The cardioid crossed pair shown in Figure 16.10 is angled at approximately 131°, although angles of between 90° and 180° may be used to good effect depending on the width of the sound stage to be covered. At an angle of 131° a center source is 65.5° off-axis from each capsule, resulting in a 3 dB drop in level compared with the maximum on-axis output (the cardioid mic response is equivalent to $0.5 (1 + \cos \theta)$, where θ is the angle off-axis of the source, and thus the output at 65.5° is $\sqrt{2}$ times that at 0°). A departure from the theoretically correct angle is often necessary in practical situations, and it must be remembered that the listener will not necessarily be aware of the 'correct' location of each source, neither may it matter that

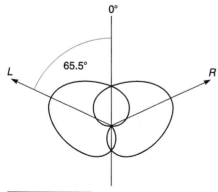

FIGURE 16.10 *A coincident pair of cardioid microphones should theoretically be angled at 131°, but deviations either side of this may be acceptable in practice.*

the true and perceived positions are different. A pair of
'back-to-back' cardioids has often been used to good effect
(see Figure 16.11), since it has a simple MS equivalent
of an omni and a figure-eight, and has no out-of-phase
region. Although the maximum level difference between
the channels is at 90° off-center there will in fact be a satisfactory level difference for a phantom image to appear
fully left or right at a substantially smaller angle than this.

XY or LR coincident pairs in general have the possible disadvantage that central sounds are off-axis to both
mics, perhaps considerably so in the case of crossed cardioids. This may result in a central signal with a poor
frequency response and possibly an unstable image if the
polar response is erratic. Whether or not this is important depends on the importance of the central image in
relation to that of offset images, and will be most impor-

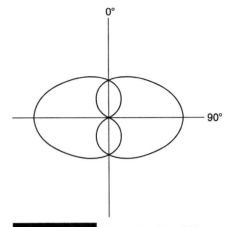

FIGURE 16.11 *Back-to-back cardioids have been found to work well in practice and should have no out-of-phase region.*

tant in cases where the main source is central (such as in television, with
dialog). In such cases the MS technique described in the next section is
likely to be more appropriate, since central sources will be on-axis to the
M microphone. For music recording it would be hard to say whether central sounds are any more important than offset sources, so either technique
may be acceptable.

Using MS processing on coincident pairs

Although some stereo microphones are built specifically to operate in the
MS mode, it is possible to take any coincident pair capable of at least one
capsule being switched to figure-eight, and orientate it so that it will produce suitable signals. The S component (being the difference between left
and right signals) is always a sideways-facing figure-eight with its positive lobe facing left. The M (middle) component may be any polar pattern facing to the center-front, although the choice of M pattern depends
on the desired equivalent pair, and will be the signal that a mono listener
would hear. True MS mics usually come equipped with a control box that
matrixes the MS signals to LR format if required. A control for varying S
gain is often provided as a means of varying the effective acceptance angle
between the equivalent LR pair.

MS signals are not suitable for direct stereo monitoring, they are sum
and difference components and must be converted to a conventional loudspeaker format at a convenient point in the production chain. The advantages of keeping a signal in the MS format until it needs to be converted

will be discussed below, but the major advantage of pickup in the MS format is that central signals will be on-axis to the M capsule, resulting in the best frequency response. Furthermore, it is possible to operate an MS mic in a similar way to a mono mic which may be useful in television operations where the MS mic is replacing a mono mic on a pole or in a boom.

To see how MS and LR pairs relate to each other, and to draw some useful conclusions about stereo width control, it is informative to consider a coincident pair of figure-eight mics again. For each MS pair there is an LR equivalent. The polar pattern of the LR equivalent to any MS pair may be derived by plotting the level of (M + S)/2 and (M − S)/2 for every angle around the pair. Taking the MS pair of figure-eight mics shown in Figure

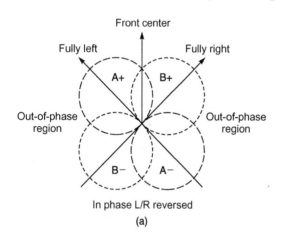

16.12, it may be seen that the LR equivalent is simply another pair of figure-eights, but rotated through 45°. Thus the correct MS arrangement to give an equivalent LR signal where both 'capsules' are oriented at 45° to the center-front (the normal arrangement) is for the M capsule to face forwards and the S capsule to face sideways.

A number of interesting points arise from a study of the LR/MS equivalence of these two pairs, and these points apply to all equivalent pairs. First, fully left or right in the resulting stereo image occurs at the point where S = M (in this case at 45° off-center). This is easy to explain, since the fully left point is the point at which the output from the right capsule is zero. Therefore M = L + 0, and S = L − 0, both of which equal L. Second, at angles of incidence greater than 45° off-center in either direction the two channels become out-of-phase, as was seen above, and this corresponds to the region in which S is greater than M. Third, in the rear quadrant where the signals are in phase again, but left–right reversed, the M signal is greater than S again. The relationship between S and M levels, therefore, is an excellent guide to the phase relationship between the equivalent LR signals. If S is lower than M, then the LR signals will be in phase. If S = M, then the source is

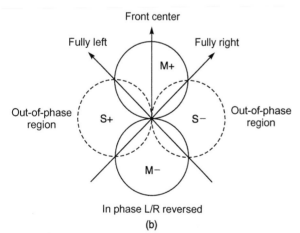

FIGURE 16.12 *Every coincident pair has an MS equivalent. The conventional left–right arrangement is shown in (a), and the MS equivalent in (b).*

either fully left or right, and if S is greater than M, then the LR signals will be out-of-phase.

To show that this applies in all cases, and not just that of the figure-eight pair, look at the MS pair in Figure 16.13 together with its LR equivalent. This MS pair is made up of a forward-facing cardioid and a sideways-facing figure-eight (a popular arrangement). Its equivalent is a crossed pair of hypercardioids, and again the extremes of the image (corresponding to the null points of the LR hypercardioids) are the points at which S equals M. Similarly, the signals go out-of-phase in the region where S is greater than M, and come back in phase again for a tiny angle round the back, due to the rear lobes of the resulting hypercardioids. Thus the angle of acceptance (between fully left and fully right) is really the frontal angle between the two points on the MS diagram where M equals S.

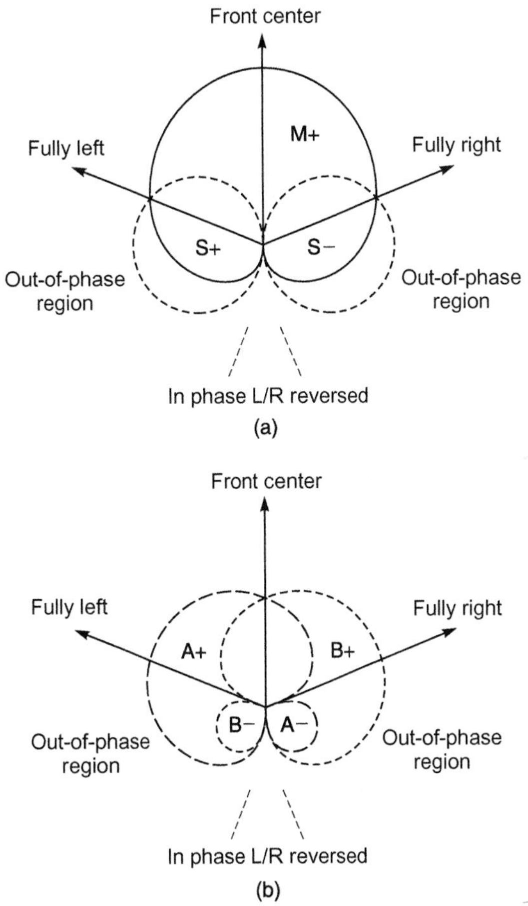

Now, consider what would happen if the gain of the S signal was raised (imagine expanding the lobes of the S figure-eight). The result of this would be that the points where S equaled M would move inwards, making the acceptance angle smaller. As explained earlier, this results in a wider stereo image, since off-center sounds will become closer to the extremes of the image, and is equivalent to increasing the angle between the equivalent LR capsules. Conversely, if the S gain is reduced, the points at which S equals M will move further out from the center, resulting in a

FIGURE 16.13 *The MS equivalent of a forward facing cardioid and sideways figure-eight, as shown in (a), is a pair of hypercardioids whose effective angle depends on S gain, as shown in (b).*

narrower stereo image, equivalent to decreasing the angle between the equivalent LR capsules. This helps to explain why Blumlein-style shufflers work by processing the MS equivalent signals of stereo pairs, as one can change the effective stereo width of pairs of signals, and this can be made frequency dependent if required.

This is neatly exemplified in a commercial example, the Neumann RSM 191i, which is an MS mic in which the M capsule is a forward-facing short shotgun mic with a polar pattern rather like a hypercardioid. The polar pattern of the M and S capsules and the equivalent LR pair is shown in Figure 16.14 for three possible gains of the S signal with relation to

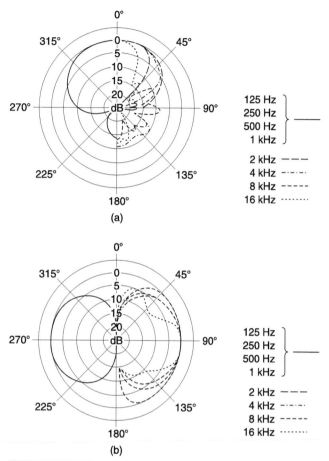

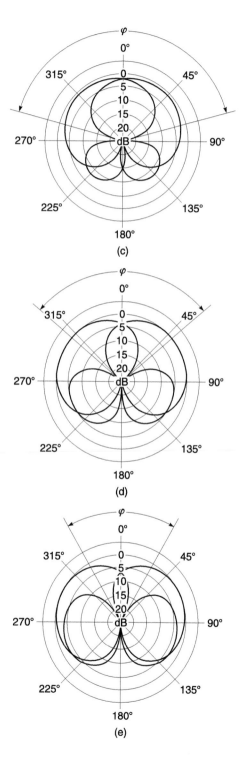

FIGURE 16.14   *Polar patterns on the Neumann RSM191i microphone. (a) M capsule, (b) S capsule, (c) LR equivalent with − 6dB S gain, (d) 0dB S gain, (e) 6dB S gain.*

M(−6 dB, 0 dB and +6 dB). It will be seen that the acceptance angle (M) changes from being large (narrow image) at −6 dB, to small (wide image) at +6 dB. Changing the S gain also affects the size of the ear lobes of the LR equivalent. The higher the S gain, the larger the rear lobes. Not only does S gain change stereo width, it also affects rear pickup, and thus the ratio of direct to reverberant sound.

Any stereo pair may be operated in the MS configuration, simply by orientating the capsules in the appropriate directions and switching them to an appropriate polar pattern, but certain microphones are dedicated to MS operation simply by the physical layout of the capsules (see Fact File 16.7).

FACT FILE 16.7 END-FIRE AND SIDE-FIRE CONFIGURATIONS

There are two principal ways of mounting the capsules in a coincident stereo microphone, be it MS or LR format: either in the 'end-fire' configuration where the capsules 'look out' of the end of the microphone, such that the microphone may be pointed at the source (see the diagram), or in the 'side-fire' configuration where the capsules 'look out' of the sides of the microphone housing. It is less easy to see the direction in which the capsules are pointing in a side-fire microphone, but such a microphone makes it possible to align the capsules vertically above each other so as to be time-coincident in the horizontal plane, as well as allowing for the rotation of one capsule with relation to the other. An end-fire configuration is more suitable for the MS capsule arrangement (see diagram below), since the S capsule may be mounted sideways behind the M capsule, and no rotation of the capsules is required. There is a commercial example of an LR end-fire microphone for television ENG (electronic news gathering) use which houses two fixed cardioids side by side in an enlarged head.

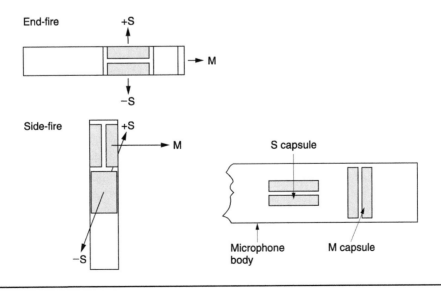

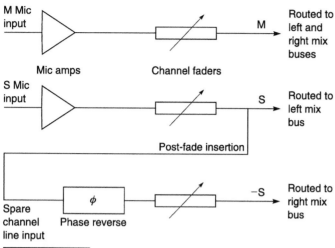

FIGURE 16.15 *An LR mix with variable width can be derived from an MS microphone connected to three channels of a mixer as shown. The S faders should be ganged together and used as a width control.*

Operational considerations with coincident pairs

The control of S gain is an important tool in determining the degree of width of a stereo sound stage, and for this reason the MS output from a microphone might be brought (unmatrixed) into a mixing console, so that the engineer has control over the width. This in itself can be a good reason for keeping a signal in MS form during the recording process, although M and S can easily be derived at any stage using a conversion matrix.

Although some mixers have MS matrixing facilities on board, the diagram in Figure 16.15 shows how it is possible to derive an LR mix with variable width from an MS microphone using three channels on a mixer without using an external MS matrix. M and S outputs from the microphone are fed in phase through two mixer channels and faders, and a post-fader feed of S is taken to a third channel line input, being phase-reversed on this channel. The M signal is routed to both left and right mix buses (panned centrally), whilst the S signal is routed to the left mix box (M + S = 2 L) and the 2 S signal (the phase-reversed version) is routed to the right mix bus (M − S = 2 R). It is important that the gain of the 2 S channel is matched very closely with that of the S channel. (A means of deriving M and S from an LR format input is to mix L and phase-reversed R together to get S, and without the phase reverse to get M.)

Outdoors, coincident pairs will be susceptible to wind noise and rumble, as they incorporate velocity-sensitive capsules which always give more problems in this respect than omnis. Most of the interference will reside in the S channel, since this always has a figure-eight pattern, and thus would not be a problem to the mono listener. Similarly, physical handling of the stereo microphone, or vibration picked up through a stand, will be much more noticeable than with pressure microphones. Coincident pairs should not generally be used close to people speaking, as small movements of their heads can cause large changes in the angle of incidence, leading to considerable movement in their apparent position in the sound stage.

Near-coincident microphone configurations

'Near-coincident' pairs of directional microphones introduce small additional timing differences between the channels which may help in the localization of transient sounds and increase the spaciousness of a recording, and which at the same time remain nominally coincident at low frequencies, giving rise to suitable amplitude differences between the channels. Headphone compatibility is also quite good owing to the microphone spacing being similar to ear spacing. The family of near-coincident (or closely spaced) techniques relies on a combination of time and level differences between the channels which can be traded off for certain widths of sound stage and microphone pattern.

Subjective evaluations often seem to show good results for such techniques. One comprehensive subjective assessment of stereo microphone arrangements, performed at the University of Iowa, consistently resulted in the near-coincident pairs scoring amongst the two favored performers for their sense of 'space' and realism. Critics have attributed these effects to 'phasiness' at high frequencies (which some people may like, nonetheless), and argued that truly coincident pairs were preferable.

A number of examples of near-coincident pairs exist as 'named' arrangements, although there is a whole family of possible near-coincident arrangements using combinations of spacing and angle. Some near-coincident pairs of different types, based on the 'Williams curves' (see Fact File 16.3) are given in Table 16.1. The so-called 'ORTF pair' is an arrangement of two cardioid mics, deriving its name from the organization which first adopted it (the Office de Radiodiffusion-Télévision Française). The two mics are spaced apart by 170 mm, and angled at 110°. The 'NOS' pair (Nederlande Omroep Stichting, the Dutch Broadcasting Company) uses cardioid mics spaced apart by 300 mm and angled at 90°. Figure 16.16 illustrates these two pairs, along with a third pair of figure-eight microphones spaced apart by 200 mm, which

Table 16.1	Some Near-Coincident Pairs Based on the 'Williams curves'			
Designation	**Polar Pattern**	**Mic Angle**	**Spacing**	**Recording Angle**
NOS	Cardioid	±45°	30 cm	80°
RAI	Cardioid	±50°	21 cm	90°
ORTF	Cardioid	±55°	17 cm	95°
DIN	Cardioid	±45°	20 cm	100°
–	Omni	0°	50 cm	130°
–	Omni	0°	35 cm	160°

FIGURE 16.16
Near-coincident pairs
(a) ORTF, (b) NOS,
(c) Faulkner.

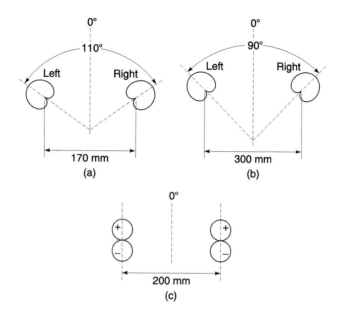

(a)

(b)

(c)

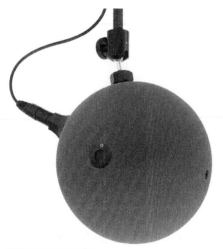

FIGURE 16.17 *The Schoeps KFM6U microphone consists of two presssure microphones
mounted on the surface of a sphere. (Courtesy of
Schalltechnik Dr. -Ing. Schoeps GmbH).*

has been called a 'Faulkner' pair, after the British recording engineer who first adopted it (this is not strictly based
on the Williams curves). This latter pair has been found to
offer good image focus on a small-to-moderate-sized central
ensemble with the mics placed further back than would normally be expected.

Pseudo-binaural techniques

Binaural techniques could be classed as another form of near-
coincident technique. The spacing between the omni microphones in a dummy head is not great enough to fit any of the
Williams models described above for near-coincident pairs,
but the shadowing effect of the head makes the arrangement
more directional at high frequencies. Low-frequency width is
likely to need increasing to make the approach more loud-
speaker-compatible, as described earlier, unless one adheres
to Theile's association theory of stereo in which case little
further processing is required except for equalization.

The Schoeps KFM6U microphone, pictured in Figure 16.17, was
designed as a head-sized sphere with pressure microphones mounted on
the surface of the sphere, equalized for a flat response to frontal incidence
sound and suitable for generating signals that could be reproduced on

loudspeakers. This is in effect a sort of dummy head without ears. Dummy heads also exist that have been equalized for a reasonably natural timbral quality on loudspeakers, such as the Neumann KU100. The use of unprocessed dummy head techniques for stereo recording intended for loudspeakers has found favor with some recording engineers because they claim to like the spatial impression created, although others find the stereo image somewhat unfocused or vague.

Spaced microphone configurations

Spaced arrays have a historical precedent for their usage, since they were the first to be documented (in the work of Clement Ader at the Paris Exhibition in 1881), were the basis of the Bell Labs stereo systems in the 1930s, and have been widely used since then. They are possibly less 'correct' theoretically, from a standpoint of soundfield representation, but they can provide a number of useful spatial cues that give rise to believable illusions of natural spaces. Many recording engineers prefer spaced arrays because the omni microphones often used in such arrays tend to have a flatter and more extended frequency response than their directional counterparts, although it should be noted that spaced arrays do not have to be made up of omni mics (see below).

Spaced arrays rely principally on the precedence effect. The delays that result between the channels tend to be of the order of a number of milliseconds. With spaced arrays the level and time difference resulting from a source at a particular left–right position on the sound stage will depend on how far the source is from the microphones (see Figure 16.18), with a more distant source resulting in a much smaller delay and level difference. In order to calculate the time and level differences that will result from a particular spacing it is possible to use the following two formulae:

$$\Delta t = (d_1 - d_2)/c \qquad \Delta L = 20\log_{10}(d_1/d_2)$$

where Δt is the time difference and ΔL the pressure level difference which results from a source whose distance is d_1 and d_2 respectively from the two microphones, and c is the speed of sound (340 m/s).

When a source is very close to a spaced pair there may be a considerable level difference between the microphones, but this will become small once the source is more than a few meters distant. The positioning of spaced microphones in relation to a source is thus a matter of achieving a compromise between

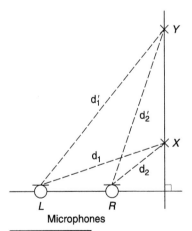

FIGURE 16.18 *With spaced omnis a source at position X results in path lengths d_1 and d_2 to each microphone respectively, whilst for a source in the same LR position but at a greater distance (source Y) the path length difference is smaller, resulting in smaller time difference than for X.*

closeness (to achieve satisfactory level and time differences between channels), and distance (to achieve adequate reverberant information relative to direct sound). When the source is large and deep, such as a large orchestra, it will be difficult to place the microphones so as to suit all sources. It may therefore be found necessary to raise the microphones somewhat so as to reduce the differences in path length between sources at the front and rear of the orchestra.

Spaced microphone arrays do not stand up well to theoretical analysis when considering the imaging of continuous sounds, the precedence effect being related principally to impulsive or transient sounds. Because of the phase differences between signals at the two loudspeakers created by the microphone spacing, interference effects at the ears at low frequencies may in fact result in a contradiction between level and time cues at the ears. It is possible in fact that the ear on the side of the earlier signal may not experience the higher level, thus producing a confusing difference between the cues provided by impulsive sounds and those provided by continuous sounds. The lack of phase coherence in spaced-array stereo is further exemplified by phase inverting one of the channels on reproduction, an action which does not always appear to affect the image particularly, as it would with coincident stereo, showing just how uncorrelated the signals are. (This is most noticeable with widely spaced microphones.)

Accuracy of phantom image positioning is therefore lower with spaced arrays, although many convincing recordings have resulted from their use. It has been suggested that the impression of spaciousness that results from the use of spaced arrays is in fact simply the result of phasiness and comb-filtering effects. Others suggest that there is a place for the spaciousness that results from spaced techniques, since the highly decorrelated signals which result from spaced techniques are also a feature of concert hall acoustics.

Griesinger has often claimed informally that spacing the mics apart by at least the reverberation radius (critical distance) of a recording space gives rise to adequate decorrelation between the microphones to obtain good spaciousness, and that this might be a suitable technique for ambient sound in surround recording. Mono compatibility of spaced pairs is variable, although not always as poor in practice as might be expected.

The so-called 'Decca Tree' is a popular arrangement of three spaced omnidirectional mics. The name derives from the traditional usage of this technique by the Decca Record Company, although even that company did not adhere rigidly to this arrangement. A similar arrangement is described by Grignon (1949). Three omnis are configured according to the diagram in

Figure 16.19, with the center microphone spaced so as to be slightly forward of the two outer mics, although it is possible to vary the spacing to some extent depending on the size of the source stage to be covered. The reason for the center microphone and its spacing is to stabilize the central image which tends otherwise to be rather imprecise, although the existence of the center mic will also complicate the phase relationships between the channels, thus exacerbating the comb-filtering effects that may arise with spaced pairs. The advance in time experienced by the forward mic will tend to solidify the central image, due to the precedence effect, avoiding the hole-in-the-middle often resulting from spaced pairs. The outer mics are angled outwards slightly, so that the axes of best HF response favor sources towards the edges of the stage whilst central sounds are on-axis to the central mic.

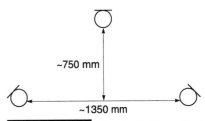

FIGURE 16.19 *The classic 'Decca Tree' involved three omnis, with the center microphone spaced slightly forward of the outer mics.*

A pair of omni outriggers are often used in addition to the tree, towards the edges of wide sources such as orchestras and choirs, in order to support the extremes of the sound stage that are some distance from the tree or main pair (see Figure 16.20). This is hard to justify on the basis of any conventional imaging theory, and is beginning to move toward the realms of multi-microphone pickup, but

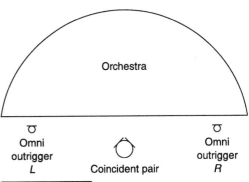

FIGURE 16.20 *Omni outriggers may be used in addition to a coincident pair or Decca Tree, for wide sources.*

can be used to produce a commercially acceptable sound. Once more than around three microphones are used to cover a sound stage one has to consider a combination of theories, possibly suggesting conflicting information between the outputs of the different microphones. In such cases the sound balance will be optimized on a mixing console, subject to the creative control of the recording engineer.

Spaced microphones with either omnidirectional or cardioid patterns may be used in configurations other than the Decca Tree described above, although the 'tree' has certainly proved to be the more successful arrangement in practice. The precedence effect begins to break down for delays greater than around 40 ms, because the brain begins to perceive the two arrivals of sound as being discrete rather than integrated. It is therefore reasonable to assume that spacings between microphones which give rise to greater delays than this between channels should be avoided. This maximum delay, though, corresponds to a mic spacing of well over 10 meters. Such extremes have not proved to work well in practice due to the great distance of central sources from either microphone compared with the

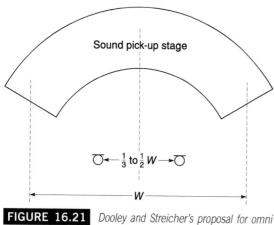

Sound pick-up stage

$\frac{1}{3}$ to $\frac{1}{2}$ W

W

FIGURE 16.21 *Dooley and Streicher's proposal for omni spacing.*

FIGURE 16.22 *B&K omni microphones mounted on a stereo bar that allows variable spacing.*

closeness of sources at the extremes, resulting in a considerable level drop for central sounds and thus a hole in the middle.

Dooley and Streicher have shown that good results may be achieved using spacings of between one-third and one-half of the width of the total sound stage to be covered (see Figure 16.21), although closer spacings have also been used to good effect. Bruel and Kjaer manufacture matched stereo pairs of omni microphones together with a bar which allows variable spacing, as shown in Figure 16.22, and suggest that the spacing used is smaller than one-third of the stage width (they suggest between 5 cm and 60 cm, depending on stage width). Their principal rule is that the distance between the microphones should be small compared with the distance from microphones to source.

BINAURAL RECORDING AND 'DUMMY HEAD' TECHNIQUES

Whilst it is possible to use a real human head for binaural recording (generally attached to a live person), it can be difficult to mount high-quality microphones in the ears and the head movements and noises of the owner can be obtrusive. Sometimes heads are approximated by the use of a sphere or a disc separating a pair of microphones, and this simulates the shadowing effect of the head but it does not give rise to the other spectral filtering effects of the outer ear. Recordings made using such approaches have been found to have reasonable loudspeaker compatibility as they do not have the unusual equalization that results from pinna filtering. (Unequalized true binaural recordings replayed on loudspeakers will typically suffer two stages of pinna filtering – once on recording and then again on reproduction – giving rise to distorted timbral characteristics.)

Dummy heads are models of human heads with pressure microphones in the ears that can be used for originating binaural signals suitable for measurement or reproduction. A number of commercial products exist, some of which also include either shoulders or a complete torso.

A complete head-and-torso simulator is often referred to as a 'HATS', and an example is shown in Figure 16.23. The shoulders and torso are considered by some to be important owing to the reflections that result from them in natural listening, which can contribute to the HRTF. This has been found to be a factor that differs quite considerably between individuals and can therefore be a confusing cue if not well matched to the listener's own torso reflections.

Some dummy heads or ear inserts are designed specifically for recording purposes whereas others are designed for measurement. As a rule, those designed for recording tend to have microphones at the entrances of the ear canals, whereas those designed for measurement have the mics at the ends of the ear canals, where the ear drum should be. (Some measurement systems also include simulators for the transmission characteristics of the inner parts of the ear.) The latter types will therefore include the ear canal resonance in the HRTF, which would have to be equalized out for recording/reproduction purposes in which headphones were located outside the ear canal. The ears of dummy heads are often interchangeable in order to vary the type of ear to be simulated, and these ears are modeled on 'average' or 'typical' physical properties of human ears, giving rise to the same problems of HRTF standardization as mentioned above.

FIGURE 16.23

Head and torso simulator (HATS) from B&K.

The equalization of dummy heads for recording has received much attention over the years, mainly to attempt better headphone/loudspeaker compatibility. Equalization can be used to modify the absolute HRTFs of the dummy head in such a way that the overall spatial effect is not lost, partly because the differences between the ears are maintained. Just as Theile has suggested using diffuse field equalization for headphones as a good means of standardizing their response, he and others have also suggested diffuse field equalization of dummy heads so that recordings made on such heads replay convincingly on such headphones and sound reasonably natural on loudspeakers. This essentially means equalizing the dummy head microphone so that it has a near-flat response when measured in one-third octave bands in a diffuse sound field. The Neumann KU100, pictured in Figure 16.24, is a dummy head that is designed to have good compatibility between loudspeaker and headphone reproduction, and uses equalization that is close to Theile's proposed diffuse field response.

FIGURE 16.24

Neumann KU100 dummy head. (Courtesy of Georg Neumann GmbH, Berlin).

Binaural cues do not have to be derived from dummy heads. Provided the HRTFs are known, or can be approximated for the required angle of sound incidence, signals can be synthesized with the appropriate time delays and spectral characteristics. Such techniques are increasingly used in digital signal processing applications that aim to simulate natural spatial cues, such as flight simulators and virtual reality. Accurate sets of HRTF data for all angles of incidence and elevation have been hard to come by until recently, and they are often quite closely guarded intellectual property as they can take a long time and a lot of trouble to measure. The question also arises as to how fine an angular resolution is required in the data set. For this reason a number of systems base their HRTF implementation on relative coarse resolution data and interpolate the points in between.

SPOT MICROPHONES AND TWO-CHANNEL PANNING LAWS

We have so far considered the use of a small number of microphones to cover the complete sound stage. It is also possible to make use of a large number of mono microphones or other mono sources, each covering a small area of the sound stage and intended to be as independent of the others as possible. This is the normal basis of most studio pop music recording, with the sources often being recorded at separate times using overdubbing techniques. In the ideal world, each mic in such an arrangement would pick up sound only from the desired sources, but in reality there is usually considerable spill from one to another. It is not the intention in this chapter to provide a full résumé of studio microphone technique, and thus discussion will be limited to an overview of the principles of multi-mic pickup as distinct from the more simple techniques described above.

In multi-mic recording each source feeds a separate channel of a mixing console, where levels are individually controlled and the mic signal is 'panned' to a virtual position somewhere between left and right in the sound stage. The pan control takes the monophonic signal and splits it two ways, controlling the proportion of the signal fed to each of the left and right mix buses. Typical pan control laws follow a curve which gives rise to a 3 dB drop in the level sent to each channel at the center, resulting in no perceived change in level as a source is moved from left to right (see Fact File 6.2). This has often been claimed to be due to the way signals from left and right loudspeakers sum acoustically at the listening position, which includes a diffuse field component of the room. The −3 dB pan-pot law is not correct if the stereo signal is combined electrically to mono, since the summation of two equal signal voltages would result in a 6 dB rise in level for signals panned centrally. A −6 dB law is more appropriate for mixers whose outputs will be summed to mono (e.g. radio and TV operations) as well as stereo, although this will then result in a drop in level in the center for stereo signals. A compromise law of −4.5 dB is sometimes adopted by manufacturers for this reason.

Panned mono balances rely on channel level differences, separately controlled for each source, to create phantom images on a synthesized sound stage, with relative level between sources used to adjust the prominence of a source in a mix. Time delay is hardly ever used as a panning technique, for reasons of poor mono compatibility and technical complexity. Artificial reverberation may be added to restore a sense of space to a multi-mic balance. Source distance can be simulated by the addition of reflections and reverberation, as well as by changes in source spectrum and overall level (e.g. HF roll-off can simulate greater distance).

It is common in classical music recording to use close mics in addition to a coincident pair or spaced pair in order to reinforce sources that appear to be weak in the main pickup. These close mics are panned to match the true position of the source. The results of this are variable and can have the effect of flattening the perspective, removing any depth which the image might have had, and thus the use of close mics must be handled with subtlety. David Griesinger has suggested that the use of stereo pairs of mics as spots can help enormously in removing this flattening effect, because the spill that results between spots is now in stereo rather than in mono and is perceived as reflections separated spatially from the main signal.

The recent development of cheaper digital signal processing (DSP) has made possible the use of delay lines, sometimes as an integral feature of digital mixer channels, to adjust the relative timing of spot mics in relation to the main pair. This can help to prevent the distortion of distance, and to equalize the arrival times of distant mics so that they do not exert a precedence 'pull' over the output of the main pair. It is also possible to process the outputs of multiple mono sources to simulate binaural delays and head-related effects in order to create the effect of sounds at any position around the head when the result is monitored on headphones or on loudspeakers using crosstalk canceling, as described earlier.

RECOMMENDED FURTHER READING

Alexander, R.C., 2000. The Inventor of Stereo. The Life and Works of Alan Dower Blumlein. Focal Press.

Bartlett, B., 1991. Stereo Microphone Techniques. Focal Press.

Eargle, J. (Ed.), 1986. Stereophonic Techniques – An Anthology. Audio Engineering Society.

Rayburn, R., 2011. Eargle's Microphone Book: From Mono to Stereo to Surround – A Guide to Microphone Design and Application, third edition. Focal Press.

Rumsey, F., 2001. Spatial Audio. Focal Press.

Surround Sound

CHAPTER CONTENTS

This chapter is concerned with the most commonly encountered multichannel (i.e. more than two channels) stereo reproduction configurations, most of which are often referred to as surround sound. Standards or conventions that specify basic channel or loudspeaker configurations are distinguished from proprietary systems such as Dolby Digital and DTS whose primary function is the coding and delivery of multichannel audio signals. The latter are discussed in the second part of the chapter, in which is also contained an explanation of the Ambisonic system for stereo signal representation. Surround sound standards often specify little more than the channel configuration and the way the loudspeakers should be arranged. This leaves the business of how to create or represent a spatial sound field entirely up to the user.

THREE-CHANNEL (3-0) STEREO

It is not proposed to say a great deal about the subject of three-channel stereo here, as it is rarely used on its own. Nonetheless it does form the basis of a lot of surround sound systems. It requires the use of a left (L), center (C) and right (R) channel, the loudspeakers arranged equidistantly across the front sound stage, as shown in Figure 17.1. It has some precedents in historical development, in that the stereophonic system developed by Steinberg and Snow in the 1930s used three channels (see Chapter 16). Three front channels have also been commonplace in cinema stereo systems, mainly because of the need to cover a wide listening area and because wide screens tend to result in a large distance between left and right loudspeakers. Two channels only became the norm in consumer systems for reasons of economy and convenience, and particularly because it was much more straightforward to cut two channels onto an analog disk than three.

There are various advantages of three-channel stereo. First, it allows for a somewhat wider front sound stage than two-channel stereo, if desired, because the center channel acts to 'anchor' the central image and the left and right loudspeakers can be placed further out to the sides (say ± 45°). (Note, though, that in the current five-channel surround sound standard the L and R loudspeakers are in fact placed at ± 30°, for compatibility with two-channel stereo material.) Second, the center loudspeaker enables a wider range of listening positions in many cases, as the image does not collapse quite as readily into the nearest loudspeaker. It also anchors dialog more clearly in the middle of the screen in sound-for-picture applications. Third, the center image does not suffer the same timbral modification as the center image in two-channel stereo, because it emanates from a real source.

A practical problem with three-channel stereo is that the center loudspeaker position is often very inconvenient. Although in cinema reproduction it can be behind an acoustically transparent screen, in consumer environments, studios and television environments it is almost always just where one wants a television monitor or a window. Consequently the center channel has to be mounted above or below the object in question, and possibly made smaller than the other loudspeakers.

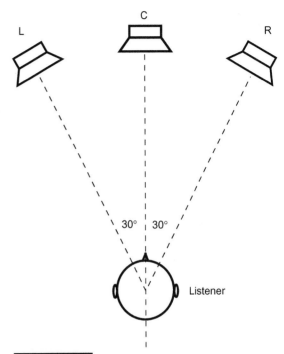

FIGURE 17.1   *Three-channel stereo reproduction usually involves three equally spaced loudspeakers in front of the listener. The angle between the outer loudspeakers is 60° in the ITU standard configuration, for compatibility with two-channel reproduction, but the existence of a center loudspeaker makes wider spacings feasible if compatibility is sacrificed.*

FOUR-CHANNEL SURROUND (3-1 STEREO)

In this section the form of stereo called '3-1 stereo' in some international standards, or 'LCRS surround' in some other circles, is briefly described. Proprietary encoding and decoding technology from Dolby relating to this format is described later.

'Quadraphonic' reproduction using four loudspeakers in a square arrangement is not covered further here (it was mentioned in the Introduction), as it has little relevance to current practice.

Purpose of four-channel systems

The merits of three front channels have already been introduced in the previous section. In the 3-1 approach, an additional 'effects' channel or 'surround' channel is added to the three front channels, routed to a loudspeaker or loudspeakers located behind (and possibly to the sides) of listeners. It was developed first for cinema applications, enabling a greater degree of audience involvement in the viewing/listening experience by providing a channel for 'wrap-around' effects. This development is attributed to 20th Century Fox in the 1950s, along with wide-screen Cinemascope viewing, being intended to offer effective competition to the new television entertainment.

There is no specific intention in 3-1 stereo to use the effects channel as a means of enabling 360° image localization. In any case, this would be virtually impossible with most configurations as there is only a single audio channel feeding a larger number of surround loudspeakers, effectively in mono.

Loudspeaker configuration

Figure 17.2 shows the typical loudspeaker configuration for this format. In the cinema there are usually a large number of surround loudspeakers fed from the single S channel ('surround channel', not to be confused with the 'S' channel in sum-and-difference stereo), in order to cover a wide audience area. This has the tendency to create a relatively diffuse or distributed reproduction of the effects signal. The surround speakers are sometimes electronically decorrelated to increase the degree of spaciousness or diffuseness of surround effects, in order that they are not specifically localized to the nearest loudspeaker or perceived inside the head.

In consumer systems reproducing 3-1 stereo, the mono surround channel is normally fed to two surround loudspeakers located in similar positions to the 3-2 format described below. The gain of the channel is usually reduced by 3 dB so that the summation of signals from the two speakers does not lead to a level mismatch between front and rear.

Limitations of four-channel reproduction

The mono surround channel is the main limitation in this format. Despite the use of multiple loudspeakers to reproduce the surround channel, it is still not possible to create a good sense of envelopment of spaciousness without using surround signals that are different on both sides of the listener. Most of the psychoacoustic research suggests that

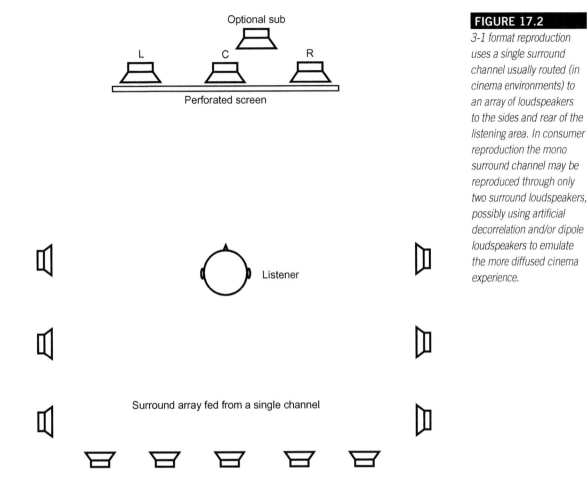

Optional sub

L   C   R

Perforated screen

Listener

Surround array fed from a single channel

FIGURE 17.2
3-1 format reproduction uses a single surround channel usually routed (in cinema environments) to an array of loudspeakers to the sides and rear of the listening area. In consumer reproduction the mono surround channel may be reproduced through only two surround loudspeakers, possibly using artificial decorrelation and/or dipole loudspeakers to emulate the more diffused cinema experience.

the ears need to be provided with decorrelated signals to create the best sense of envelopment and effects can be better spatialized using stereo surround channels.

5.1-CHANNEL SURROUND (3-2 STEREO)

This section deals with the 3-2 configuration that has been standardized for numerous surround sound applications, including cinema, television and consumer applications. Because of its wide use in general parlance, the term '5.1 surround' will be used below. Whilst without doubt a compromise, it has become widely adopted in professional and consumer circles and is likely to form the basis for consumer surround sound for the foreseeable future.

Various international groups have worked on developing recommendations for common practice and standards in this area, and some of the information below is based on the effort of the AES Technical Committee on Multichannel and Binaural Audio Technology to bring together a number of proposals.

Purpose of 5.1-channel systems

Four-channel systems have the disadvantage of a mono surround channel, and this limitation is removed in the 5.1-channel system, enabling the provision of stereo effects or room ambience to accompany a primarily front-orientated sound stage. This front-orientated paradigm is a most important one as it emphasizes the intentions of those that finalized this configuration, and explains the insistence in some standards on the use of the term '3-2 stereo' rather than 'five-channel surround'. Essentially the front three channels are intended to be used for a conventional three-channel stereo sound image, whilst the rear/side channels are only intended for generating supporting ambience, effects or 'room impression'. In this sense, the standard does not directly support the concept of 360° image localization, although it may be possible to arrive at recording techniques or signal processing methods that achieve this to a degree.

The front–rear distinction is a conceptual point often not appreciated by those that use the format. Two-channel stereo can be relatively easily modeled and theoretically approached in terms of localization vectors, etc. for sounds at any angle between the loudspeakers. It is more difficult, though, to come up with such a model for the five-channel layout described below, as it has unequal angles between the loudspeakers and a particularly large angle between the two rear loudspeakers. It is possible to arrive at gain and phase relationships between these five loudspeakers that are similar to those used in Ambisonics for representing different source angles, but the varied loudspeaker angles make the imaging stability less reliable in some sectors than others. For those who do not have access to the sophisticated panning laws or psychoacoustic matrices required to feed five channels accurately for all-round localization it may be better to treat the format in 'cinema style' – in other words with a three-channel front image and two surround effect channels. With such an approach it is still possible to create very convincing spatial illusions, with good envelopment and localization qualities.

One cannot introduce the 5.1 surround system without explaining the meaning of the '.1' component. This is a dedicated low-frequency effects (LFE) channel or sub-bass channel. It is called '.1' because of its limited bandwidth. Strictly, the international standard nomenclature for 5.1 surround should be '3-2-1', the last digit indicating the number of LFE channels.

International standards and configurations

The loudspeaker layout and channel configuration is specified in the ITU-R BS.775 standard. This is shown in Figure 17.3 and Fact File 17.1. A display screen is also shown in the figure for sound with picture

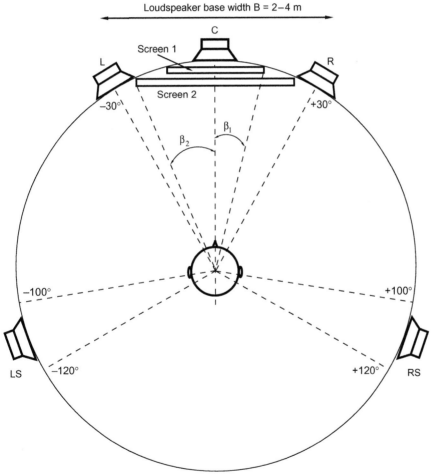

FIGURE 17.3

3-2 format reproduction according to the ITU-R BS.775 standard uses two independent surround channels routed to one or more loudspeakers per channel.

Screen 1: Listening distance = 3H ($2\beta_1$ = 33°) (possibly more suitable for TV screen)
Screen 2: Listening distance = 2H ($2\beta_2$ = 48°) (more suitable for projection screen)
H: Screen height

FACT FILE 17.1 TRACK ALLOCATIONS IN 5.1

Standards recommend the track allocations to be used for 5.1 surround on eight-track recording formats, as shown in the table below. Although other configurations are known to exist there is a strong move to standardize on this arrangement (see also the notes below the table).

Track[1]	Signal	Comments		Color[2]
1	L	Left		Yellow
2	R	Right		Red
3	C	Center		Orange
4	LFE	Low frequency enhancement	Additional sub-bass and effects signal for subwoofer, optional[3]	Gray
5	LS	Left surround	−3 dB in the case of mono surround	Blue
6	RS	Right surround	−3 dB in the case of mono surround	Green
7	Free use in program exchange[4]		Preferably left signal of a 2/0 stereo mix	Violet
8	Free use in program exchange[4]		Preferably right signal of a 2/0 stereo mix	Brown

[1]The term 'track' is used to mean either tracks on magnetic tape or virtual tracks on other storage media where no real tracks exist.
[2]This color coding is only a proposal of the German Surround Sound Forum at present, and not internationally standardized.
[3]Preferably used in film sound, but is optional for home reproduction. If no LFE signal is being used, track 4 can be used freely, e.g. for commentary. In some regions a mono surround signal MS = LS + RS is applied, where the levels of LS and RS are decreased by 3 dB before summing.
[4]Tracks 7 and 8 can be used alternatively, for example for commentary, for additional surround signals, or for half-left/half-right front signal (e.g. for special film formats), or rather for the matrix format sum signal Lt/Rt.

applications, and there are recommendations concerning the relative size of the screen and the loudspeaker base width shown in the accompanying table. The left and right loudspeakers are located at ± 30° for compatibility with two-channel stereo reproduction. In many ways this need for compatibility with 2/0 is a pity, because the center channel unavoidably narrows the front sound stage in many applications, and the front stage could otherwise take advantage of the wider spacing facilitated by three-channel reproduction. It was nonetheless considered crucial for the same loudspeaker configuration to be usable for all standard forms of stereo reproduction, for reasons most people will appreciate.

The surround loudspeaker locations, at approximately ±110°, are placed so as to provide a compromise between the need for effects panning behind the listener and the lateral energy important for good envelopment. In this respect they are more like 'side' loudspeakers than rear loudspeakers, and in many installations this is an inconvenient location causing people to mount them nearer the rear than the standard suggests. (Some have said that a 150° angle for the rear loudspeakers provides a more exciting surround effect.) In the 5.1 standard there are normally no loudspeakers directly behind the listener, which can make for creative difficulties. This has led to a Dolby proposal called EX (described below) that places an additional speaker at the center-rear location. (This is not part of the current standard, though.) The ITU standard allows for additional surround loudspeakers to cover the region around listeners, similar to the 3-1 arrangement described earlier. If these are used then they are expected to be distributed evenly in the angle between ±60° and ±150°.

Surround loudspeakers should be the same as front loudspeakers where possible, in order that uniform sound quality can be obtained all around. That said, there are arguments for use of dipole loudspeakers in these positions. Dipoles radiate sound in more of a figure-eight pattern and one way of obtaining a diffuse surround impression is to orient these with the nulls of the figure-eight towards the listening position. In this way the listener experiences more reflected than direct sound and this can give the impression of a more spacious ambient soundfield that may better emulate the cinema listening experience in small rooms. Dipoles make it correspondingly more difficult to create defined sound images in rear and side positions, though.

The LFE channel and use of subwoofers

The low-frequency effects channel is a separate sub-bass channel with an upper limit extending to a maximum of 120 Hz (see Fact File 17.2). It is intended for conveying special low-frequency content that requires greater sound pressure levels and headroom than can be handled by the main channels. It is not intended for conveying the low-frequency component of the main channel signals, and its application is likely to be primarily in sound-for-picture applications where explosions and other high-level rumbling noises are commonplace, although it may be used in other circumstances.

In consumer audio systems, reproduction of the LFE channel is considered optional. Because of this, recordings should normally be made so that they sound satisfactory even if the LFE channel is not reproduced. The EBU (European Broadcasting Union) comments on the use of the LFE channel as follows.

When an audio program originally produced as a feature film for theatrical release is transferred to consumer media, the LFE channel is often derived from the dedicated theatrical subwoofer channel. In the cinema, the dedicated subwoofer channel is always reproduced, and thus film mixes may use the subwoofer channel to convey important low frequency

FACT FILE 17.2 BASS MANAGEMENT IN 5.1

It is a common misconception that any sub-bass or sub-woofer loudspeaker(s) that may be used on reproduction must be fed directly from the LFE channel in all circumstances. Whilst this may be the case in the cinema, bass management in the consumer reproducing system is not specified in the standard and is entirely system dependent. It is not mandatory to feed low-frequency information to the LFE channel during the recording process, neither is it mandatory to use a subwoofer, indeed it has been suggested that restricting extreme low-frequency information to a monophonic channel may limit the potential for low-frequency spaciousness in balances. In music mixing it is likely to be common to send the majority of full-range LF information to the main channels, in order to retain the stereo separation between them.

In practical systems it may be desirable to use one or more subwoofers to handle the low-frequency content of a mix on reproduction. The benefit of this is that it enables the size of the main loudspeakers to be correspondingly reduced, which may be useful practically when it comes to finding places to put them in living rooms or sound control rooms. In such cases crossover systems split the signals between main loudspeakers and subwoofer(s) somewhere between 80 Hz and 160 Hz. In order to allow for reproduction of the LFE channel and/or the low-frequency content from the main channels through subwoofer loudspeakers, a form of bass management akin to that shown below is typically employed.

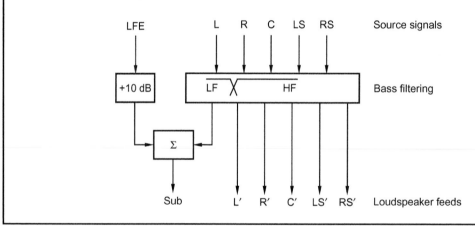

program content. When transferring programs originally produced for the cinema over television media (e.g. DVD), it may be necessary to re-mix some of the content of the subwoofer channel into the main full bandwidth channels. It is important that any low frequency audio which is very significant to the integrity of the program content is not placed into the LFE channel. The LFE channel should be reserved for extreme low frequency, and for very high level, 120 Hz program content which, if not reproduced, will not compromise the artistic integrity of the program.

With cinema reproduction the in-band gain of this channel is usually 10 dB higher than that of the other individual channels. This is achieved by a level increase of the reproduction channel, not by an increased recording level. (This does not mean that the broadband or weighted SPL of the LFE loudspeaker should measure 10 dB higher than any of the other channels – in fact it will be considerably less than this as its bandwidth is narrower.)

Limitations of 5.1-channel reproduction

The main limitations of the 5.1 surround format are first, that it was not intended for accurate 360° phantom imaging capability, as explained above. Whilst it may be possible to achieve a degree of success in this respect, the loudspeaker layout is not ideally suited to it. Second, the front sound stage is narrower than it could be if compatibility with 2/0 reproduction was not a requirement. Third, the center channel can prove problematic for music balancing, as conventional panning laws and coincident microphone techniques are not currently optimized for three loudspeakers, having been designed for two-speaker stereo. Simple bridging of the center loudspeaker between left and right signals has the effect of narrowing the front image compared with a two-channel stereo reproduction of the same material. This may be resolved over time as techniques suited better to three-channel stereo are resurrected or developed. Fourth, the LS and RS loudspeakers are located in a compromise position, leading to a large hole in the potential image behind the listener and making it difficult to find physical locations for the loudspeakers in practical rooms.

These various limitations of the format, particularly in some people's view for music purposes, have led to various non-standard uses of the five or six channels available on new consumer disc formats such as DVD-A (Digital Versatile Disc – Audio) and SACD (Super Audio Compact Disc). For example, some are using the sixth channel (which would otherwise be LFE) in its full bandwidth form on these media to create a height channel. Others are making a pair out of the 'LFE' channel and the center channel so as to feed a pair of front-side loudspeakers, enabling the rear loudspeakers to be further back. These are non-standard uses and should be clearly indicated on any recordings.

Signal levels in 5.1 surround

In film sound environments it is the norm to increase the relative recording level of the surround channels by 3 dB compared with that of the front channels. This is in order to compensate for the −3 dB acoustic alignment of each surround channel's SPL with respect to the front that takes place in dubbing stages and movie theaters. It is important to be aware of this discrepancy between practices, as it is the norm in music mixing and broadcasting to align all channels for equal level both on recording media and for acoustical monitoring. Transfers from film masters to consumer or broadcast media may require 3 dB alteration in the gain of the surround channels.

OTHER MULTICHANNEL CONFIGURATIONS

Although the 5.1 surround standard is becoming widely adopted as the norm for the majority of installations, other proposals and systems exist, typically involving more channels to cover a large listening area more accurately. It is reasonable to assume that the more real loudspeakers exist in different locations around the listener, the less one has to rely on the formation of phantom images to position sources accurately, and the more freedom one has in listener position. The added complication of mixing for such larger numbers of channels must be considered as a balancing factor.

The reader is also referred to the discussion of Ambisonics, as this system can be used with a wide range of different loudspeaker configurations depending on the decoding arrangements used.

7.1-channel surround

Deriving from widescreen cinema formats, the 7.1-channel configuration normally adds two further loudspeakers to the 5.1-channel configuration, located at center-left (CL) and center-right (CR), as shown in Figure 17.4. This is not a format primarily intended for consumer applications, but for large cinema auditoria where the screen width is such that the additional channels are needed to cover the angles between the loudspeakers satisfactorily for all the seats in the auditorium. Sony's SDDS cinema system is a common proprietary implementation of this format, as is the original 70 mm Dolby Stereo format (see below), although the original 70 mm analog format only used one surround channel.

Lexicon and Meridian have also implemented a seven-channel mode in their consumer surround decoders, but the recommended locations for the loudspeakers are not quite the same as in the cinema application. The additional channels are used to provide a wider side-front component and allow the rear speakers to be moved round more to the rear than in the 5.1 arrangement.

10.2-channel surround

Tomlinson Holman has spent considerable effort promoting a 10.2-channel surround sound system as 'the next step' in spatial reproduction, but this has not yet been adopted as standard. To the basic five-channel array

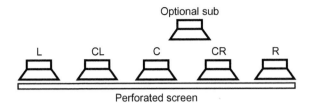

FIGURE 17.4

Some cinema sound formats for large auditorium reproduction enhance the front imaging accuracy by the addition of two further loudspeakers, center-left and center-right.

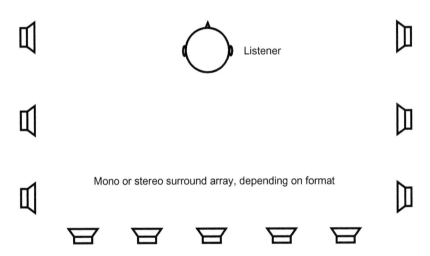

he adds wider side-front loudspeakers and a center-rear channel to 'fill in the holes' in the standard layout. He also adds two height channels and a second LFE channel. The second LFE channel is intended to provide lateral separation of decorrelated low bass content to either side of the listening area, as suggested by Griesinger, to enhance low-frequency spaciousness.

SURROUND SOUND SYSTEMS

This part of the chapter concerns what will be called surround sound 'systems', which includes proprietary formats for the coding and transfer of surround sound. These are distinguished from the generic configurations and international standards discussed already. Most of the systems covered here are the subject of patents and intellectual property rights. In some proprietary systems the methods of signal coding or matrixing for storage and delivery are defined (e.g. Dolby Stereo), whilst others define a full source-receiver signal representation system (e.g. Ambisonics).

MATRIXED SURROUND SOUND SYSTEMS

Whilst ideally one would like to be able to transfer or store all the channels of a surround sound mix independently and discretely, it may be necessary to make use of existing two-channel media for compatibility with other systems. The systems described in the following sections all deal with multichannel surround sound in a matrixed form (in other words, using an algorithm that combines the channels in such a way that they can be subsequently extracted using a suitable decoder). By matrixing the signals they can be represented using fewer channels than the source material contains. This gives rise to some side-effects and the signals require careful dematrixing, but the approach has been used widely for many years, mainly because of the unavailability of multichannel delivery media in many environments.

Dolby stereo, surround and prologic

Dolby Labs was closely involved with the development of cinema surround sound systems, and gradually moved into the area of surround sound for consumer applications.

The original Dolby Stereo system involved a number of different formats for film sound with three to six channels, particularly a 70 mm film format with six discrete tracks of magnetically recorded audio, and a 35 mm format with two optically recorded audio tracks onto which were matrixed four audio channels in the 3-1 configuration (described above). The 70 mm format involved L, LC, C, RC, R and S channels, whereas the 35 mm format involved only L, C, R and S. Both clearly only involved mono surround information. The four-channel system is the one most commonly known today as Dolby Stereo, having found widespread acceptance in the cinema world and used on numerous movies. Dolby Surround was introduced in 1982 as a means of emulating the effects of Dolby Stereo in a consumer environment. Essentially the same method of matrix decoding was used, so movies transferred to television formats could be decoded in the home in a similar way to the cinema. Dolby Stereo optical sound tracks for the cinema were Dolby A noise-reduction encoded and decoded, in order to improve the signal-to-noise ratio, but this is not a feature of consumer Dolby Surround (more recent cinema formats have used Dolby SR-type noise reduction, alongside a digital soundtrack).

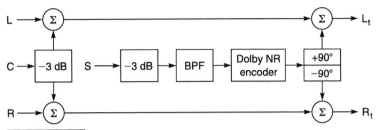

FIGURE 17.5 *Basic components of the Dolby Stereo matrix encoding process.*

The Dolby Stereo matrix (see Figure 17.5) is a form of '4-2-4' matrix that encodes the mono surround channel so that it is added out of phase into the left and right channels (+90° in one channel and −90° in the other). The center channel signal is added to left and right in phase. The resulting sum is called L_t/R_t (left total and right total). By doing this the surround signal can be separated from the front signals upon decoding by summing the L_t/R_t signals out of phase (extracting the stereo difference signal), and the center channel can be extracted by summing L_t/R_t in phase. In consumer systems using passive decoding the center channel is not always fed to a separate loudspeaker but can be heard as a phantom image between left and right. A decoder block diagram for the consumer version (Dolby Surround) is shown in Figure 17.6. Here it can be seen that in addition to the sum-and-difference-style decoding, the surround channel is subject to an additional delay, band-limiting between 100 Hz and 7 kHz and a modified form of Dolby B noise reduction. The low-pass filtering and the delay are both designed to reduce matrix side-effects that could otherwise result in front signals appearing to come from behind. Crosstalk between channels and effects of any misalignment in the system can cause front signals to 'bleed' into the surround channel, and this can be worse at high frequencies than low. The delay (of the order of 20–30 ms in consumer systems, depending on the distance of the rear speakers) relies on the precedence effect (see Chapter 2) to cause the listener to localize signals according to the first arriving wavefront which will now be from the front rather than the rear of the sound stage. The rear signal then becomes psychoacoustically better separated from the front and localization of primary signals is biased more towards the front. The modified B-type NR reduces surround channel noise and also helps to reduce the effects of decoding errors and interchannel crosstalk, as some distortions introduced between encoding and decoding will be reduced by B-type decoding.

A problem with passive Dolby Surround decoding is that the separation between adjacent channels is relatively modest, although the separation of left/right and center/surround remains high. When a signal is panned fully left it will tend to appear only 3 dB down in the center, and also in the surround, for example. The effects of this can be ameliorated in passive

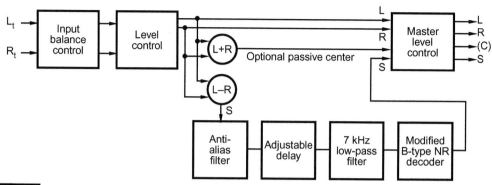

FIGURE 17.6 *Basic components of the passive Dolby surround decoder.*

consumer systems by the techniques described above (phantom center and surround delay/filtering). Dolby's ProLogic system, based on principles employed in the professional decoder, attempts to resolve this problem by including sophisticated 'steering' mechanisms into the decoder circuit to improve the perceived separation between the channels. A basic block diagram is shown in Figure 17.7. This enables a real center loudspeaker to be employed. Put crudely, ProLogic works by sensing the location of 'dominant' signal components and selectively attenuating channels away from the dominant component. (A variety of other processes are involved as well as this.) So, for example, if a dialog signal is predominantly located in the center, the control circuit will reduce the output of the other channels (L, R, S) in order that the signal comes mainly from the center loudspeaker. (Without this it would also have appeared at quite high level in left and right as well.) A variety of algorithms are used to determine how quickly the system should react to changes in dominant signal position, and what to do when no signal appears dominant.

Dolby has recently introduced an enhancement to ProLogic, entitled ProLogic 2, that adds support for full-bandwidth stereo rear channels, with various options that make it more suitable for music programs. It is also claimed to be effective in the up-conversion of unencoded two-channel material to five-channel surround.

Mixes that are to be matrix encoded using the Dolby system should be monitored via the encode–decode chain in order that the side-effects of the process can be taken into account by the balance engineer. Dolby normally licenses the system for use on a project, and will assist in the configuration and alignment of their equipment during the project.

Dolby Stereo/Surround can be complemented by the THX system, as described in Fact File 17.3.

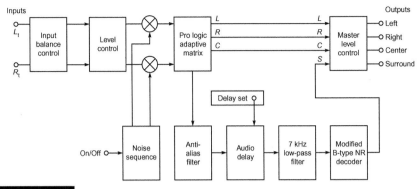

FIGURE 17.7   *Basic components of the active Dolby ProLogic decoder.*

FACT FILE 17.3 WHAT IS THX?

The THX system was developed by Tomlinson Holman at Lucasfilm (THX is derived from 'Tomlinson Holman Experiment'). The primary aim of the system was to improve the sound quality in movie theaters and make it closer to the sound experienced by sound mixers during post-production. It was designed to complement the Dolby Stereo system, and does not itself deal with the encoding or representation of surround sound. In fact THX is more concerned with the acoustics of cinemas and the design of loudspeaker systems, optimizing the acoustic characteristics and noise levels of the theater, as well as licensing a particular form of loudspeaker system and crossover network. THX licenses the system to theaters and requires that the installation is periodically tested to ensure that it continues to meet the specification.

Home THX was developed, rather like Dolby Surround, in an attempt to convey the cinema experience to the home. Through the use of a specific controller, amplifiers and speakers, the THX system enhances the decoding of Dolby Surround and can also be used with digital surround sound signals. The mono surround signal of Dolby Surround is subject to decorrelation of the signals sent to the two surround loudspeakers in order that the surround signal is made more diffuse and less 'mono'. It is claimed that this has the effect of preventing surround signals from collapsing into the nearest loudspeaker. Signals are re-equalized to compensate for the excessive high-frequency content that can arise when cinema balances are replayed in small rooms, and the channels are 'timbre matched' to compensate for the spectral changes that arise when sounds are panned to different positions around the head.

In terms of hardware requirements, the Home THX system also specifies certain aspects of amplifier performance, as well as controlling the vertical and horizontal directivity of the front loudspeakers. Vertical directivity is tightly controlled to increase the direct sound component arriving at listeners, whilst horizontal directivity is designed to cover a reasonably wide listening area. Front speakers should have a frequency response from 80 Hz to 20 kHz and all speakers must be capable of radiating an SPL of 105 dB without deterioration in their response or physical characteristics. The surround speakers are unusual in having a bipolar radiation pattern, arranged so that the listener hears reflected sound rather than direct sound from these units. These have a more relaxed frequency response requirement of 125 Hz to 8 kHz. A subwoofer feed is usually also provided.

Circle Surround

Circle Surround was developed by the Rocktron Corporation (RSP Technologies) as a matrix surround system capable of encoding stereo surround channels in addition to the conventional front channels. They proposed the system as more appropriate than Dolby Surround for music applications, and claimed that it should be suitable for use on material that had not been encoded as well as that which had.

The Circle Surround encoder is essentially a sum and difference L_t/R_t process (similar to Dolby but without the band limiting and NR encoding of the surround channel). One incarnation of this involves 5-2 encoding, intended for decoding back to five channels (the original white paper on the system described a 4-2 encoder). Amongst other methods, the Circle decoder steers the rear channels separately according to a split-band technique that steers low- and high-frequency components independently from each other. In this way they claim to avoid the broad-band 'pumping' effects associated with some other systems. They also decode the rear channels slightly differently, using L–R for the left rear channel and R–L for the right rear channel, which it is claimed allows side-images to be created on either side. They avoid the use of a delay in the rear channels for the 'Music' mode of the system and do not band-limit the rear channels as Dolby Surround does.

Lexicon logic 7

Logic 7 is another surround matrix decoding process that can be used as an alternative for Dolby Surround decoding. Variants on this algorithm (such as the so-called Music Logic and Music Surround modes) can also be used for generating a good surround effect from ordinary two-channel material. Lexicon developed the algorithm for its high-end consumer equipment, and it is one of a family of steered decoding processes that distributes sound energy appropriately between a number of loudspeakers depending on the gain and phase relationships in the source material. In this case seven loudspeaker feeds are provided rather than five, adding two 'side' loudspeakers to the array, as shown in Figure 17.8. The rear speakers can then be further to the rear than would otherwise be desirable. The side loudspeakers can be used for creating an enhanced envelopment effect in music modes and more accurate side panning of effects in movie sound decoding.

In Logic 7 decoding of Dolby matrix material the front channel decoding is almost identical to Dolby ProLogic, with the addition of a variable center channel delay to compensate for non-ideal locations of the center

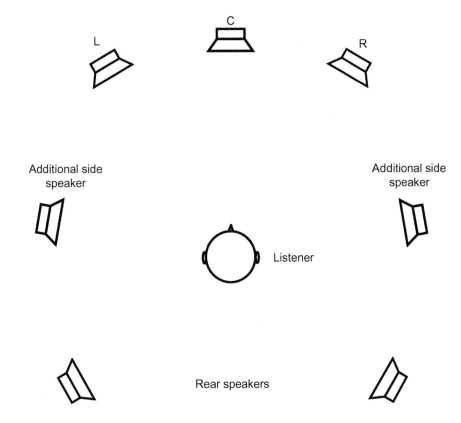

FIGURE 17.8

Approximate loudspeaker layout suitable for Lexicon's Logic 7 reproduction. Notice the additional side loudspeakers that enable a more enveloping image and may enable rear loudspeakers to be placed further to the rear.

speaker. The rear channels operate differently depending on whether the front channel content is primarily steered dialog/effects or music/ambience. In the former case the front signals are canceled from the rear channels and panned effects behave as they would with ProLogic, with surround effects panned 'full rear' appearing in mono on both rear channels. In the latter case the rear channels work in stereo, but reproducing the front left and right channels with special equalization and delay to create an enveloping spatial effect. The side channels carry steered information that attempts to ensure that effects which pan from left to rear pass through the left side on the way, and similarly for the right side with right-to-rear pans.

It is claimed that by using these techniques the effect of decoding a 3-1 matrix surround version of a 3-2 format movie can be brought close to that of the original 3-2 version. Matrix encoding of five channels to L_t/R_t is also possible with a separate algorithm, suitable for decoding to five or more loudspeakers using Logic 7.

Dolby EX

In 1998 Dolby and Lucasfilm THX joined forces to promote an enhanced surround system that added a center rear channel to the standard 5.1-channel setup. They introduced it, apparently, because of frustrations felt by sound designers for movies in not being able to pan sounds properly to the rear of the listener – the surround effect typically being rather diffuse. This system was christened 'Dolby Digital – Surround EX', and apparently uses matrix-style center channel encoding and decoding between the left and right surround channels of a 5.1-channel mix. The loudspeakers at the rear of the auditorium are then driven separately from those on the left and right sides, using the feed from this 'rear-center' channel, as shown in Figure 17.9.

FIGURE 17.9

Dolby EX adds a center-rear channel fed from a matrix-decoded signal that was originally encoded between left and right surround channels in a manner similar to the conventional Dolby Stereo matrix process.

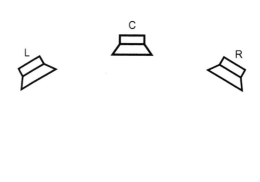

Matrix derived
rear centre speaker

DIGITAL SURROUND SOUND FORMATS

Data-reduced digital encoding has largely replaced analog matrix encoding for surround sound and is now covered in Chapter 8.

AMBISONICS

Principles

The Ambisonic system of directional sound pickup and reproduction is discussed here because of its relative thoroughness as a unified system, being based on some key principles of psychoacoustics. It has its theoretical basis in work by Gerzon, Barton and Fellgett in the 1970s, as well as work undertaken earlier by Cooper and Shiga.

Ambisonics aims to offer a complete hierarchical approach to directional sound pickup, storage or transmission and reproduction, which is equally applicable to mono, stereo, horizontal surround sound, or full 'periphonic' reproduction including height information. Depending on the number of channels employed it is possible to represent a lesser or greater number of dimensions in the reproduced sound. A number of formats exist for signals in the Ambisonic system, as detailed in the next section. A format known as UHJ ('Universal HJ', 'HJ' simply being the letters denoting two earlier surround sound systems) is also used for encoding multichannel Ambisonic information into two or three channels whilst retaining good mono and stereo compatibility for 'non-surround' listeners. Thus, Ambisonically-encoded material can be released as a conventional two-channel stereo recording and if required a UHJ decoder can be used to convert it into surround sound.

Ambisonic sound should be distinguished from quadraphonic sound, since quadraphonics explicitly requires the use of four loudspeaker channels, and cannot be adapted to the wide variety of pickup and listening situations that may be encountered. Quadraphonics generally works by creating conventional stereo phantom images between each pair of speakers and, as Gerzon states, conventional stereo does not perform well when the listener is off-center or when the loudspeakers subtend an angle larger than 60°. Since in quadraphonic reproduction the loudspeakers are angled at roughly 90° there is a tendency towards a hole-in-the-middle, as well as there being the problem that conventional stereo theories do not apply correctly for speaker pairs to the side of the listener. Ambisonics, however, encodes sounds from all directions in terms of pressure and velocity components, and decodes these signals to a number of loudspeakers, with psychoacoustically optimized shelf filtering above 700 Hz to correct for the shadowing effects of the head. It also incorporates an amplitude matrix that determines the correct levels for each speaker for the layout chosen, and can therefore be decoded correctly for 5.1 speaker layouts, for instance.

Ambisonics might thus be considered as the theoretical successor to coincident stereo on two loudspeakers, since it is the logical extension of Blumlein's principles to surround sound.

The source of an Ambisonic signal may be an Ambisonic microphone such as the Calrec Soundfield, or it may be an artificially panned mono signal, split into the correct B-format components (see below) and placed in a position around the listener by adjusting the ratios between the signals.

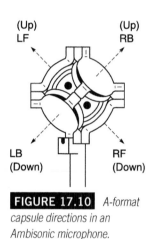

FIGURE 17.10 *A-format capsule directions in an Ambisonic microphone.*

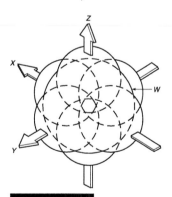

FIGURE 17.11 *The B-format components W, X, Y and Z in Ambisonics represent an omnidirectional pressure component and three orthogonal velocity (figure-eight) components of the sound field respectively.*

Signal formats

As indicated above, there are four basic signal formats for Ambisonic sound: A, B, C and D. The A-format consists of the four signals from a microphone with four sub-cardioid capsules orientated as shown in Figure 17.10 (or the pan-pot equivalent of such signals). These are capsules mounted on the four faces of a tetrahedron, and correspond to left-front (LF), right-front (RF), left-back (LB) and right-back (RB), although two of the capsules point upwards and two point downwards. Such signals should be equalized so as to represent the soundfield at the center of the tetrahedron, since the capsules will not be perfectly coincident.

The B-format consists of four signals that between them represent the pressure and velocity components of the sound field in any direction, as shown in Figure 17.11. It can be seen that there is a similarity with the sum and difference format of two channel stereo, described in the previous chapter, since the B-format is made up of three orthogonal figure-eight components (X, Y and Z), and an omni component (W). All directions in the horizontal plane may be represented by scalar and vector combinations of W, X and Y, whilst Z is required for height information. X is equivalent to a forward-facing figure-eight (equivalent to M in MS stereo), Y is equivalent to a sideways-facing figure-eight (equivalent to S in MS stereo). The X, Y and Z components have a frontal, sideways or upwards gain of 3 dB or $\sqrt{2}$ with relation to the W signal (0 dB) in order to achieve roughly similar energy responses for sources in different positions. B-format signals may also be created directly by arranging capsules or individual microphones in the B-format mode (two or three figure-eights at 90° plus an omni). The Z component is not necessary for horizontal information. If B-format signals are recorded instead of speaker feeds (D-format), subsequent manipulation of the soundfield is possible, and the signal will be somewhat more robust to interchannel errors.

The C-format consists of four signals L, R, T and Q, which conform to the UHJ hierarchy, and are the signals used for mono- or stereo-compatible transmission or recording. The C-format is, in effect, a useful consumer matrix format. L is a two-channel-compatible left channel, R is the corresponding right channel, T is a third channel which allows more accurate horizontal decoding, and Q is a fourth channel containing height information. The proportions of B-format signals which are combined to make up a C-format signal have been carefully optimized for the best compatibility with conventional stereo and mono reproduction.

Two, three, or four channels of the C-format signal may be used depending on the degree of directional resolution required, with a two-and-a-half channel option available where the third channel (T) is of limited bandwidth. For stereo compatibility only L and R are used. The UHJ or C-format hierarchy is depicted graphically in Figure 17.12.

D-format signals are those distributed to loudspeakers for reproduction, and are adjusted depending on the selected loudspeaker layout. They may be derived from either B- or C-format signals using an appropriate decoder, and the number of speakers is not limited in theory, nor is the layout constrained to a square. Four speakers give adequate surround sound, whilst six provide better immunity against the drawing of transient and sibilant signals towards a particular speaker, and eight may be used for full periphony with height. The decoding of B- and C-format components into loudspeaker signals is too complicated and lengthy a matter to go into here, and is the subject of several patents that were granted to the NRDC (the UK National Research and Development Council, as was). It is sufficient to say that the principle of decoding involves the passing of two or more UHJ signals via a phase-amplitude matrix, resulting in B-format signals that are subjected to

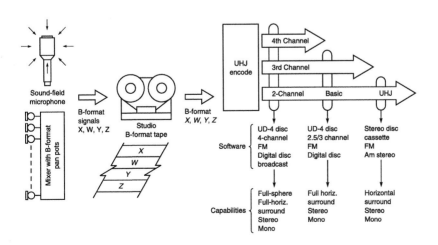

FIGURE 17.12

The C-format or UHJ hierarchy enables a variety of matrix encoding forms for stereo signals, depending on the amount of spatial information to be conveyed and the number of channels available.

shelf filters (in order to correct the levels for head-related transfer functions such as shadowing and diffraction). These are passed through an amplitude matrix which feeds the loudspeakers (see Figure 17.13). A layout control is used to vary the level sent to each speaker depending on the physical arrangement of speakers. See also Fact File 17.4. AES Convention Paper 5788, available on-line, provides further information about this surround format, including higher order Ambisonics.

FIGURE 17.13

C-format signals are decoded to provide D-format signals for loudspeaker reproduction.

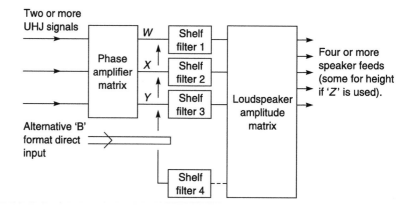

FACT FILE 17.4 LOUDSPEAKER MOUNTING

In many studios it is traditional to mount the monitor loudspeakers flush with the front wall. This has the particular advantage of avoiding the reflection that occurs with free-standing loudspeakers from the wall behind the loudspeaker, causing a degree of cancelation at a frequency where the spacing is equal to one-quarter of the radiated wavelength. It also improves the low-frequency radiation conditions if the front walls are hard. Nonetheless, it is hard to find places to mount five large loudspeakers in a flush-mounted configuration, and such mounting methods can be expensive. Furthermore the problems noted above, of detrimental reflections from rear loudspeakers off a hard front wall or speaker enclosure, can arise, depending on the angle of the rear loudspeakers. For such reasons, some sources recommend making the surfaces around the loudspeakers reflective at low frequencies and absorbent at mid and high frequencies.

The problem of low-frequency cancelation notches with free-standing loudspeakers can be alleviated but not completely removed. The perceived depth of the notch depends on the absorption of the surface and the directivity of the loudspeaker. By adjusting the spacing between the speaker and the wall, the frequency of the notch can be moved (downwards by making the distance greater), but the distance needed is often too great to be practical. If the speaker is moved close to the wall the notch position rises in frequency. This can be satisfactory for large loudspeakers whose directivity is high enough at middle frequencies to avoid too much rear radiation, but is a problem for smaller loudspeakers.

The use of a 5.1-channel monitoring arrangement (rather than five full-bandwidth loudspeakers), with proper bass management and crossovers, can in fact ameliorate the problems of free-standing loudspeakers considerably. This is because a subwoofer can be used to handle frequencies below 80–120 Hz and it can be placed in the corner or near a wall where the cancelation problem is minimized. Furthermore, the low-frequency range of the main loudspeakers can then be limited so that the cancelation notch mentioned above occurs below their cut-off frequency.

FURTHER DEVELOPMENTS

Other surround formats will be described presently, and with the number of replay channels increasing as this area of activity develops, it is pertinent here to consider how many channels would be necessary for convincing, comprehensive coverage. Research indicates that the lateral directional resolution of the hearing mechanism can achieve an accuracy of about four degrees under ideal conditions, an exact figure being impossible to quote because it depends upon the frequency and transient content of a sound. It also depends upon different combinations of subject and environment. Under more general conditions, lateral resolution is probably little better than 10 degrees. This can be compared with the positioning resolution of the eyes, which is a fraction of a degree. Few bow and arrow shots and spear throws would reach their moving targets if they depended upon the ears alone, resolution being a function of wavelength of sound that the ears can respond to. Bats of course use ultrasonic frequencies with their very short wavelengths to achieve sufficient resolution to enable them to forage and avoid dangers in darkness. A format such as Dolby Atmos with its 64 replay speaker capability may therefore be capable of delivering directional information with a degree of resolution comparable with the hearing's ability to resolve it. What one sees with one's eyes on the projection screen, as in real life, is what supplies the pin-point homing in to fix the direction of an object precisely. The fact that the ear's ability to resolve height information is rather less acute than in the lateral plain suggests that a comparatively small number of loudspeakers could give adequate representation of height.

A brief description of recent surround formats follows.

Auro-3D

The Auro-3D system has been developed for with-height surround sound in 9.1 channels, and for compatibility reasons it is embedded in the standard 5.1 format. Height information is created 30 degrees above the four main front-left, front-right, rear-left and rear-right channels with respect to a central listening position. A full Auro-3D system is specified as 13.1, the 11.1 format being regarded as the ideal format for cinema replay. Extra channels are added by applying the Auro-3D Octopus codec: it first reduces information of existing channels by forming data subsets and equating of adjacent samples. This makes room for corresponding samples of additional channels to be added, and seed samples of the original data are embedded such that two composite channels can be separated on replay by the application of a mathematical algorithm. The system designers point out that the least significant bits of a 24-bit system, a routine production bit-rate, cannot in practice be used for replay because the dynamic range of 24 bits, theoretically 144 dB, is far too large to reproduce. The codec therefore continuously monitors the signal during encoding and reduces its dynamic range so that it can be accommodated within 18 bits or fewer, freeing the lower bits which can now be used to carry decoding information.

Dolby Atmos

Dolby Atmos is a cinema surround sound system, providing up to 128 discrete audio input tracks feeding up to 64 separate loudspeaker feeds including overhead channels. A specific channel format can be supplied to a particular cinema which is optimized for its setup and replay capability. Figure 17.14 shows the basic processing chain. Dolby Atmos supports 'beds' – channel-based sub-mixes or systems which contain a variety of background atmospheric sounds, and combines these with objects – and specific foreground sounds such as dialog and story-telling effects, to produce an Atmos object and bed combination. A 'print master' is created during mastering which contains bed and object audio data together with metadata; this contains the Dolby Atmos mix along with Dolby Surround 7.1 and 5.1 mixes as needed. Material Exchange Format (MXF) wrapping techniques are used to deliver it to the digital cinema packaging facility via the standard Digital Cinema Project (DCP) format. The Dolby Atmos-equipped cinema server recognizes the format and processes it for rendering; an Ethernet connection between the server and cinema processor allows the audio to be identified and synchronized. Cinema servers without the appropriate decoding simply ignore it and reproduce the standard 5.1 or 7.1 information which exists alongside.

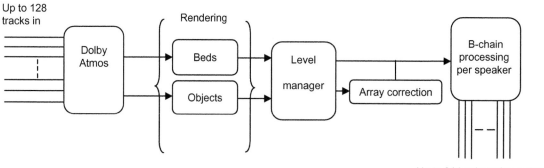

FIGURE 17.14 *Basic Dolby Atmos processing chain.*

A complex installation is required for full rendering, and Dolby Labs supply a setup service which includes comprehensive room analysis with equalization and level matching of loudspeakers. Dolby also recognizes that the system must be compatible with existing cinema layouts and future updated systems which still fall some way short of the ultimate 64-speaker implementation, and their setup is configured accordingly.

NHK 22.2

The NHK 22.2 or Hamasaki 22.2 format, named after its Japanese inventor, is a surround sound system to partner Super Hi-Vision, a television format having sixteen times the resolution of HDTV. It is worth mentioning here that Super Hi-Vision sound mixing recognizes that distance perception depends not only upon volume of a sound but also upon tonal quality and the direct-to-reverberation ratio as perceived from the listening position, and so mixer channels are equipped with two faders: one controls level in the conventional manner, the other controls the direct/reflected sound ratio of the object, thus controlling perceived distance.

22.2 supports 24 loudspeakers and two subwoofers arranged essentially in two layers, or heights:

CHANNELS 1–6: are equivalent to the familiar 5.1 layout.
CHANNELS 7–12: are a further five listener-level channels: FL centre, FR centre, back centre, side L, side R, plus a second subwoofer.
CHANNELS 13–21: are nine upper level channels, eight arranged around the periphery of the listening room, one directly overhead in the centre.
CHANNELS 22–24: are three lower front channels.

An additional low frequency effects channel (LFE) is also available. It is unusual for loudspeakers to be positioned at a height lower than the listening positions in surround sound formats, and it is more commonly encountered in live theater work where speakers can be positioned under rows of seats and beneath raised stages and stage traps, but these positions tend to be show specific. Cinema seating tends to be tiered, and one anticipates that future surround formats may use the voids underneath to house loudspeakers firing upwards through an acoustically transparent floor, giving complete envelopment.

For live transmission, NHK 22.2 experiments have used the MPEG-2 coding scheme, the video data being compressed to less than 600 Mbit/sec. from the original data rate of 24 GB/sec. A 32-channel digital audio signal, with 48 K, 24 bit resolution, was embedded in the MPEG-2 stream.

Wave field synthesis

Wave Field Synthesis (WFS) is based on the Huygens-Fresnel principle which was originally developed for the analysis of light wave propagation. Christiaan Huygens (1629–95) argued that light consisted of waves, but Isaac Newton's particle theory was the one generally accepted because of the latter's prestige in the scientific community. Augustin-Jean Fresnel (1788–1827) however established by both theory and experiment that light was a wave phenomenon. The principle states that a light (or acoustical) wavefront can be regarded as the result of a superposition of a multitude of elementary spherical waves: each point of the wavefront can be regarded as the starting point of an elementary wave. Figure 16.1 illustrates early spatial recording and reproduction ideas for film which involved the use of a large number of microphones arranged in a line across the front of a stage feeding the same number of loudspeakers in a line in front of the listeners in the listening room. The resulting wavefront, created by the array of essentially hemispherical point sources of sound, created the information necessary for the ears to perceive directional information regarding the original sound field. Because the wavefront is created from many sound sources rather than just the two of conventional stereo, perceived positioning of sound sources is rather more consistent regardless of the lateral position of the listener. Conventional stereo is prone to image shift with comparatively little movements of the head because of the precedence effect, and also because of the various amplitude differences between the two speakers as heard from different listening positions.

The Kirchhoff-Helmholtz integral, too complex for brief description here (see AES Convention Paper 5788 on-line, which also provides information on higher order Ambisonics) indicates that the wave front consists of a continuous distribution of secondary sources, each emanating from both velocity (pressure gradient) and pressure primary source components. Thus, the ideal microphone array should consist of both pressure (omni) and pressure gradient (figure of eight) types, and ideally the reproducing loudspeakers should also consist of both dipole and essentially conventional hemispherical types.

A practical appreciation of how the system can work may be gleaned by first imagining a stone being dropped into a pond of water. The waves that are produced by the disturbance radiate uniformly in concentric circles towards the edge of the pond. As a wave approaches, one can draw an imaginary tangent and a perpendicular line where the tangent just touches the wave. The perpendicular line will point to the place where the stone entered the water, giving an accurate indication of the direction of the original disturbance without the observer necessarily having witnessed it. Figure 17.15a shows how this is accomplished with sound. Initially, the sound waves radiate from the source and are picked up by a large number of microphones in a row which 'sample' the wave front. These signals are conveyed to an equally large number of loudspeakers which are shown occupying effectively the same positions in space as the microphones, the latter used during the recording process, the former during listening. The loudspeakers can be seen to recreate the original wave front in the listening space as if it has passed through the dividing wall, supplying the ears with the necessary information for the perception of the direction of the original sound. Returning to the pond analogy, if one now moves to a different place at the edge of the pond, one can again use the approaching wave to deduce the direction of the original disturbance just as before, and so it is with the sound wave-front as illustrated in Figure 17.15b. Whatever

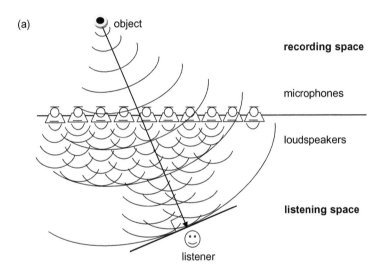

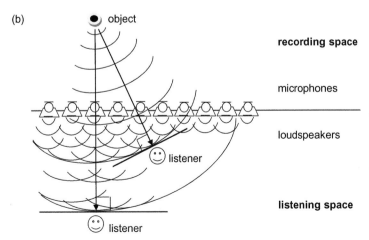

FIGURE 17.15

a) In Wave Field Synthesis, a multitude of loudspeaker outputs in the listening space reproduce the wave front created by the original sound source in the recording space. b) The sound source remains in apparently the same position for a variety of listening positions.

the listening position, the sound source will appear to come from the same direction, and this phenomenon gives WFS a distinct advantage over stereo and other surround sound formats which rely upon amplitude and to a lesser extent time of arrival differences between loudspeakers to give directional clues. In these systems, changes in listening positions result in different amplitude relationships between loudspeakers which result in image shifting, something which WFS avoids.

An additional factor is present in the practical implementation of the technique. As well as the listener hearing the output from the loudspeaker which lies on a direct line (the shortest path) between sound source and listening position, an ideal which would fix the sound source firmly in place, output from other speakers in its near vicinity will also be heard. This would tend to produce blurred or over-wide images if it were not for the fact that the speaker lying on the direct line delivers the sound to the ears *first*, the path length from sound source to listener via any other combination of microphone and loudspeaker being longer. The precedence effect therefore ensures that a firm image is perceived, and each listening position will have a particular loudspeaker lying on a direct line between listener and sound source which helps to ensure a firm image.

The Wave Field Synthesis technique can also produce images in front of the speakers in the listening space by creating a concave wave front. For instance speakers to the left and right can reproduce the sound of a central source ahead of the sound from the central speakers (in practice, the signals sent to the central speakers are delayed), creating in–the-room or even in-the-head images. It is the equivalent of the sound from an object in the listening space being reflected back to the listener by a concave surface.

A complete implementation of Wave Field Synthesis would involve the use of a large number of microphones positioned both in a line in front of the sound stage, and also at different heights and around the walls of the room to capture height and reverberation information accurately. But there is more than one possible approach to the placing of microphones. For instance, one can place them in positions so that they envelop the entire listening area, the performers being outside of this, so that reproduction again places the listeners in that same area. An alternative is to envelop the performers in an array of microphones, reproduction then placing the listeners among the musicians. In the listening room a similar number of loudspeakers would need to be placed in positions corresponding to the microphone positions. An approximation to the ideal was achieved at a concert in 2008 in Cologne cathedral, reproduced in a lecture hall by the Technical University of Berlin, where 2700 loudspeakers fed by 832 independent channels comprised the replay system. This emphasizes the necessity to develop rationalized systems for practical implementation of the idea capable of general adoption. The use of cardioid microphones with their tendency towards omni at low frequencies and a narrower pickup at high frequencies, rather than a combination of omnis and figure-of-eights, together with conventional loudspeakers with their tendency towards an omni polar pattern at low frequencies and a somewhat narrower dispersion at HF, has been found to complement the system.

Alternatively, rather fewer microphones can be used close up to the sound sources, their signals processed to give appropriate amplitude and time delays for allocation to the replay channels for the required perceived positions during replay. Imagine in Figure 17.15a that the object is very close miced, the signal from it being conveyed to the loudspeakers with the appropriate delays and amplitude levels, *as if* the large array of microphones were picking up the sound. This gives complete control over both the sound balance and the natural acoustics, and would be appropriate for non-classical music recording and close-miced live events. It also introduces flexibility and creativity during the recording and post-production stages. Speakers can be conventional types with essentially hemispherical or fan-like dispersion, arranged around the periphery of the listening room to give good horizontal directional information rather than full surround sound with height. A problem with a simplified replay system is that too few speakers can cause aliasing. At certain frequencies speakers will be spaced at half or whole wavelengths or multiples thereof, and a particular speaker can therefore give information that is apparently the same as that from an adjacent speaker. This, and missing information that would otherwise be present from a continuous array of speakers, can result in ambiguous directional information to be presented.

SURROUND SOUND MONITORING

This section is principally concerned with monitoring environments and configurations for 5.1-channel surround sound, although many of the principles may be found to apply in other configurations. The Audio Engineering Society has published an information document on this topic, containing more detailed guidelines (see 'Recommended further reading' at the end of this chapter).

Differences between two-channel and surround mixing rooms

There is a gradual consensus building around the view that rooms for surround monitoring should have an even distribution of absorbing and diffusing material. This is so that the rear loudspeakers function in a similar acoustic environment to the front loudspeakers. This is contrary to a number of popular two-channel control room designs that have one highly absorbtive end and the other end more reflective.

The effects of the acoustics of non-ideal control room acoustics on surround channels may be ameliorated if a distributed array of surround loudspeakers is used, preferably with some form of decorrelation between them to avoid strong comb filtering effects. (Appropriate gain/EQ modification should also be applied to compensate for the acoustic summing of their outputs.) This is more akin to the film sound situation, though, and may only be possible in larger dubbing stages. In smaller control rooms used for music and broadcasting mixing the space may not exist for such arrays. The ITU standard allows for more than one surround loudspeaker on either side and recommends that they are spaced equally on an arc from 60 to 150° from the front.

One of the difficulties of installing loudspeaker layouts according to the ITU standard, with equal spacing from the listening position and the surrounds at 110° ± 10°, is the required width of the space. This arrangement often makes it appropriate for the room to be laid out 'wide' rather than 'long' (as it might be for two-channel setups). If the room is one that was previously designed for two-channel stereo the rotation of the axis of symmetry may result in the acoustic treatment being inappropriately distributed. Also the location of doors and windows may make the modification of existing rooms difficult. If building a new room for surround monitoring then it is obviously possible to start from scratch and make the room wide enough to accommodate the surround loudspeakers and absorption in more suitable places. See also Fact File 17.4.

Front loudspeakers in general

As a rule, front loudspeakers can be similar to those used for two-channel stereo, although noting the particular problems with the center loudspeaker described in the next section. It has been suggested that low-directivity front loudspeakers may be desirable when trying to emulate the effect of a film mixing situation in a smaller surround control room. This is because in the large rooms typical of cinema listening the sound balancer is often well beyond the critical distance where direct and reflected sound are equal in level, and using speakers with low directivity helps to emulate this scenario in smaller rooms. Film mixers generally want to hear what the large auditorium audience member would hear, and this means being further from the loudspeakers than for small room domestic listening or conventional music mixing.

What to do with the center loudspeaker

One of the main problems encountered with surround monitoring is that of where to put the center loudspeaker in a mixing room. Ideally it should be of the same type or quality as the rest of the channels and this can make such speakers quite large. In 5.1 surround setups there is an increasing tendency to use somewhat smaller monitors for the five main channels than would be used for two-channel setups, handling the low bass by means of a subwoofer or two. This makes it more practical to mount a center loudspeaker behind the mixing console, but its height will often be dictated by a control room window or video monitor (see below). The center loudspeaker should be on the same arc as that bounding the other loudspeaker positions, as shown in the ITU layout above, otherwise the time delay of its direct sound at the listening position will be different from that of the other channels. If the center speaker is closer than the left or right channels then it should be delayed slightly to put it back in the correct place acoustically.

The biggest problem with the center loudspeaker arises when there is a video display present. A lot of 5.1 surround work is carried out in conjunction with pictures and clearly the display is likely to be in exactly the same place as one wants to put the center speaker. In cinemas this is normally solved by making the screen acoustically 'transparent' and using front projection, although this transparency is never complete and usually requires some equalization. In smaller mixing rooms the display is often a flat-screen plasma monitor or a CRT display and these do not allow the same arrangement.

With modestly sized solid displays for television purposes it can be possible to put the center loudspeaker underneath the display, with the display raised slightly, or above the display angled down slightly. The presence of a mixing console may dictate which of these is possible, and care should be taken to avoid strong reflections from the center loudspeaker off the console surface. Neither position is ideal and the problem may not be solved easily. Dolby suggests that if the center loudspeaker has to be offset height-wise it could be turned upside down compared with the left and right channels to make the tweeters line up, as shown in Figure 17.16.

Interestingly, the flat-panel loudspeaker company, NXT, has shown large flat-panel loudspeakers that can double as projection display screens, which may be one way forward if the sound quality of the flat panel speakers can be made high enough.

Surround loudspeakers

The standard recommendations for professional setups suggest that the surround loudspeakers should be of the same quality as the front ones.

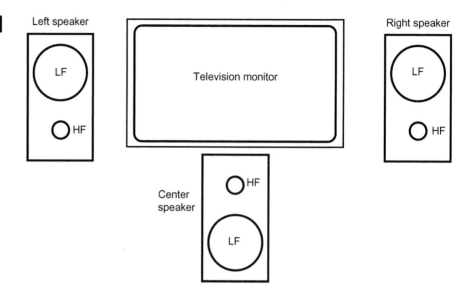

FIGURE 17.16
Possible arrangement of the centre loudspeaker in the presence of a TV screen, aligning HF units more closely.

This is partly to ensure a degree of inter-system compatibility. In consumer environments this can be difficult to achieve, and the systems sold at the lower end of the market often incorporate much smaller surround loud-speakers than front. As mentioned above, the use of a separate loudspeaker to handle the low bass (a so-called 'subwoofer') may help to ameliorate this situation, as it makes the required volume of all the main speakers quite a lot smaller. Indeed Bose has had considerable success with a consumer system involving extremely small satellite speakers for the mid–high-frequency content of the replay system, mountable virtually anywhere in the room, coupled with a low-frequency driver that can be situated somewhere unobtrusive.

The directivity requirements of the surround loudspeakers have been the basis of some considerable disagreement in recent years. The debate centers around the use of the surround loudspeakers to create a diffuse, enveloping soundfield – a criterion that tends to favor either decorrelated arrays of direct radiators (speakers that produce their maximum output in the direction of the listener) or dipole surrounds (bi-directional speakers that are typically oriented so that their main axes do not point towards the listener). If the creation of a diffuse, enveloping rear and side soundfield is the only role for surround loudspeakers then dipoles can be quite suitable if only two loudspeaker positions are available. If, on the other hand, attempts are to be made at all-round source localization (which, despite the evidence in some literature, is not entirely out of the question), direct radiators are considered more suitable. Given the physical restrictions in the majority of control rooms it is likely that conventional loudspeakers will be more practical to install than dipoles (for the reason that dipoles, by their nature, need to be free-standing, away from the walls) whereas conventional speakers can be mounted flush with surfaces.

A lot depends on the application, since film sound mixing has somewhat different requirements from some other forms of mixing, and is intended for large auditoria. Much music and television sound is intended for small-room listening and is mixed in small rooms. This was also the primary motivation behind the use of dipoles in consumer environments – that is, the translation of the large-room listening experience into the small room. In large rooms the listener is typically further into the diffuse field than in small rooms, so film mixes made in large dubbing stages tend not to sound right in smaller rooms with highly directional loudspeakers. Dipoles or arrays can help to translate the listening experience of large-room mixes into smaller rooms.

Subwoofers

Low-frequency interaction between loudspeakers and rooms has a substantial bearing on the placement of subwoofers or low-frequency loudspeakers. There appears to be little agreement about the optimum location for a single subwoofer in a listening room, although it has been suggested that a corner location for a single subwoofer provides the most extended, smoothest, low-frequency response. In choosing the optimum locations for subwoofers one must remember the basic principle that loudspeakers placed in corners tend to give rise to a noticeable bass boost, and couple well to most room modes (because they have antinodes in the corners). Some subwoofers are designed specifically for placement in particular locations whereas others need to be moved around until the most subjectively satisfactory result is obtained. Some artificial equalization may be required to obtain a reasonably flat overall frequency response at the listening position. Phase shifts or time-delay controls are sometimes provided to enable some correction of the time relationship of the subwoofer to other loudspeakers, but this will necessarily be a compromise with a single unit. A subwoofer phase shift can be used to optimize the sum of the subwoofer and main loudspeakers in the crossover region for a flat response.

There is some evidence to suggest that multiple low-frequency drivers generating decorrelated signals from the original recording create a more natural spatial reproduction than monaural low-frequency reproduction from a single driver. Griesinger proposes that if monaural LF content is reproduced it is better done through two units placed to the sides of the listener, driven 90° out of phase, to excite the asymmetrical lateral modes more successfully and improve LF spaciousness.

Others warn of the dangers of multiple low-frequency drivers, particularly the problem of mutual coupling between loudspeakers that takes place when the driver spacing is less than about half a wavelength. In such situations the outputs of the drivers couple to produce a level greater than would be predicted from simple summation of the powers. This is due to the way in which the drivers couple to the impedance of the air and the effect that one unit has on the radiation impedance of the other. The effect of this coupling will depend on the positions to which sources are panned between drivers, affecting the compatibility between the equalization of mixes made for different numbers of loudspeakers.

SURROUND SOUND RECORDING TECHNIQUES

This section deals with the extension of the conventional two-channel recording technique to multiple channels for surround sound applications, concentrating on standard 5(.1)-channel reproduction. Many of the concepts described here have at least some basis in conventional two-channel stereo, although analysis of the psychoacoustics of 5.1 surround has been nothing like as exhaustively investigated to date. Consequently a number of the techniques described below are at a relatively early stage of development and are still being evaluated.

The section begins with a review of microphone techniques that have been proposed for the pickup of natural acoustic sources in surround, followed by a discussion of multichannel panning and mixing techniques, mixing aesthetics and artificial reverberation, for use with more artificial forms of production such as pop music. Film sound approaches are not covered in any detail as they are well established and not the main theme of this book.

Principles of surround sound microphone technique

Surround sound microphone technique, as discussed here, is unashamedly biased towards the pickup of sound for 5.1 surround, although Ambisonic techniques are also covered because they are well documented and can be reproduced over five-channel loudspeaker systems if required, using suitable decoders. The techniques described in this section are most appropriate for use when the spatial acoustics of the environment are as important as those of the sources within, such as in classical music and other 'natural' recording. These microphone techniques tend to split into two main groups: those that are based on a single array of microphones in reasonably close proximity to each other, and those that treat the front and rear channels separately. The former are usually based on some theory that attempts to generate phantom images with different degrees of accuracy around the full 360° in the horizontal plane. (The problems of this are outlined in Fact File 17.5.) The latter usually have a front array providing reasonably accurate phantom images in the front, coupled with a separate means of capturing the ambient sound of the recording space (often for feeding to all channels in varying degrees). It is rare for such microphone techniques to provide a separate feed for the LFE channel, so they are really five-channel techniques not 5.1-channel techniques.

The concept of a 'main array' or 'main microphone configuration' for stereo sound recording is unusual to some recording engineers, possibly being a more European than American concept. The traditional European approach has tended to involve starting with a main microphone technique of some sort that provides a basic stereo image and captures the spatial effect of the recording environment in an aesthetically satisfactory way, and then supporting this subtly to varying degrees with spot mics as necessary. It has been suggested by some that many balances in fact end up with more sound coming from the spot mics than from the main array in practice, and that in this case it is the spatial treatment of the spot mics and any artificial reverberation that will have most effect on the perceived result. This is covered in the next section and the issue is open to users for further experimentation.

One must accept also that the majority of consumer systems will have great variability in the location and nature of the surround loudspeakers, making it unwise to set too much store by the ability of such systems to enable accurate soundfield reconstruction in the home. Better, it seems, would be to acknowledge the limitations of such systems and to create recordings that work best on a properly configured reproduction arrangement but do not rely on 100% adherence to a particular reproduction alignment and layout, or on a limited 'hot spot' listening position. Surround sound provides an opportunity to create something that works over a much wider range of listening positions than two-channel stereo, does not collapse rapidly into the nearest loudspeaker when one moves, and enhances the spatial listening experience.

Five-channel 'main microphone' arrays

Recent interest in five-channel recording has led to a number of variants on a common theme involving fairly closely spaced microphones (often

FACT FILE 17.5 SURROUND IMAGING

It is difficult to create stable phantom images to the sides of a listener in a standard 5.1 surround configuration, using simple pairwise amplitude or time differences. If the listener turns to face the speaker pair then the situation may be improved somewhat, but the subtended angle of about 80° still results in something of a hole in the middle and the same problem as before then applies to the front and rear pairs. Phantom sources can be created between the rear speakers but the angle is again quite great (about 140°), leading to a potential hole in the middle for many techniques, with the sound pulling towards the loudspeakers. This suggests that those techniques attempting to provide 360° phantom imaging may meet with only limited success over a limited range of listening positions, and might imply that one would be better off working with two- or three-channel stereo in the front and decorrelated ambient signals in the rear.

There is no escaping the fact that it is easiest to create images where there are loudspeakers, and that phantom images between loudspeakers subtending wide angles tend to be unstable or 'hole-in-the-middle'. Given this unavoidable aspect of surround sound psychoacoustics, one should always expect imaging in standard five-channel replay systems to be best between the front loudspeakers, only moderate to the rear, and highly variable to the sides, as shown below. Since the majority of material one listens to tends to conform to this paradigm in any case (primary sources in front, secondary content to the sides and rear), the problem is possibly not as serious as it might seem.

Good phantom images
between left, center and right
loudspeakers

Typically poor and
unstable phantom images
between front and
surround loudspeakers

Typically poor and
unstable phantom images
between front and
surround loudspeakers

Only moderately satisfactory
phantom images between rear
loudspeakers, with a tendency
towards a 'hole in the middle'

cardioids) configured in a five-point array. The basis of most of these arrays is pairwise time–intensity trading, usually treating adjacent microphones as pairs covering a particular sector of the recording angle around the array. The generic layout of such arrays is shown in Figure 17.17. Cardioids or even supercardioids tend to be favored because of the increased direct-to-reverberant pickup they offer, and the interchannel level differences created for relatively modest spacings and angles, enabling the array to be mounted on a single piece of metalwork. The center microphone is typically spaced slightly forward of the L and R microphones thereby introducing a useful time advance in the center channel for center-front sources.

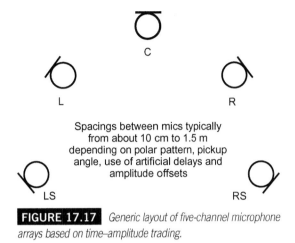

Spacings between mics typically from about 10 cm to 1.5 m depending on polar pattern, pickup angle, use of artificial delays and amplitude offsets

FIGURE 17.17 *Generic layout of five-channel microphone arrays based on time–amplitude trading.*

The spacing and angles between the capsules are typically based on the so-called 'Williams curves', based on time and amplitude differences required between single pairs of microphones to create phantom sources in particular locations. (In fact the Williams curves were based on two-channel pairs and loudspeaker reproduction in front of the listener. It is not necessarily the case that the same technique can be applied to create images between pairs at the sides of the listener, or that the same level and time differences will be suitable. There is some evidence that different delays are needed between side and rear pairs than those used between front pairs, and that inter-microphone crosstalk can affect the accuracy of stereo imaging to varying degrees depending on the array configuration and microphone type.) One possible configuration of many that satisfy Williams' psychoacoustic criteria is pictured in Figure 17.18. To satisfy the requirements for this particular array the front triplet is attenuated by 2.4 dB in relation to the back pair.

Some success has also been had by the author's colleagues using omni microphones instead of cardioids, with appropriate adjustments to the spacings according to 'Williams-style' time–amplitude trading curves (also with modifications to correct for different inter-loudspeaker angles and spacings to the sides and rear). These tend to give better overall sound quality but (possibly unsurprisingly) poorer front imaging. Side imaging has proved to be better than expected with omni arrays.

The closeness between the microphones in these arrays is likely to result in only modest low-frequency decorrelation between the channels.

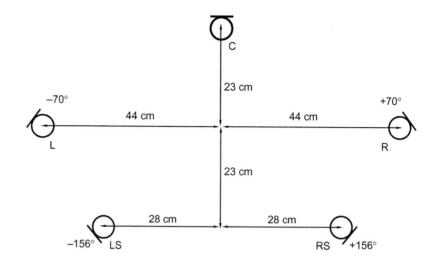

Good LF decorrelation is believed to be important for creating a sense of spaciousness, so these 'near-coincident' or 'semi-correlated' techniques will be less spacious than more widely spaced microphone arrays. Furthermore, the strong dependence of these arrays on precedence effect cues for localization makes their performance quite dependent on listener position and front–rear balance.

The INA (Ideale Nieren Anordnung) or 'Ideal Cardioid Array' (devised by Hermann and Henkels) is a three-channel front array of cardioids (INA-3) coupled with two surround microphones of the same polar pattern (making it into an INA-5 array). One configuration of this is shown in Figure 17.19, and a commercial implementation by Brauner is pictured in Figure 17.20. Table 17.1 shows some possible combinations of microphone spacing and recording angle for the front three microphones of this proposed array. In the commercial implementation the capsules can be moved and rotated and their polar patterns can be varied. The configuration shown in Figure 17.20 is termed an 'Atmokreuz' (atmosphere cross) by the authors. Its large front recording angle of 180° means that to use it as a main microphone it would have to be placed very close to the source unless all the sources were to appear to come from near the center. This might make it less well placed for the surroundings. Such a configuration may be more suitable for general pickup slightly further back in the hall.

Separate treatment of front imaging and ambience

Many alternative approaches to basic microphone coverage for 5.1 surround treat the stereo imaging of front signals separately from the capture

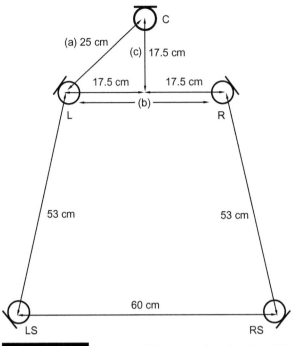

INA-5 cardioid array configuration. (see Table 17.1)

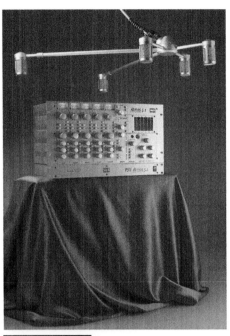

FIGURE 17.20 *SPL Atmos 5.1 Surround Recording System. (Courtesy of Sound Performance Lab.)*

Table 17.1 Dimensions and angles for the front three cardioid microphones of the INA array (see Figure 17.20). Note that the angle between the outer microphones should be the same as the recording angle

Recording angle (0)°	Microphone spacing (a) cm	Microphone spacing (b) cm	Array depth (c) cm
100	69	126	29
120	53	92	27
140	41	68	24
160	32	49	21
180	25	35	17.5

of a natural-sounding spatial reverberation and reflection component, and some are hybrid approaches without a clear theoretical basis. Most do this by adopting a three-channel variant on a conventional two-channel technique for the front channels, as introduced in the previous chapter (sometimes optimized for more direct sound than in a two-channel array), coupled with a more or less decorrelated combination of microphones in a different location for capturing spatial ambience (sometimes fed just to the surrounds, other times to both front and surrounds). Sometimes the front microphones also contribute to the capture of spatial ambience, depending on the proportion of direct to reflected sound picked up, but the essential point here is that the front and rear microphones are not intentionally configured as an attempt at a 360° imaging array.

The so-called 'Fukada Tree', shown in Figure 17.21, is based on a Decca Tree, but instead of using omni mics it mainly uses cardioids. The reason for this is to reduce the amount of reverberant sound pickup by the front mics. Omni outriggers are sometimes added as shown, typically panned between L–LS and R–RS, in an attempt to increase the breadth of orchestral pickup and to integrate front and rear elements. The rear mics are also cardioids and are typically located at approximately the critical distance of the space concerned (where the direct and reverberant components are equal). They are sometimes spaced further back than the front mics by nearly 2 meters, although the dimensions of the tree can be varied according to the situation, distance, etc. (Variants are known that have the rear mics quite close to the front ones, for example.) The spacing between the mics more closely fulfills requirements for the decorrelated microphone signals needed to create spaciousness, depending on the critical distance of the space in which they are used. (Mics should be separated by at least the room's critical distance for adequate decorrelation.) The front imaging of such an array would be similar to that of an ordinary Decca Tree (not bad, but not as precise as some other techniques).

FIGURE 17.21

The so-called 'Fukada Tree' of five spaced microphones for surround recording.

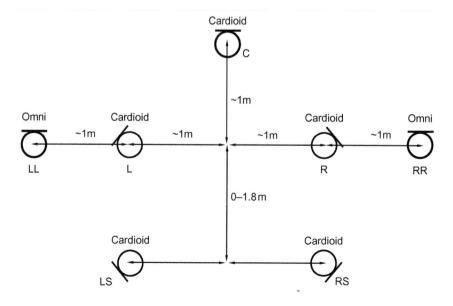

The Dutch recording company Polyhymnia International has developed a variant on this approach that uses omnis instead of cardioids, to take advantage of their better sound quality. Using an array of omnis separated by about 3 meters between left–right and front–back they achieve a spacious result where the rear channels are well integrated with the front. The center mic is placed slightly forward of left and right. It is claimed that placing the rear omnis too far away from the front tree makes the rear sound detached from the front image, so one gets a distinct echo or repeat of the front sound from the rear.

Hamasaki of NHK (the Japanese broadcasting company) has proposed an arrangement based on near-coincident cardioids (30 cm) separated by a baffle, as shown in Figure 17.22. Here the center cardioid is placed slightly forward of left and right, and omni outriggers are spaced by about 3 meters.

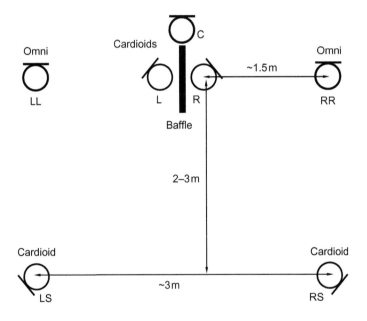

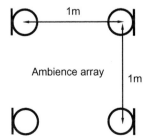

FIGURE 17.22

A surround technique proposed by Hamasaki (NHK) consisting of a cardioid array, omni outriggers and separate ambience matrix.

These omnis are low-pass filtered at 250 Hz and mixed with the left and right front signals to improve the LF sound quality. Left and right surround cardioids are spaced about 2–3 meters behind the front cardioids and 3 meters apart. An ambience array is used further back, consisting of four figure-eight mics facing sideways, spaced by about 1 meter, to capture lateral reflections, fed to the four outer channels. This is placed high in the recording space.

Theile proposes a front microphone arrangement shown in Figure 17.23. Whilst superficially similar to the front arrays described in the previous section, he reduces crosstalk between the channels by the use of supercardioid microphones at ±90° for the left and right channels and a cardioid for the center. (Supercardioids are more directional than cardioids and have the highest direct/reverberant pickup ratio of any first-order directional microphone. They have a smaller rear lobe than hypercardioids.) Theile's rationale behind this proposal is the avoidance of crosstalk between the front segments. He proposes to enhance the LF response of the array by using a hybrid microphone for left and right, which crosses over to omni below 100 Hz, thereby restoring the otherwise poor LF response. The center channel is high-pass filtered above 100 Hz. Furthermore, the response of the supercardioids should be equalized to have a flat response to signals at about 30° to the front of the array (they would normally sound quite colored at this angle). Schoeps has developed a prototype of this array, and it has been christened 'OCT' for 'Optimum Cardioid Triangle'.

For the ambient sound signal, Theile proposes the use of a crossed configuration of microphones, which has been christened the 'IRT cross' or 'atmo-cross'. This is shown in Figure 17.24. The microphones are either cardioids or omnis, and the spacing is chosen according to the degree of correlation desired between the channels. Theile suggests 25 cm for cardioids and about 40 cm for omnis, but says that this is open to experimentation. Small spacings are appropriate for more accurate imaging of reflection

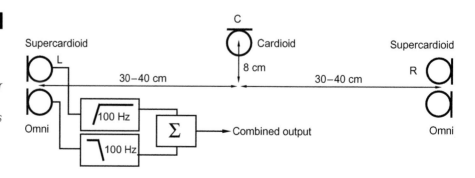

sources at the hot spot, whereas larger spacings are appropriate for providing diffuse reverberation over a large listening area. The signals are mixed in to L, R, LS and RS channels, but not the center.

A 'double MS' technique has been proposed by Curt Wittig and others, shown in Figure 17.25. Two MS pairs (see previous chapter) are used, one for the front channels and one for the rear. The center channel can be fed from the front M microphone. The rear pair is placed at or just beyond the room's critical distance. S gain can be varied to alter the image width in either sector, and the M mic's polar pattern can be chosen for the desired directional response (it would typically be a cardioid). Others have suggested using a fifth microphone (a cardioid) in front of the forward MS pair, to feed the center channel, delayed to time align it with the pair. If the front and rear MS pairs are co-located it may be necessary to delay the rear channels somewhat (10–30 ms) so as to reduce perceived spill from front sources into rear channels. In a co-located situation the same figure-eight microphone could be used as the S channel for both front and back pairs.

In general, the signals from separate ambience microphones fed to the rear loudspeakers may often be made less obtrusive and front–back 'spill' may be reduced by rolling off the high-frequency content of the rear channels. Some additional delay may also assist in the process of integrating the rear channel ambience. The precise values of delay and equalization can only really be arrived at by experimentation in each situation.

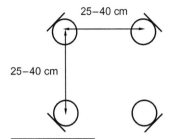

FIGURE 17.24 *The IRT 'atmo-cross' designed for picking up ambient sound for routing to four loudspeaker channels (omitting the centre). Mics can be cardioids or omnis (wider spacing for omnis).*

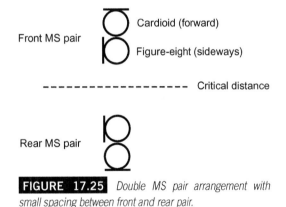

FIGURE 17.25 *Double MS pair arrangement with small spacing between front and rear pair.*

Pseudo-binaural techniques

As with two-channel stereo, some engineers have experimented with pseudo-binaural recording techniques intended for loudspeaker reproduction. Jerry Bruck adapted the Schoeps 'Sphere' microphone (described earlier) for surround sound purposes by adding bi-directional (figure-eight) microphones near to the 'ears' (omni mics) of the sphere, with their main axis front–back, as pictured in Figure 17.26. This microphone is now manufactured by Schoeps as the KFM360. The figure-eights are mounted just below the sphere transducers so as to affect their frequency response in as benign a way as possible for horizontal sources. The outputs from the figure-eight and the omni at each side of the sphere are MS matrixed

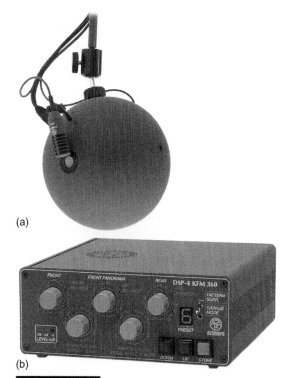

(a)

(b)

FIGURE 17.26 (a) Schoeps KFM360 sphere microphone with additional figure-eights near the surface-mounted omnis. (b) KFM360 control box. (Courtesy of Schalltechnik Dr.-Ing. Schoeps GmbH.)

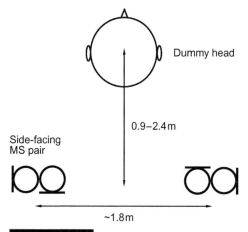

FIGURE 17.27 Double MS pairs facing sideways used to feed the side pairs of channels combined with a dummy head facing forwards to feed the front image.

to create pairs of roughly back-to-back cardioids facing sideways. The size of the sphere creates an approximately ORTF spacing between the side-facing pairs. The matrixed output of this microphone can be used to feed four of the channels in a five-channel reproduction format (L, R, LS and RS). A Schoeps processing unit can be used to derive an equalized center channel from the front two, and enables the patterns of front and rear coverage to be modified.

Michael Bishop of Telarc has reportedly adapted the 'double MS' technique described in the previous section by using MS pairs facing sideways, and a dummy head some 1–2.5 m in front, as shown in Figure 17.27. The MS pairs are used between side pairs of channels (L and LS, R and RS) and line-up is apparently tricky. The dummy head is a model equalized for a natural response on loudspeakers (Neumann KU100) and is used for the front image.

Multimicrophone techniques

Most real recording involves the use of spot microphones in addition to a main microphone technique of some sort, indeed in many situations the spot microphones may end up at higher levels than the main microphone or there may be no main microphone. The principles outlined in the previous chapter still apply in surround mixing, but now one has the issue of surround panning to contend with. The principles of this are covered in more detail in 'Multichannel panning techniques', below.

Some engineers report success with the use of multiple sphere microphones for surround balances, which is probably the result of the additional spatial cues generated by using a 'stereo' spot mic rather than a mono one, avoiding the flatness and lack of depth often associated with panned mono sources. Artificial reverberation of some sort is almost always helpful when trying to add spatial enhancement to panned mono sources, and some engineers prefer to use amplitude-panned signals to create a good balance in the front image, plus artificial reflections and reverberation to create a sense of spaciousness and depth.

Ambisonic or 'SoundField' microphone principles

The so-called 'SoundField' microphone, pictured in Figure 17.28, is designed for picking up full periphonic sound in the Ambisonic A-format (see 'Signal formats', above), and is coupled with a control box designed for converting the microphone output into both the B-format and the D-format. Decoders can be created for using the output of the SoundField microphone with a 5.1-channel loudspeaker array, including that recently introduced by SoundField Research. The full periphonic effect can only be obtained by reproduction through a suitable periphonic decoder and the use of a tetrahedral loudspeaker array with a height component, but the effect is quite stunning and worth the effort.

The SoundField microphone is capable of being steered electrically by using the control box, in terms of azimuth, elevation, tilt or dominance, and as such it is also a particularly useful stereo microphone for two-channel work.

(a)

(b)

(c)

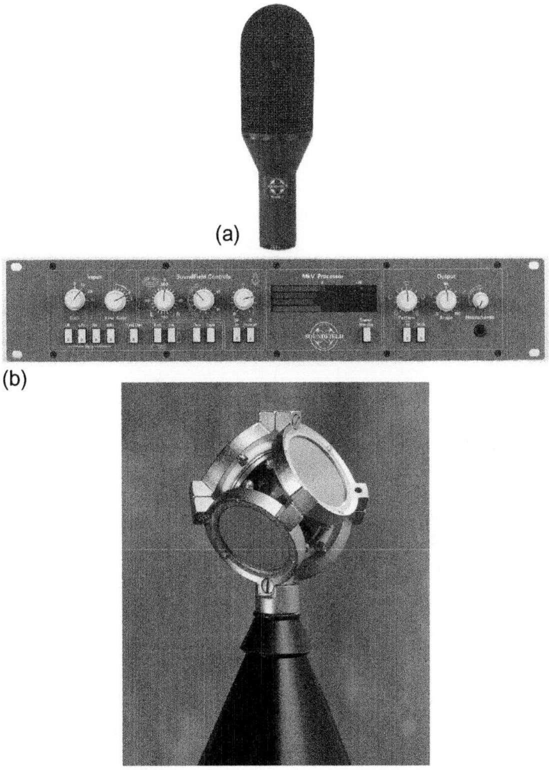

FIGURE 17.28 *(a) The SoundField microphone, (b) accompanying control box, (c) capsule arrangement. (Courtesy of SoundField Ltd.)*

The microphone encodes directional information in all planes, including the pressure and velocity components of indirect and reverberant sounds.

Figure 17.28c shows the physical capsule arrangement of the microphone, which was shown diagrammatically in Figure 17.10. Four capsules with sub-cardioid polar patterns (between cardioid and omni, with a response equal to $2 + \cos\theta$) are mounted so as to face in the A-format directions, with electronic equalization to compensate for the inter-capsule spacing, such that the output of the microphone truly represents the soundfield at a point (true coincidence is maintained up to about 10 kHz). The capsules are matched very closely and each contributes an equal amount to the B-format signal, thus resulting in cancelation between variations in inherent capsule responses. The A-format signal from the microphone can be converted to B-format according to the equations given in 'Signal formats', above.

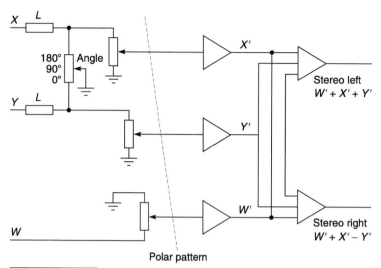

FIGURE 17.29 *Circuit used for controlling stereo angle and polar pattern in SoundField microphone. (Courtesy of Ken Farrar.)*

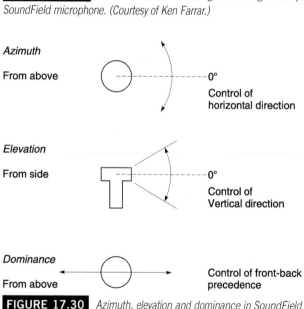

FIGURE 17.30 *Azimuth, elevation and dominance in SoundField microphone.*

The combination of B-format signals in various proportions can be used to derive virtually any polar pattern in a coincident configuration, using a simple circuit as shown in Figure 17.29 (two-channel example). Crossed figure-eights are the most obvious and simple stereo pair to synthesize, since this requires the sum-and-difference of X and Y, whilst a pattern such as crossed cardioids requires that the omni component be used also, such that:

$$Left = W + (X/2) + (Y/2)$$
$$Right = W + (X/2) - (Y/2)$$

From the circuit it will be seen that a control also exists for adjusting the effective angle between the synthesized pair of microphones, and that this works by varying the ratio between X and Y in a sine/cosine relationship.

The microphone may be controlled, without physical reorientation, so as to 'point' in virtually any direction (see Figure 17.30). It may also be electrically inverted, so that it may be used upside-down. Inversion of the microphone is made possible by providing a switch which reverses the phase of Y and Z components. W and X may remain unchanged since their directions do not change if the microphone is used upside-down.

MULTICHANNEL PANNING TECHNIQUES

The panning of signals between more than two loudspeakers presents a number of psychoacoustic problems, particularly with regard to appropriate energy distribution of signals, accuracy of phantom source localization, off-center listening and sound timbre. A number of different solutions have been proposed, in addition to the relatively crude pairwise approach used in much film sound, and some of these are outlined below. The issue of source distance simulation is also discussed.

Here are Michael Gerzon's criteria for a good panning law for surround sound:

> *The aim of a good panpot law is to take monophonic sounds, and to give each one amplitude gains, one for each loudspeaker, dependent on the intended illusory directional localisation of that sound, such that the resulting reproduced sound provides a convincing and sharp phantom illusory image. Such a good panpot law should provide a smoothly continuous range of image directions for any direction between those of the two outermost loudspeakers, with no 'bunching' of images close to any one direction or 'holes' in which the illusory imaging is very poor.*

Pairwise amplitude panning

Pairwise amplitude panning is the type of pan control most recording engineers are familiar with, as it is the approach used on most two-channel mixers. As described in the previous chapter, it involves adjusting the relative amplitudes between a pair of adjacent loudspeakers so as to create a phantom image at some point between them. This has been extended to three front channels and is also sometimes used for panning between side loudspeakers (e.g. L and LS) and rear loudspeakers. The typical sine/cosine panning law devised by Blumlein for two-channel stereo is often simply extended to more loudspeakers. Most such panners are constructed so as to ensure constant power as sources are panned to different combinations of loudspeakers, so that the approximate loudness of signals remains constant.

Panning using amplitude or time differences between widely spaced side loudspeakers is not particularly successful at creating accurate phantom images. Side images tend not to move linearly as they are panned and tend to jump quickly from front to back. Spectral differences resulting from differing HRTFs of front and rear sound tend to result in sources appearing to be spectrally split or 'smeared' when panned to the sides.

In some mixers designed for five-channel surround work, particularly in the film domain, separate panners are provided for L–C–R, LS–RS, and front-surround. Combinations of positions of these amplitude panners enable sounds to be moved to various locations, but some more successfully than others. For example, sounds panned so that some energy is emanating from all loudspeakers (say, panned centrally on all three pots) tend to sound diffuse for center listeners, and in the nearest loudspeaker for those sitting off-center. Joystick panners combine these amplitude relationships under the control of a single lever that enables a sound to be 'placed' dynamically anywhere in the surround soundfield. Moving effects made possible by these joysticks are often unconvincing and need to be used with experience and care.

Research undertaken by Jim West at the University of Miami showed that, despite the limitations of constant power 'pairwise' panning, it proved to offer reasonably stable images for center and off-center listening positions, for moving and stationary sources, compared with some other more esoteric algorithms. Front–back confusion was noticed in some cases, for sources panned behind the listener.

'Ambisonic' panning laws

A number of variations of panning laws loosely based on Ambisonic principles have been attempted. These are primarily based on the need to optimize psychoacoustic localization parameters according to low- and high-frequency models of human hearing. Gerzon proposed a variety of psychoacoustically optimal panning laws for multiple speakers that can theoretically be extended to any number of speakers. Some important features of these panning laws are:

- There is often output from multiple speakers in the array, rather than just two.
- They tend to exhibit negative gain components (out-of-phase signals) in some channels for some panning positions.
- The channel separation is quite poor.

A number of authors have shown how this type of panning could be extended to five-channel layouts according to the standards of interest in this book. McKinnie proposed a five-channel panning law based on similar principles, suitable for the standard loudspeaker angles. It is shown in Figure 17.31. Moorer also proposed some four- and five-channel panning laws, pictured in Figure 17.32 (only half the circle is shown because

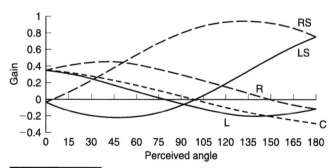

FIGURE 17.31 *Five channel panning law based on Gerzon's psychoacoustic principles. (Courtesy of Douglas McKinnie.)*

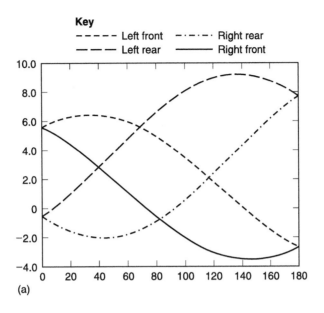

(a)

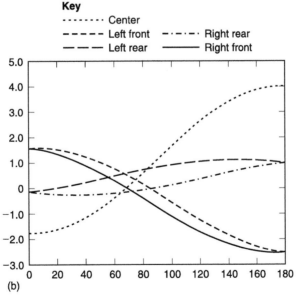

(b)

FIGURE 17.32 *Two panning laws proposed by Moorer designed for optimum velocity and energy vector localization with 2nd spatial harmonics constrained to zero. (a) Four-channel soundfield panning. The front speakers are placed at 30° angles left and right, and the rear speakers are at 110° left and right. (b) This shows an attempt to perform soundfield panning across five speakers where the front left and right are at 30° angles and the rear left and right are at 110° angles. Note that at zero degrees, the center speaker is driven strongly out of phase. At 180°, the center speaker is driven quite strongly, and the front left and right speakers are driven strongly out of phase. At low frequencies, the wavelengths are quite large and the adjacent positive and negative sound pressures will cancel out. At higher frequencies, their energies can be expected to sum in an RMS sense. (Courtesy of James A. Moorer.)*

the other side is symmetrical). They differ because Moorer has chosen to constrain the solution to first order spatial harmonics (a topic beyond the scope of this book). He proposes that the standard ±30° angle for the front loudspeakers is too narrow for music, and that it gives rise to levels in the center channel that are too high in many cases to obtain adequate L–R decorrelation, as well as giving rise to strong out-of-phase components. He suggests at least ±45° to avoid this problem. Furthermore, he states that the four-channel law is better behaved with these particular constraints and might be more appropriate for surround panning.

Head-related panning

Horbach of Studer has proposed alternative panning techniques based on Theile's 'association model' of stereo perception. This uses assumptions similar to those used for the Schoeps 'sphere' microphone, based on the idea that 'head-related' or pseudo-binaural signal differences should be created between the loudspeaker signals to create natural spatial images. It is proposed that this can work without crosstalk canceling, but that crosstalk canceling can be added to improve the full 3D effect for a limited range of listening positions.

In creating his panning laws, Horbach chooses to emulate the response of a simple spherical head model that does not give rise to the high-frequency peaks and troughs in response typical of heads with pinnae. This is claimed to create a natural frequency response for loudspeaker listening, very similar to that which would arise from a sphere microphone used to pick up the same source. Sources can be panned outside the normal loudspeaker angle at the front by introducing a basic crosstalk canceling signal into the opposite front loudspeaker (e.g. into the right when a signal is panned left). Front–back and center channel panning are incorporated by conventional amplitude control means. He also proposes using a digital mixer to generate artificial echoes or reflections of the individual sources, routed to appropriate output channels, to simulate the natural acoustics of sources in real spaces, and to provide distance cues.

RECOMMENDED FURTHER READING

AES, 2001. Proceedings of the 19th International Conference: Surround Sound – Techniques, Technology and Perception. Audio Engineering Society.

AES, 2001. Technical document ESTD1001.0.01-05: multichannel surround sound systems and operations. Available from website: http://www.aes.org.

Gerzon, M., 1992. Panpot laws for multispeaker stereo. Presented at 92nd AES Convention, Vienna. Preprint 3309. Audio Engineering Society.

Holman, T., 1999. 5.1 Surround Sound: Up and Running. Focal Press.

ITU-R, 1993. Recommendation BS 755: Multi-channel stereophonic sound system with or without accompanying picture. International Telecommunications Union.

Rayburn, R., 2011. Eargle's Microphone Book: From Mono to Stereo to Surround – A Guide to Microphone Design and Application, third edition. Focal Press.

Rumsey, F., 2001. Spatial Audio. Focal Press.

USEFUL WEBSITES

www.dolby.com: Contains an informative history of cinema surround sound systems, and information about current Dolby formats.

Sound Quality

This chapter is intended as an introduction to the topic of sound quality – what it means and what affects it. Examples are taken from both the analog and digital domains, and there is also an introduction to the topic of listening tests, including some of the related international standards. Finally there is an introduction to the question of how to measure sound quality and the use of perceptual models.

WHAT IS SOUND QUALITY?

It is possible to talk about sound quality in physical or technical terms, and in perceptual terms. In physical terms it generally relates to certain desirable measured characteristics of audio devices, transmission channels or signals. In perceptual terms, however, it relates to what is heard, interpreted and judged by human listeners. In an ideal world one domain could be related or mapped directly to the other. However, there may be aspects of sound quality that can be perceived, even though they cannot be measured, and some that can be measured but not perceived. One of the goals of perceptual model research, discussed later on, is to find better ways of measuring those aspects of audio signals that predict perceived quality.

Tomasz Letowski distinguishes between sound quality and sound character in his writings on the topic, the former having to do with a value judgment about the superiority of one sound over another and the latter being purely descriptive. Jens Blauert, in his Audio Engineering Society masterclass on sound quality (see 'Recommended further reading', below), refers to perceived sound quality in a broader sense as being related to a judgment about the perceived character of a sound in terms of its suitability for a particular task, expectation or pragmatic purpose. In his model it requires comparison of the sound to a reference set defined for the context in question, because otherwise the concept of sound quality 'floats' on a scale that has no fixed points and no meaning. The choice of the reference is therefore crucially important in defining the context for sound quality evaluation. Blauert also refers to different levels of abstraction when talking about sound quality, where the low levels are closely related to features of audio signals themselves, while the higher levels are related to ideas, concepts and meanings of sound (see Figure 18.1, which was inspired by Blauert and Jekosch, but not identical to their representation). This suggests that it is important to be careful what one means when talking about sound quality. One should be clear about the conceptual level at which one is talking – whether about signal quality, some descriptive attribute of the sound, or some more abstracted aspect of its overall cognition or perception.

Objective and subjective quality

Sound quality can be defined both in 'objective' and 'subjective' terms, but this categorization can be misleading and workers in the field have interpreted the meaning of these words differently. The term objective is often related to measurable aspects of sound quality, whereas the term subjective is often related to perceived aspects of sound quality. However, the term objective, in one sense of the word, means 'free of any bias or prejudice caused by personal feelings' and it has been shown that some descriptive attributes of sound quality can be evaluated by experienced listeners

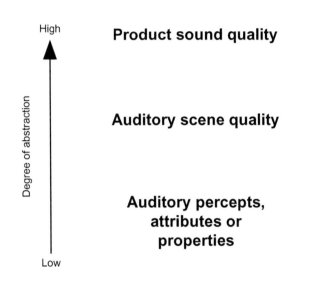

High

Degree of abstraction

Low

Product sound quality

Auditory scene quality

Auditory percepts, attributes or properties

Evaluation of meaning, signs, functionalities, plausibility

Integrative or gestalt judgments of e.g. fidelity, authenticity, quality enhancement

Psychoacoustical measurements of low-level attributes

FIGURE 18.1

Sound quality can be considered at different levels of abstraction from low to high. The lower levels are closely related to audio signals themselves and to associated auditory percepts whereas the higher levels have more cognitive complexity and relate to the meaning or signification of sound. (Adapted from Blauert.)

in a reliable and repeatable way that conforms to this definition. This can result in a form of perceptual measurement that is almost as 'objective' as a technical measurement. For this reason it may be more appropriate to differentiate between physically measured and perceived attributes of quality, reserving the term 'subjective' for questions such as liking and preference (discussed below). Figure 18.2, adapted from Bech and Zacharov's *Perceptual Audio Evaluation* (see Recommended further reading, below) shows this in a graphical form.

Sound quality attributes

Perceived sound quality is a many-faceted concept, and a term often used to describe this is 'multi-dimensional'. In other words, there are many perceptual dimensions or attributes making up human judgments about sound quality. These may be arranged in a hierarchy, with an integrative judgment of quality at the top, and judgments of individual descriptive attributes at the bottom (see Figure 18.3). According to Letowski's model this 'tree' may be divided broadly into spatial and timbral attributes, the spatial attributes referring to the three-dimensional features of sounds such as their location, width and distance, and the timbral attributes referring to aspects of sound color. Effects of non-linear distortion and noise are also sometimes put in the timbral group. The higher one goes up the tree, the more one is usually talking about the acceptability or suitability of the sound for some purpose and in relation to some frame of reference, whereas at the lower levels one may be able to evaluate the

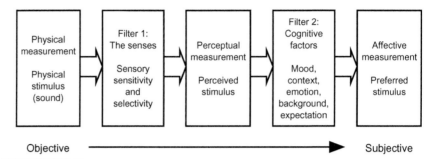

Objective ——————————————————————————→ Subjective

FIGURE 18.2 *Sound quality evaluation can be modeled as shown, involving perceptual and cognitive 'filters' at different stages of the process. Processes to the right can be said to be more 'subjective' than those towards the left, as they are likely to vary more between individuals and contexts. (Adapted from Bech and Zacharov.)*

FIGURE 18.3

Overall sound quality can be subdivided into timbral and spatial quality domains. Each of these consists of contributions from a number of relevant low-level attributes.

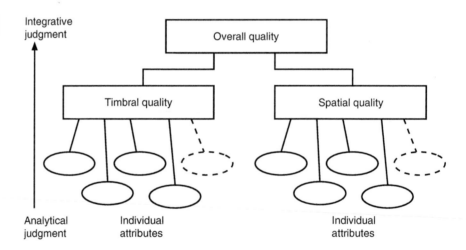

attributes concerned in value-free terms. In other words a high-level judgment of quality is an integrative evaluation that takes into account all of the lower-level attributes and weighs up their contribution. As mentioned above, the nature of the reference, the context, and the definition of the task govern the way in which the listener decides which aspects of the sound should be taken into consideration. One conception of such a model suggests that it may be possible to arrive at an overall prediction of quality by some weighted combination of ratings of individual low-level attributes. However, it is highly likely that such weightings are strongly context and task-dependent. (For example, whilst accurately located sources may be crucial for flight simulators, they may be less important in home entertainment systems.)

Quality and fidelity

The concept of fidelity has been of fundamental importance in defining the role of sound recording and reproduction. Fidelity can be defined variously as relating to faithfulness, as well as to accuracy in the description or reporting of facts and their details. In sound recording, it concerns the extent to which technical equipment is capable of accurately capturing, storing and reproducing sounds. The fidelity of two reproductions should really be a measure of their similarity. However, there has been a tendency in the history of sound quality evaluation either explicitly or implicitly to include a value judgment in concepts of fidelity.

Floyd Toole, for example, describes his concept of fidelity in a paper on listening tests, and states that in addition to rating various aspects of sound quality 'listeners conclude with an overall "fidelity rating" intended to reflect the extent to which the reproduced sound resembles an ideal. With some music and voice the ideal is a recollection of live sound, with other source material the ideal must be what listeners imagine to be the intended sound' (Toole, 1982: 440). Fidelity is thus defined in relation to a memorized or imagined ideal reference. Toole's fidelity scale, shown in Figure 18.4, is really a hybrid of a value judgment (e.g. using terms like 'worse') and a faithfulness judgment (e.g. 'faithful to the ideal'). There is not the option of 'less faithful' meaning 'better' (although this appears to be an unlikely combination it highlights the assumption implicit in this scale). It assumes that listeners know what is correct reproduction, and that what is correct is good. Gabrielsson and Lindström, on the other hand, define fidelity as 'the similarity of the reproduction to the original sound{...}the music sounds exactly as you heard it in the same room where it was originally performed' (1985: 52), but acknowledge the difficulty in judging this when listeners do not know what the music sounded like in reality (such as in studio-manufactured pop music).

Quality and naturalness

Naturalness is another term that crops up frequently in human descriptions of sound quality. It may be related to a higher-level cognitive response whereby the 'plausibility' of lower-level factors is somehow weighed in relation to a remembered experience of natural listening conditions. Naturalness and liking or preference are often found to be highly correlated, suggesting

```
10–
9–Excellent
8–
7–Good
6–
5–Fair
4–
3–Poor
2–
1–Bad
0–
```

The number 10 denotes a reproduction that is perfectly faithful to the ideal. No improvement is possible.

The number 0 denotes a reproduction that has no similarity to the ideal. A worse reproduction cannot be imagined.

FIGURE 18.4 *Toole's fidelity scale from 1982 includes elements of value judgment as well as similarity to an ideal.*

that listeners have a built-in preference for 'natural' sounds. This suggests that auditory cues in reproduced sound that contradict those encountered in the natural environment, or that are combined in an unnatural way, will be responded to negatively by listeners. Whether this phenomenon will continue to be noticed as people are exposed to more and more artificial sounds is open to question.

Quality and liking

One of the biggest mistakes that can be made when talking about sound quality is to confuse liking with correctness. One cannot automatically assume that the sound with the least distortion or flattest frequency response will be the most liked, although in many cases it is so. Some people actively like the sound of certain kinds of distortion, and this may have something to do with the preference shown by some for analog recording systems or vinyl LP records, for example. There is nothing wrong with liking distorted recordings, but liking should not be confused with fidelity. Learning and familiarity also play an important role in determining preferred sound quality. In one early experiment, for example, students who had spent a period of time listening to restricted frequency range audio systems demonstrated a preference for those compared with full range systems.

In audio engineering there is a long tradition of the concept of fidelity – the assumption that there is a correct way to reproduce a recording and it is the job of a sound system to reproduce as faithfully as possible an artist's intention or a natural sound. The idea of a reference reproduction is well embedded in the hearts of most audio engineers. If you like something that is different from what is correct, you are entitled to your opinion but basically you are wrong, so the argument runs. There is not space here to expound this argument in more than a basic way, but an alternative view is inspired by the food and beverage industry, which attempts to optimize products primarily for consumer preference. The most liked product is *de facto* the best, there being little room for a notion of correct reproduction of an ideal product. There is no 'correct' bottle of Shiraz wine, for example, just one that is most liked by a certain group of consumers. There may be no real conflict here in reality because in most cases sound systems are conduits for a creative product (e.g. a recorded song) that someone wants to convey to a consumer. It stands to reason that this conduit should be as transparent as possible, rather as the wine bottle and cork should taint the wine as little as possible. When talking about sound design on the other hand, perhaps there is more of a parallel with the food and beverage industry, as one is likely to be more interested in how to optimize different aspects of the sonic product so that people like the result. In the end it probably depends on defining clearly what the 'product' is, and who has control over it. This will certainly become more important in an age of interactive sonic products that can be altered or created on the fly by consumers who have no notion of a correct version of that product, or where multiple ways of rendering the product are available.

It is interesting in this respect to consider an experiment conducted by the author and his colleagues that attempted to discover which aspects of fidelity contributed most to listener preference in surround sound reproduction. It was found that untrained listeners simply did not care about precise stereo imaging in the frontal arc, whereas trained listeners regarded it as highly important. The untrained listeners mainly liked a broadly surrounding effect in the reproduction. This suggests that it may matter whose opinion is sought when evaluating liking for different aspects of sound reproduction. In the food industry, for example, it is common to use subjects representative of different populations to evaluate consumer preference because it is known that this depends on demographic and socio-economic factors. Sean Olive (2003), however, found that there is a fairly consistent relationship between measured loudspeaker performance and listener preference, for both trained and untrained listeners. This suggests that there are at least some contexts in which there is a relationship between technical fidelity of audio products and listeners' hedonic judgments about sound quality.

METHODS OF SOUND QUALITY EVALUATION

This section is intended as an introduction to methods of sound quality evaluation, although it is by no means a comprehensive exposition. It includes an introduction to the topic of listening tests and also to that of perceptual models in sound quality prediction. Those wishing to study the topic in more detail are referred to the excellent book by Bech and Zacharov (see 'Recommended further reading', below).

Listening tests

Listening tests are the most widely used formal method for sound quality evaluation. When properly structured and carried out in a scientific manner with experienced listeners they can provide reliable and repeatable information about a number of perceived aspects of sound quality, as introduced earlier in this chapter. In order to carry out reliable listening tests it is important to be aware of the various difficulties inherent in evaluating sensory information using human subjects.

Listeners

Inevitably human judgment carries with it a degree of uncertainty and variability, and it is the goal of most experimenters to reduce this to a minimum by using experienced listeners who are carefully trained to recognize and grade the quality attributes in question. On the other hand, it can be important to recognize that variability between humans is not always to be regarded as annoying statistical 'noise', but may represent a real difference in the way that people interpret what they hear. It should be clear whether one is looking for the *opinions* or *sentiments* of different potential user groups, or for a clearly defined objective *description* of sound quality features. In the former case it may be necessary to engage a pool of listeners representing different consumer types, and quite large numbers are usually needed in order to observe trends. In the latter case sufficient training, experience or familiarization is needed on behalf of a smaller group of subjects to enable them to employ a common language, understanding and reliable scaling method. In some types of listening test it is also common to employ so-called 'expert listeners' to detect and rate changes in overall sound quality in a consistent and reliable manner. In such cases the aim is not so much the profiling of consumer opinion but the accurate rating of sound quality changes caused by audio processing.

There is, however, no escaping the fact that a listener's background, experience and training will bias his or her judgment of sound quality, so it is important to recognize the limitations in generalizability of results.

Blind tests

It is widely accepted that knowing what you are listening to affects your judgment. In an interesting series of experiments by Toole and Olive, for example, in which they compared listener preferences for loudspeakers in both blind and sighted conditions, it was found that being aware of the model of loudspeaker under test affected the results as much as other variables such as listening position or program material. People's responses were so strongly biased by their expectations of certain brands or models that their judgments of sound quality were sometimes radically altered from their ratings under blind conditions. This applied to experienced listeners as well as to more naïve ones, suggesting that even people who 'know what sounds good' can have their opinions altered by their expectation or visual bias. 'Obviously', said the authors, 'listeners' opinions were more attached to the products that they could see, than they were to the differences in sound associated with program' (Toole and Olive, 1994: 13).

Because of this tendency, controlled listening tests employ 'blind' evaluation, which means that the listener is not allowed to know the identity of the stimuli concerned. 'Blind' does not necessarily mean blindfolded, unless there is a specific intention to ensure the removal of a visual bias, but it does mean that listeners should not be told the identity of the things they are to evaluate. Sometimes loudspeakers are hidden behind a curtain, for example. This aims to ensure that the evaluation is based only on what is heard and not biased by expectations, prejudices or other information that might change the result of the test. In the case of double-blind experiments, the test supervisor is also supposed to be unaware of the exact stimuli being presented to the listener at any one time in order that there is no chance of inadvertent biasing.

Types of listening test

A number of types of formal listening test are commonly employed, depending on the aim of the investigation. It is not intended to describe all of them, but only the most commonly encountered. Common to nearly all is the need to define exactly what is to be evaluated, whether this is an overall quality judgment supposed to include all aspects of sound quality, or whether it is one or more specific attributes of sound quality (such as 'brightness' or 'spaciousness'). It is also very common to employ a reference of some sort, which is a signal against which others are compared. This is most likely in cases involving the evaluation of impairments in quality, such as caused by audio codecs or processing devices. Human listeners tend to be quite poor at judging sound quality attributes when they have nothing to compare with, but can be quite reliable when there is one or more 'anchor' points on the scale concerned.

Most listening tests require the listener to rate a quality attribute on a scale of some kind, and in some cases the scale is labeled with words or numbers at intervals. The words may describe some level of quality or magnitude of some attribute in terms that are meaningful to listeners (e.g. 'excellent' and 'bad', or perhaps 'bright' and 'dark'). There is considerable debate about the problems arising from the use of scale labels as they may not have a universal meaning, neither may the words represent equal perceptual intervals. There is not space here to go into the matter further, but it is sufficient to note that these problems have led some to propose the use of alternative indirect scaling methods, or alternatives involving label-free scales, 'audible labels', or even approaches involving the observation of listener behavior when exposed to sound stimuli with different levels of quality.

The method of evaluation often involves presenting listeners with a number of stimuli to evaluate, either in pairs, triplets or multiples, since comparative judgment tends to be much more reliable than one-off grading. The ability to switch freely between stimuli also aids reliable comparison. It is also common practice to employ hidden reference signals in the case of some tests, in order to enable the experimenter to determine whether or not the listener is able to identify any reference signal consistently. This is one way of weeding out unreliable listeners.

The two primary evaluation methods employed in the international standards that deal with listening tests are the 'triple stimulus with hidden reference' approach used in ITU-R BS.1116, and the 'multiple stimulus with hidden reference and anchors' approach used in ITU-R BS.1534. The latter is commonly known as 'MUSHRA' and both are intended for the evaluation of quality impairments, mainly those caused by low bit rate codecs, although they can be adapted for other purposes. The BS.1116 approach presents the listener with three stimuli at a time – A, B and C – where the original unprocessed reference is always A and the other two are randomly assigned to the reference or the impaired signal. The listener has to decide which of B or C is the same as the reference and rate it '5' on the 5-point scale, rating the other stimulus according to its perceived quality. The MUSHRA method presents the listener with a number of stimuli together on the same screen, enabling him to switch between them at will and enter gradings for each one. Hidden among the stimuli are anchor signals representing specific quality levels, and a reference signal. The listener is expected to rate at least one of the stimuli (the hidden anchor, one hopes) at the top of the scale. Examples of scales for these test types are shown in Figure 18.5. Scales range from 'imperceptible' to 'very annoying' (when rating quality impairment) or from 'excellent' to 'bad' (when rating quality). BS.1116 is intended when the impairments involved are relatively small and hard to detect, whereas BS.1534 is intended for intermediate levels of quality.

An alternative to these types of test is the so-called 'ABX' test, which is intended for determining only whether a difference can be reliably perceived between a pair of stimuli. In an ABX test one or more listeners is presented with a pair of stimuli and asked to identify whether X is the same as A or the same as B. X is randomly assigned to being one or the other. This process is repeated a large number of times. If there is no perceivable difference between the two then on average, over a large number of trials, half of the answers should be correct (simply what could be arrived at by chance). If there is truly a difference between the stimuli then the

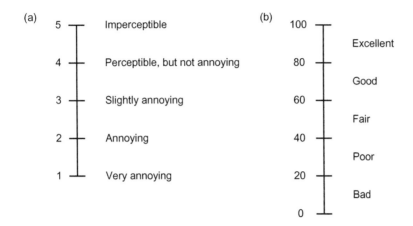

(a)

5 — Imperceptible

4 — Perceptible, but not annoying

3 — Slightly annoying

2 — Annoying

1 — Very annoying

(b)

100 —

    Excellent

80 —

    Good

60 —

    Fair

40 —

    Poor

20 —

    Bad

0 —

FIGURE 18.5

Scales used in ITU-R listening tests. (a) BS.1116 impairment scale, (b) BS.1534 quality scale. Note that in the latter the labels denote the meaning of intervals rather than points on the scale.

number of correct answers will be significantly higher than half, and the proportion gives information about the strength of the difference. This is one approach that can be used when trying to find out whether some supposedly inaudible process, such as an audio watermark, can be heard. Or it could be used, for example, if trying to discover whether listeners really can tell the difference between two loudspeaker cables.

Perceptual models for sound quality

Listening tests are time consuming and resource intensive, so there is a strong motivation for the development of perceptual models that aim to predict the human response to different aspects of sound quality. As opposed to making the relatively basic conventional measurements of frequency response and distortion, such models tend to be based on measurements or 'metrics' corresponding to a range of dynamic and static features of the audio signals in question, which are known to have an effect on perceived quality. Either by employing a calibrated statistical model or a trained neural network, relationships are established between these metrics and the results of listening tests on selected signals so that predictions can be made about the perceived quality of other signals. Such models either base their predictions on measurements made using program material or they make use of specially prepared test items or probe signals with characteristics designed to stress the perceptual range of interest. A standard model exists for predicting the overall quality of low bit rate audio codecs (commercially known as 'PEAQ' – for Perceptual Evaluation of Audio Quality) and another is available for speech coding quality prediction (commercially known as 'PESQ' – for Perceptual Evaluation of Speech Quality). Others have attempted to build models for predicting spatial quality, and for individual quality attributes.

A typical perceptual model for sound quality is calibrated in a similar way to that shown in Figure 18.6. Audio signals, usually consisting of a set of reference and impaired versions of chosen program items, are scaled in standard listening tests to generate a database of 'subjective' grades. In parallel with this a set of audio features is defined and measured, leading to a set of metrics representing perceptually relevant aspects of the audio signals concerned. These are sometimes termed 'objective metrics'. The statistical model or neural network is then calibrated or trained based on these data so as to make a more or less accurate prediction of the quality ratings given by the listeners. The output of the prediction is commonly referred to as an 'objective' measurement of sound quality, although this may be misleading. It is only as objective as the 'subjective' data used in its calibration, the appropriateness of the metrics and test signals used, and the sophistication of the mathematical model employed. The generalizability of such models needs wide testing.

Such models are typically 'double-ended' in that they have access to both the original unimpaired reference signal and the processed, impaired signal (see Figure 18.7). As such they are primarily used as quality impairment models, predicting perceived reductions in quality compared with a known 'correct' reference. It is much more difficult to predict quality without a known signal for comparison, although it may be possible to predict individual descriptive quality attributes in this way if the scales are calibrated using known anchor stimuli.

One further use for predictive quality models is in preference mapping. This derives from methods used in the food and beverage industry where it is often wished to know the relationship between certain descriptive features of quality and consumer preference. In such cases a set of

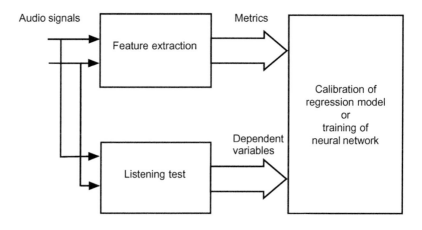

A typical approach used in the calibration of sound quality prediction models.

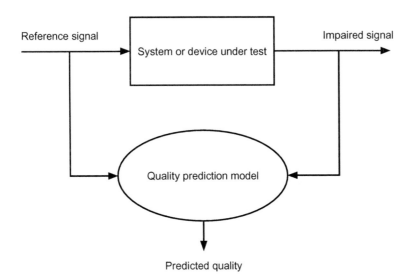

FIGURE 18.7

Most predictive models for sound quality, such as those standardized by the ITU, compare a reference (unimpaired) and impaired version of the same audio signal, leading to a prediction of the difference in the perceived quality.

expert-derived quality metrics relating to calibrated sensory attributes of the food are mapped onto preference ratings obtained from different consumer groups. In this way it is attempted to 'engineer' the sensory attributes of food products (relating to taste, texture and smell) so that they are maximally preferred by the target consumer. As discussed earlier in this chapter, there is still considerable debate about the relevance of such approaches to audio engineering.

ASPECTS OF AUDIO SYSTEM PERFORMANCE AFFECTING SOUND QUALITY

In the following sections a number of basic aspects of audio systems will be discussed in relation to sound quality. In particular the technical performance of devices and systems will be related where possible to audible artefacts.

Frequency response

The most commonly quoted specification for a piece of audio equipment is its frequency response. It is a parameter that describes the frequency range handled by the device – that is, the range of frequencies that it can pick up, record, transmit or reproduce. To take a simple view, for high-quality reproduction the device would normally be expected to cover the whole audio-frequency range, which was defined earlier in this book as being from 20 Hz to 20 kHz, although some have argued that a response which extends above the human hearing range has audible benefits. It is not enough, though, simply to consider the range of frequencies reproduced, since this says nothing about the relative levels of different frequencies or the amplitude of signals at the extremes of the range. If further qualification is not given then a frequency response specification of 20 Hz–20 kHz is not particularly useful.

The ideal frequency response for transparent transmission is one which is 'flat' – that is with all frequencies treated equally and none amplified more than others. Technically, this means that the gain of the system should be the same at all frequencies, and this could be verified by plotting the amplitude of the output signal on a graph, over the given frequency range, assuming a constant-level input signal. An example of this is shown in Figure 18.8, where

FIGURE 18.8

Plot of a flat frequency response from 20 Hz to 20 kHz. (b) Examples of two non-flat responses.

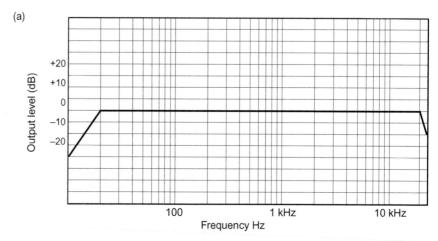

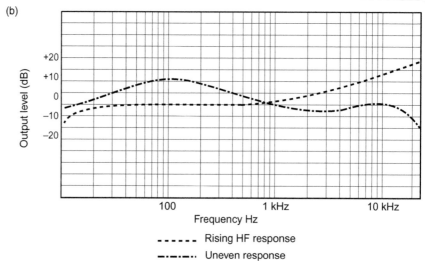

it will be seen that the graph of output level versus frequency is a straight horizontal line between the limits of 20 Hz and 20 kHz – that is, a flat frequency response. Also shown in Figure 18.8 are examples of non-flat responses, and it will be seen that these boost some frequencies and cut others, affecting the balance between different parts of the sound spectrum. Some discussion of the effects of this on sound quality is provided in Fact File 18.1.

Concerning the effects of very low and very high frequencies, close to the limits of human hearing, it can be shown that the reproduction of sounds below 20 Hz does sometimes offer an improved listening experience, since it can cause realistic vibrations of the surroundings. Also, the ear's frequency response does not cut off suddenly at the extremes, but gradually decreases, and thus it is not true that one hears nothing below 20 Hz and above 20 kHz – one simply hears much less and the amplitude has to be exceptionally high to create a sensation. Recent research suggests that the sound pressure level has to be well above 70 dB for any response to be detected above 20 kHz in the auditory brain stem, for example. Similarly, extended HF responses can sometimes help sound quality, mainly because a gentle HF roll-off above 20 kHz usually implies less steep filtering of the signal, which may have the by-product of improved quality for other reasons. The sensitivity of humans to frequency extremes is also

FACT FILE 18.1 EFFECTS OF FREQUENCY RESPONSE ON SOUND QUALITY

Deviations from a flat frequency response will affect perceived sound quality. If the aim is to carry through the original signal without modifying it, then a flat response will ensure that the original amplitude relationships between different parts of the frequency spectrum are not changed. Some forms of modification to the ideal flat response are more acceptable than others. For example, a gentle roll-off at the high-frequency (HF) end of the range is often regarded as quite pleasant in some microphones. FM radio receivers, for example, tend to have an upper limit of around 15 kHz, but are relatively flat below this and thus do not sound unpleasant. Frequency responses that deviate wildly from flat over the audio-frequency range, on the other hand, sound much worse, even if the overall range of reproduction is wider than that of FM radio. Middle frequency peaks and troughs in the response are particularly objectionable, sounding very 'colored' and sometimes with a ringing effect. However, extremely narrow troughs in loudspeaker responses may not be noticed under some circumstances. If the frequency response of a system rises at high frequencies then the sibilant components of the sound will be emphasized, music will sound very 'bright' and 'scratchy', and any background hiss will be emphasized. If the response is down at high frequencies then the sound will become dull and muffled, and any background hiss may appear to be reduced. If the frequency response rises at low frequencies then the sound will be more 'boomy', and bass notes will be emphasized. If low frequencies are missing, the sound will be very 'thin' and 'tinny'. A rise in the middle frequency range will result in a somewhat 'nasal' or 'honky' sound, perhaps having a rather harsh quality, depending on the exact frequency range concerned.

level dependent, as discussed in Fact File 2.2, which affects the perceived frequency balance of reproduced sound. For example, a mix that is made at a very high listening level may seem lacking in bass when played back more quietly. This is one reason why 'loudness' controls are provided on some consumer reproduction systems – they boost the bass (and sometimes treble) to compensate for the reduced sensitivity to these regions when replay is quiet. It also helps to explain why even small differences in reproduction level can give rise to noticeable differences in perceived sound quality.

Electronic devices and digital recording systems tend to have a flatter response than analog recording systems or electro-acoustical transducers (microphones or loudspeakers). An amplifier is an example of the former case, and it is unusual to find a well-designed power amplifier, say, that does not have a flat frequency response these days – flat often to within a fraction of a decibel from 5 Hz up to perhaps 100 kHz. (This does not, however, imply that the full power of the amplifier is necessarily available over this whole range, making the frequency response of power amplifiers a potentially misleading specification.) Essentially, however, a flat frequency response is relatively easy to engineer in most electronic audio systems today.

Transducers are the most prone of audio devices to frequency response errors, and some poor loudspeakers exhibit deviations of 10 dB or more from 'flat'. Since such devices are also affected by the acoustics of rooms it is difficult to divorce a discussion of their own response from a discussion of the way in which they interact with their surroundings. The room in which a loudspeaker is placed has a significant effect on the perceived response, since the

FACT FILE 18.2 EFFECTS OF HARMONIC DISTORTION ON SOUND QUALITY

Harmonic distortion is not always unpleasant, indeed many people find it quite satisfying and link it with such subjective parameters as 'warmth' and 'fullness' in reproduced sound, calling sound which has less distortion 'clinical' and 'cold'. Since the distortion is harmonically related to the signal that caused it, the effect may not be unmusical and may serve to reinforce the pitch of the fundamental in the case of even-harmonic distortion.

Because distortion tended to increase gradually with increasing recording level on analog tape recordings, the onset of distortion was less noticeable than it is when an amplifier or digital recording system 'clips', for example. Many old analog tape recordings contain quite

high percentages of odd harmonic distortion that have been deemed acceptable (e.g. 1–5%). Amplifier or digital system clipping, on the other hand, is very sudden and results in a 'squaring-off' of the audio waveform when it exceeds a certain level, at which point the harmonic distortion becomes severe. This effect can be heard when the batteries are going flat on a transistor radio, or when a loudspeaker is driven exceedingly hard from a low-powered amplifier. It sounds like a buzzing or 'fuzz-like' breaking up of the sound on peaks of the signal. It is no accident that this sounds like the fuzz box effect on a rock electric guitar, as this effect is created by overdriving an amplifier circuit.

room will resonate at certain frequencies, creating pressure peaks and troughs throughout the room. Depending on the location of the listener, some frequencies may be emphasized more than others. (For further in-depth discussion of this topic see Floyd Toole's book, *Sound Reproduction: Loudspeakers and Rooms*, listed in 'Recommended further reading', below.) A loudspeaker's response can be measured in so-called 'anechoic' conditions, where the room is mostly absorbent and cannot produce significant effects of its own, although other methods now exist which do not require the use of such a room. These alternative methods, such as time-delay spectrometry and maximum length sequence analysis, enable the system to exclude the portion of the time response that includes reflections, concentrating only on the direct sound. This has limitations in terms of the low-frequency range that can be measured correctly, depending on the delay before the first reflection.

A good loudspeaker will have a response that covers the majority of the audio-frequency range, with a tolerance of perhaps ± 3 dB, but the LF end is less easy to extend than the HF end because it requires large cabinet volume and driver displacement. Smaller loudspeakers will only have a response that extends to perhaps 50 or 60 Hz. The off-axis response of a loudspeaker has also been found to have an important influence on the perceived quality, since this often governs the frequency content of the reflected sound that makes up a large proportion of what is heard. An interesting paper by Sean Olive relates measured parameters of the loudspeaker response to listener preference ratings, showing the relative contributions of various on- and off-axis measurements (see 'Recommended further reading', below).

Some typical frequency response specifications of devices are shown in Table 18.1.

Harmonic distortion

Harmonic distortion is another common parameter used in the specifications of audio systems. Such distortion is the result of so-called 'non-linearity' within a device – in other words, when there is not a 1:1 relationship between what comes out of the device and what went in, when looked at over the whole signal amplitude range. In Chapter 1 it was shown that only simple sinusoidal waveforms are completely 'pure', consisting of a single frequency without harmonics. More complex repetitive waveforms can be analyzed into a set of harmonic components based on the fundamental frequency of the wave. Harmonic distortion in audio equipment arises when the shape of the sound waveform is changed slightly between input and output, such that harmonics are introduced into the signal which were not originally present, thus modifying the sound to some extent (see Figure 18.9). It is virtually impossible to avoid a small

Table 18.1	Examples of typical frequency responses of audio system
Device	**Typical frequency response**
Telephone system	300 Hz–3 kHz
AM radio	50 Hz–6 kHz
Consumer cassette machine	40 Hz–15 kHz (± 3 dB)
Professional analog tape recorder	30 Hz–25 kHz (± 1 dB)
CD player	20 Hz–20 kHz (± 0.5 dB)
Good-quality small loudspeaker	60 Hz–20 kHz (± 3 dB)
Good-quality large loudspeaker	35 Hz–20 kHz (± 3 dB)
Good-quality power amplifier	6 Hz–100 kHz (+ 0, − 1 dB)
Good-quality omni microphone	20 Hz–20 kHz (± 3 dB)

FIGURE 18.9

A sine wave input signal is subject to harmonic distortion in the device under test. The waveform at the output is a different shape to that at the input, and its equivalent line spectrum contains components at harmonics of the original sine wave frequency.

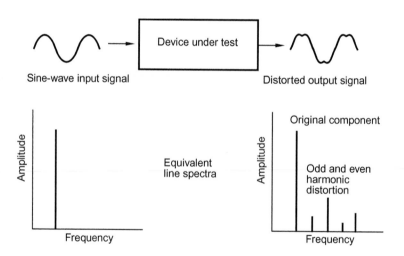

amount of harmonic distortion, since no device carries through a signal entirely unmodified, but it can be reduced to extremely low levels in modern electronic audio equipment.

Harmonic distortion is normally quoted as a percentage of the signal which caused it (e.g. THD 0.1% @ 1 kHz), but, as with frequency response, it is important to be specific about what type of harmonic distortion is being quoted, and under what conditions. One should distinguish, for example, between third-harmonic distortion and total harmonic distortion, and unfortunately both can be abbreviated to 'THD' (although THD most often refers to total harmonic distortion). Total harmonic distortion is the sum of the contributions from all the harmonic components introduced by the device, assuming that the fundamental has been filtered out, and is normally measured by introducing a 1 kHz sine wave into the device and measuring the resulting distortion at a recognized input level. The level and frequency of the sine wave used depends on the type of device and the test standard used. Third-harmonic distortion, on the other hand, is a measurement of the amplitude of the third harmonic of the input frequency only, and was commonly used in analog tape recorder tests. It can be important to be specific about the level and frequency at which the distortion specification is made, since in many audio devices distortion varies with these parameters.

Typical examples of harmonic distortion in audio products are given in Table 18.2.

Intermodulation (IM) distortion

IM distortion results when two or more signals are passed through a non-linear device. Since all audio equipment has some non-linearity there will always be small amounts of IM distortion, but these can be very low. Low IM distortion figures are an important mark of a high-quality system, since such distortion is a major contributor to poor sound quality, but it is less often quoted than THD. Unlike harmonic distortion, IM distortion may not be harmonically related to the frequency of the signals causing the distortion, and thus it is audibly more unpleasant. If two sine-wave tones are passed through a non-linear device, sum and difference tones may arise between them (see Figure 18.10). For example, a tone at $f_1 = 1100\,\text{Hz}$ and a tone at $f_2 = 1000\,\text{Hz}$ might give rise to IM products at $f_1 - f_2 = 100\,\text{Hz}$, and also at $f_1 + f_2 = 2100\,\text{Hz}$, as well as subsidiary products at $2 f_1 - f_2$ and so on. The dominant components will depend on the nature of the non-linearity.

Table 18.2 Typical THD percentages

Device	% THD
Good power amplifier @ rated power	<0.05% (20 Hz–20 kHz)
16 bit digital recorder (via own convertors)	<0.05% (−15 dB input level)
Loudspeaker	<1% (25 W, 200 Hz)
Professional analog tape recorder	<1% (ref. level, 1 kHz)
Professional capacitor microphone	<0.5% (1 kHz, 94 dB SPL)

FIGURE 18.10

Intermodulation distortion between two input signals in a non-linear device results in low-level sum-and-difference frequency components in the output signal.

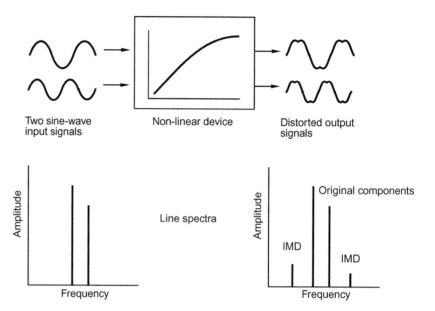

Two sine-wave input signals

Non-linear device

Distorted output signals

Line spectra

Amplitude

Frequency

Amplitude

Original components

IMD

IMD

Frequency

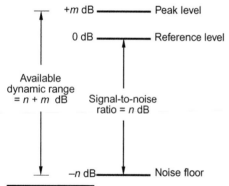

+*m* dB —— Peak level

0 dB —— Reference level

Available dynamic range = *n* + *m* dB

Signal-to-noise ratio = *n* dB

−*n* dB —— Noise floor

FIGURE 18.11 *Signal-to-noise ratio is often quoted as the number of decibels between the reference level and the noise floor. Available dynamic range may be greater than this, and is often quoted as the difference between the peak level and the noise floor.*

Dynamic range and signal-to-noise ratio

Dynamic range and signal-to-noise (S/N) ratio are often considered to be interchangeable terms for the same thing. This depends on how the figures are arrived at. S/N ratio is normally considered to be the number of decibels between the 'reference level' and the noise floor of the system (see Figure 18.11). The noise floor may be weighted according to one of the standard curves which attempts to account for the potential 'annoyance' of the noise by amplifying some parts of the frequency spectrum and attenuating others (see Fact Files 1.4 and 18.3). Dynamic range may be the same thing, or it may be the number of decibels between the peak level and the noise floor, indicating the 'maximum-to-minimum' range of signal levels that may be handled by the system. Either parameter quoted without qualification is difficult to interpret. It is also difficult to compare S/N ratios between devices measured using different weighting curves. In analog tape recorders, for example, dynamic range is sometimes quoted as the number of decibels between the 3% MOL and the weighted noise floor. This gives an idea of the available recording 'window', since the MOL is often well above the reference level. In digital recorders, the peak recording level is really also the

FACT FILE 18.3 NOISE WEIGHTING CURVES

Weighting filters are used when measuring noise, to produce a figure that more closely represents the subjective annoyance value of the noise. Some examples of regularly used weighting curves are shown in the diagram, and it will be seen that they are similar but not the same. Here 0 dB on the vertical axis represents the point at which the gain of the filter is 'unity', that is where it neither attenuates nor amplifies the signal. The 'A' curve is not normally used for measuring audio equipment noise, since it was designed for measuring acoustic background noise in buildings. The various DIN and CCIR curves are more commonly used in audio equipment specifications.

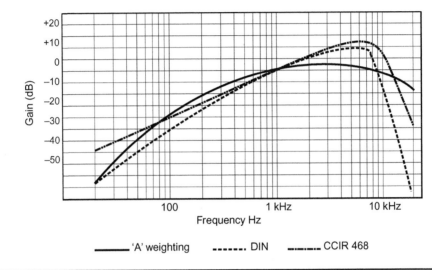

reference level, since there is no point in recording above this point due to the sudden clipping of the signal.

In general a constant level of background hiss is much less subjectively annoying than intermittent noises that draw attention to themselves. Mains-induced hum or buzz is also potentially annoying because of its tonal quality and the harmonics of the fundamental frequency that are often present. Noise that is modulated in some way by the audio signal, such as the digital codec noise described below, can also be subjectively quite annoying.

Wow, flutter and jitter

Wow and flutter are names used to describe speed variations of an analog tape machine or turntable, which are usually translated into frequency fluctuations in the audio signal. Wow means slow variations in speed while flutter means faster variations in speed. A weighting filter (usually to the DIN standard) is used when measuring to produce a figure that closely correlates with one's perception of the annoyance of speed variations. A machine with poor W&F results will sound most unpleasant, with either uncomfortable deviations in the pitch of notes or a 'roughness' in the sound and possibly some intermodulation distortion. Digital systems, on the other hand, tend to suffer more from 'jitter' as discussed in Fact File 18.4.

Sound quality in the digital signal chain

A number of basic aspects of digital conversion that affect sound quality were already mentioned in Chapter 8, such as the relationship between quantizing resolution and dynamic range, and between sampling frequency and frequency response. In this section some operational and systems issues will be considered.

Typical audio systems today have a very wide dynamic range that equals or exceeds that of the human hearing system. The distortion and noise inherent in the recording or processing of audio are at exceptionally low levels owing to the use of high resolution A/D convertors, up to 24 bit storage, and wide range floating-point signal processing. The sound quality achievable with modern audio devices is therefore exceptionally high. Of course there will always be those for whom improvements can be made, but technical performance of digital audio systems is no longer really a major issue today.

If one accepts the foregoing argument, the maintenance of sound quality in the digital signal chain comes down more to understanding the operational areas in which quality can be compromised. These include things like ensuring as few A/D and D/A conversions as possible, maintaining audio resolution at 20 bits or more throughout the signal chain (assuming

FACT FILE 18.4 JITTER

Happily the phenomenon of wow and flutter is now consigned almost entirely to history, as modern digital recorders do not suffer from it owing to the fact that transport or disk speed variations can be ironed out by reclocking the audio signal before reproduction. The modern equivalent, however, is jitter. Jitter is the term used to describe clock speed or sample timing variations in digital audio systems, and can give rise to effects of a similar technical nature to wow and flutter, but with a different spectral spread and character. It typically only affects sound quality when it interferes with the A/D or D/A conversion process. Because of the typical frequency and temporal characteristics of jitter it tends to manifest itself as a rise in the noise floor or distortion content of the digital signal, leading to a less 'clean' sound when jitter is high. If an A/D convertor suffers from jitter there is no way to remove the distortion it creates from the digital signal subsequently, so it pays to use convertors with very low jitter specifications.

this is possible), redithering appropriately at points where requantizing is done, and avoiding sampling frequency conversions. The rule of thumb should be to use the highest sampling frequency and resolution that one can afford, but no higher than strictly necessary for the purpose, otherwise storage space and signal processing power will be squandered. The scientific merits of exceptionally high sampling frequencies are dubious, although the marketing value may be considerable.

The point at which quality can be affected in a digital audio system is at A/D and D/A conversion. In fact the quality of an analog signal is irretrievably fixed at the point of A/D conversion, so this should be done with the best equipment available. There is very little that can be done afterwards to improve the quality of a poorly converted signal. At conversion stages the stability of timing of the sampling clock is crucial, because if it is unstable the audio signal will contain modulation artefacts that give rise to increased distortions and noise of various kinds. This so-called clock jitter (see Fact File 18.4) is one of the biggest factors affecting sound quality in convertors and high-quality external convertors usually have much lower jitter than the internal convertors used on PC sound cards.

The quality of a digital audio signal, provided it stays in the digital domain, is not altered unless the values of the samples are altered. It follows that if a signal is recorded, replayed, transferred or copied without altering sample values then the quality will not have been affected, despite what anyone may say. Sound quality, once in the digital domain, therefore depends entirely on the signal processing algorithms used to modify the program. There is little a user can do about this except choose high quality plug-ins and other software, written by manufacturers that have a good reputation for DSP that takes care of rounding errors, truncation, phase errors and all the other nasties that can arise in signal processing. This is really no different from the problems of choosing good-sounding analog equipment. Certainly not all digital equalizer plug-ins sound the same, for example, because this depends on the filter design. Storage of digital data, on the other hand, does not affect sound quality at all, provided that no errors arise and that the signal is stored at full resolution in its raw PCM form (in other words not using some form of lossy coding).

The sound quality the user hears when listening to the output of a professional system in the studio is not necessarily what the consumer will hear when the resulting program is issued on the release medium. One reason for this is that the sound quality depends on the quality of the D/A convertors used for monitoring. The consumer may hear better or worse, depending on the convertors used, assuming the bit stream is delivered without modification. One hopes that the convertors used in professional environments are better than those used by consumers, but this is not always the case. High-resolution audio may be mastered at a lower resolution for consumer release (e.g. 96 kHz, 24 bit recordings reduced to 44.1 kHz, 16 bits for release on CD), and this can affect sound quality. It is very important that any down-conversion of master recordings be done using the best dithering and/or sampling frequency conversion possible, especially when sampling frequency conversion is of a non-integer ratio.

When considering the authoring of interactive media such as games or virtual reality audio, there is a greater likelihood that the engineer, author, programmer and producer will have less control over the ultimate sound quality of what the consumer hears. This is because much of the sound material may be represented in the form of encoded 'objects' that will be rendered at the replay stage, as shown in Figure 18.12. Here the quality

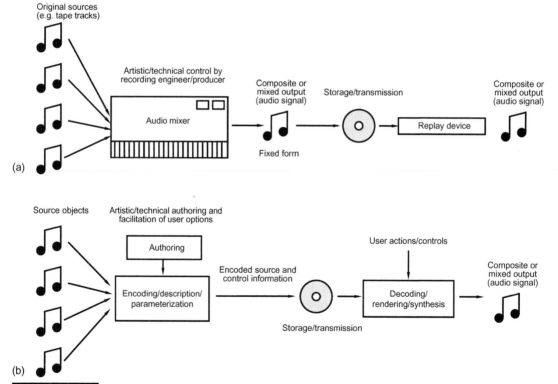

FIGURE 18.12 *(a) In conventional audio production and delivery sources are combined and delivered at a fixed quality to the user, who simply has to replay the signal. The quality is limited by the resolution of the delivery link. (b) In some virtual and synthetic approaches the audio information is coded in the form of described objects that are rendered at the replay stage. Here the quality is strongly dependent on the capabilities of the rendering engine and the accuracy of description.*

depends more on the quality of the consumer's rendering engine, which may involve resynthesis of some elements, based on control data. This is a little like the situation that arises when distributing a song as a MIDI sound file, using General MIDI voices. The audible results, unless one uses downloadable sounds (and even then there is some potential for variation), depends on the method of synthesis and the precise nature of the voices available at the consumer end of the chain.

Sound quality in audio codecs

It is increasingly common for digital audio signals to be processed using one or more perceptual audio codecs (see Chapter 8 for further details). This is typically done in order to limit the bit rate for transmission, internet delivery, or storage on portable devices such as phones and iPods. In mobile telephony and communications audio coding is often used to transmit speech signals at very low bit rates. Such coding normally has an effect on sound quality because the systems involved achieve the reduction in bit rate by allowing an increase in noise and distortion. However, the effect is a dynamic one and depends on the current signal and the perceptual processing algorithms employed to optimize the bit rate. Therefore the perceived effects on sound quality are often more difficult to describe, evaluate and measure than when a signal is affected by the simpler processes described in the previous section.

The aim of perceptual audio coding is to achieve the highest possible perceived quality at the bit rate in question. At the highest bit rates it is possible to achieve results that are often termed 'transparent' because they appear to convey the audio signal without audible changes in noise or distortion. As the bit rate is reduced there is an increasing likelihood that some of the effects of audio coding will be noticed perceptually, and these effects are normally referred to as coding artefacts. In the following section some of the most common coding artefacts will be considered, along with suggestions about how to keep them under control. Some tips to help minimize coding artefacts are included in Fact File 18.5.

Although the nature of coding artefacts depends to some extent on the type of codec, there are enough similarities between codecs to be able to generalize about this to some degree. This makes it possible to say that the most common artefacts can normally be grouped into categories involving coding noise of various types, bandwidth limitation, temporal smearing and spatial effects. Coding noise is generally increased quantizing noise that is modulated by various features of the audio signal. At high bit rates it is often possible to ensure that most or all of the noise resulting

FACT FILE 18.5 MINIMIZING CODING ARTEFACTS

As a rule coding artefacts can be minimized by using a high-quality codec at as high a bit rate as possible. Joint stereo coding is one way of helping to reduce the impact of coding artefacts when using MP3 codecs at lower bit rates, but it does this by simplifying the spatial information. It is therefore partly a matter of trading off one type of distortion against another. (It seems to be generally true that timbral distortions are perceived as more objectionable than spatial distortions by the majority of listeners.).

Avoidance of long chains of codecs ('cascading') will also help to maintain high audio quality, as each generation of coding and decoding will add artefacts of its own that are compounded by the next stage. This danger is particularly prevalent in broadcasting systems where an audio signal may have been recorded in the field on a device employing some form of perceptual encoding,

transmitted to the broadcasting center over a link employing another codec, and then coded again for final transmission.

Limiting the bandwidth of audio signals to be encoded is another way of reducing annoying coding artefacts, although many codecs tend to do this automatically as the first 'tool in the armory' when trying to maximize quality at low bit rates. (A fixed limitation in bandwidth is generally perceived as less annoying than time-varying coding noises.) Under conditions of very low bit rate, such as when using streaming codecs for mobile or Internet applications, the audio signal may require some artificial enhancement such as compression or equalization to make it sound acceptable. This should be considered as a form of mastering that aims to optimize sound quality, taking into account the limitations of the transmission medium.

from requantization of signals is constrained so that it lies underneath the masking threshold of the audio signal, but at lower bit rates there may be regions of the spectrum where it becomes audible at certain times. The 'swooshing' effect of coding noise is the result of its changing spectrum, as shaped by the perceptual model employed, and this can be easily heard if the original uncoded signal is subtracted from a version that has been coded and decoded. It is remarkable to hear just how much noise has been added to the original signal by the encoding process when performing this revealing operation, yet to realize that the majority of it remains masked by the audio signal.

Bandwidth limitation can be either static or dynamic and gives rise to various perceivable artefacts. Static bandwidth limitation is sometimes used to restrict the frequency range over which a codec has to operate, in order to allow the available bits to be used more efficiently. Some codecs therefore operate at reduced overall bandwidths when low bit rates are needed, leading to the slightly muffled or 'dull' sound that results from a loss of very high frequencies. This is often done in preference to allowing the more unpleasant coding artefacts that would result from maintaining full audio bandwidth, because a lot of audio program material, particularly speech, does not suffer unduly from having the highest frequencies removed. The so-called 'birdies' artefact, on the other hand, is related to a dynamic form of spectral change that sounds like twittering birds in the high-frequency range. It tends to result from the changing presence or lack of energy in individual frequency bands, as determined by the perceptual algorithms employed in the codec.

Temporal smearing is mainly the result of audio being coded in blocks of a number of milliseconds at a time. This can have the effect of spreading noise over a period of time so that it becomes more audible, particularly at transients (when the signal energy rises or falls rapidly), or causing pre- and post-echoes (small repetitions of the signal either before or after its intended time). The main perceptual effect of this tends to be a blurring of transients and a dulling or 'fuzzing' of attacks in musical signals, which is most noticeable on percussive sounds. A well-known test signal that tends to reveal this effect is a recording of solo castanets, which can be found on the EBU SQAM (Sound Quality Assessment Material) test disk (see 'Recommended further reading', below).

Many codecs act on stereophonic signals involving two or more channels, and often the bits available are shared between the channels using a variety of algorithms that minimize the information needed to transmit spatial information. These algorithms include intensity panning, M-S coding (see Chapter 16), and parametric spatial audio coding. Intensity panning and parametric spatial coding both simplify the spatial information used for representing source positions and diffuse spatial impression by boiling it down to a set of instructions about the interchannel relationships in terms of time, intensity or correlation. A simplified 'downmix', either to mono or two-channel stereo, can be transmitted and accompanied by the parameters needed to enable the original interchannel relationships to be reconstructed, as shown in Figure 18.13. This is the basic principle of MPEG spatial audio coding, for example. The degree to which this succeeds perceptually depends on the rate at which the spatial information can be transmitted, and the accuracy with which the interchannel relationships

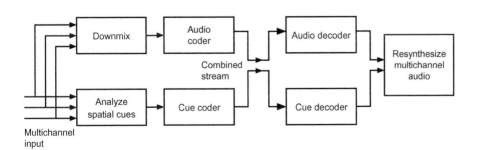

FIGURE 18.13

Block diagram of a typical spatial audio codec.

can be reconstructed. When the bit rate is low compromises have to be made, leading to potential distortions of the perceived spatial scene, which can include narrowing of the stereo image, blurring or movement of source locations, and a reduction in the sense of 'spaciousness' or envelopment arising from reverberation or other diffuse background sounds.

RECOMMENDED FURTHER READING

AES, 2002. Perceptual audio coders: what to listen for. A tutorial CD-ROM on coding artefacts. Available from: http://www.aes.org/publications.

Blauert, J., 2008. Masterclass: Concepts in Sound Quality [online]. Audio Engineering Society, New York. Available at: www.aes.org/tutorials (Accessed 10 November 2008).

Bech, S., Zacharov, N., 2006. Perceptual Audio Evaluation: Theory, Method and Application. John Wiley.

EBU Tech. 3253, 1988. Sound Quality Assessment Material (SQAM). European Broadcasting Union, Geneva. (Most of the audio material on the SQAM disk can be downloaded freely at: http://www.ebu.ch/en/technical/publications/tech3000_series/tech3253/index.php).

Gabrielsson, A., Lindström, B., 1985. Perceived sound quality of high fidelity loudspeakers. J. Audio. Eng. Soc. 33(1), 33–53.

ITU-R Recommendation BS.1116-1, 1994. Methods for Subjective Assessment of Small Impairments in Audio Systems Including Multichannel Sound Systems. International Telecommunications Union, Geneva.

ITU-R Recommendation BS.1534-1, 2003. Method for the Subjective Assessment of Intermediate Quality Level of Coding Systems. International Telecommunications Union, Geneva.

Letowski, T., 1989. Sound quality assessment: cardinal concepts. Presented at the 87th Audio Engineering Society Convention, New York. Preprint 2825.

Olive, S., 2003. Differences in performance and preference of trained versus untrained listeners in loudspeaker tests: a case study. J. Audio. Eng. Soc. 51(9), 806–825.

Toole, F., 1982. Listening tests: turning opinion into fact. J. Audio. Eng. Soc. 30(6), 431–445.

Toole, F., Olive, S.., 1994. Hearing is believing vs. believing is hearing: blind vs. sighted listening tests, and other interesting things. Presented at the 97th Audio Engineering Society Convention, San Francisco. Preprint 3894.

Toole, F., 2008. Sound Reproduction: Loudspeakers and Rooms. Focal Press.

Appendix: Record Players

Record players are obsolete technology, but it is felt that information should remain in the literature describing how to get the best out of the format, not least because inadequate or inept setting up of equipment can cause permanent damage to records, and because it seems likely that the LP will persist for some time yet. Also, there is still valuable material to be found here which is not always available on CD.

PICKUP MECHANICS

The replay stylus motion should describe an arc offset from the vertical by 20°, as shown in Figure A1.1. This will be achieved if the arm height at the pivot is adjusted such that the arm tube is parallel to the surface of the record when the stylus is resting in the groove. The stylus tip should have a cone angle of 55°, as shown in Figure A1.2 The point is rounded such that the tip makes no contact with the bottom of the groove. Stylus geometry is discussed further in Fact File A1.1.

The arm geometry is arranged so that a line drawn through the cartridge body, front to back, forms a tangent to the record groove at a point where the stylus rests in the groove, at two points across the surface of the record: the outer groove and the inner position just before the lead-out groove begins. Figure A1.3 illustrates this. Note that the arm tube is bent to achieve the correct geometry. Alternatively, the arm tube can be straight

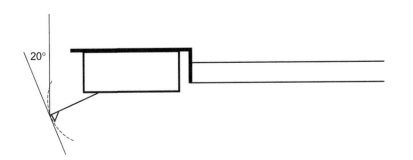

FIGURE A1.1

Stylus vertical tracking geometry.

with the cartridge headshell set at an offset angle which achieves the same result. The arc drawn between the two stylus positions shows the horizontal path of the stylus as it plays the record. Due to the fact that the arm has a fixed pivot, it is not possible for the stylus to be exactly tangential to the groove throughout its entire travel across the record's surface, but setting up the arm to meet this ideal at the two positions shown gives a good compromise, and a correctly designed and installed arm can give less than $\pm 1°$ tracking error throughout the whole of the playing surface of the disc.

Alignment protractors are available which facilitate the correct setting up of the arm. These take the form of a rectangular piece of card with a hole towards one end which fits over the center spindle of the turntable when it is stationary. It has a series of parallel lines marked on it (tangential to the record grooves) and two points corresponding to the outer and inner groove extremes. The stylus is lowered on to these two points in turn and the cartridge and arm are set up so that a line drawn through the cartridge from front to back is parallel to the lines on the protractor.

The original cutting stylus is driven across the acetate in a straight line towards the center, using a carriage which does not

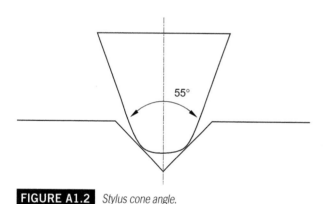

FIGURE A1.2 *Stylus cone angle.*

FACT FILE A1.1 STYLUS PROFILE

Two basic cross-sectional shapes exist for a replay stylus – conical and elliptical, as shown in the diagram. The elliptical profile can be seen to have a smaller contact area with the wall of the groove, and this means that for a given tracking weight (the downforce exerted by the arm on to the record surface) the elliptical profile exerts more force per unit area than does the conical tip. To compensate, elliptical styli have a specified tracking force which is less than that for a conical tip. The smaller contact area of the elliptical tip enables it to track the small, high-frequency components of the signal in the groove walls, which have short wavelengths, more faithfully. This is particularly advantageous towards the end of the side of the record where the

groove length per revolution is shorter and therefore the recorded wavelength is shorter for a given frequency. Virtually all high-quality styli have an elliptical profile or esoteric variation of it, although there are still one or two high-quality conical designs around. The cutting stylus is, however, always conical.

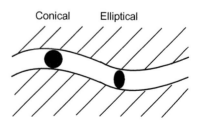

have a single pivot point like the replay arm, and it can therefore be exactly tangential to the groove all the way across the disc. The cutting lathe is massively engineered to provide an inert, stable platform. There are some designs of record player which mimic this action so that truly zero tracking error is achieved on replay. The engineering difficulties involved in implementing such a technique are probably not justified since a well-designed and well-set-up arm can achieve excellent results using just a single conventional pivot.

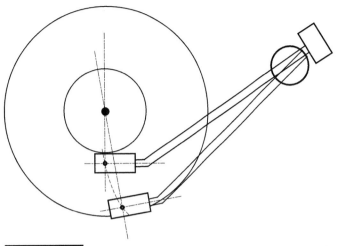

FIGURE A1.3 *Ideal lateral tracking is achieved when a line through the head-shell forms a tangent to the groove.*

A consequence of the pivoted arm is that a side thrust is exerted on the stylus during play which tends to cause it to skate across the surface of the record. This is a simple consequence of the necessary stylus overhang in achieving low tracking error from an arm which is pivoted at one end. Consider Figure A1.4. Initially it can be considered that the record is not rotating and the stylus simply rests in the groove. Consider now what happens when the record rotates in its clockwise direction. The stylus has an immediate tendency to drag across the surface of the record in the arrowed direction towards the pivot rather than along the record groove. The net effect is that the stylus feels a force in a direction towards the center of the record causing it to bear harder on the inner wall of the groove than on the outer wall. One stereo channel will therefore be tracked more securely than the other, and uneven wear of the groove and stylus will also result. To overcome this, a system of bias compensation or 'anti-skating' is employed at the pivot end of the arm which is arranged so that a small outward force is exerted on the arm to counteract its natural inward tendency. This can be implemented in a variety of ways, including a system of magnets; or a small weight and thread led over a pulley which is contrived so as to pull the arm outwards away from the center of the record; or a system of very light springs. The degree of force which is needed for this bias compensation varies with different stylus tracking forces but it is in the order of one-tenth of that value (see Fact File A1.2).

Although every cartridge will fit into every headshell apart from one or two special types, one must be aware of certain specifications of both the arm and the cartridge in order to determine whether the two are

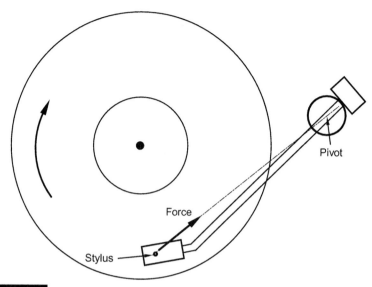

Pivot

Force

Stylus

FIGURE A1.4 *The rotation of the disc can create a force which pulls the arm towards the centre of the disc.*

FACT FILE A1.2 TRACKING WEIGHT

The required weight varies from cartridge to cartridge. A small range of values will be quoted by the manufacturer such as '1 gram ± 0.25 grams' or '1–2 grams' and the exact force must be determined by experiment in conjunction with a test record. First, the arm and cartridge must be exactly balanced out so that the arm floats in free air without the stylus moving either down towards the record surface or upwards away from it, i.e. zero tracking force. This is generally achieved by moving the counterweight on the end of the arm opposite to the cartridge either closer to or further away from the pivot until an exact balance point is found. The counterweight is usually moved by rotating it along a thread about the arm, or alternatively a separate secondary weight is moved. This should be carried out with bias compensation off.

When a balance has been achieved, a tracking weight should be set to a value in the middle of the cartridge manufacturer's values. Either the arm itself will have a calibrated tracking force scale, or a separate stylus

balance must be used. The bias compensation should then be set at the appropriate value, which again will have either a scaling on the arm itself or an indication in the setting up instructions. A good way to set the bias initially is to lower the stylus towards the play-in groove of a rotating record such that the stylus initially lands mid-way between these widely spaced grooves on an unused part of the surface of the record. Too much bias will cause the arm to move outwards before dropping into the groove. Too little bias will cause the arm to move towards the center of the record before dropping into the groove. Just the right amount will leave the arm stationary until the relative movement of the groove itself eventually engages the stylus. From there, the optimum tracking and bias forces can then be determined using the test record according to the instructions given. In general, a higher tracking force gives more secure tracking but increases record wear. Too light a tracking force, though, will cause mistracking and damage to the record grooves.

compatible. In order for the stylus to move about in the groove, the cantilever must be mounted in a suitable suspension system so that it can move to and fro with respect to the stationary cartridge body. This suspension has compliance or springiness and is traditionally specified in (cm/dyne) × 10^{-6}, abbreviated to cu ('compliance units'). This is a measure of how many centimeters (in practice, fractions of a centimeter!) the stylus will deflect when a force of 1 dyne is exerted on it. A low-compliance cartridge will have a compliance of, say, 8 cu. Highest compliances reach as much as 45 cu. Generally, values of 10–30 cu are encountered, the value being given in the maker's specification.

RIAA EQUALIZATION

The record groove is an analog of the sound waves generated by the original sources, and this in itself caused early pioneers serious problems. In early electrical cutting equipment the cutter stylus velocity remained roughly constant with frequency, for a constant input voltage (corresponding to a falling amplitude response with frequency) except at extreme LF where it became of more constant amplitude. Thus, unequalized, low frequencies would cause stylus movements of considerably greater excursion per cycle for a given stylus velocity than at high frequencies. It would be difficult for a pickup stylus and its suspension system inside the cartridge body to handle these relatively large movements, and additionally low frequencies would take up relatively more playing surface or 'land' on the record curtailing the maximum playing time. Low-frequency attenuation was therefore used during cutting to restrict stylus excursions.

In modern record cutting, a standard known as RIAA equalization has been adopted which dictates a recorded velocity response, no matter what the characteristics of the individual cutting head. Electrical equalization is used to ensure that the recorded velocity corresponds to the curve shown in Figure A1.5a. A magnetic replay cartridge will have an output voltage proportional to stylus velocity (its unequalized output would rise with frequency for a constant amplitude groove) and thus its output must be electrically equalized according to the curve shown in Figure A1.5b in order to obtain a flat voltage–frequency response.

The treble pre- and de-emphasis of the RIAA replay curve have the effect of reducing HF surface noise. An additional recommendation is very low 20 Hz bass cut on replay (time constant 7960 μs) to filter out subsonic rumble and non-program-related LF disturbance. The cartridge

FIGURE A1.5

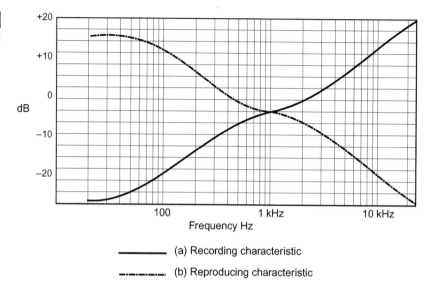

RIAA recording and reproducing characteristics.

(a) Recording characteristic

(b) Reproducing characteristic

needs to be plugged into an input designed for this specific purpose; the circuitry will perform the above discussed replay equalization as well as amplification.

CARTRIDGE TYPES

The vast majority of cartridges in use are of the moving-magnet type, meaning that the cantilever has small powerful magnets attached which are in close proximity to the output coils. When the stylus moves the cantilever to and fro, the moving magnets induce current in the coils to generate the output. The DC resistance of the coils tends to be several hundred ohms, and the inductance several hundred millihenries (mH). The output impedance is therefore 'medium', and rises with frequency due to the inductance. The electrical output level depends upon the velocity with which the stylus moves, and thus for a groove cut with constant deviation the output of the cartridge would rise with frequency at 6 dB per octave. The velocity of the stylus movements relative to an unmodulated groove is conveniently measured in $cm\,s^{-1}$, and typical output levels of moving magnet cartridges are in the order of $1\,mV\,cm^{-1}\,s^{-1}$.

The average music program produces cartridge outputs of several millivolts, and an upper limit of 40 or 50 mV will occasionally be encountered at mid frequencies. Due to the RIAA recording curve, the output will be less at low frequencies but not necessarily all that much more at high frequencies owing to the falling power content of music with rising frequency. A standard input impedance of an RIAA input of 47 k has been adopted, and around 40 dB of gain (× 100) is needed at mid frequencies to bring the signal up to line level.

Another type of cartridge which is much less often encountered but has a strong presence in high-quality audio circles is the moving-coil cartridge. Here, the cantilever is attached to the coils rather than the magnets, the latter being stationary inside the cartridge body. These cartridges tend to give much lower outputs than their moving-magnet counterparts because of the need to keep coil mass low by using a small number of turns, and they have a very low output impedance (a few ohms up to around a hundred) and negligible inductance. They require 20–30 dB more gain than moving magnets do, and this is often provided by a separate head amplifier or step-up transformer, although many high-quality hi-fi amplifiers provide a moving-coil input facility. The impedance of such inputs is around 100 ohms or so.

CONNECTING LEADS

Owing to the inductive nature of the output impedance of a moving-magnet cartridge, it is sensitive to the capacitance present in the connecting leads and also that present in the amplifier input itself. This total capacitance appears effectively in parallel with the cartridge output, and thus forms a resonant circuit with the cartridge's inductance. It is the high-frequency performance of the cartridge which is affected by this mechanism, and the total capacitance must be adjusted so as to give the best performance. Too little capacitance causes a frequency response which tends to droop several decibels above about 5 kHz, sharply rising again to a 2–3 dB peak with respect to the mid band at around 18–20 kHz, which is the result of the resonant frequency of the stylus/record interface, the exact value depending upon the stylus tip mass. Adding some more capacitance lifts the 5–10 kHz trough and also curtails the tip mass resonant peak to smooth out the frequency response. Too much capacitance causes attenuation of the highest frequencies giving a dull sound.

Around 300–400 pF total is the usual range of capacitance to be tried. Adding capacitance is most conveniently carried out using special in-line plugs containing small capacitors for this purpose. Assume that 200 pF is present already, and try an extra 100 pF, then 200 pF. Alternatively, small polystyrene capacitors can be purchased and soldered between signal and earth wires of the leads inside the plugs or sockets. Sometimes solder tags are present in the base of the record player itself which are convenient.

FIGURE A1.6 *The SME Series V pickup arm. (Reprinted with permission of the Society of Manufacturing Engineers, USA.)*

NEVER solder anything to the cartridge pins. The cartridge can very easily be damaged by doing this.

Moving-coil cartridges have a very low output impedance and negligible inductance, and their frequency response is therefore not affected by capacitance.

ARM CONSIDERATIONS

The effective mass of the pickup arm (Figure A1.6), which is the inertial mass of the arm felt by the stylus, coupled with the cartridge's suspension compliance, together form a resonant system, the frequency of which must be contrived such that it is low enough not to fall within the audio band but high enough to avoid coinciding with record warp frequencies and other LF disturbances which would continually excite the resonance causing insecure tracking and even groove jumping. Occasionally, large, slow excursions of the cones of the speaker woofers can be observed when the record is being played, which is the result of non-program-related, very low-frequency output which results from an ill-matched arm/cartridge combination.

A value of 10–12 Hz is suitable, and a simple formula exists which enables the frequency to be calculated for a given combination of arm and cartridge:

$$f = 1000/(2\pi\sqrt{(MC)})$$

where f = resonant frequency in hertz, M = effective mass of the arm 1 mass of the cartridge 1 mass of hardware (nuts, bolts, washers) in grams, C = compliance of the cartridge in compliance units.

For example, consider a cartridge weighing 6 g, having a compliance of 25 cu; and an arm of effective mass 20 g, additional hardware a further 1 g. The resonant frequency will therefore be 6.2 Hz. This value is below the optimum, and such a combination could give an unsuitable performance due to this resonance being excited by mechanical vibrations such as people walking across the floor, record warps and vibrations emanating from the turntable main bearing. Additionally, the 'soft' compliance of the cartridge will have difficulty in coping with the high effective mass of the arm, and the stylus will be continually changing its position in the groove somewhat as the arm's high inertia tends to flex the cartridge's suspension and dominate its performance.

If the same cartridge in an arm having an effective mass of 8 g is considered, then f = 8.4 Hz. This is quite close to the ideal, and would be acceptable. It illustrates well the need for low-mass arms when high compliances are encountered. The resonance tends to be high Q, and this sharp resonance underlines the need to get the frequency into the optimum range. Several arms provide damping in the form of a paddle, attached to the arm, which moves in a viscous fluid, or some alternative arrangement. This tends to reduce the amplitude of the resonance somewhat, which helps to stabilize the performance. Damping cannot, however, be used to overcome the effects of a non-optimum resonant frequency, which must still be carefully chosen.

LASER PICKUPS

The idea of reading a record groove with a laser beam rather than a stylus has been mooted for quite some time, and in 1990 a player using such a technique finally appeared. It is a very attractive proposition for record libraries due to the fact that record wear becomes a thing of the past. However, a laser beam does not push particles of dust aside as does a stylus, and the commercial system needs to be fed with discs which are almost surgically clean, otherwise signal drop-outs occur. Two entirely separate laser beams are in fact used, one reading the information from each wall of the groove. Error concealment circuitry is built in, which suppresses the effects of scratches. Towards the center of the record, short wavelengths occupy a proportionately smaller area of the groove than at the perimeter of the disc, and the width of the laser beam means that difficulties in reading these high-amplitude HF signals occur. The frequency response of the player therefore droops by around 10 dB if the highest frequencies towards the end of the side are of a high amplitude.

CD-type features are offered such as pause, track repeat and track search, which are very useful. The player does not suffer from traditional record player ills such as LF arm/cartridge resonance, rumble, and wow and flutter. Costing not much under £10000, the player obviously has a limited appeal, the professional user being the main potential customer.

RECOMMENDED FURTHER READING

AES, 1981. Disk Recording – An Anthology. Audio Engineering Society.

BS 7063. British Standards Office.

Earl, J., 1973. Pickups and Loudspeakers. Fountain Press.

Roys, H.E. (Ed.), 1978. Disk Recording and Reproduction. Hutchinson and Ross, Dowde.

Glossary of Terms

AAC	Advanced Audio Coding.
ABR	Auxiliary Bass Radiator.
AC	Alternating Current.
A/D	Analog-to-digital conversion.
AES/EBU	Audio Engineering Society/European Broadcasting Union: typically refers to a standard digital audio interface.
AF	Audio Frequency.
AFL	After Fade Listen.
AGC	Automatic Gain Control.
Aliasing	The generation of in-band spurious frequencies caused by using a sampling rate which is inadequate for the chosen frequency range, i.e. it is less than twice the highest frequency present in the signal.
AM	Amplitude Modulation.
Amp	The unit of electrical current, named in honor of André Marie Ampère (1775–1836).
Anechoic Chamber	A highly absorbtive room; the walls, floor and ceiling of which are virtually non-reflective. Used for acoustical measurements of devices such as microphones and loudspeakers.
Antinode	The part of a waveform where its propagating medium has maximum velocity and minimum acceleration.
ASK	Amplitude Shift Keying.
ASW	Auditory Source Width.
ATRAC	Adaptive Transform Acoustic Coding.
Bandwidth	The range of frequencies over which a device will operate; formally, the measurement is taken from the points at which the response is 3 dB down at the frequency extremes compared with mid-frequencies.
BCD	Binary Coded Decimal.
Bell Curve	A narrow-band EQ curve shaped like a bell when displayed on a frequency response graph.
Bias	An ultrasonic frequency added to the audio frequencies sent to an analog tape recorder's record head to bias the tape into a more linear part of its operating range.
BPF	Band Pass Filter.
CD	Compact Disc.
CMRR	Common Mode Rejection Ratio.

Compansion	The complete process of compression during the recording or transmission stages followed by reciprocal expansion during the replay or reception stages.
Compliance	'Springiness'.
CRC	Cyclic Redundancy Check.
Crossover	A device in a loudspeaker system for splitting the audio signal into frequency bands which can then be fed to appropriate speaker drive units (e.g. low frequency, mid range and high frequency). A passive crossover is placed between the power amplifier and the speaker drivers. An active crossover splits the frequencies at line level ahead of the power amplifiers.
CU	Compliance Unit.
D/A	Digital-to-Analog conversion.
Damping factor	In power amplifiers, the impedance of the speaker it is driving divided by the amplifier's output impedance. It is an indication of the ability of an amplifier to control spurious movements of the speaker cones (particularly at low frequencies) which can be caused by resonances and energy storage in the drivers' suspension systems.
DASH	Digital Audio Stationary Head. An open-reel digital format.
DAT	Digital Audio Tape.
dBm	Signal level in decibels, referred to 1 milliwatt (0 dBm = 775 mV across 600 ohms).
dBu	Signal level in decibels, referred to 0.775 V with an unspecified impedance value (775 mV = 0 dBu).
dBV	Signal level in decibels, referred to 1 volt (0 dBV = 1 volt).
dBv	The same as dBu, sometimes used in the USA.
DC	Direct Current.
DCC	Digital Compact Cassette.
DCLD	Downmix Channel Level Difference.
DCP	Digital Cinema Project.
Decibel (dB)	In audio, the unit used to denote the logarithm of the ratio between two quantities, e.g. voltages or power levels. Also used to denote acoustical sound pressure level. Named in honor of Alexander Graham Bell (1847–1922).
De-emphasis	Reciprocal treble cut (see also Pre-emphasis) during a replay or reception process.
DHCP	Dynamic Host Configuration Protocol.
DI	Direct Inject.
Directivity	Defines the angle of coverage of a loudspeaker's output.
Directivity Factor	The number which denotes the ratio between a sound source's output on its axis of maximum radiation and its

	output if it were perfectly omnidirectional and its total acoustical output were to be evenly spread all around it.
Directivity Index	Directivity Factor expressed in dB.
Dispersion	See Directivity.
Dither	A continuous low-level noise signal added to the program prior to A/D conversion and quantization or during signal processing.
DML	Distributed-Mode Loudspeaker.
Drive unit	See Driver.
Driver	The component of a speaker system which actually vibrates or 'drives' the air, e.g. a speaker cone or tweeter.
DSD	Direct Stream Digital.
DSP	Digital Signal Processing.
DST	Direct Stream Transfer.
DVD	Digital Versatile Disc.
EBU	European Broadcasting Union.
Eigentone	A standing wave in a room which is set up when half the wavelength of a sound or a multiple of it is equal to one of the dimensions of the room (height, width, length).
EIN	Equivalent Input Noise.
EQ	Equalization.
FET	Field Effect Transistor.
FIR	Finite Impulse Response.
FLAC	Free Lossless Audio Coding.
FM	Frequency Modulation.
FSK	Frequency Shift Keying.
FX	Effects.
Haas Effect	If two sound sources emit similar sounds but one is delayed with respect to the other (up to about 50 mS) the ears perceive the non-delayed sound source to be the louder of the two, and the sound appears to come from a direction close to the non-delayed source, the exact location depending on the amount of delay between them. Beyond the 50 mS time difference, the ears tend to perceive the sounds as coming from two distinct sources.
Harmonics	(Also known as Overtones or Partials.) Components of a waveform which are multiples of the fundamental frequency; together with the starting transient they contribute much to the character or tone color of a sound.
HDTV	High Definition Television.
Hertz (Hz)	The unit of vibration in cycles per second, named in honor of Heinrich Rudolf Hertz (1857–1894).
HPF	High Pass Filter.
HRTF	Head Related Transfer Function.

ICS	Internet Connection Sharing.
IIR	Infinite Impulse Response.
Impedance	Measured in ohms, it is a device's opposition to the flow of AC current. Reactive devices such as loudspeakers, capacitors and inductors exhibit impedances which vary with frequency.
IOC	Inter-Object cross Coherence.
IP	Internet Protocol.
ITD	Interaural Time Difference.
LED	Light Emitting Diode.
Longitudinal wave	A wave in which the to and fro movement of the wave carrier is in the same plane as the wave's travel. A sound wave is an example of this, the source of sound pushing and pulling in the direction of the wave.
LP	Long Playing gramophone record.
LPF	Low Pass Filter.
MADI	Multichannel Audio Digital Interface.
Masking	A psychacoustic phenomenon, whereby quiet sounds in the presence of loud sounds and/or sounds with a similar frequency content will be rendered less audible.
MD	MiniDisc.
MDCT	Modified Discrete Cosine Transform.
MFM	Miller Frequency Modulation.
MIDI	Musical Instrument Digital Interface.
MLP	Meridian Lossless Packing.
MMC	MIDI Machine Control.
MOL	Maximum Output Level.
MOSFET	Metal Oxide Semiconductor Field Effect Transistor.
MP3	Short for MPEG-1 Layer 3.
MPEG	Moving Pictures Expert Group. ('Empeg'.)
MPX	Multiplex.
MS	Main (or Middle)-Side.
MTC	MIDI Time Code.
MXF	Material Exchange Format.
NAT	Network Address Translation.
NC	Noise Criterion.
Node	The part of a waveform where its propagating medium has maximum acceleration and minimum velocity.
Noise shaping	The technique of reducing noise in perceptually sensitive regions of the audio band at the expense of increasing it above the audio band, or in less sensitive regions, thereby improving the perceived signal-to-noise ratio. It is used in digital systems such as that used in SACD and over-sampling CD players.

NR	Noise Reduction. (Also Noise Rating.)
nWb/m	Nanowebers per metre. The unit of flux along a magnetic recording medium, named in honor of Wilhelm Eduard Weber (1804–1891).
Nyquist frequency	Half the sampling frequency in a digital system.
OCA	Open Control Architecture.
Ohm	The unit of resistance and impedance, named in honor of Georg Simon Ohm (1789–1854).
OLD	Object Level Difference.
Overtones	See Harmonics.
PA	Public Address. (Also Power Amplifier.)
Pad	An input attenuator.
PAM	Pulse Amplitude Modulation.
Partials	See Harmonics.
PCM	Pulse Code Modulation.
PD	ProDigi. An open-reel digital format.
PDM	Pulse Density Modulation.
PFL	Pre-Fade Listen.
Phase	Two waves of the same frequency are 'in phase' when their positive and negative half-cycles coincide exactly in time. For example, two loudspeakers are in phase if, when fed by the same source signal, their cones move backwards and forwards in step with each other, and their acoustical outputs reinforce. If they are out-of-phase, cancellation of sound results. Electrical signals can similarly be in or out-of-phase, or in any intermediate relationship.
Phase response	A measure of phase lead or lag of signals across the frequency range as they pass through an electrical circuit or device.
Phon	A unit denoting the subjective loudness of a sound, the scale derived from research data.
Pink noise	Noise which has equal energy per octave. Its frequency spectrum is therefore flat when the usual logarithmic horizontal frequency scale is used.
Power bandwidth	Superficially similar to Bandwidth (qv) for a power amplifier, but it is the range of frequencies over which the amplifier can deliver full power, with −3 dB being allowed at the frequency extremes. A power amplifier's frequency response is normally wider than its power bandwidth.
PPM	Peak Programme Meter.
Pre-emphasis	Treble boost applied during a recording or transmission process. See also De-emphasis.
PWM	Pulse Width Modulation.
PZM	Pressure Zone Microphone.

Q	Historically, the 'quality' of a tuned radio-frequency receiving circuit. A good sharp tuning which centered on the station of interest, greatly attenuating the unwanted frequencies to either side, was said to be of high quality or Q, and it could be quantified as set out below. In the audio industry it can denote a number of things, including: (1) the bandwidth or 'sharpness' of an EQ curve. The Q is defined as the center frequency divided by the bandwidth. See Principal EQ bands, Chapter 5. (2) In a loudspeaker system: at its low-frequency resonant point, Q is the ratio between the speaker's output level here and its output level over the nominally flat part of its frequency range, expressed as a number. For example, if the output is 3 dB down at a speaker's LF resonant frequency (which is fairly typical), it has a Q of 0.707 ($20 \log Q = -3$ dB). If it is 6 dB down, it has a Q of 0.5 ($20 \log Q = -6$ dB).
QPSK	Quadrature Phase Shift Keying.
Quantization	After sampling a waveform, the quantization process assigns a numerical value to each sample according to its amplitude. For example, a 16-bit digital system can assign one of a possible 65 536 values (2^{16}) to a particular sample, with no 'in-between' values being permitted.
RAID	Redundant Array of Independent Disks.
RAM	Random Access Memory.
R-DAT	Rotary-head Digital Audio Tape (same as DAT).
Resonance	This takes place in a system at frequencies where the balance of its moving mass and its compliance gives rise to regions where a relatively small amount of input energy is required to produce vibrations of large amplitude compared with that required at most other frequencies.
RF	Radio Frequency.
RIAA	Recording Industry Association of America.
RMS	Root-Mean-Square. The RMS heating power of a sine wave is $0.707 \times$ the value of its peak-to-peak measurement.
ROM	Read Only Memory.
SACD	Super Audio Compact Disc.
Sampling	(1) The process of encoding a signal digitally by registering it as discrete values of level at specified intervals of time (the sampling frequency), in contrast to analog recording which registers the waveform continuously. (2) The process of recording sounds into a 'sampler' which can then be edited and processed in various ways.
SAOC	Spatial Audio Object Coding.
SCSI	Small Computer Systems Interface. ('Scuzzy'.)

S-DAT	Stationary-head Digital Audio Tape.
Sensitivity	For present purposes, sensitivity effectively denotes the efficiency with which a transducer converts electrical energy into acoustic energy (e.g. a loudspeaker) or vice versa (e.g. a microphone).
Shelving	Low- or high-frequency boost or cut with a gentle curve up to a 'shelf'.
Signal-to-noise ratio	The ratio in dB between the wanted signal and the unwanted noise in a system.
Sine wave	A wave which is made up of one single frequency.
Sinusoidal waveform	See Sine wave.
Slew rate	The maximum rate of change in volts/microsecond of which a circuit's output is capable.
SMART	System Managed Audio Resource Technique.
SMPTE	Society of Motion Pictures and Television Engineers. ('Simpty'.)
Solo	On a mixer, pressing 'solo' on a channel routes its post-fade output to the monitor output. It is the same as AFL.
SPDIF	Sony-Philips Digital Interface.
SPL	Sound pressure level.
SPMIDI	Scalable Polyphonic Musical Instrument Digital Interface.
SPP	Song Position Pointers.
SR	(1) Sound Reinforcement. (2) Spectral Recording (Dolby).
Standing wave	A standing wave is the result of reflections from room boundaries reinforcing each other at certain frequencies to create points where the sound pressure level is very high, and other points where it is very low.
SWR	Standing wave ratio.
Sync	Synchronization.
THD	Total Harmonic Distortion.
TOA	Time of arrival.
Transducer	A device which converts one form of energy into another form of energy. For example, a loudspeaker converts electrical energy into acoustical energy.
Transverse wave	A wave in which the device or particles creating it move at right angles to the direction of the wave's travel, the carrying medium also oscillating at right angles to the wave's travel. An example is electromagnetic radiation, created by the electrons' up-and-down motion along the length of a transmitting aerial.
UHF	Ultra-high frequency.
UHJ	Universal HJ. The Ambisonics surround sound encoding and decoding system, the H and J denoting earlier systems.

VCA	Voltage controlled amplifier.
VCO	Voltage controlled oscillator.
VHF	Very high frequency.
VITC	('Vitcee') Vertical Interval Time Code.
Volt	The unit of electrical pressure, named in honor of Alessandro Volta (1745–1827).
VTR	Video Tape Recorder.
VU	Volume Unit.
WAP	Wireless Application Protocol.
Watt	The unit of electrical power, named in honor of James Watt (1736–1819).
WFS	Wave Field Synthesis.
White noise	Noise which has equal energy per Hz of frequency. Its frequency spectrum therefore rises by 3 dB per octave when the usual logarithmic horizontal frequency scale is used.
WMA	Windows Media Audio.
WORM	Write Once Read Many times.
XLR	Originally a part code for the ITT-Canon company's professional audio connector, the most familiar of which is the three-pin microphone and balanced line XLR-3.

GENERAL FURTHER READING

Alkin, G., 1989. Sound Techniques for Video and TV, second edition. Focal Press.

Alkin, G., 1996. Sound Recording and Reproduction, third edition. Focal Press.

Ballou, G. (Ed.), 2008. Handbook for Sound Engineers, fourth edition. Focal Press.

Borwick, J. (Ed.), 1996. Sound Recording Practice, fourth edition. Oxford University Press.

Eargle, J., 2005. Handbook of Recording Engineering, fourth edition. Springer.

Huber, D.M., 2013. Modern Recording Techniques, eighth edition. Focal Press.

Izhaki, R., 2011. Mixing Audio: Concepts, Practices and Tools. Focal Press.

Leonard, J.A., 2001. Theatre Sound. A. and C. Black.

Nisbett, A., 2003. The Sound Studio, seventh edition. Focal Press.

Owzinski, B., 2006. The Mixing Engineer's Handbook. Course Technology Inc..

Owzinski, B., 2009. The Recording Engineer's Handbook. Delmar.

Index

Note: Page numbers in *italics* indicate figures; those in **bold** indicate tables.